Cation binding by humic substances

Humic substances are highly abundant organic compounds formed in soils and sediments by the decay of dead plants, microbes and animals. This book focuses on the important binding properties of these compounds that regulate the chemical reactivity and bioavailability of hydrogen and metal ions in the natural environment.

Topics covered include the physico-chemical properties of humic matter and interactions of protons and metal cations with weak acids and macromolecules. Experimental laboratory methods are also discussed, together with mathematical modelling. Finally the author looks at how the results of this research can be used to interpret environmental phenomena in soils, waters and sediments.

This comprehensive account of cation binding by humic matter is a valuable resource for advanced undergraduate and graduate students, environmental scientists, ecologists and geochemists.

EDWARD TIPPING works at the Natural Environment Research Council Centre for Ecology and Hydrology in Cumbria, England, and is a Visiting Professor at the University of Lancaster. After receiving his PhD from the University of Manchester in 1973, he spent five years at the Middlesex Hospital Medical School researching cancer biochemistry. He has spent short periods in laboratories in Finland, Sweden, the USA, Norway and the Netherlands, and published over 120 papers in international journals.

CAMBRIDGE ENVIRONMENTAL CHEMISTRY SERIES

Series Editors:
P. G. C. Campbell, *Institut National de la Recherche Scientifique, Université du Québec à Québec, Canada*
R. M. Harrison, *School of Chemistry, University of Birmingham, England*
S. J. de Mora, *International Atomic Energy Agency – Marine Environment Laboratory, Monaco*

Other books in the series:
A. C. Chamberlain *Radioactive Aerosols*
M. Cresser, K. Killham and A. Edwards *Soil Chemistry and its Applications*
R. M. Harrison and S. J. de Mora *Introductory Chemistry for the Environmental Sciences* Second Edition
S. J. de Mora *Tributyltin: Case Study of an Environmental Contaminant*
T. D. Jickells and J. E. Rae *Biogeochemistry of Intertidal Sediments*
P. Brimblecombe *Air Composition and Chemistry* Second Edition
S. J. de Mora, S. Demers and M. Vernet *The Effects of UV Radiation in the Marine Environment*

Cation binding by humic substances

EDWARD TIPPING

Centre for Ecology and Hydrology, Windermere, UK

CAMBRIDGE
UNIVERSITY PRESS

CAMBRIDGE UNIVERSITY PRESS
Cambridge, New York, Melbourne, Madrid, Cape Town, Singapore, São Paulo

Cambridge University Press
The Edinburgh Building, Cambridge CB2 2RU, UK

Published in the United States of America by Cambridge University Press, New York

www.cambridge.org
Information on this title: www.cambridge.org/9780521621465

First published 2002
This digitally printed first paperback version 2005

A catalogue record for this publication is available from the British Library

Library of Congress Cataloguing in Publication data

Tipping, Edward, 1948–
Cation binding by humic substances / Edward Tipping. – 1st ed.
p. cm. – (Cambridge environmental chemistry series)
Includes bibliographical references and index.
ISBN 0 521 62146 1
1. Humic acid – Derivatives. 2. Cations. 3. Ligand binding (Biochemistry) – Mathematical models I. Title II. Series.
QD341.A2 T625 2002
572′.33–dc21 2001043867

ISBN-13 978-0-521-62146-5 hardback
ISBN-10 0-521-62146-1 hardback

ISBN-13 978-0-521-67565-9 paperback
ISBN-10 0-521-67565-0 paperback

Contents

Preface

The study of the interactions of cations with humic substances has been going on for half a century or more, and given the wide and continuing interest in the subject it is perhaps surprising that there has not been a text dedicated to the subject. With this volume, I attempt to fill the gap. My aims are to provide a basic account of the interactions, discuss how they can be represented in models, and relate the results to the natural environment. I hope the book will be of interest and use to students of environmental chemistry, to individuals beginning research projects, and to established research workers, not only chemists but also those in other environmental disciplines.

I am grateful to Peter Campbell (INRS-Eau, Canada), who first suggested that I write this book, and to John Hamilton-Taylor (University of Lancaster) who kindly read and commented upon draft chapters. Lisa Baldwin's cheerful help in obtaining references, maintaining my filing system, and obtaining permissions to reproduce published figures, is greatly appreciated. Thanks are due to James Alberts, David Kinniburgh and Erwin Temminghoff for supplying copies of original diagrams. The present copyright laws require permission to be obtained only from the publisher of a work to reproduce figures and tables (although some publishers require the author(s) to agree as well). However, I sought the approval of all the authors anyway, and was much heartened to receive the willing agreement of all who replied, and by the many good wishes expressed.

Over the years, I have worked with numerous colleagues on the subjects covered in this book, and I thank them all for their help and ideas. In particular, I would like to mention Margaret Ohnstad, Clare Backes, Colin Woof, Dudley Thompson, Margaret Hurley, Jan Mulder, John Hamilton-Taylor, Stephen Lofts and Stephen Bryan. I also offer my deep and lasting gratitude to the three people who set me on the path of scientific research – Malcolm Jones, Brian Ketterer and the late Hank Skinner. Finally, I dedicate the book to my wife Pam, and our dear children Miles and Rose, with all my love.

The Ferry House
Windermere

1

Introduction

Our immediate environment – the atmosphere, soils, surface water and groundwater – is host to a vast number of chemical reactions, even leaving aside the biochemical reactions of living organisms. Environmental chemists aim to define and quantify these reactions, make analytical determinations of reactants and products, construct predictive models, and relate their findings to the functioning of ecosystems. A major part of the research effort is to take account of natural organic matter and mineral particulates, because these abundant, poorly defined, components exert a powerful and ubiquitous chemical influence on the environment. This book focuses on the most chemically significant fraction of natural organic matter, humic substances.

Humic substances are complex, acidic organic molecules formed by the decomposition of plant, animal and microbial material. They are abundant and persistent in the biosphere and immediate subsurface, being present in particulate and dissolved forms in soils, waters and sediments. They interact with a variety of solutes, adsorb at surfaces, and are photochemically active. Humic substances first came to prominence in agriculture, because of their positive influence on the structure, water retention properties, and nutrient status of soils. More negatively, they pose problems to the water supply industry, requiring removal to minimise water colour, and giving rise to potentially mutagenic by-products as a result of chlorination. There is interest in using humic substances to recover metals from wastewaters and from seawater. Schnitzer (1991) lists a variety of uses of humic matter, in agriculture, industry, environmental engineering, and medicine. However, over and above these practical and utilitarian matters, the greatest present interest in humic substances concerns their rôles in the natural environment. They are recognised to be important in a range of issues (Table 1.1), and consequently there is a considerable worldwide research effort aimed at characterising them and understanding their environmental behaviour.

Pre-eminent among the interactions of humic substances with solutes are those involving cations, i.e. protons (H^+) and metallic cations such as Na^+, Al^{3+}, Ca^{2+}, Cu^{2+}, Pb^{2+} and Am^{3+}. The interactions involve both immobile organic matter, in the solid phases of soils and sediments, and dissolved humic substances, which may be mobile and transport the ions. Cation–humic interactions exert control on the reactivity of the cation, including its bioavailability. But this is a two-way process, since cation binding influences the physico-chemical state of the humic matter, and thereby its interactions with other components of the environmental system. Therefore it is important to describe not only how humic substances influence cations, but also how cations influence humic substances.

Table 1.1. Environmental issues involving humic substances. Those for which cation binding is recognised to be an important factor are marked with an asterisk.

Issue	Role of humic substances
Carbon cycling	Major C pool, transformations, transport and accumulation
Light penetration into waters	Absorption and attenuation of light by humic chromophores
Soil warming	Absorption of solar radiation by soil humic matter
Soil and water acidification*	Binding of protons, aluminium and base cations in soils and waters
Nutrient source	Reservoir of carbon, nitrogen, phosphorus and sulphur
Nutrient control*	Binding of iron and phosphate
Microbial metabolism	Substrate for microbes
Weathering*	Enhancement of mineral dissolution rates
Soil formation (podzolisation)*	Translocation of dissolved humic substances and associated metals (Al, Fe)
Properties of fine sediments*	Adsorption at surfaces and alteration of colloidal properties
Soil structure*	Aggregating effect on soil mineral solids
Photochemistry	Mediation of light-driven reactions
Heavy metals*	Binding, transport, influence on bioavailability, redox reactions
Pesticides, xenobiotics*	Binding, transport, influence on bioavailability
Radioactive waste disposal*	Binding and transport of radionuclide ions in groundwaters
Ecosystem buffering*	Control of proton and metal ion concentrations, persistence

The aim of the present volume is to draw together current knowledge about cation interactions with humic matter, and to show how it can be used to understand and predict the natural environment. The book is structured as follows. Chapter 2 reviews information about humic substances, emphasising those aspects of most relevance to cation–humic interactions. The reviews of environmental and solution chemistry in Chapters 3–5 define the chemical territory within which cation–humic interactions take place. Chapter 6 surveys methods available for determining the extents of interaction of cations with humic substances. Chapter 7 reviews available quantitative information, and Chapter 8 discusses knowledge about the physico-chemical natures of cation–humic complexes. Chapters 9–11 address the mathematical modelling of cation–humic reactions, which is a key requirement if information from laboratory studies is to be applied to the natural environment. Chapter 12 looks at how cation binding influences other environmentally significant physico-chemical interactions. Chapters 13 and 14 examine the implications of cation–humic interactions in natural waters and soils respectively. Finally, Chapter 15 considers research needs and wider applications.

2

○ ○ ○ ○ ○ ○ ○ ○ ○ ○ ○ ○ ○ ○ ○ ○ ○ ○ ○

Humic substances – a brief review

An appreciation of the formation and properties of humic substances is needed to understand their interactions with cations, and the influence of those interactions within the natural environment. This chapter presents a summary of current knowledge, emphasising those aspects of most relevance to cation binding. More detailed, wider-ranging, information can be found in the books of Aiken *et al.* (1985a), Hayes *et al.* (1989), Averett *et al.* (1989), Beck *et al.* (1993), Stevenson (1994), Piccolo (1995) and Gaffney *et al.* (1996). Jones & Bryan (1998) reviewed the colloidal properties of humic substances. Thurman (1985) is an essential text on the organic matter of natural waters, and includes much information on humic substances.

2.1 Natural organic matter and humic substances

Most (99.95%) of the Earth's carbon is held in sedimentary rocks, to and from which it cycles on geological time scales. The remaining carbon (about 4×10^{19} g) is in the biosphere or shallow subsurface, 90% of it in the form of carbonate dissolved in seawater, and about 9% in organic forms (Table 2.1). In soils, natural waters and their sediments, the majority of the organic carbon is present in humic matter.

Stevenson (1994) considers natural organic matter in soils to consist principally of litter (macroorganic material lying on the soil surface), the light fraction (plant residues within the soil proper), soil biomass (predominantly microorganisms living in the soil), and stable humus. Of these pools, the last predominates in most agricultural, forest and moorland soils. In peatlands, plant residues may constitute most of the organic matter, but there are nonetheless large amounts of humic substances. Humus consists of humic substances and non-humic substances. The latter comprise identifiable classes of biochemicals (amino acids, carbohydrates, fats, waxes, resins, low molecular weight organic acids etc.). Stevenson defines humic substances as: 'a series of relatively high-molecular-weight, yellow to black colored substances formed by secondary synthesis reactions'.

Natural waters contain a range of identifiable solutes, together with humic substances. Thurman (1985) lists the main classes of identifiable compounds as carboxylic acids, phenols, amino acids, carbohydrates and hydrocarbons. There are also many trace compounds, of both natural and anthropogenic origin. A similar range of compounds is found in soil waters (Qualls & Haines, 1991). Thurman (1985) points out the difficulty of defining 'in some limited yet useful way' what aquatic humic substances are, and he offers:

> (Aquatic humic substances) are colored, polyelectrolytic, organic acids isolated from water on XAD resins, weak base ion-exchange resins, or a comparable procedure. They are nonvolatile and range in molecular weight from 500 to 5000; their elemental composition is approximately 50 percent carbon, 4 to 5 percent hydrogen, 35 to 40 percent oxygen, 1 to 2 percent nitrogen, and less than 1 percent for sulfur plus phosphorus. The major functional groups include: carboxylic acids, phenolic hydroxyl, carbonyl, and hydroxyl groups. Within aquatic humic substances there are two fractions, which are humic and fulvic acid. Humic acid is that fraction that precipitates at pH 2.0 or less, and fulvic acid is that fraction that remains in solution at pH 2.0 or less.

The suspended and bed sediments of rivers consist chiefly of material

Table 2.1. Pools of organic carbon in the Earth's surface and subsurface.

Pool	$g \times 10^{-15}$
Seawater	1700
Soils	1500
Terrestrial plants	560
Groundwater	15
Surface freshwater	0.5

The data are mainly taken from the compilation by Schlesinger (1997), also from Aiken *et al.* (1985b).
The approximate value for groundwater is based on an assumed average concentration of 1 mgC dm^{-3} and a total groundwater volume of 1.5×10^{16} m^3 (Schlesinger, 1997). The value for freshwater assumes an average concentration of 5 mg dm^{-3} and a total volume of 10^{14} m^3 (Smith, 1981; Schlesinger, 1997).

derived from the land, and the organic matter is therefore related to soil material. Humic substances are the principal forms of organic carbon in marine and lake sediments. In both cases they can be derived from both terrestrial and aquatic sources (Vandenbroucke *et al.*, 1985; Ishiwatari, 1985).

Two definitions of humic substances for soils and waters have already been given. To emphasise that definition is difficult, a third attempt, that of Aiken *et al.* (1985b), can be mentioned: '(Humic substances are) naturally occurring, biogenic, heterogeneous organic substances that can generally be characterized as being yellow to black in color, of high molecular weight and refractory'. For the most part, the three definitions are summaries of the properties of compounds obtained when natural samples (waters, soils, sediments) are subjected to defined extraction procedures, the principal ones being (a) solubilisation with base and (b) adsorption or precipitation from acid solution (see Section 2.2). The definitions do not include any recognition of function, i.e. they do not say what the rôle(s) of humic substances might be. A contrast can be drawn with biochemical definitions of enzymes, which are primarily defined in terms of the reaction that they catalyse. Much of the research on humic substances aims to determine their functions under different environmental circumstances, and the study of cation binding is an important part of that research effort.

2.2 Isolation and classification of humic substances

Traditionally, humic substances have been considered to comprise three main fractions, distinguished by their solubility and adsorption properties. Humic acids are soluble in base but insoluble in acid, while fulvic acids are soluble in both base and acid. A third fraction, humin, is readily soluble in neither acid nor base. Humin may consist of humic acid in strong association with mineral matter, highly condensed insoluble humic matter, fungal melanins, and paraffinic substances (Stevenson, 1994). Broadly speaking, humic acids and humin occur mostly in soils and sediments as part of the solid phase, while fulvic acids are more mobile and account for a major part of the dissolved organic matter in natural waters. The word 'humic', used generically, encompasses all humic substances.

2.2.1 Isolation and fractionation procedures

Isolation procedures have been discussed fully by Hayes (1985), Aiken (1985) and Stevenson (1994). Figures 2.1 and 2.2 show outline schemes of the most commonly used methods for the isolation of humic and fulvic acids from soils and waters. To a large extent, the procedures depend upon controlling the electrical charge of the humic matter. When the molecules are highly charged (high pH), they are more soluble. When the charge is decreased (low pH), precipitation and adsorption to hydrophobic surfaces are promoted.

Hatcher *et al.* (1985) considered humin to be the organic residue after successive extraction of sediments by benzene + methanol (to remove lipids), 1 mol dm^{-3} HCl, and 0.5 mol dm^{-3} NaOH. In marine sediments, further treatment with HF + HCl was needed to remove mineral matter.

Within the humic substances research community, there is some unease about how representative isolated humic substances are of the materials in their natural states. The unease arises mainly because the extraction conditions are rather extreme, involving high concentrations of acid and base; if such conditions were used in the isolation of enzymes, for example, most or all of their biological activity would be lost. A number of authors

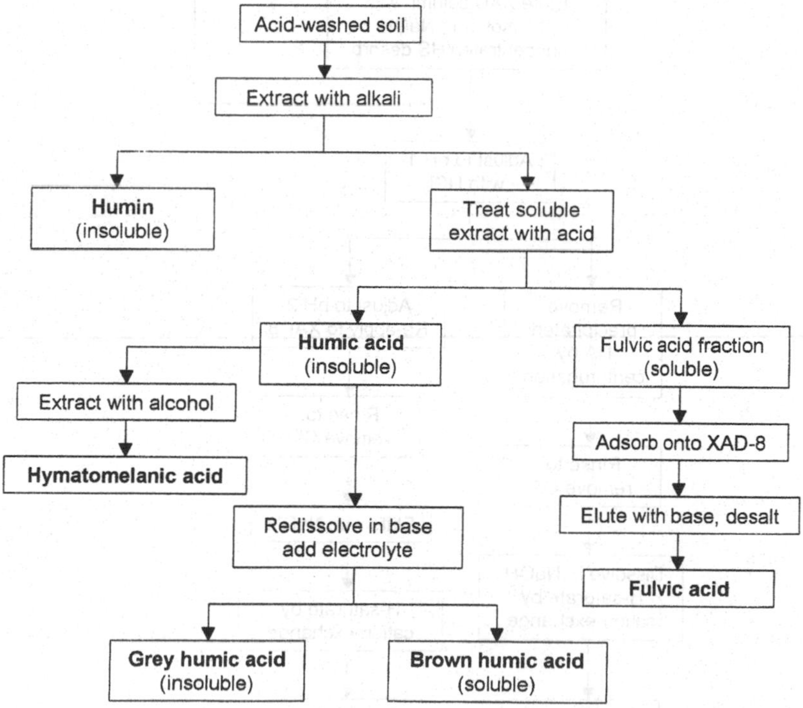

Figure 2.1. Scheme for the isolation of humic substances from soils. [Adapted from Stevenson, F.J. (1994), *Humus Chemistry: Genesis, Composition, Reactions,* 2nd edn., by permission of John Wiley & Sons, Inc.]

have argued that the standard methods of isolation introduce artefacts. According to Hayes (1985), 'the classical extraction procedures use aqueous alkaline solvents and these give rise to partially oxidised and slightly degraded humic artefacts'. He argued that milder, but still efficient, methods should be sought. Shuman (1990) speculated that the isolation procedure for aquatic humic substances tends to select molecules with certain properties, leading to an unrepresentative, but relatively

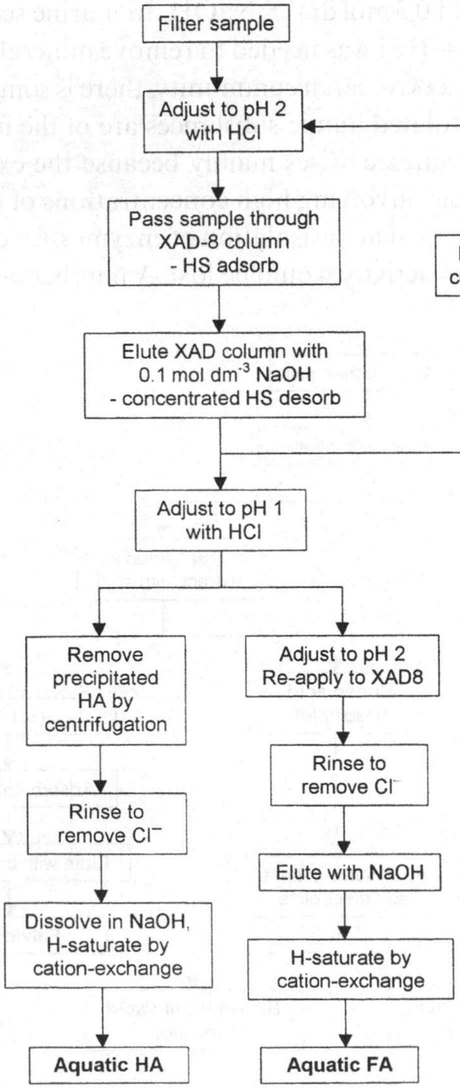

Figure 2.2. Scheme for the isolation of humic substances from waters, based on Aiken (1985).

uniform, isolated material. De Haan (1992) claimed that the extremes of pH to which aquatic humic substances are subjected during isolation (Fig. 2.2) cause artefacts. These concerns have led to alternative isolation methods, including extraction of soil humic matter with acidic dimethylsulphoxide (Hayes, 1985), and acid pyrophosphate (Gregor & Powell, 1987). Reverse osmosis, a purely physical method, has been used to concentrate dissolved organic matter from natural waters (Sun *et al.*, 1995), but other (unwanted) solutes are also concentrated.

Fractionation of humic acid can be achieved by extraction with alcohol, to obtain hymatomelanic acid, and by treatment of its solution in base with electrolyte, to obtain grey humic acid (precipitated) and brown humic acid (not precipitated). Fractionation has also been carried out successfully by gel chromatography, which separates the material by molecular weight (Swift, 1985). More extensive fractionation schemes have been developed for dissolved organic matter, the most well-established being that of Leenheer (1981), in which chromatographic separations are used to divide the sample into hydrophobic and hydrophilic fractions, each sub-divided into acidic, basic and neutral fractions. Table 2.2 shows suggestions for the components of the different fractions.

2.2.2 Buffle's classification scheme for natural organic matter

Buffle (1984, 1988) proposed a classification of natural organic matter that is ecologically orientated, being based on the identification of 'homologous compound groups', which correspond more closely to specific ecosystem functions than do fractions obtained from the analytically based separation schemes described above. Buffle defined natural organic matter as all the organic matter other than living organisms and synthetic compounds. The classification scheme is summarised in Table 2.3. Buffle (1988) explained in detail the natures of the different classes, and showed how they are related to the chemically fractionated humic substances. Buffle's scheme is arguably more logical, less restrictive, and more useful than the classification of humic substances (and non-humic substances) that currently holds sway. As yet there is not much evidence that other workers are adopting it in their research. However, at the very least it serves as a reminder that two samples of humic matter isolated by the same means may not have the same functional properties, and that differences may be related to the origins of the materials.

2.3 Formation and decomposition of humic substances

2.3.1 Formation

In most environments, the majority of dead biomass is converted to carbon dioxide, in the process of mineralisation. However, mineralisation is rarely complete, and therefore organic matter accumulates in the form of humic substances. The extent of accumulation depends upon turnover times, climate, vegetation, parent mineral material, topography and cropping (Stevenson, 1994). Humic substances are especially prevalent in

Table 2.2. Fractions obtained from the Leenheer (1981) fractionation scheme, and some of their likely non-anthropogenic constituents in surface and soil waters, based on Thurman (1985) and Qualls & Haines (1991).

	Hydrophobics	Hydrophilics
Acids	Fulvic acid Humic acid Humic-bound carbohydrates Simple aromatic acids Oxidised polyphenols Long-chain fatty acids Tannins Flavonoids Polyphenols Vanillin Proteins Phospholipids	Hydrophilic acids Simple carboxylic acids Oxidised carbohydrates with COOH groups Proteins Amino acids Adenosine di- and tri-phosphates Inositol and other sugar phosphates
Neutrals	Hydrocarbons Chlorophyll Carotenoids Phospholipids Low-acidity humic compounds Proteins Alcohols Ketones Ethers	Simple neutral sugars Polysaccharides Alcohols Ketones Ethers Proteins Amino acids Urea
Bases	Peptides Proteins	Amino-sugar polymers Pyrimidines Purines Amino acids Peptides Proteins Low molecular weight amines

peat soils. The formation of humic substances (humification) does not appear to be governed by controlled reactions of the type that combine within living organisms. Instead, a collection of undirected reactions, some chemical, some microbially mediated, produces mixtures of complex organic compounds. Information about the processes of humification has been obtained from a large number of analytical and experimental studies. Only a brief review of current knowledge is possible here. The reader is referred to Tate (1987) and Stevenson (1994) for accounts of humification in soils. Information on other environments is given in Aiken *et al.* (1985a), and by Thurman (1985).

In soils, the main source of humic matter is terrestrial plant material. Other contributors are animal and microbial remains and microbially synthesised products. Mechanisms proposed for the formation of humic substances are summarised in Fig. 2.3. Pathways 1, 2 and 3 involve the breakdown of the plant precursors to small molecules, followed by their reassembly into larger entities. Pathway 4 is the 'lignin theory'. Stevenson (1994) stated that humic substances in soil may be formed by all four of the pathways, but considered the condensation pathways 2 and 3, involving polyphenols and quinones, to dominate in most soils. Between them, these two pathways comprise the 'polyphenol theory'.

Table 2.3. Classification of natural organic matter (NOM) according to Buffle (1984, 1988).

Abbreviation	Name	Origin
SOM	Soil organic matter	Compounds produced in soil, especially by decomposition of higher plants
TOM	Peat organic matter	Compounds resulting from degradation less advanced than SOM
	Aquatic organic matter	NOM accumulated in water as POM and AOM
POM	Pedogenic organic matter	Originating from leaching of SOM
AOM	Aquagenic organic matter	Originating from metabolism and degradation of plankton and aquatic bacteria
SedOM	Sediment organic matter	Compounds resulting from accumulation and transformation of POM, AOM and organism debris in sediments
AntOM	Anthropogenic organic matter	Compounds resulting from human activity

According to Ishiwatari (1985), most of the humic matter in lacustrine sediments is derived from sedimented algal material, but with some contributions from terrestrial plant remains, washed into the lakes from their catchments. He proposed that the major pathways for humic substance formation are non-enzymic 'browning' reactions involving amines and reducing compounds (e.g. carbohydrates), or amino–carbonyl reactions. Algal lipids are also involved. The formation is supposed to be rapid, being almost completed in decaying algae, and/or in surface sediments. With burial in the sediment, slower transformations occur to produce humin. Formation of humic matter in marine sediments results from condensation reactions that involve breakdown products of cellular constituents of dead organisms, either algae or land plants, depending upon the characteristics of the sedimentary basin. Marine sediments also contain soil humic substances nearly in their original form (Vandenbroucke *et al.*, 1985).

Humic substances form in the water columns of lakes and oceans (Thurman, 1985). Harvey & Boran (1985) proposed that in seawater the formation involves the free radical cross-linking of unsaturated lipids released from algae.

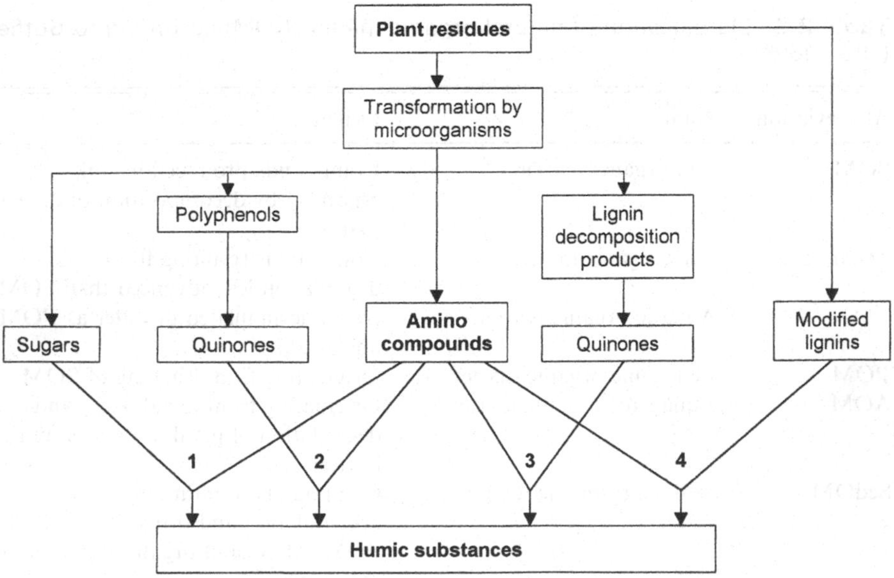

Figure 2.3. Formation pathways of humic substances. [Redrawn from Stevenson, F.J. (1994), *Humus Chemistry: Genesis, Composition, Reactions*, 2nd edn., by permission of John Wiley & Sons, Inc.]

2.3.2 Decomposition

Once formed, humic substances resist degradation, i.e. they are refractory. According to MacCarthy & Rice (1991), this is due to their chemical and physical heterogeneity, which discourages the evolution of degrading enzymes. Physical and chemical protection mechanisms also operate; organic compounds, including humic substances, tend to be less susceptible to breakdown when adsorbed by minerals, when aggregated, or when complexed with metal ions, especially Al^{3+} (McKeague *et al.*, 1986). Nonetheless, breakdown does occur. A number of microorganisms with the capability to metabolise humic substances have been isolated from soils. Tate (1987) discussed the mechanisms and ecological significance of such metabolism. Jones (1992) concluded that, in humic-rich lakes, the metabolism of humic matter by bacterioplankton is a significant process, made possible by the photochemical breakdown of biologically refractory humic substances to yield lower molecular weight biologically labile organic products. Such utilisation of humic substances by bacteria is the basis of a food chain distinct from the 'conventional' one that is driven by phytoplanktonic primary production.

2.4 Chemical and physical properties of humic substances

2.4.1 Elemental composition

Some representative elemental compositions of humic compounds are shown in Table 2.4. The carbon content is typically just greater than 50%, although some soil fulvic acids have less. Hydrogen contents of soil and freshwater samples centre around 5%, but marine samples have higher levels, reflecting their aliphatic nature. Oxygen contents are in the range 30–40%, with groundwater values being the lowest. The ^{13}C and ^{14}C contents of humic substances provide information on the origin and age of the materials. For example, ^{13}C data are consistent with the formation of marine fulvic acid from algae, and of soil humic matter from terrestrial plants. Examples of average ages estimated from ^{14}C data are 30 years for Suwannee River fulvic acid (Thurman, 1985), 100–500 years for soil fulvic acid, 700–1600 years for soil humic acid, and 100–2400 years for soil humin (Jenkinson, 1981). These data attest to the refractory nature of humic substances already mentioned. Another use of ^{13}C is in nuclear magnetic resonance (NMR) studies (next section).

2.4.2 Molecular structure

Here we are concerned with the groupings of atoms that comprise the isolated materials, and how they are joined together. Information on structure has come from degradative techniques and from spectroscopy. Methods and results are extensively reviewed in Aiken *et al.* (1985a), Hayes *et al.* (1989) and Stevenson (1994). The application of ^{13}C-NMR has been especially revealing (e.g. Wilson, 1987; Malcolm, 1989; Steelink *et al.*, 1989). An NMR spectrum is shown in Fig. 2.4 to illustrate the groupings that can be identified and quantified. Table 2.5 summarises results for different humic materials. Six main classes account for nearly all of the organic carbon.

Approximately 80% of the hydrogen in humic matter is bonded to carbon, the rest to oxygen. Since hydrogen can only dissociate from bonds with oxygen, there is an upper limit of approximately 10 meq g^{-1} on the

Table 2.4. Elemental compositions of humic substances, corrected for ash content.

Humic compound	*n*	C	H	O	N	S
Soil HA	many	52.8–58.7	2.2–6.2	32.8–38.3	0.8–4.3	0.1–1.5
Soil FA	many	40.7–50.7	2.8–7.0	39.7–49.8	0.9–2.3	0.1–2.6
Groundwater HA	5	65.5	5.2	24.8	2.4	1.0
Groundwater FA	5	60.4	6.0	32.0	0.9	0.7
Seawater FA	1	51.8	7.0	37.7	6.6	0.5
River water HA	15	52.2	4.9	41.7	2.1	—
River water FA	15	52.7	5.1	40.9	1.1	0.6
Lake water FA	3	54.8	5.5	41.1	1.4	1.1
Soil humin	2	55.9	5.8	32.8	4.9	—
Peat humin	2	56.3	5.1	36.5	2.1	—
Marine humin	2	56.2	7.0	31.7	5.2	—

The data are taken from the compilations of Steelink (1985), Thurman (1985) and Hatcher *et al.* (1985).
The number of samples considered is denoted by *n*.

Table 2.5. Distributions (%) of organic carbon in different humic compounds, adapted from Malcolm (1990).

	n	Unsaturated aliphatics	N-alkyl methoxyl	Carbo-hydrate	Aromatic	Carboxyl	Ketonic
Soil humic acid	8	17–30	4–9	12–18	24–42	12–18	4–7
Soil fulvic acid	1	22	5	20	26	24	4
Aquatic humic acid	4	23–30	5–6	9–21	29–36	14–17	6–8
Aquatic fulvic acid	4	30–40	5–7	10–18	14–18	16–19	5–11

The number of samples considered is denoted by *n*.

content of protons that can dissociate. With regard to the binding of cations, the proton-dissociating groups are of prime importance. Data from acid–base titrations and other analytical methods, notably NMR spectroscopy and infra-red (IR) spectroscopy, have established that humic substances possess a range of weak acid groups, including carboxylic and phenolic OH. The total content of proton-dissociating groups in fulvic acid is 6–10 meq g^{-1}, while the range for humic acid is 4–6 meq g^{-1}. Most of the groups dissociating at pH < 7 are carboxylic. The natures of the proton-dissociating groups are discussed in more detail in Chapters 7 and 8.

Nitrogen may play an important rôle in the creation of strong binding sites for certain metals (Chapter 8). Acid hydrolysis of humic substances releases various forms of nitrogen, notably in ammonia, amino acids and amino sugars. However, an appreciable amount of the released nitrogen is in unidentified compounds, while approximately half of the nitrogen is not released by acid hydrolysis. Schnitzer (1985) and Stevenson (1994) have reviewed knowledge in this area. Another element that may be significant in the creation of strong binding sites is sulphur, which is especially favoured by polarisable metals such as mercury.

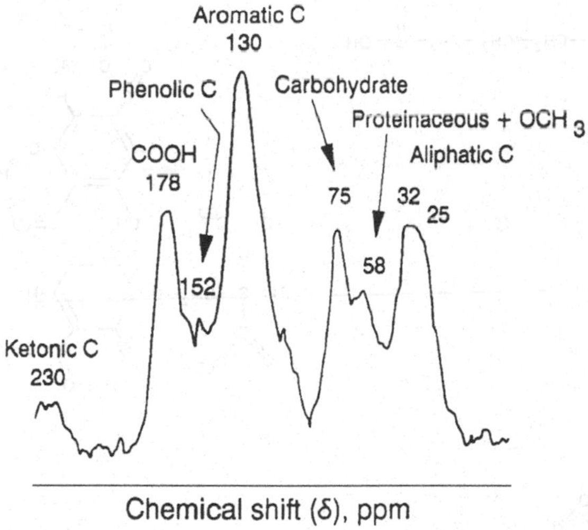

Figure 2.4. The ^{13}C-NMR spectrum of a soil humic acid, showing chemical groupings and their chemical shifts. [Reproduced from the adaptation in Stevenson, F.J. (1994), *Humus Chemistry: Genesis, Composition, Reactions*, 2nd edn., of a diagram published by Schnitzer, M. (1990), Selected methods for the characterisation of soil humic substances, in *Humic Substances in Soil and Crop Sciences: Selected Readings*, ed. P. MacCarthy, C.E. Clapp, R.L. Malcolm & P.R. Bloom, pp. 65–89, by permission of John Wiley & Sons, Inc., the American Society of Agronomy, and the Soil Science Society of America.]

A variety of structural models for humic matter have been suggested, and these have been reviewed by Stevenson (1994). As a result of intensive research into fulvic acid from the Suwannee River (USA), Leenheer *et al.* (1989) proposed several structures that are consistent with elemental composition, NMR and titration data (Fig. 2.5). The compounds were assigned a molecular weight of approximately 800, which corresponds to the number-average value, and is therefore biased towards smaller molecules (see Section 2.4.3). Larger, more complex, molecules are therefore also to be expected, although based on the same molecular groupings. A recent suggestion for the structure of a fragment of humic

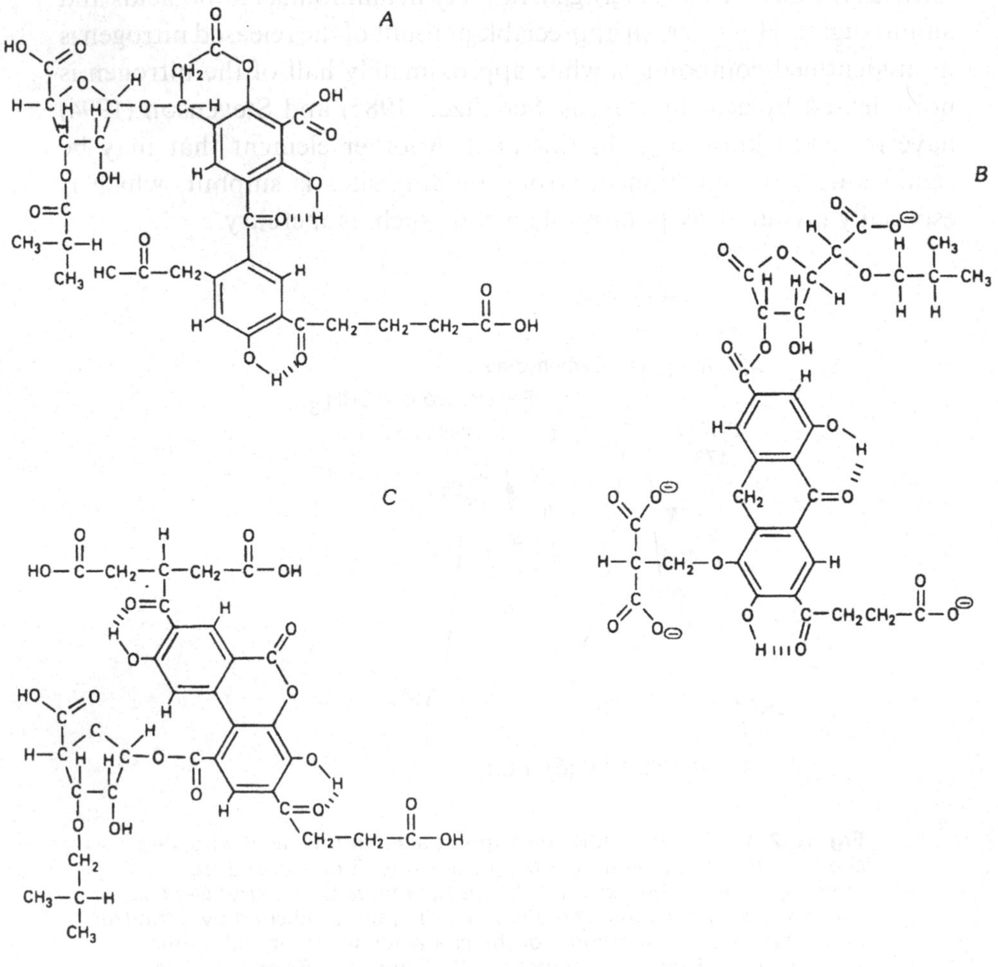

Figure 2.5. Proposed molecular structures of Suwannee River fulvic acid. [Reproduced from Leenheer *et al.* (1989).]

acid by Schulten (1996) contains many aromatic residues, a considerable degree of cross-linking, and in three dimensions a fairly open structure. The carbon content of the model material is greater (67%) and the oxygen content lower (25%) than the typical values given in Table 2.4. Del Rio & Hatcher (1996) have questioned the 'network' structure of the model. It must be appreciated that any proposed structure of humic matter can only be regarded as in some way representative or typical of the different compounds making up the sample.

The structures of Fig. 2.5 show a number of possibilities for cation-binding sites. The most abundant are single carboxylic and phenolic groups, but it can be envisaged that folding of the molecules could bring such groups into proximity and generate multidentate sites, allowing metal ions to interact with more than one grouping. As already noted, small amounts of nitrogen and sulphur, not included in the model structures, may be important in the generation of strong binding sites (see Chapter 8).

2.4.3 Molecular weight

Since humic substances are mixtures, molecular weight values are averages. The two most commonly used are number-average (M_n) and weight-average (M_w), defined by

$$M_n = \frac{\sum_i n_i M_i}{\sum_i n_i} \qquad (2.1)$$

and

$$M_w = \frac{\sum_i c_i M_i}{\sum_i c_i} \qquad (2.2)$$

Here n_i and c_i are respectively the number and weight concentrations of the different humic fractions. Polydispersity is the ratio of the weight and number averages. For a pure compound, M_n and M_w are equal and the polydispersity is unity. Swift (1989a) has discussed polydispersity in humic substances in detail.

From measurements on a variety of samples by different techniques, there is general agreement that fulvic acids have molecular weights of at most a few thousand. Some results are shown in Table 2.6, from which

Table 2.6. Some reported molecular weights of fulvic acid (FA), humic acid (HA) and dissolved organic matter (DOM).

Sample	M_n	M_w	Method(s)	References
Freshwater FA	600–1400	1000–2300	Cry, FFF, HPSEC, UC, VPO	6–10, 12, 14
Freshwater HA		4000–8000	UC	12
Freshwater DOM	500–1300	800–2200	Abs, HPSEC, UC	13, 14
Marine FA	500–800		VPO	7
Soil water DOM	850–1800	1200–3300	HPSEC	11
Soil FA	600–1000		Cry, VPO	2–4
Aldrich Chemicals HA	1500–3100	4000–20 000	FFF, HPSEC, Visc	10, 14
Soil HA		25 000–200 000 +	DGU, UC, Visc	1, 2, 5
Peat HA		8000–17 000	UC	12

Key to methods: Abs, molar absorptivity correlation; Cry, cryoscopy; DGU, density gradient ultracentrifugation; FFF, field flow fractionation; HPSEC, high-pressure size-exclusion chromatography; UC, ultracentrifugation; Visc, viscosity; VPO, vapour pressure osmometry.

References: 1 Piret *et al.* (1960); 2 Visser (1964); 3 DeBorger & DeBacker (1968); 4 Hansen & Schnitzer (1969); 5 Cameron & Posner (1974); 6 Wilson & Weber (1977); 7 Gillam & Riley (1981); 8 Reuter & Perdue (1981); 9 Aiken & Malcolm (1987); 10 Beckett *et al.* (1987); 11 Berdén & Berggren (1990); 12 Reid *et al.* (1990); 13 Wilkinson *et al.* (1993); 14 Chin *et al.* (1994).

representative values for M_n and M_w of 800 and 1600 respectively can be derived, so that the polydispersity is about two. Published molecular weights of unfractionated humic acids vary from a few thousand to a few hundred thousand (Table 2.6). When both number and weight average values are available, high polydispersity is apparent. This is consistent with the results of Cameron *et al.* (1972), who fractionated a soil humic acid by gel filtration and then determined molecular weights by ultracentrifugation. The range of fractions they obtained had weight-average molecular weights between 2400 and 1 360 000, the most abundant fractions having values around 100 000. Stevenson (1994) suggested that the highest values may have been affected by aggregation, which is plausible in view of the high concentrations of material (*c.* 1 g dm^{-3}) used in the centrifugation experiments.

Wershaw (1986) proposed that humic molecules are not inherently large, and that the apparently high molecular weights reflect aggregation of small monomer units. There is as yet no experimental proof of this idea, nor have values been suggested for the molecular weights of the monomer units, although to judge from schematic diagrams presented by Wershaw they must be of the order of a few hundred. Conte & Piccolo (1999) claimed, on the basis of high-pressure size-exclusion chromatographic experiments, that soil humic substances comprise primary units of low molecular weight. Their estimates of M_w for fulvic acids were in the range 1500–3500, which are actually similar to or larger than those quoted in Table 2.6. The estimates for humic acids were between 2200 and 11 000, which are generally lower than the values of Table 2.6.

Some determinations have been made of the molecular weights of humic substances dissolved in natural waters. Berdén & Berggren (1990) studied soil solutions by high-pressure size-exclusion chromatography, and obtained values of 850–1800 for M_n and 1200–3300 for M_w. Wilkinson *et al.* (1993) reported a value of M_w of 1400 for humic matter dissolved in a softwater lake. These values are in reasonable agreement with values for isolated fulvic acid (Table 2.6), which would be expected to dominate the natural solutions.

2.4.4 Size and shape

The key aspects of size and shape are the degree of solvation, i.e. how much solvent (water) is associated with the humic 'particle' in solution, and conformation, which is how different parts of the molecule are positioned relative to one another. A familiar instance of the importance of conformation is in enzymes, the conformations of which are crucial in determining the geometry of the active site; this can involve the coming together of functional groups that would be far apart on the extended polypeptide chain. A full understanding of cation binding by humic substances should take into account the sizes and shapes of the molecules. While this goal has not been achieved, or even approached, mechanistic models of the interactions should at least not violate knowledge of structures.

The partial specific volume ($cm^3\,g^{-1}$) of a material is a measure of the space it occupies. Different atoms and molecular fragments make appreciably different contributions to the overall value. For example, a phenyl group has a partial specific volume of $0.8363\,cm^3\,g^{-1}$, while the values for methylene and carbonyl groups are 1.1621 and $0.5177\,cm^3\,g^{-1}$ respectively (Birkett *et al.*, 1997). Thus it can be seen that the partial specific volume of humic material depends on its molecular structure. Measured partial specific volumes of humic substances lie in the range $0.45–0.71\,cm^3\,g^{-1}$ (Jones *et al.*, 1995).

In solution, solvent molecules are present within the domain occupied by the humic molecule, and consequently the hydrodynamic volume – the effective volume governing transport behaviour – of the 'particle' comprising the solute and associated solvent is greater than that of the solute alone. The hydrodynamic volume v_h is given by

$$v_h = \frac{M_w}{N}(v_2 + \delta_1 v_1^0) \tag{2.3}$$

where N is Avogadro's number ($6.023 \times 10^{23}\,mol^{-1}$), v_2 is the partial specific volume of the humic material, δ_1 is the number of grams of solvent per gram of dry humic material, and v_1^0 is the specific volume ($cm^3\,g^{-1}$) of the solvent. According to Buffle (1988), the value of δ_1 for water associated with humic matter should be between 0.8 and 1.5, based on the attraction of water for hydrophilic groups. Taking $0.6\,cm^3\,g^{-1}$ as a typical value for v_2, these two values correspond to the solvent occupying 57% and 71% respectively of the total particle volume. The value is probably greatest when the molecules bear their highest charges, i.e. at high pH, where most of the protons are dissociated. Other authors have postulated extended structures for humic substances, containing greater amounts of water, up to 90% or more by volume (see below).

For fulvic acids with $M_w = 1600$, application of equation (2.3) gives values of v_h of $2.7–5.6\,nm^3$, depending on the value of δ_1 chosen. Values of v_h can also be estimated from measured intrinsic viscosities of fulvic acid solutions, again on the assumption that the particles are spherical (Clapp *et al.*, 1989). From data compiled by Clapp *et al.* (1989) for fulvic acid samples in water and 0.001–0.1 mol dm^{-3} NaCl solutions at various pH values, a range of v_h of 2.2–5.9 nm^3 is obtained for an assumed molecular weight of 1600. By the same means, an intrinsic viscosity of 2.62 cm^3 g^{-1} in 0.5 mol dm^{-3} NaCl for an aquatic fulvic acid (Birkett *et al.*, 1997) translates to $v_h = 2.9$ nm^3. Thus the viscosity data are in accord with the assumption of spherical particles and the degree of solvation suggested by Buffle (1988).

The situation with humic acids is less clear than that for fulvic acids, because of the doubts about molecular weights discussed in the previous section. Swift (1989b) presented a discussion of the size and shape of humic acids, based largely on the frictional coefficient data of Cameron *et al.* (1972) for fractionated soil humic acids. He argued that humic acids are linear polymers having an open 'random-coil' conformation, the particles being perfused throughout with solvent, the mass being greatest at the centre and decreasing towards the outer limits. Such a picture of humic acids in solution is accepted in a number of texts (Buffle, 1988; Hayes *et al.*, 1989; Clapp *et al.*, 1993; Stevenson, 1994). It also accords with the view of Marinsky (1987) that humic substances are gels, i.e. macromolecules that have large amounts of solvent trapped within them. However, the data of Cameron *et al.* (1972) for the smaller humic acid fractions can be explained reasonably well in terms of solvated spheres with δ_1 in the range 0.3 to 1.5, corresponding to water contents of *c.* 30–70%. Alternatively, the water content could be more constant, and there could be modest deviations from sphericity.

In summary, the physical data suggest that dissolved fulvic acids are approximately spherical, with up to 75% of the space of the 'particle' being occupied by water. Humic acids of molecular weight less than about 50 000 may be similar, while larger humic acid molecules could tend towards more open structures, with greater amounts of associated solvent. The smaller molecules undergo modest changes in shape and size as their degree of charging varies, and as electrolyte concentrations are changed. Larger molecules may expand and contract to a greater extent. Given the fact that humic molecules are heterogeneous, molecules with differing conformational properties may exist.

While there are difficulties in understanding humic substances in solution, uncertainty is still greater when it comes to solid-phase humic substances. For humic acid, the solid phase – either precipitated by protons and metal ions, or adsorbed to mineral surfaces – is the commonest state in the environment. According to Swift (1989a), the solid state would be little different from the solution state if humic substances have 'an essentially condensed molecular conformation'. If, however, the random-coil model is adopted, rather large changes in conformation are expected. The binding of ions neutralises humic charges, reducing intramolecular charge repulsion. The macromolecule starts to collapse, water is expelled and different parts of the molecule are forced together to interact by hydrogen bonds and hydrophobic interactions. Swift envisaged that the radius may decrease to less than half its value in solution, implying a fall in volume of about an order of magnitude.

It must be expected that molecular size and shape influence, and in turn are influenced by, cation binding. Changes in conformation associated with proton dissociation have already been mentioned, as have the precipitating effects of bound multivalent ions. The folding of the molecules may bring together ligand groups to form multidentate sites, and the binding of one metal ion may cause a change in conformation that influences the binding of subsequent ones (cooperativity). If the structures are quite rigid, then the question arises as to whether all the cation-binding sites can be located at or near the outer surfaces of the molecules. The electrostatic field associated with the humic molecule – which can markedly influence cation binding – will depend upon size and shape, the location of charged groups, and the extent to which solvent pervades the structure. The uncertainty about the physical chemistry of humic matter, as discussed in the previous paragraphs, limits our ability to explain cation-binding behaviour. However, it may be possible to draw conclusions about size and shape from information on cation binding.

2.4.5 Specific surface area of HS

Specific surface area refers to solids, and is usually expressed in units of $m^2\,g^{-1}$. For example, spheres of diameter 1 μm and density $2\,g\,cm^{-3}$ have a specific surface area of $3\,m^2\,g^{-1}$. Chiou *et al.* (1990) reported a value of $1\,m^2\,g^{-1}$ for humic acid, from determinations of nitrogen adsorption (BET method). This value is controversial, and its validity has been debated (Pennel & Rao, 1992; Chiou *et al.*, 1992). It is interesting to compute the specific surface area of a humic acid of given particle size for comparison; for a humic acid of molecular weight 50 000 and $\delta_1 = 1.5$, the value is *c.* $2000\,m^2\,g^{-1}$, which might be considered the effective surface area of the dissolved or hydrated form. For a typical fulvic acid ($M_w = 1600$, $\delta_1 = 1$), the value is *c.* $5000\,m^2\,g^{-1}$. The $1\,m^2\,g^{-1}$ value of Chiou *et al.* (1990) is no doubt applicable to the dried material, but is unlikely to have environmental relevance, except for situations where severe drying takes place. Effective surface area may be used in computing the accumulation of counterions around humic molecules, and is therefore highly relevant to the binding of cations (Chapter 9).

2.4.6 Optical properties

Humic materials absorb light in the ultraviolet and visible ranges (MacCarthy & Rice, 1985; Bloom & Leenheer, 1989; Stevenson, 1994). The spectra are without major features, the absorbance declining approximately exponentially with increasing wavelength (Fig. 2.6). The spectra result from mixtures of chromophores, the most important of which are believed to be aromatic, quinone and conjugated structures. Charge-transfer bands and light scattering may be significant at wavelengths longer than 400 nm. Because of the absorption in the visible region, humic substances are coloured yellow to brown. Typically, a 10 mg dm^{-3} solution of fulvic acid in a 1 cm path length cell has an absorbance of about 0.1 at 300 nm. Humic acids tend to have higher absorbances. Molar

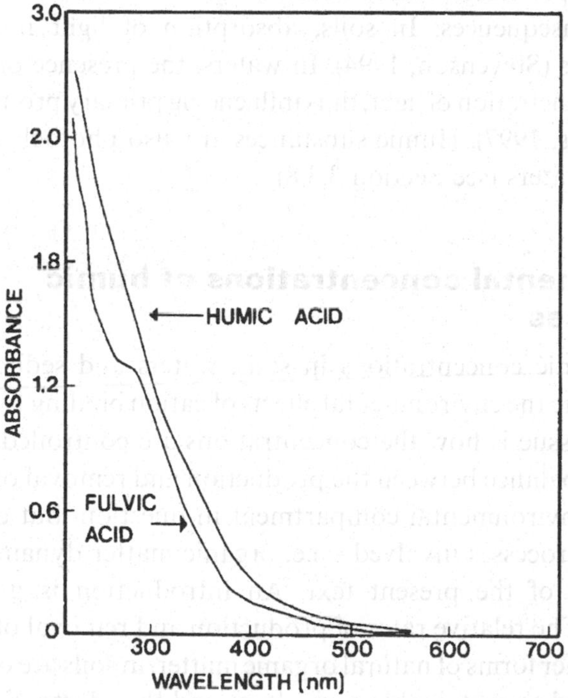

Figure 2.6. UV–visible spectra of humic and fulvic acids. [Reproduced from Bloom, P.R. & Leenheer, J.A. (1989), Vibrational, electronic, and high-energy spectroscopic methods for characterizing humic substances. In *Humic Substances II. In Search of Structure*, ed. M.H.B. Hayes, P. MacCarthy, R.L. Malcolm & R.S. Swift, pp. 409–446, by permission of John Wiley & Sons, Inc.]

absorptivity at 280 nm correlates strongly with aromaticity (Chin *et al.*, 1994).

Humic absorption spectra are influenced by interactions of the humic matter with cations. As pH is raised and protons dissociate, the absorbance tends to increase. The binding of some metals, notably iron, intensifies colour. The absorbance properties are useful in field studies as a simple means of estimating concentrations of dissolved humic substances, and they provide insight into sources of dissolved organic matter (Buffle, 1988; Tipping *et al.*, 1988a).

Humic substances are fluorescent. Bloom & Leenheer (1989) suggest that there are two main fluorophores. One, excited in the wavelength range 315–390 nm, is probably a carboxyphenol, while the other, excited in the range 415–470 nm, remains to be identified. The fluorescence tends to increase with pH, as proton dissociation takes place. Bound metal ions quench the fluorescence, a phenomenon which has been used both to estimate binding and to obtain information about the natures of the metal–humic interactions (see Chapters 6 and 8).

The optical properties of humic substances have several important environmental consequences. In soils, absorption of light influences warming properties (Stevenson, 1994). In waters, the presence of humic matter affects the penetration of light, thus influencing primary productivity (Schindler & Curtis, 1997). Humic substances are also photochemically active in natural waters (see Section 3.3.8).

2.5 Environmental concentrations of humic substances

Knowledge of humic concentrations in soils, waters and sediments is needed to appreciate the environmental effects of cation binding by humic matter. A related issue is how the concentrations are controlled, which depends upon the balance between the production and removal of humic matter from the environmental compartment in question, but detailed discussion of the processes involved – i.e. organic matter dynamics – is beyond the scope of the present text. An introduction is given by Stevenson (1994). The relative rates of production and removal of humic substances, and other forms of natural organic matter, in soils are of major concern with regard to sustainable agriculture and the effects of climate change (Jenkinson, 1981; Parton *et al.*, 1987; Schlesinger, 1997). The turnover of organic matter in sediments, which is important in relation to the formation of fossil fuels, has been discussed by Ishiwatari (1985), Vandenbroucke *et al.* (1985) and Huc (1988). Dissolved organic matter dynamics, in surface waters, groundwaters and marine waters, has been discussed by Buffle (1984, 1988), Malcolm (1985) and Thurman (1985), and in soil waters by Zsolnay (1996) and Tipping (1998a).

Because humic substances are mixtures, not defined by a single chemical characteristic, their concentrations are difficult to measure. Strictly, all that can be done is to isolate, quantitatively, the material in a given sample by the operational methods that define the humic compounds, and for soils and sediments this is the usual method. For waters, where repeated measurements on numerous occasions are often required, full scale isolation from each sample is impractical. The most useful humic-related determinand is dissolved organic carbon (DOC). Although not a proper measure of humic matter, DOC concentrations provide a useful indication, since much of the DOC is usually in humic substances. Optical absorbance may provide a useful adjunct to DOC in estimating humic concentrations. An interesting approach is that of Quentel *et al.* (1987) who determined humic matter in seawater by polarography, the method being based on the adsorption of the humic material on a mercury electrode. In terms of cation binding, 'active concentrations' can be estimated by determining buffer capacity or the binding of metals (see Chapters 13 and 14).

2.5.1 Soil contents of humic substances

The organic matter content of soils varies from 1% or less in the mineral horizons of some tropical and sandy soils to nearly 100% in peats. In most mineral soils, the content is greatest in the upper parts of the profile, due to the input of plant litter. Some examples are shown in Fig. 2.7. The fraction of the organic matter that is humic substances can be high; according to Stevenson (1994) up to 80% of the organic matter in mineral soils is extractable with NaOH, and therefore is classified as humic and fulvic acids. Kononova (1961) presented data for a number of Russian soil types in which the humic acid plus fulvic acid contents were in the range 30–60%. Some or all of the material that is not extractable with base is humin. The amount of base-extractable organic matter from peats and the organic horizons of podzols may be comparatively low, 25% or less, if the plant litter is decomposing slowly (Mathur & Farnham, 1985). The relative amounts of the three main humic fractions vary among soils. For example, in mollisols (grassland soils), fulvic acid, humic acid and humin are present in approximately equal amounts, whereas in spodosol B horizons fulvic acid comprises approximately 60% of the total carbon, humic acid and humin about 20% each (Stevenson, 1994).

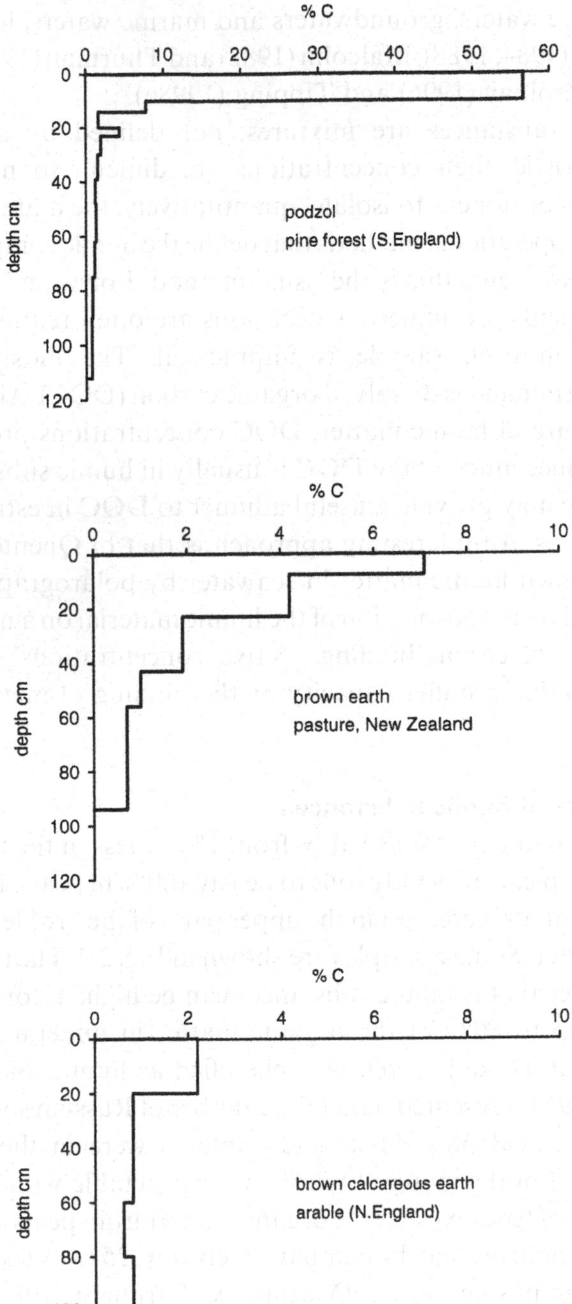

Figure 2.7. Carbon profiles in three soils: podzol, Gooddy *et al.* (1995); brown earth, O'Brien & Stout (1978); brown calcareous earth, Jarvis *et al.* (1984). Note that Gooddy *et al.* (1995) reported their results with the boundary between the FH layer (top 10 cm) and the underlying soil as the zero reference depth.

To consider cation binding by humic substances in soils, concentrations can usefully be expressed in terms of mass of humic matter per unit volume of soil water. For example, in a peat which is 80% by weight water, with 20% of the solid organic material being humic substances, the effective concentration of the humic substances is 50 g dm^{-3}. In B horizons of forest soils studied by Lofts *et al.* (2001a), the average content of humic matter involved in binding ions is estimated to be *c.* 40 mg g^{-1}. For a typical water content of 40% by weight, this translates to an effective humic matter concentration of 60 g dm^{-3}. Since there are 5–10 × 10^{-3} moles g^{-1} of proton-dissociating groups in humic substances, the concentrations of these groups are approximately 0.25–0.5 mol dm^{-3} for the peat and 0.3–0.6 mol dm^{-3} for the B horizons. Thus it can be seen that the humic matter in soils is present at very high effective concentrations, and will therefore exert a powerful influence on the solid-solution partitioning of cations. This is not to say that humic matter dominates cation binding in all soils. For example, in B horizons oxide phases may also contribute significantly, while in other soil types clay minerals may be important. It also has to be borne in mind that (a) soil water is not homogeneous, so that effective humic concentrations may vary spatially within the soil, and (b) not all of the humic binding sites may be exposed to soil water.

2.5.2 Concentrations of humic substances in subsurface waters

According to Thurman (1985) and Herbert & Bertsch (1995), concentrations of DOC in soil waters vary from 1 to 70 mg dm^{-3}. However, in organic-rich soils, the upper limit can be much higher. For example, van Hees *et al.* (2000a) reported a value of nearly 400 mg dm^{-3} for the O horizon of a Swedish forest podzol. The values obtained depend upon the method used for sampling, because of the inhomogeneity of soil water. Thus, tensionless lysimeters collect freely moving water that has not made full contact with all soil surfaces, whereas if pore water is isolated by centrifugation the concentrations tend to be higher. The issue of soil pore sizes and DOC has been discussed in detail by Zsolnay (1996). He suggested that pores belong to three classes. In Class I (< 0.2 μm) the DOC is subject only to diffusive transport, and is relatively unavailable for metabolism. In Class II (0.2–6 μm) transport is still mainly diffusive but microbes are able to access the DOC. In Class III (> 6 μm) transport is mainly convective and subject to a wider range of biotic metabolism. He estimated that in agricultural soils the DOC is distributed approximately in the ratio 2:5:3 in Class I, II and III pores respectively.

Herbert & Bertsch (1995) reviewed information on the dissolved organic matter of soil waters. They concluded that about 50% consists of hydrophobic (fulvic and humic) acids, and about 30% of macromolecular hydrophilic acids. The remainder is due to identifiable compounds such as carbohydrates, carboxylic acids, amino acids and hydrocarbons. The acid–base properties of the hydrophilic acid fraction are generally similar to those of the hydrophobic acids (see Chapter 7). Concentrations of DOC in deep groundwaters are generally quite low. Thurman (1985) gives a range of 0.2 to 15 mg dm^{-3}, and a median value of 0.7 mg dm^{-3}. The low concentrations reflect long residence times, adsorption on aquifer material, and low supply from aquifer solids. However, there are some groundwaters with much higher concentrations. At the Gorleben salt dome in Germany, some groundwater contains up to 100 mg dm^{-3} of humic DOC due to the leaching of 'brown coal' sand and clay (Dearlove *et al.*, 1991). Even more extreme is groundwater formed when 'trona water', formed by the dissolution of sodium bicarbonate, leaches kerogen from oil shales, giving rise to DOC concentrations of up to 40 g dm^{-3} (Thurman, 1985).

2.5.3 Concentrations of humic substances in surface waters

Concentrations of DOC in freshwater streams, rivers and lakes vary from less than 0.5 mg dm^{-3} to as high as 100 mg dm^{-3}, depending on the nature of the catchment, including the climate, the trophic status of the water body, and pollutant inputs. The highest values are found for water draining catchments containing large amounts of peat. The average value for the major rivers of the world is *c.* 5 mg dm^{-3} (Thurman, 1985). The majority of the dissolved organic matter comprises fulvic and humic acids, together with hydrophilic acids (e.g. Malcolm, 1993). Concentrations of DOC, and therefore of humic substances, in freshwaters vary in time, more so than in other environmental compartments. In temperate regions, a seasonal variation is commonly observed, as illustrated by Fig. 2.8. The variation appears to be due to the accumulation of potentially soluble organic matter in soil during the summer, when microbial activity is highest, followed by leaching in autumn (Scott *et al.*, 1998).

The concentration of DOC in the open ocean is low, *c.* 1 mg dm^{-3} in the surface water, and *c.* 0.5 mg dm^{-3} at depth. Estuaries and coastal waters may have higher concentrations, due to the influence of local terrestrial sources. Humic substances are formed from marine algal debris (Section 2.3.1). Terrestrially derived humic matter may also contribute significantly (Mantoura & Woodward, 1983), although recent work

(Hedges *et al.*, 1997) suggests that it may have a relatively short turnover time in the oceans.

2.5.4 Contents of humic substances in aquatic sediments

Organic matter is present in suspended sediments in lakes, rivers, estuaries and coastal seawaters, originating for the most part from eroded soils. The humic matter present in suspension is therefore largely terrestrial in origin. In the world's major rivers, the concentration of particulate organic matter is similar to that of the DOC (Thurman, 1985), but the relative amounts vary depending on the amount and nature of the suspended material. Humic matter plays a significant rôle in cation binding by riverine suspended sediments (Lofts & Tipping, 1998).

Ishiwatari (1985) reported that in six Japanese lakes the average total organic matter content of the near-surface sediment was 44 mg g^{-1}, of which 47–67% (depending upon the extraction time) was humin, 11–25% fulvic acid, and 17–22% humic acid. Kemp & Johnston (1979) found that 70–80% of the organic matter of near-surface sediments from the North American Great Lakes was present as 'nonbiochemicals', i.e. compounds not identifiable and therefore principally humic substances. At greater depths in the sediment of Lake Biwa, the organic matter content is lower than in surface layers, falling to 10 mg g^{-1} at a depth in the sediment of 100–200 m. The extractable humic substances content falls also, so that at

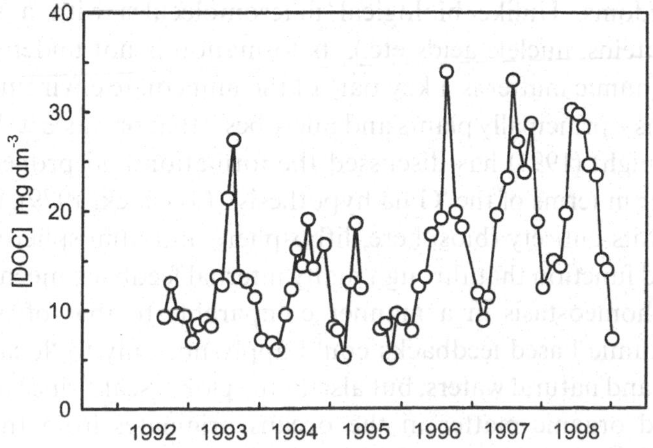

Figure 2.8. Temporal variation in the concentration of dissolved organic carbon (DOC) in a shallow peaty pool in the Pennines (N. England). The gap in the record in the summer of 1995 indicates a period of drought.

100–200 m it is only 16% of the total organic matter (Ishiwatari, 1985). According to Vandenbroucke *et al.* (1985), the organic content of marine sediments is typically 100 mg g^{-1}, of which 90% is kerogen, consisting of humin (60%), humic acid (25%) and fulvic acid (15%).

As with soils (see Section 2.5.1), the effective concentrations of sedimentary humic substances can be expressed in terms of mass per unit volume of porewater. The contents quoted for surface Japanese lake sediments in the previous paragraph correspond to a concentration of humic substances (including humin) of *c.* 2 g dm^{-3}, on the assumption that the water content of the sediments is 95%. At this concentration, the humic substances will have a substantial influence on the distributions of cations between the sediment solids and the porewaters.

Concentrations of DOC in the interstitial waters of sediments depend on whether the sediment is oxic or anoxic. Thurman (1985) quotes a range of 4–20 mg dm^{-3} for the former and 10–390 mg dm^{-3} for the latter. Much of the dissolved material may be due to humic substances.

2.6 Humic substances – accident or design?

Environmental functions of humic substances were described in Chapter 1, and the rest of the book is about one of the main functions, i.e. cation binding. Given these important rôles of humic matter in both terrestrial and aquatic ecosystems, the question arises as to whether selection of the functions has occurred in the course of evolution. Humic matter occupies a unique position in the natural environment, straddling the biotic and abiotic kingdoms. Unlike biological macromolecules with a specific purpose (proteins, nucleic acids etc.), its formation is not under genetic control. Yet humic matter is a key part of the immediate environment of the organisms – principally plants and microbes – that produce it (Wetzel, 1975). McKnight (1991) has discussed the formation and properties of humic matter in terms of the 'Gaia hypothesis' (Lovelock, 1979), i.e. that the Earth in its entirety (biosphere, lithosphere and atmosphere) has a structure and function that through many internal feedback mechanisms maintain a homeostasis in a manner comparable to that of a single organism. Humic-based feedbacks could apply not only to 'local' units such as soils and natural waters, but also at the global scale, since much of the dissolved organic matter in the oceans originates from the land, implying an influence of terrestrial organisms on the marine ecosystem.

An inescapable feature of humic matter is its heterogeneity, and this will be a recurring theme throughout our discussion of cation binding.

Buffle *et al.* (1990) have propounded the view that heterogeneity in the cation-binding sites of humic matter, and other 'homologous complexants' (e.g. mineral and biological surfaces), leads to the buffering of environmental systems against variability. Thus, these materials have exerted control over the chemical media in which organisms have developed and evolved. MacCarthy & Rice (1991) have also argued that the heterogeneity of humic substances confers stability and persistence, because microoganisms cannot develop a sufficient array of enzymes to bring about complete and rapid decomposition. Several authors (e.g. Dubach & Mehta, 1963; Gjessing, 1976) have speculated that no two humic molecules are exactly the same; MacCarthy & Rice (1991) conclude that this is probably true, and state that 'in the case of humus the mixture is the message'. This is worth bearing in mind when considering the molecular-level interactions of humic substances with cations, not least to stop ourselves complaining that sample heterogeneity makes it difficult to interpret the results of our experiments.

3

Environmental solution and surface chemistry

This chapter reviews briefly a number of basic topics in the field of physical, colloid and surface chemistry. The aim is to prepare the ground for consideration of cation–humic interactions *per se*, and to show their place in the chemistry of the aqueous natural environment (surface and groundwaters, soils and sediments). Fuller general accounts of the concepts and phenomena covered may be found in chemistry textbooks, and in texts on colloid and surface chemistry (e.g. Hiemenz, 1977; Shaw, 1978), and environmental chemistry, soil chemistry and aqueous geochemistry (e.g. Sposito, 1989; Morel & Hering, 1993; Schwarzenbach *et al.*, 1993; Brezonik, 1994; Stumm & Morgan, 1996; Harrison & De Mora, 1996; Langmuir, 1997).

3.1 Solutions and solutes

A solution is a homogeneous mixture of two or more substances. Usually the most abundant is called the solvent, while the others are solutes. The solubility of a substance is its dissolved concentration at saturation, i.e. when the solution is in equilibrium with undissolved material, at a given temperature and pressure. Solutions may involve all kinds of substances, and may be formed by mixtures of liquids, solids and gases, but we are exclusively concerned with aqueous solutions.

3.1.1 Factors governing aqueous solubility

The solubility of a compound in water depends upon the polarities of its chemical groupings, and whether it can ionise. Common examples of polar groups are —OH, —NH_2 and >C=O, which confer solubility by the formation of hydrogen bonds with water molecules. For example, the aqueous solubility of glucose is very high due to its five hydroxyl groups and one carbonyl group, all of which form hydrogen bonds. A special property of water, and one that makes it a good solvent, is its ability to act as both donor and acceptor in the formation of hydrogen bonds.

Simple hydrocarbons are the most obvious examples of apolar compounds, i.e. they lack polarisable groups. Thus they do not interact strongly with the polar water molecules, and are poorly soluble in water. However, even tiny quantities of some substances can exert significant biological or ecological effects, and therefore very low concentrations may be highly significant in the natural environment.

Ionic solutions are formed when the solute dissociates into ions. Examples are KCl (dissociation of an ionic solid) and HCl (ionisation of a neutral molecule). The strong attraction of the ions for water molecules (solvation) – which is considerably greater than that of simply polar groupings – provides the major driving force for dissolution. Some solutes dissociate completely in water, for example $NaNO_3$, while others only do so partially, for example acetic acid (CH_3COOH).

3.1.2 Amphiphiles

Generally, the balance among solubility attributes will determine the solubility of a given compound. A good illustration is provided by *amphiphiles*, organic compounds that consist of a hydrophobic part and a hydrophilic one. Detergents consist of a hydrocarbon chain and a polar (often ionised) head group. In water they form micelles, comprising a hydrocarbon core surrounded by the solvated polar groups. Phospholipid bilayers form similarly. Amphiphiles also accumulate at the water–air interface and adsorb to hydrophobic surfaces. The key feature of these reactions is the removal of the hydrophobic moieties from water, within which they impose ordered local structure. The entropy-increasing abolition of this order drives the reaction, and is known as the *hydrophobic effect* (Tanford, 1980).

With regard to humic substances, the information on molecular groupings (Table 2.5), together with the hypothetical structures of Fig. 2.5, show that hydrophobic aliphatic and aromatic hydrocarbon moieties comprise substantial fractions of these materials. There are also many polar and charged groups, which are hydrophilic. Thus, humic substances can be classified as amphiphiles, although they are clearly more complex and heterogeneous than detergents and phospholipids. The amphiphilic properties of humic substances, especially in relation to cation binding, are discussed in Chapter 12.

3.1.3 Solubilities of macromolecules

Most biological and synthetic macromolecules are polymeric in nature, consisting of subunits joined by repeating covalent bonds, usually of a single type. Examples are polystyrene, proteins (polypeptides), and nucleic acids. Humic substances are the most abundant macromolecular compounds on the planet, but there is no good evidence that they are polymeric, in the sense of well-defined repeating subunits joined by one type of bond.

Macromolecular solubility depends upon the chemical properties of the subunits (polarity, degree of ionisation) and also on the degree to which they are exposed to solvent water. For example, globular proteins tend to fold so that the majority of the apolar groupings form a central core, surrounded by the more hydrophilic moieties. Hydrogen bonding is especially significant when polar groups find themselves in hydrophobic regions, for example in the interiors of globular proteins, and this may well also apply to humic matter. Thus the molecular conformation (tertiary structure) plays a significant rôle in determining the solubility properties of macromolecules. Tanford (1961) and Richards (1980) discussed in detail the properties of macromolecules, including solubility.

3.1.4 Solutes in natural waters

The multitude of solutes in natural waters can broadly be divided into major (present in high concentrations) and minor (present in trace concentrations). Most of the major solutes are ions. Solute concentrations in natural waters are governed by a variety of chemical, biological and physical processes, giving rise to a wide range of compositions, illustrated by the examples of Table 3.1. As discussed in Chapter 2, the dissolved organic carbon (DOC) is predominantly in humic substances, but also in identifiable organic compounds of both natural and anthropogenic origin. Sizes of solutes are compared in Fig. 3.1.

3.2 Natural particulate matter

Suspended particulate matter is always present in surface waters, with concentrations ranging from 1 mg dm^{-3} or less in some freshwaters and the open ocean, to tens of grams per dm^3 in rivers draining regions with high erosion rates (e.g. the Yangtse River). High concentrations are also encountered in estuaries where the energy of water movement is sufficient to suspend coarser material (Eisma, 1993). In soils, sediments and groundwaters, the solution is in intimate contact with immobile solids, the

Table 3.1. Compositions of natural waters (major solutes).

	pH	Na	Mg	Al	K	Ca	NH_4	Cl	NO_3	SO_4	ΣCO_3	Si	DOC
Rain water N. England[a]	4.5	0.090	0.010	0	0.003	0.006	0.030	0.10	0.020	0.030	0	—	0.5
Upland stream Gaitscale Gill, N. England[a]	5.1	0.13	0.019	0.008	0.008	0.009	0	0.13	0.039	0.029	0	0.033	0.7
Bog water Pennines, N. England[a]	4.7	0.13	0.020	0.015	0.008	0.072	0.018	0.17	0.001	0.062	0	0.012	34
High altitude continental lake Loch Vale, Colorado[b]	6.3	0.026	0.011	—	0.005	0.038	0.003	0.006	0.017	0.019	0.047	0.15	—
Tropical river Rio Solimões, Amazon basin[c]	6.9	0.10	0.05	0.002	0.023	0.18	—	0.087	—	—	0.56	0.14	13
Tropical river Rio Negro, Amazon basin[c]	5.1	17	4.8	0.004	8.4	5.3	—	0.048	—	—	0.14	0.071	11
Industrially influenced river River Rhine, German–Dutch border[d]	8	4.3	0.45	—	0.19	2.1	0.085	5.0	0.23	0.81	2.5	0.092	6
A-horizon forest soil solution Crowthorne, S. England[e]	3.0	1.3	0.084	0.085	0.29	0.27	0.012	0.95	0.050	0.79	0	0.63	260
Carbonate groundwater Florida aquifer[f]	7.7	0.34	0.50	—	0.026	1.4	—	0.34	—	0.55	2.6	—	—
Seawater[g]	8	468	53	0	10	10	0.001	545	0.01	28	2.4	0.04	1
Saline surface water Mono Lake, California[h]	9.6	930	1.4	—	30	0.11	—	380	—	77	260	0.23	—

Concentrations are in mmol dm^{-3}, except for DOC (mg dm^{-3}).
A zero indicates less than 0.001, a dash indicates no reported value.
[a]Unpublished data from the author's laboratory. [b]Baron & Bricker, 1987. [c]Furch, 1984. [d]Zobrist & Stumm, 1981. [e]Goody *et al.*, 1995. [f]Freeze & Cherry, 1979. [g]Riley & Chester, 1971. [h]Eugster & Hardie, 1978.

smallest and/or most porous of which contribute a large surface area. Chemical interactions between the fine particulate matter and the aqueous phase can exert a powerful control on solution composition, by dissolution and adsorption reactions (Section 3.3).

3.2.1 Particle composition

Natural fine particulate materials consist of inorganic (mineral) and organic components. The mineral matter includes rock fragments, alteration products of chemical weathering, such as oxides and aluminosilicates, and precipitated oxides, carbonates and sulphides. Organic materials include living organisms, predominantly bacteria, protozoa, algae and fungi, and decomposition products, of which the main form is humic matter. The humic matter may be present as insoluble material *per se*, or as soluble forms adsorbed to solid phases.

3.2.2 Particle size

Particle size is important because it governs the physical behaviour of particulate matter, notably sedimentation, erosion and permeability. Fine particulates can be physically fractionated into clay (equivalent diameter < 2 μm), silt (2–60 μm) and sand (60 μm – 2 mm). Size is also significant in

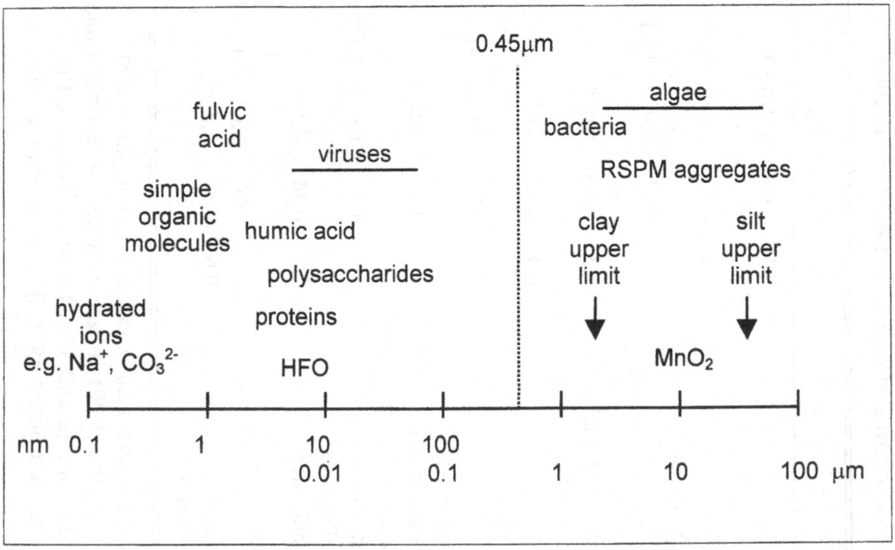

Figure 3.1. Approximate sizes of chemical and biological components of natural waters, based on diagrams in Buffle (1988) and Stumm & Morgan (1996). HFO, hydrous ferric oxide (primary particles); RSPM, riverine suspended particulate matter at moderate flow; the MnO_2 particle size refers to lake water precipitates. The 0.45 μm line indicates the size of filter conventionally used to distinguish 'dissolved' from 'particulate' components.

terms of chemical reactivity: the more finely divided is a solid the more reactive it is per unit mass. Figure 3.1 shows typical sizes of entities significant in natural water chemistry.

3.2.3 Colloids

Colloids are materials with dimensions in the range 1 nm – 1 μm (Fig. 3.1). They can have very large surface areas. For example, when freshly precipitated, the iron oxyhydroxide ferrihydrite exists as approximately spherical particles of diameter approximately 5 nm, with a specific surface area of approximately 600 $m^2 g^{-1}$ (Dzombak & Morel, 1990). Classically, colloidal dispersions in water are regarded as being either hydrophobic (finely divided insoluble material) or hydrophilic (chiefly dissolved macromolecules).

In research on natural waters, suspended solids are commonly separated by passage of the water sample through filters. The most commonly used filter has an average pore size of 0.45 μm. Material that passes such filters is conventionally referred to as dissolved (Fig. 3.1). However, it will usually include small insoluble colloids, which may, under some circumstances, account for most of a 'dissolved' element or chemical species.

3.3 Physico-chemical interactions in environmental aqueous systems

Our main interest is in reactions that are reversible, i.e. do not involve the formation or breaking of strong bonds, such as the C—H bond. Cation–humic interactions fall into this category, along with reactions involving other solutes, colloids and surfaces. The study of these reactions accounts for a large part of environmental chemical research. The following sections provide a synopsis of the most important types of interaction, and possible involvements of humic substances are pointed out. The subdivision of processes is somewhat artificial, and extra complexity arises from multiple or simultaneous interactions.

Knowledge of these interactions is a key requirement for the understanding and prediction of transport, retention and bioavailability of solutes and other components in natural systems. The chemical knowledge has to be combined with knowledge about physical and biological processes to describe dynamic aspects.

3.3.1 Complexation of metal ions and protons

The interaction of dissolved metal species with other types of solute is known as complexation. A simple example is the reaction of cadmium ions with chloride ions

$$Cd^{2+} + Cl^{-} = CdCl^{+} \tag{3.1}$$

The non-metal in these interactions (Cl^{-} in the present case) is known as the ligand. Protons may undergo similar reactions with ligands, for example

$$H^{+} + CO_3^{2-} = HCO_3^{-} \tag{3.2}$$

Many ligands exist, including simple inorganic species like Cl^{-}, HCO_3^{-}, SO_4^{2-}, and NH_3, and low molecular weight organic compounds such as oxalic and citric acids. Dissolved humic substances are often the dominant complexing agents in natural waters.

Figure 3.2 shows how the concentration of Cd^{2+} is predicted to depend upon the concentration of Cl^{-}, and how Cu^{2+} binds to fulvic acid. The cadmium is appreciably complexed at concentrations of Cl^{-} greater than 0.01 mol dm^{-3}. Even low concentrations of fulvic acid are sufficient to complex Cu significantly, and at concentrations greater than 0.01 g dm^{-3} nearly all the metal is in the form of fulvic acid complexes. Thus, a solution of copper containing fulvic acid has a much lower concentration of Cu^{2+} than a similar solution lacking the organic material. The concentration of Cu^{2+} is a measure of the tendency of the metal to enter into other interactions, and so the general effect of fulvic acid is to attenuate the reactivity of the metal, including its participation in other chemical

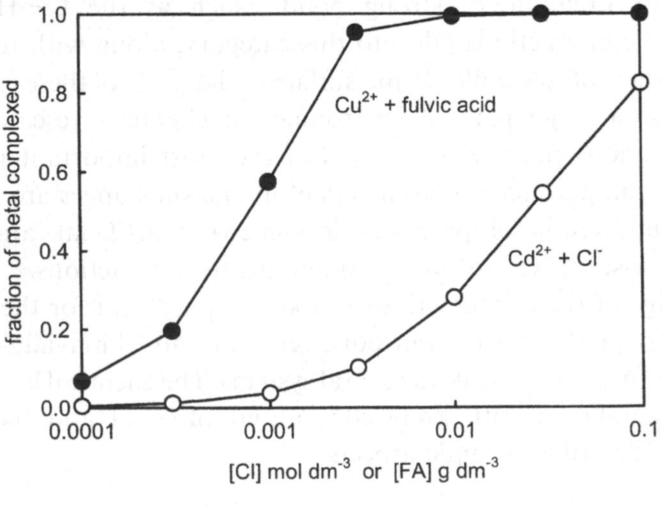

Figure 3.2. Complexation of Cd^{2+} by Cl^{-} and of Cu^{2+} by fulvic acid (FA) at pH 7, calculated with WHAM (Tipping, 1994).

reactions, and its uptake by biota. The principles of complexation are explored in more detail in Chapters 4 and 5.

Although ligands are classically regarded as dissolved species, the natural environment is not so well-ordered. Thus, when it comes to dealing with the binding of ions by solid-phase organic matter (mainly humic substances), it is helpful to treat the interactions in the same way as for the dissolved materials. A similar approach has been adopted for interactions of ions with mineral surfaces (Section 3.3.2). Bacteria and algae also possess appreciable contents of functional groups on their surfaces, able to interact with ions.

3.3.2 Adsorption of ions by mineral particles

Adsorption is the accumulation of chemical species at a surface. In contact with water, the oxides of Al, Si, Mn and Fe develop surfaces with exposed hydroxyl groups that can bind and release protons, in reactions that can be represented by

$$\equiv SOH + H^+ = \equiv SOH_2^+ \tag{3.3}$$

$$\equiv SOH = \equiv SO^- + H^+ \tag{3.4}$$

They also interact with metal cations, inorganic anions like phosphate, and organic ligands

$$\equiv SO^- + Pb^{2+} = \equiv SOPb^+ \tag{3.5}$$

$$\equiv SOH + HPO_4^{2-} = \equiv SOPO_3H^- + OH^- \tag{3.6}$$

$$\equiv SOH + RCOO^- = \equiv SOOCR + OH^- \tag{3.7}$$

This type of reaction is referred to as surface complexation (Schindler & Stumm, 1987). The interactions closely parallel those in solution involving dissolved species (Section 3.3.1). They are more complicated than these simple representations suggest, because of the complexity of the solid surface, and electrostatic effects. They are of widespread importance, and their understanding and description is an active area of research. Figure 3.3 shows examples of the adsorption of cations and anions. Detailed accounts are given by Stumm (1992) and van Riemsdijk & Koopal (1992).

The surface charges of oxides are determined by the binding and dissociation of ions, especially protons, and oxides are said to possess variable charge surfaces. Other mineral surfaces bear permanent charges. For example, the isomorphous substitution of Si^{4+} in silicate structures by Al^{3+} produces fixed negative charges. In both cases, the surface charge

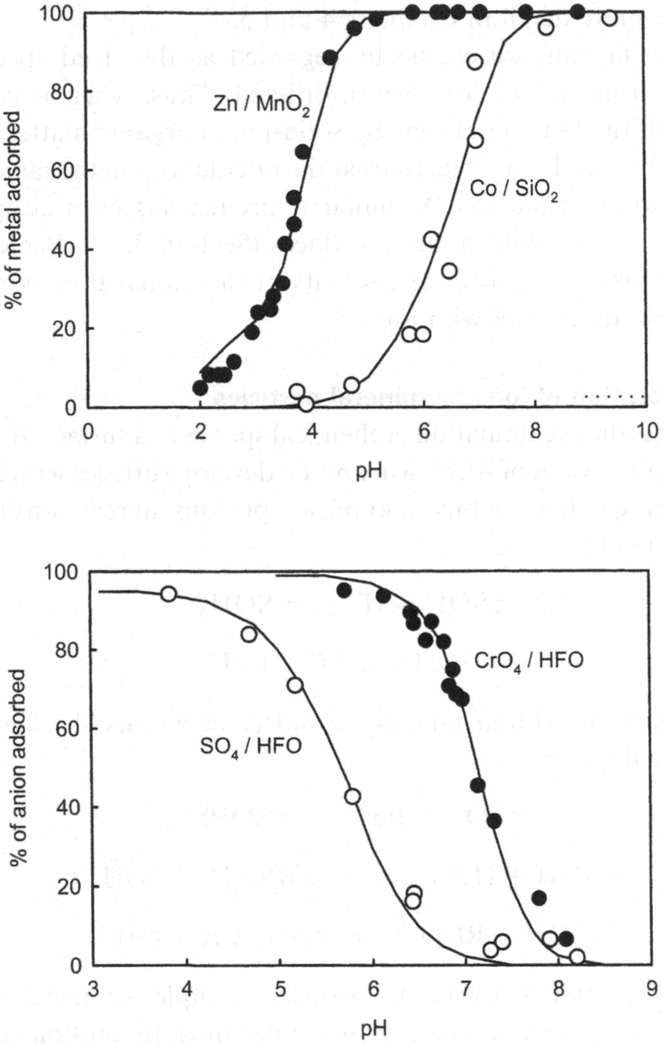

Figure 3.3. Adsorption of cations and anions by oxides and hydroxides. The points are experimental observations, the lines are fits with surface complexation models. [The data are from the following sources: James & Healy (1972), Co/silica; Catts & Langmuir (1986), Zn/manganese dioxide; Davis & Leckie (1980), sulphate and chromate/hydrous ferric oxide. The plots in the upper panel are redrawn from Lofts, S. & Tipping, E. (1998), An assemblage model for cation-binding by natural particulate matter, *Geochim. Cosmochim. Acta.* **62**, 2609–2625, with permission from Elsevier Science. Those in the lower panel are from Dzombak, D.A. & Morel, F.M.M. (1990), *Surface Complexation Modeling; Hydrous Ferric Oxide,* by permission of John Wiley & Sons, Inc.]

is balanced by counterions, to maintain local electroneutrality. Generally, two approaches to describing the distributions of counterions have been taken. One is based on the concept of an electrical double layer (see Section 4.6). The other treats the exchanger as a reaction of the form

$$X\text{–}(Na)_2 + Ca^{2+} = X\text{–}Ca + 2Na^+ \quad (3.8)$$

In soils, cation exchange capacity is regarded as a key indicator of the general chemical state. It is due principally to finely divided aluminosilicate minerals, and to natural organic matter (chiefly humic substances). Cation exchange in soils is discussed in detail by Bolt (1982).

3.3.3 Mineral dissolution and precipitation

The dissolution of minerals – chemical weathering – supplies solutes to natural waters and soils. The most important contributors are the abundant silicate minerals and carbonates. Rates of dissolution are dependent upon the composition of the mineral and its surface chemistry, and on the composition of the solution in contact with the surface. Detailed accounts are given by Sverdrup (1990) and in the book edited by White & Brantley (1995). Humic substances may affect weathering by adsorption to the dissolving surface and by complexing dissolved cations (Section 12.6).

Precipitation is akin to complexation and occurs when the solubility of the solid is exceeded. For example, a common phenomenon in natural waters is the formation of calcium carbonate

$$Ca^{2+} + CO_3^{2-} = CaCO_3(s) \quad (3.9)$$

The solution concentrations of aluminium, manganese and iron may be controlled by oxide formation in oxic environments, while those of iron and a number of trace metals may be controlled by sulphide formation under anoxic conditions. Complexation reactions in solution, perhaps involving humic substances, increase the total dissolved concentration of a metal in equilibrium with a dissolving solid. The interactions of humic matter with mineral precipitates can affect the rate of further precipitation, and transformations from one crystal form to another (Section 12.7).

3.3.4 Reduction and oxidation

These reactions involve changes in oxidation state, and exchanges of electrons. They are known as redox reactions. An example, taken from Stumm & Morgan (1996), is the reaction of PbO_2 with Mn^{2+}

$$2Mn^{2+} + 5PbO_2 + 4H^+ = 2MnO_4^- + 5Pb^{2+} + 2H_2O \quad (3.10)$$

Here, the Pb has been reduced from oxidation state IV to II, while the Mn has been oxidised, from oxidation state II to oxidation state VII. In the reaction as written, 10 electrons have been transferred from the Mn to the Pb.

In the natural environment, most redox reactions are biologically mediated. Non-photosynthesising organisms make use of the energy yielded from redox reactions involving the products of photosynthesis. Thus, bacteria and fungi draw energy from the decomposition reactions of plant litter involving the sequential reductions of O_2, NO_3^-, iron oxide and SO_4^{2-}. Many metals can exist in more than one oxidation state, and interconversions occur in the natural environment. Examples include Cr(III)/Cr(VI), Mn(II)/Mn(IV), Fe(II)/Fe(III) and Hg(0)/Hg(II).

Humic matter has redox properties, with a reduction potential of 0.5–0.7 V (Skogerboe & Wilson, 1981). For example, it can reduce Cr(VI) to Cr(III), Fe(III) to Fe(II), Hg(II) to Hg(0), and Pu(VI) to Pu(IV), and it can oxidise Cu(I) to Cu(II). Humic hydroquinone and phenolic groups have been implicated in the reactions. Humic substances play an important rôle in photochemical redox reactions in natural waters (see Section 3.3.8).

3.3.5 Sorption reactions of organic chemicals

The behaviours and fates of many anthropogenic organic compounds, including solvents, polycyclic aromatic hydrocarbons, polychlorinated biphenyls and pesticides, are influenced by their interactions with natural organic matter (especially humic substances) and mineral surfaces. The strengths of interaction of neutral molecules with natural organic matter are strongly correlated with their solubility in organic solvents, indicating the importance of the hydrophobic effect (Section 3.1.2). There is some debate as to whether the interactions of organic compounds with humic substances are due to adsorption, implying reaction at specific sites on the molecular surface, or absorption, by which is meant non-specific 'dissolution' in the three-dimensional structure of the humic molecule (Beck & Jones, 1993). Other processes contributing to binding by natural organic matter include cation exchange (when the organic compound is positively charged), hydrogen bonding, formation of charge-transfer complexes, metal cation bridging (when the organic compound is a ligand), dipole–dipole forces, van der Waals forces, and covalent bonding (Senesi, 1993). The chemicals may bind to natural organic matter in the solid or dissolved states.

Mineral particles also adsorb organic compounds, and this becomes especially important when natural organic matter is present at low concentrations, for example in groundwaters. Even polar mineral surfaces such as that of silica can bind neutral molecules (Schwarzenbach *et al.*, 1993).

3.3.6 Adsorption of humic substances by mineral surfaces

Humic substances adsorb to the surfaces of mineral particles, oxides and clays. The extent of interaction depends strongly upon the nature and concentrations of the cations that are present, and this is considered in detail in Chapter 12.

3.3.7 Aggregation

Two kinds of aggregation are commonly distinguished. The aggregation of organic molecular species due to the hydrophobic effect was described in Section 3.1.2. The second form of aggregation involves hydrophobic colloids (Section 3.2.3), dispersions of which are thermodynamically unstable, so that irreversible aggregation must occur. However, the rate of such aggregation may be very slow, in which case the dispersion is said to be colloidally stable. A stable dispersion of a hydrophobic colloid may behave like a solution in that the particles may pass through filters of small pore size, and not settle under the influence of gravity, because thermal motion is sufficient to maintain a homogeneous mixture. Colloid stability depends strongly upon the surface chemistry of the particles. For example it may be conferred by the adsorption of surface-active organic compounds (notably humic substances in natural waters and soils) or abolished by the addition of salt. Aggregation of colloidal matter may result in the formation of silt-sized aggregates, much more influenced by gravity. However, the aggregates may still have a high surface area and be chemically reactive. Thus, entities in a larger size class can have chemical properties characteristic of smaller primary particles.

Buffle *et al.* (1998) have discussed interactions among colloidal particles in natural waters. They considered the systems to comprise three types of component: compact inorganic colloids, large rigid biopolymers, and humic substances, mainly fulvic-type material. The influence of humic substances on colloid stability is discussed in Chapter 12.

3.3.8 Photochemical reactions

Photosynthesis is the prime example of a chemical reaction involving sunlight, but there are many photochemical reactions that do not involve living cells. Direct photolysis involves the absorption of light energy by a chemical. For example, an important reaction in the atmosphere is the photolysis of NO_2 to yield NO and O (atomic oxygen), while in surface waters the photolysis of organic pollutants is a significant phenomenon. A second class of photochemical reaction involves the generation of unstable products (photoreactants). Dissolved organic matter (chiefly humic substances) plays a major rôle in this process, its interaction with light energy leading to the generation of photoreactants such as peroxy radicals, superoxide anions, hydrated electrons, hydrogen peroxide and hydroxyl radicals (Cooper *et al.*, 1989; Hoigné *et al.*, 1989). These species then enter into reactions with other chemical entities.

3.4 Equilibrium and kinetics

In a closed system under constant conditions, chemical reactions reach equilibrium given sufficient time, and the concentrations of reactants and products become constant. However, natural systems are open, and receive continual inputs of energy. Many reactions are far from equilibrium in the natural environment, perhaps the most obvious examples being those between oxygen and nitrogen in the atmosphere, and between oxygen and organic compounds, including living matter. Life on Earth is made possible by the slowness of these reactions.

In solution and surface chemistry, the reactions can be fast enough that, under some conditions, equilibrium is achieved sufficiently rapidly that the reaction is essentially instantaneous. The importance or otherwise of reaction rates should therefore be assessed in terms of other (physical, biochemical) processes. If these are slow compared to the chemical reactions, then chemical equilibrium can often be assumed. Buffle (1988) classified environmentally significant processes according to their rates as follows:

I fast homogeneous chemical reactions
II slow homogeneous chemical reactions
III heterogeneous chemical reactions
IV physical mixing processes

Table 3.2 gives examples of these processes and their characteristic time scales. The time scales of Classes II and III can overlap with those of Class IV.

In principle, we should try to deal with environmental chemistry – including reactions involving humic matter – in terms of kinetics. However, equilibria are more easily dealt with, both experimentally and mathematically, because history does not have to be taken into account, and simultaneous reactions can more easily be considered. The assumption of equilibrium allows the researcher to draw on the enormous amount of available equilibrium (thermodynamic) data, determined for well-defined reactions under controlled conditions, to calculate the extents of the different reactions. Therefore, whenever the approximation of chemical equilibrium can be justified, it is made. Furthermore, even if the chemical reactions are slow, equilibrium concepts are useful – if not essential – to establish the 'chemical boundaries' of the system (Stumm & Morgan, 1996). Detailed accounts of the applications of kinetics in aquatic and soil chemistry respectively are given by Brezonik (1994) and Sparks (1989).

3.5 Chemical speciation

It will be evident from the foregoing sections that there are many reactants and reactions in aqueous environmental systems. In a given system, an element may therefore exist in a variety of forms, and the distribution among those forms is known as the *chemical speciation* of the element. Table 3.3 shows an example for Mn, which as well as interacting in the Mn(II) form with ligands and surfaces, also exists in other oxidation states. Chemical speciation is the key to understanding and predicting much environmental chemistry, including the transport, retention and bioavailability of solutes, and the chemical states of particles and natural organic matter. Much of the research into cation–humic reactions has as

Table 3.2. Characteristic rates of some important environmental processes, based on a more comprehensive classification in Buffle (1988).

Class	Timescale (s)	Process
I	10^{-10}–10^{-3}	Water exchange among solutes Protonation/deprotonation Formation of outer-sphere complexes
II	10^{-2}–10^{4}	Formation and dissociation of chelates
III	10^{2}–10^{8}	Redox reactions Adsorption Precipitation Uptake of chemicals by organisms
IV	10^{3}–10^{10}	Water mixing in lakes and oceans Water flow in streams and rivers Particle settling Biological cycles Flow of ground water Diffusion in sediments Physical and chemical weathering Ocean residence of water

its goal the understanding and prediction of proton and metal binding under different circumstances, and thus contributes significantly to the development of chemical speciation knowledge. Therefore, the subject of this book can be considered to be a particular, but very important, part of environmental chemical speciation.

3.6 Calculation of equilibrium concentrations

Under equilibrium conditions, the concentrations of the chemical species in solution are related to each other via equilibrium constants, mass balances and charge balances. Algebraic manipulation of the relations allows us to calculate the concentrations of individual species, and will be used frequently in the following chapters. Here the basic principles of such manipulations are presented. We shall mainly be concerned with two types of reaction, proton dissociation and metal binding. Examples for a simple ligand, L, are

$$LH = L^- + H^+ \tag{3.11}$$

$$M^{2+} + L^- = ML^+ \tag{3.12}$$

It would be perfectly possibly to write equation (3.11) as an association reaction, or (3.12) as a dissociation reaction, to achieve consistency between protons and metals. In this book however, the more traditional system is followed. The equilibrium constants, in terms of concentrations (mol dm^{-3}) of reactants, are given by

$$K_H = \frac{[L^-][H^+]}{[LH]} \tag{3.13}$$

$$K_M = \frac{[ML^+]}{[M^{2+}][L^-]} \tag{3.14}$$

Because of the large range of values, logarithms of equilibrium constants

Table 3.3. Possible chemical forms of Mn in natural waters and soils. Oxidation states are indicated by II, III and IV.

Class of species	Examples
Aquo ion	$Mn(II)^{2+}$
Inorganic complexes	$Mn(II)SO_4{}^0$, $Mn(II)HCO_3{}^+$, $Mn(II)OH^+$
Organic complexes	$CH_3COOMn(II)^+$, Mn(II)-fulvate, Mn(II)-humate
Adsorption complexes	$Mn(II)\text{-}Fe(OH)_3$, $Mn(II)\text{-}SiO_2$, Mn(II)-clay
Precipitates	$Mn(II)CO_3(s)$, Mn(II)S(s), $Mn(III)_2O_3(s)$, $Mn(IV)O_2(s)$

are convenient to report and discuss. They are also more directly related to free energy changes.

Some authors use a highly systematic nomenclature for equilibrium constants, with fully consistent application of subscripts and net charge. This confers precision and avoids ambiguity, but can reduce clarity. In this book, the simplest possible nomenclature is adopted in the treatment of a given topic, and the reader is therefore respectfully asked to tolerate some inconsistencies.

3.6.1 Mass and charge balances

A solution mass balance is an expression of the Law of Conservation of Mass. For example, the simple acid LH when dissolved in water will give rise to two species, LH and L^-, and the sum of their concentrations will equal the total concentration, T_L

$$T_L = [LH] + [L^-] \tag{3.15}$$

If a metal is also present, there will be mass balances for the ligand and metal as follows

$$T_L = [LH] + [L^-] + [ML^+] \tag{3.16}$$

$$T_M = [M^{2+}] + [ML^+] \tag{3.17}$$

The charge balance expresses the fact that an aqueous solution is electrically neutral, i.e. the total cationic charge equals the total anionic charge. For example, in a solution of the neutral acid LH, to which has been added some NaOH, the charged species are H^+, Na^+, OH^-, and L^-, and the charge balance can be expressed as

$$[H^+] + [Na^+] - [OH^-] - [L^-] = 0 \tag{3.18}$$

It is assumed that Na^+ does not form complexes with the ligand. If the salt (MA_2) of a divalent metal that does form complexes with the ligand is present then the charge balance is

$$[H^+] + [Na^+] + 2[M^{2+}] + [ML^+] - [OH^-] - [A^-] - [L^-] = 0 \tag{3.19}$$

3.6.2 Concentrations and activities

Interactions among ions mean that they are non-ideal solutes. Therefore their activities differ from their concentrations, except at infinite dilution. Activities (a) are related to concentrations by activity coefficients (γ). Thus for the species X

$$a_X = \gamma_X[X] \tag{3.20}$$

Values of γ_X for ions have been deduced from mean activity coefficients of dissolved salts determined experimentally (see e.g. Robinson & Stokes, 1959). For all compounds, γ_X is unity at infinite dilution. As the concentration of X increases, γ_X first decreases, due to interactions among ions, then increases again or remains nearly constant. For dilute mixed ionic solutions, non-specific interactions can be described in terms of the ionic strength (I) defined by

$$I = \tfrac{1}{2}\sum_X [X]\, z_X^2 \tag{3.21}$$

where z_X is the charge of species X. The Debye–Hückel theory leads to an approximate expression for ion activity coefficients, based on a simplified picture of ions as point charges. For an ion of charge z

$$\log_{10}\gamma = -A\, z^2 \sqrt{I} \tag{3.22}$$

(A is approximately 0.5 at room temperature.) Equation (3.22) is a limiting law, and provides good estimates of activity coefficients for singly and doubly charged ions only at low ionic strengths ($<0.005\,\mathrm{mol\,dm^{-3}}$). Several improved versions have been developed to extend the range of ionic strength over which reliable calculations can be made. These take into account differences in size among ions, and the lesser dependence on I at higher I. They include the extended Debye–Hückel, Guntelberg, Davies, and Guggenheim equations (see Turner, 1995). In all the representations, the activity coefficient of an uncharged species is unity, irrespective of ionic strength.

The equations based on the Debye–Hückel theory are less satisfactory for more concentrated ionic solutions (Fig. 3.4). Activity coefficients at high ionic strengths depend upon interactions other than those among ions. Firstly, there are effects leading to non-ideality, which apply to all solutes, charged or otherwise. Secondly, there are effects peculiar to ions, notably the incorporation of solvent water into their hydration shells (Robinson & Stokes, 1959). Turner (1995) has discussed approaches for concentrated natural waters (seawater and brines). The ion interaction model of Pitzer (1991) takes account of interactions among ions of both opposite and like charge, as well as three-ion interactions.

As the ionic strength becomes very small, the activity coefficient tends to unity, at least as far as electrostatic effects are concerned. In the limiting case, the activities and concentrations become equal. For equilibrium

(3.11), equation (3.13) then becomes

$$K_{\mathrm{H},0} = \frac{[\mathrm{L}^-]_0\ [\mathrm{H}^+]_0}{[\mathrm{LH}]_0} = \frac{a_{\mathrm{L}^-}\ a_{\mathrm{H}^+}}{a_{\mathrm{LH}}} \tag{3.23}$$

where the subscript 0 indicates zero ionic strength. By making use of equation (3.20), the value of K_H at any ionic strength can be written as

$$K_{\mathrm{H}} = \frac{[\mathrm{L}^-][\mathrm{H}^+]}{[\mathrm{LH}]} = \frac{(a_{\mathrm{L}^-}/\gamma_{\mathrm{L}^-})\ (a_{\mathrm{H}^+}/\gamma_{\mathrm{L}^+})}{(a_{\mathrm{LH}}/\gamma_{\mathrm{LH}})} = K_{\mathrm{H},0}\frac{\gamma_{\mathrm{LH}}}{\gamma_{\mathrm{L}^-}\gamma_{\mathrm{H}^+}} \tag{3.24}$$

Since LH is uncharged, γ_{LH} is unity. Therefore at low to moderate ionic strengths, where γ_{L^-} and γ_{H^+} are less than one, K_H is greater than $K_{H,0}$. If the ionic strength is increased by the addition of neutral salt, the equilibrium (3.11) moves further to the right, i.e. dissociation increases. A simple physical explanation is that the electrostatic attraction between the reactants is shielded by the medium ions, allowing greater independence.

Equation (3.24) shows that K_H can be expressed in terms of $K_{H,0}$ (a true constant at a given temperature) and the collection of activity coefficients. Since the activity coefficients can be calculated from equation (3.22) or its relatives, the variation of K_H with ionic strength is obtained, an approach which can be applied to all ionic interactions at low ionic strengths. Thus the use of activity coefficients calculated from ionic strength provides a flexible and efficient means of taking into account the effects of complex

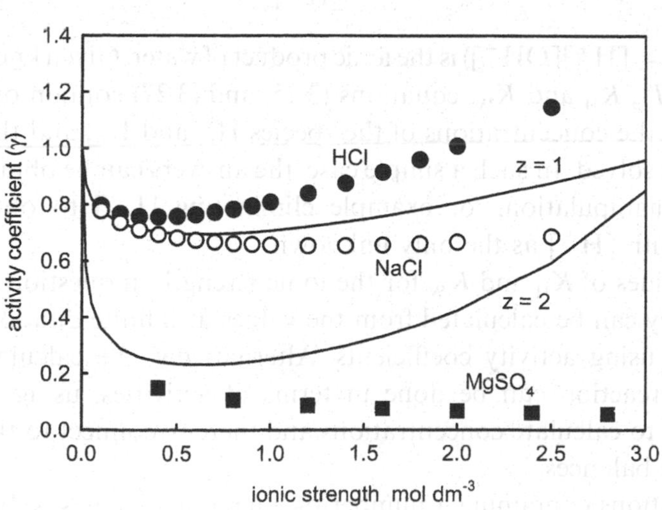

Figure 3.4. Activity coefficients as a function of ionic strength. The points are the data presented by Robinson & Stokes (1959). The lines were computed with the Davies equation.

electrolyte media, as encountered in many natural aqueous systems.

3.6.3 Solving the equations

By solving collections of the above equations, the equilibrium concentrations and activities of all the chemical species in a solution can be found, given the appropriate total concentrations and equilibrium constants. The general approach is to define a minimum number of master species (e.g. H^+, L^-) the concentrations of which can be used in equilibrium expressions to define the concentrations of all the other species. Consider for example a solution containing the ligand L, some strong acid (HA) and some strong base (BOH). The mass balance for the ligand (equation 3.15) can be written

$$T_L = [L^-]\left(1 + \frac{[H^+]}{K_H}\right) \tag{3.25}$$

and the charge balance is

$$T_B - T_A + [H^+] - [OH^-] - [L^-] = 0 \tag{3.26}$$

where T_B and T_A are respectively the total concentrations of strong base and strong acid. Equation (3.26) can be written

$$T_B - T_A + [H^+] - \frac{K_W}{[H^+]} - [L^-] = 0 \tag{3.27}$$

where $K_W (= [H^+][OH^-])$ is the ionic product of water. Given knowledge of T_B, T_A, T_L, K_H and K_W, equations (3.25) and (3.27) contain only two unknowns, the concentrations of the species H^+ and L^-, and therefore they can be solved. In such a simple case, the answers can be obtained by algebraic manipulation, for example eliminating $[L^-]$ to obtain an equation with $[H^+]$ as the only unknown.

If the values of K_H and K_W for the ionic strength in question are not known, they can be calculated from the values at infinite dilution ($K_{H,0}$ and $K_{W,0}$) using activity coefficients. Alternatively, the calculation of extents of reaction can be done in terms of activities, using activity coefficients to calculate concentrations and thereby connect to the mass and charge balances.

For solutions containing a number of interacting solutes, solving the simultaneous equations by substitution becomes impractical. Instead, numerical methods are used, involving the continued refinement of trial values until the mass and charge balances are satisfied to a specified tolerance. If the ionic strength is unknown at the start of the calculation, it too is found by continued refinement. Methods have been devised that allow the rapid calculation of the concentrations of hundreds of chemical species. Examples of speciation programs in common use for environmental systems include HALTAFALL (Ingri *et al.*, 1967), GEOCHEM (Mattigod & Sposito, 1979), PHREEQE (Parkhurst *et al.*, 1980), and MINTEQA2 (Allison *et al.*, 1991), and developments continue. WHAM (Tipping, 1994) was introduced to include reactions involving humic substances (see Chapters 13 and 14).

4

Proton dissociation from weak acids

As we saw in Chapter 2, humic substances possess carboxylic, phenolic, and other groups that can bind and release protons. In this chapter, reactions involving such weak acid groups are described, to provide the basis for addressing proton–humic interactions.

4.1 Acids and bases

A Brønsted acid is a substance that can donate a proton, while a Brønsted base is a proton acceptor. Thus HCl, CH_3COOH and H_3O^+ are acids, while Cl^-, CH_3COO^- and OH^- are bases. Lewis acids are defined as substances that can accept and share an electron pair, donated by a Lewis base. Lewis acids include species other than proton donors, notably metal ions (Section 5.1.3). In the present chapter we are concerned only with Brønsted acids and bases.

The strength of an acid refers to its degree of dissociation when dissolved in water. For example hydrochloric acid, a strong acid, dissociates almost completely, so that in a dilute aqueous solution the separated H^+ and Cl^- ions behave nearly independently. Thus, the equilibrium

$$HCl = H^+ + Cl^- \qquad (4.1)$$

lies far to the right, and the concentration of hydrogen ions is close to the total concentration of HCl. Weak acids dissociate only partially, and the hydrogen ion concentrations are therefore less than the total concentrations. For example, in a solution of 10^{-3} mol dm^{-3} acetic acid, the hydrogen ion concentration is *c.* 1.6×10^{-4} mol dm^{-3}, while in a 10^{-3} mol dm^{-3}

solution of phenol (ϕ-OH) it is only about 3×10^{-7} mol dm^{-3}, and the equilibrium

$$\phi-\mathrm{OH} = \phi-\mathrm{O}^- + \mathrm{H}^+ \tag{4.2}$$

lies far to the left. Now consider a monoprotic acid, represented as LH, which dissociates into its conjugate base L^- and a proton

$$\mathrm{LH} = \mathrm{L}^- + \mathrm{H}^+ \tag{4.3}$$

The equilibrium constant, in terms of concentrations (mol dm^{-3}) of reactants, is given by

$$K = \frac{[\mathrm{L}^-][\mathrm{H}^+]}{[\mathrm{LH}]} \tag{4.4}$$

Here, K is called the acid dissociation constant. From equation (4.4), it can be seen that when $[L^-]$ equals [LH], i.e. when the acid is half dissociated (and half undissociated), then $[H^+]$ equals K, as shown in Fig. 4.1. In logarithms, the pK of the acid is the pH for half-dissociation (p $= -\log_{10}$). This provides a ready scale for ranking acids (Table 4.1). The smaller is pK (the larger is K) the stronger is the acid. Thus, strong acids release their protons readily. To put it another way, the conjugate bases of strong acids bind protons weakly. The distinction between strong and weak acids is by no means absolute, and we are really dealing with a continuum of acid strengths.

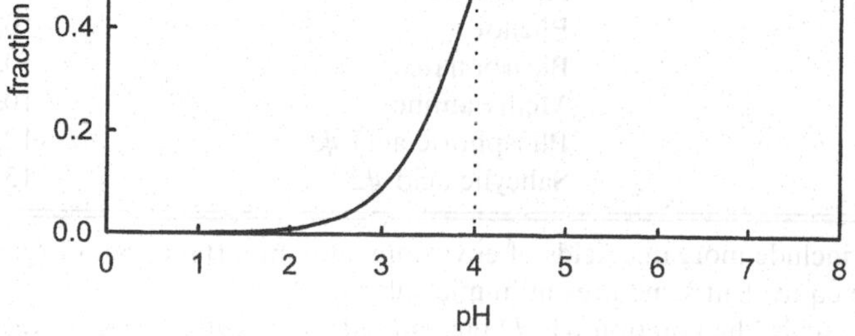

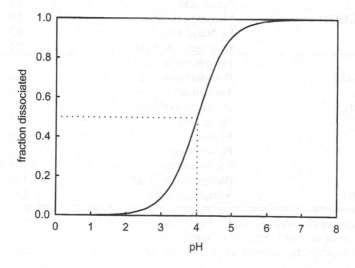

Figure 4.1. Dissociation from a monoprotic acid with pK=4. The dotted lines indicate 50% dissociation at pH 4.

In humic substances, the two main classes of acid groups are carboxylic and phenolic, both of which have —OH groups that yield protons by dissociation. Molecular structure influences the pK values of these groups chiefly through polar substituent (inductive) and resonance effects (Perdue, 1985). For example, as shown in Table 4.1, formic acid (HCOOH) is a stronger acid than acetic acid (CH_3COOH) because the hydrogen atom attracts electrons more strongly than the methyl group. This causes greater secondary displacement of the electrons in the —OH group, in the direction of the O atom, making the O—H bond in formic acid weaker than that in acetic acid. The effect is especially large in oxalic acid

Table 4.1. The pK values of some simple acids, ranked in order of decreasing acid strength.

	Acid	pK
HCl	Hydrochloric acid	~ − 3
H_2SO_4	Sulphuric acid #1	~ − 3
HNO_3	Nitric acid	− 1
HOOCCOOH	Oxalic acid #1	1.3
HSO_4^-	Bisulphate	1.9
H_3PO_4	Phosphoric acid #1	2.2
$HOOCCH_2COOH$	Malonic acid #1	2.9
HOOC-ϕ-OH	Salicylic acid #1	3.0
HCOOH	Formic acid	3.8
ϕ-COOH	Benzoic acid	4.2
$^-$OOCCOOH	Oxalic acid #2	4.3
CH_3COOH	Acetic acid	4.8
$^-OOCCH_2COOH$	Malonic acid #2	5.7
$H_2CO_3^*$	Hydrated CO_2	6.3
H_2S	Hydrogen sulphide	7.1
$H_2PO_4^-$	Phosphoric acid #2	7.2
p-CH_3OCϕ-OH	*Para*-acetylphenol	8.1
NH_4^+	Ammonium	9.2
$HO(CH_2)_2SH$	*β*-Mercaptoethanol	9.7
$Si(OH)_4$	Silicic acid	9.9
ϕ-OH	Phenol	10.0
HCO_3^-	Bicarbonate	10.3
CH_3NH_2	Methylamine	10.6
HPO_4^{2-}	Phosphoric acid #3	12.4
$^-$OOC-ϕ-OH	Salicylic acid #2	13.7

The examples include inorganic acids of environmental importance, and organic acids that may have equivalent structures in humic substances.
For polyprotic acids, the notation #1, #2 etc. is used to indicate successive dissociation reactions.
Benzene rings are indicated by ϕ.
Data from Stumm & Morgan (1996), Martell & Smith (1977) and Smith *et al.* (1993).

(HOOCCOOH), due to the very strong attraction of electrons by the carboxyl group. The influence of the polar constituent diminishes as its distance from the acid group increases, as shown by the higher p*K* for the first dissociation of malonic acid ($HOOCCH_2COOH$) compared to that of oxalic acid. Whereas saturated hydrocarbon groups are electron-repelling, compared to hydrogen, unsaturated ones are electron-withdrawing.

The different effects of substituents are insufficient to explain why the OH group of acetic acid is a stronger acid than that of phenol. The major reason is the high resonance stabilisation of the carboxylate anion. Resonance stabilisation may also play an important rôle in determining the acidities of groups in substituted benzene rings. Thus, the p*K* of the phenolic—OH group in *p*-acetylphenol is lower than that in phenol itself, partly because of the inductive influence of the acetyl group, but mainly because of the delocalisation of charge by resonance in the *p*-acetylphenolate anion (Perdue, 1985). The p*K* values of compounds with more than one proton-dissociating group also depend upon statistical effects (see Section 4.5).

4.2 Buffering

An important property of weak acids is that they are buffers, resisting the change of $[H^+]$ following additions of acid or base. Buffer intensity (β) is the differential change in the concentration of added base per pH increment, i.e.

$$\beta = \frac{d[\text{base}]}{d\text{pH}} \tag{4.5}$$

Figure 4.2 compares the titration curve of a weak acid with that of water, and shows how β varies with pH. The solution containing the weak acid requires more base to raise its pH from 3 to 11, and the extra base is needed in the pH range around the p*K*. Weak acids are widely used to control solution pH in all sorts of experimental studies. Humic substances are important buffering agents in natural waters, especially in the pH range 4–6 (Chapter 13).

4.3 Kinetics

Proton dissociation and association reactions of a simple anionic ligand can be written

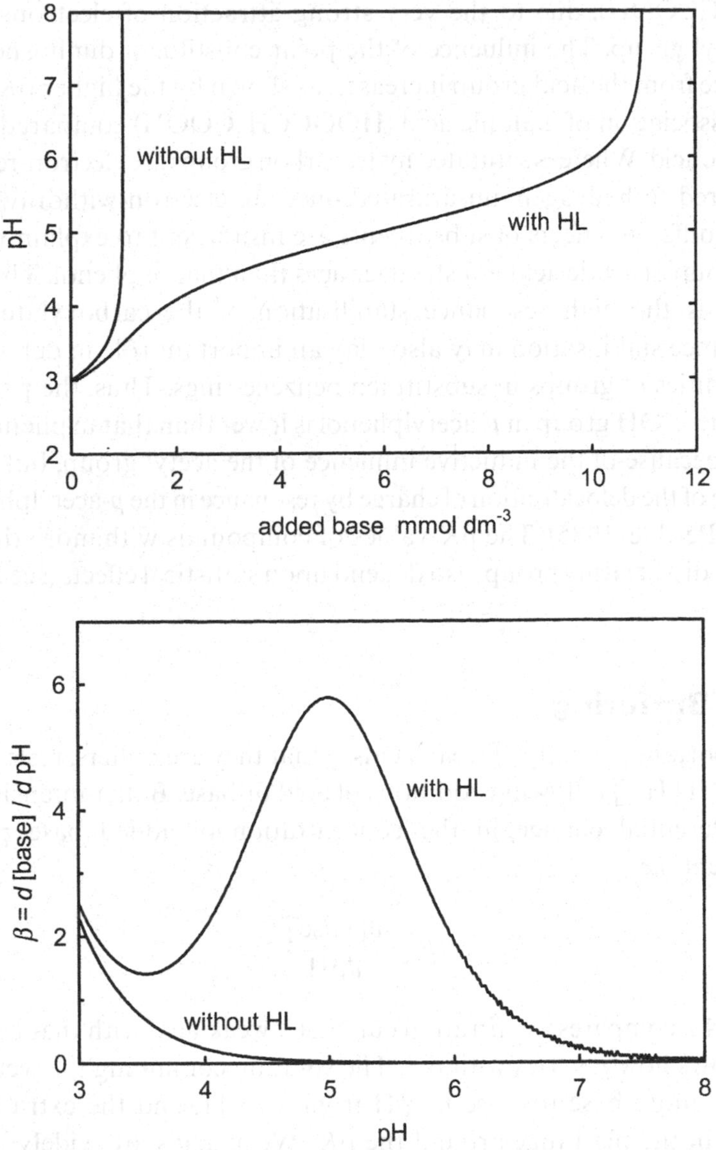

Figure 4.2. Buffering of pH by a weak acid (LH) with pK=5. Water and a solution containing 10^{-2} mol dm^{-3} LH are titrated with strong base. The upper panel shows pH response to additions of base. The lower panel shows the buffer intensity β (equation 4.5) as a function of pH.

$$\mathrm{LH} \rightarrow \mathrm{L}^- + \mathrm{H}^+ \qquad \frac{d[\mathrm{LH}]}{dt} = -k_1[\mathrm{LH}] \tag{4.6}$$

$$\mathrm{L}^- + \mathrm{H}^+ \rightarrow \mathrm{LH} \qquad \frac{d[\mathrm{LH}]}{dt} = k_2[\mathrm{L}^-][\mathrm{H}^+] \tag{4.7}$$

The rate of the association reaction (4.7) is diffusion controlled, i.e. it is governed by the frequency of contacts between the two reactants in solution, due to diffusion. Theoretical calculations give a typical value for k_1 of $4 \times 10^9\ \mathrm{dm^3\,mol^{-1}\,s^{-1}}$. At equilibrium, the rates of the two reactions are equal, and so it follows that

$$K = \frac{k_1}{k_2} \qquad k_1 = k_2 K \tag{4.8}$$

Therefore the rate constant for the dissociation reaction is proportional to K, and the stronger is the acid, the faster is the rate of dissociation. For example, an acid with pK = 4 (typical for a carboxylic acid) has a half-time for dissociation of about 1 μs, while the half-time for NH_4^+ (pK = 9.3) is about 0.1 s.

4.4 Diprotic acids

So far we have discussed the equilibria of proton dissociation from compounds with a single dissociating group. Now let us consider a diprotic acid, DH_2, for which the dissociation reactions are

$$\mathrm{DH_2} = \mathrm{DH} + \mathrm{H}^+ \qquad \mathrm{DH} = \mathrm{D} + \mathrm{H}^+ \tag{4.9}$$

and the equilibrium constants are

$$K_1 = \frac{[\mathrm{DH}][\mathrm{H}^+]}{[\mathrm{DH_2}]} \qquad K_2 = \frac{[\mathrm{D}][\mathrm{H}^+]}{[\mathrm{DH}]} \tag{4.10}$$

The average degree of dissociation per mole of DH_2 (sum of all species), r, is given by

$$r = \frac{[\mathrm{DH}] + 2[\mathrm{D}]}{[\mathrm{DH_2}] + [\mathrm{DH}] + [\mathrm{D}]} = \frac{\dfrac{K_1}{[\mathrm{H}^+]} + \dfrac{2K_1K_2}{[\mathrm{H}^+]^2}}{1 + \dfrac{K_1}{[\mathrm{H}^+]} + \dfrac{K_1K_2}{[\mathrm{H}^+]^2}} \tag{4.11}$$

Now consider the same diprotic acid, represented by HABH. The following treatment is based on those of Tanford (1961) and Perdue

(1985). At the molecular level, the first proton can dissociate by the reactions

$$\mathrm{HABH = ABH + H^+} \qquad \mathrm{HABH = HAB + H^+} \tag{4.12}$$

and the second by the reactions

$$\mathrm{ABH = AB + H^+} \qquad \mathrm{HAB = AB + H^+} \tag{4.13}$$

(Note that charges are omitted except for H^+.) The equilibrium constants for the four reactions are given by

$$K_a = \frac{[\mathrm{ABH}][\mathrm{H^+}]}{[\mathrm{HABH}]} \qquad K_b = \frac{[\mathrm{HAB}][\mathrm{H^+}]}{[\mathrm{HABH}]} \qquad K_c = \frac{[\mathrm{AB}][\mathrm{H^+}]}{[\mathrm{ABH}]}$$

$$K_d = \frac{[\mathrm{AB}][\mathrm{H^+}]}{[\mathrm{HAB}]} \tag{4.14}$$

The experimentally determined constants K_1 and K_2 are given by

$$K_1 = \frac{([\mathrm{ABH}] + [\mathrm{HAB}])[\mathrm{H^+}]}{[\mathrm{HABH}]} = K_a + K_b \tag{4.15}$$

$$K_2 = \frac{[\mathrm{AB}][\mathrm{H^+}]}{[\mathrm{ABH}] + [\mathrm{HAB}]} = \frac{K_c\, K_d}{K_c + K_d} \tag{4.16}$$

Now consider the case where the groups A and B are identical, and non-interacting, i.e. the dissociation of one does not affect the dissociation of the other. (This is somewhat hypothetical, since in most diprotic acids the dissociating groups are sufficiently close to exert appreciable effects on one another.) We then have $K_a = K_b = K_c = K_d (= K)$ and consequently $K_1 = 2K$, $K_2 = K/2$ and $K_1/K_2 = 4$. Thus, simply on statistical grounds the first dissociation constant will be four times greater than the second. The constant K is the equilibrium (dissociation) constant, which characterises the chemistry of the site, in the absence of statistical effects. The statistical effect also applies when the sites are not identical.

When the two sites are identical and non-interacting, the diprotic acid behaves exactly the same as a monoprotic acid with a dissociation equilibrium constant of K, except that it has twice the number of protons dissociating per mole. Thus if $K_1 = 2K$ and $K_2 = K/2$ are substituted into equation (4.11), it simplifies to

$$r = \frac{2K/[\mathrm{H^+}]}{1 + K/[\mathrm{H^+}]} \tag{4.17}$$

4.5 Extension to higher polyprotic acids

Consider the dissociation of protons from a molecule with n binding sites, AH_n. A series of equilibrium constants $K_1, K_2 \ldots K_n$ can be defined

$$K_1 = \frac{[AH_{n-1}][H^+]}{[AH_n]} \quad K_2 = \frac{[AH_{n-2}][H^+]}{[AH_{n-1}]} \quad K_n = \frac{[A][H^+]}{[AH]} \qquad (4.18)$$

The average extent of dissociation, r (mol mol^{-1}) is given by

$$r = \frac{[AH_{n-1}] + 2[AH_{n-2}] + \cdots n[A]}{[AH_n] + [AH_{n-1}] + [AH_{n-2}] + \cdots [A]} \qquad (4.19)$$

Combination of equations (4.18) and (4.19) then gives

$$r = \frac{\dfrac{K_1}{[H^+]} + \dfrac{2K_1K_2}{[H^+]^2} + \cdots + \dfrac{nK_1K_2 \ldots K_n}{[H^+]^n}}{1 + \dfrac{K_1}{[H^+]} + \dfrac{K_1K_2}{[H^+]^2} + \cdots + \dfrac{K_1K_2 \ldots K_n}{[H^+]^n}} \qquad (4.20)$$

The values of $K_1, K_2 \ldots K_n$ are those that would be obtained by measurement and include statistical factors as explained above for diprotic acids. The general expression for the equilibrium constants is given by (Tanford, 1961)

$$K_i = K(i)\frac{n - i + 1}{i} \qquad (4.21)$$

where $K(i)$ is the constant for the ith site, in the absence of statistical effects.

It is informative to consider how r varies with pH for different choices of values of $K(i)$. Some calculated plots for idealised hexaprotic acids with two classes of binding sites, centring on p$K(i) = 4$ and p$K(i) = 9$, are shown in Fig. 4.3. Note that these plots were obtained ignoring electrostatic effects (see Section 4.6). If there are only two different values (p$K(i) = 4$ and 9), the titration curve is well-defined, but when ranges of values around the means are introduced, the curves become relatively featureless, or 'smeared'. 'Smeared' titration curves not dissimilar to B and C in Fig. 4.3 are found for humic substances (Section 4.8).

When all the sites are equivalent, and non-interacting, equations (4.20) and (4.21) can be manipulated (Tanford, 1961) to give the simple expression

$$r = \frac{nK/[H^+]}{1 + K/[H^+]} \qquad (4.22)$$

This was demonstrated for $n = 2$ in the previous section. An alternative derivation of equation (4.22) – and of (4.17) – follows from the premise that if the sites are identical and non-interacting they must behave as a collection of individuals, so that the polyprotic acid with n sites behaves like a monoprotic acid at n times the concentration. Thus equation (4.22) is the same as that for a monoprotic acid except that the factor n is included to take account of the fact that there is more than one site per molecule.

Equation (4.22) is a very useful result if one is dealing with synthetic polymers such as polyacrylic acid, in which all the groups can reasonably be assumed to be chemically equivalent. For biopolymers such as proteins, the sites can often be apportioned into a few classes (Tanford, 1961), and the value of r can be obtained by simple addition. Thus, if there are two classes of identical non-interacting sites, A and B, we have

$$r = r_A + r_B = \frac{n_A K_A/[H^+]}{1 + K_A/[H^+]} + \frac{n_B K_B/[H^+]}{1 + K_B/[H^+]} \tag{4.23}$$

Equation (4.23) is an approximation, the validity of which depends upon the relative values of K_A and K_B. This can be seen by considering the

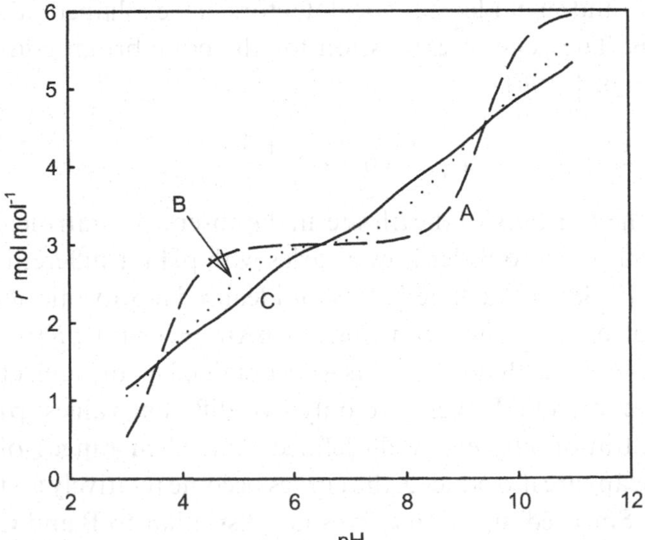

Figure 4.3. Proton dissociation from three hexaprotic acids, calculated ignoring electrostatic effects. The pK values for the six sites were set as follows:

	pK(1)	pK(2)	pK(3)	pK(4)	pK(5)	pK(6)
A	4	4	4	9	9	9
B	3	4	5	8	9	10
C	2.5	4	5.5	7.5	9	10.5

simple case where there is only one site in each class, so that it should be possible to derive an equation of the form of (4.11) from (4.23). However, the term $(K_A + K_B)/[H^+]$ then appears in the numerator and denominator instead of $K_A / [H^+]$, so the equations are equivalent only if $K_A \gg K_B$. The concept of site classes depends upon this difference in equilibrium constants. Curve A of Fig. 4.3 is an example of two classes of sites, with $K_A = 10^{-4}$ and $K_B = 10^{-9}$. The approach can be extended to any number of classes of sites, if the range of experimental data is sufficiently large to justify doing so.

Although the number of equilibrium constants required by equation (4.20) is potentially very large, it may be possible to express them in terms of a distribution that can be described by a small number of parameters. For example, Perdue & Lytle (1983) assumed that for sites in humic substances, a Gaussian distribution of dissociation constants would be found, which they parameterised with two constants, one describing the central K value and the other the width of the distribution. This approach is considered more fully in Chapter 9, in connection with the modelling of cation binding by humic substances.

4.6 Electrostatic interactions among sites

So far we have been considering algebraic expressions for more-or-less hypothetical cases in which there is no interaction among sites, i.e. the dissociation of one proton does not influence that of any other. In fact there is significant interaction, due to a number of mechanisms including changes in inductive effects, steric hindrance, induced conformational change, and – usually the most important – electrostatic interactions. Although here the focus is on protons, the principles apply to interactions involving other cations also (see Section 5.4). The treatment is based largely on Tanford (1961).

Qualitatively, electrostatic effects on proton dissociation can be understood as follows. As protons dissociate from an initially uncharged polyprotic acid, the net charge becomes progressively more negative. The first protons to dissociate therefore experience a smaller net attractive Coulombic force than subsequent ones, which means that the effective dissociation equilibrium constant decreases, i.e. pK increases, the acid becomes weaker. Tanford presented a quantitative analysis for macroions based on the electrostatic free energy, which led to the expression

$$K = K_{\text{int}} \exp(2wzZ) \qquad (4.24)$$

where Z is the average macroionic charge, z is the charge on the dissociating ion ($+1$ for a proton), K is the dissociation constant at a particular value of Z, K_{int} is the intrinsic dissociation constant, which applies when $Z = 0$, and w is a (positive) term that depends upon the nature of the macroion, the ionic strength and the temperature. Thus, as Z becomes increasingly negative, the exponential term decreases, and so does K. Given values of w, electrostatic effects on proton dissociation can be calculated.

4.6.1 Derivation of *w* from the Debye–Hückel theory

The presence of a charged particle in an electrolyte solution gives rise to a non-uniform distribution of charge in the immediate vicinity, as shown schematically in Fig. 4.4. The theory of electrostatic interactions involving mobile (unbound) ions in aqueous solutions is based on two precepts. The first is that potential, ψ, can be related to charge density (charge per unit volume), ρ, by the Poisson equation

$$\nabla^2\psi = -\frac{4\pi\rho}{D} \tag{4.25}$$

where ∇^2 is the Laplacian operator and D the dielectric constant. The second is that Boltzmann's distribution law applies. This states that the ratio of the concentration of a mobile ion experiencing a potential ψ, to its

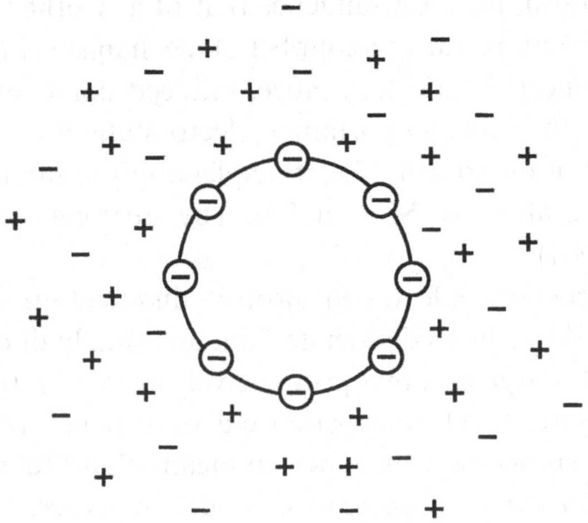

Figure 4.4. Schematic diagram (in two dimensions) of charge distribution around a central macromolecular anion. Near to the molecular surface, there is a preponderance of cations (counterions), while at distance the numbers of cations and anions (co-ions) are equal.

concentration far away, is given by $\exp(-W/kT)$, where W is the work that must be done to move the ion from infinity (potential = 0) to the point where the potential is ψ (k is Boltzmann's constant, T is temperature). For a 1:1 electrolyte of charge $\pm\varepsilon$ (coulombs per ion) the values of W are equal to $\varepsilon\psi$ and $-\varepsilon\psi$ for the positive and negative ions respectively. We therefore have

$$N_+ = N \exp\left(\frac{-\varepsilon\psi}{kT}\right) \tag{4.26}$$

$$N_- = N \exp\left(\frac{+\varepsilon\psi}{kT}\right) \tag{4.27}$$

where N_+ and N_- are the concentrations (in ions per unit volume) at the point where the potential is ψ, and N is the concentration at infinity. The charge density at this point is given by

$$\rho = N_+\varepsilon - N_-\varepsilon = N\varepsilon\left[\exp\left(-\frac{\varepsilon\psi}{kT}\right) - \exp\left(+\frac{\varepsilon\psi}{kT}\right)\right] \tag{4.28}$$

Expansion of the exponentials gives

$$\rho = -2N\varepsilon\left[\frac{\varepsilon\psi}{kT} + \frac{1}{3!}\left(\frac{\varepsilon\psi}{kT}\right)^3 + \frac{1}{5!}\left(\frac{\varepsilon\psi}{kT}\right)^5 + \cdots\right] \tag{4.29}$$

The assumption leading to the Debye–Hückel theory is that the potential is sufficiently small at all points for $(\varepsilon\psi/kT)$ to be much less than one. All but the first term in equation (4.29) can then be neglected, and we obtain

$$\rho = -\frac{2N\varepsilon^2\psi}{kT} \tag{4.30}$$

Equation (4.25) then becomes

$$\nabla^2\psi = \frac{8\pi N\varepsilon^2}{DkT}\psi = \kappa^2\psi \tag{4.31}$$

where

$$\kappa = \left(\frac{8\pi\varepsilon^2}{DkT}\right)^{\frac{1}{2}} N^{\frac{1}{2}} \tag{4.32}$$

Equation (4.31) is a version of the Poisson–Boltzmann equation. In the case of an impenetrable sphere, the equation can be solved, and the following expression is obtained for the total electrostatic free energy of a charged molecule (W_{el}), i.e. the work done in placing charges on it

$$W_{el} = \frac{Z^2\varepsilon^2}{2DR}\left(1 - \frac{\kappa R}{1 + \kappa a}\right) \tag{4.33}$$

where $Z\varepsilon$ is the (uniformly distributed) charge on the central ion, R is the radius of the central ion, and $a = (R +$ the radius of the combined ion). The variable κ is the Debye–Hückel parameter. It has units of length^{-1}, and its reciprocal is taken as a measure of the thickness of the electrical double layer. It is not restricted to 1:1 electrolytes, and can be expressed in terms of ionic strength. Thus with I in units of mol dm^{-3}

$$\kappa = \left(\frac{8\pi N_{Av}\varepsilon^2}{DkT}\right)^{\frac{1}{2}} I^{\frac{1}{2}} = 3.29 \times 10^9\, I^{\frac{1}{2}}\, \mathrm{m}^{-1} \text{ at } 25°\mathrm{C} \tag{4.34}$$

where N_{Av} is Avogadro's number. For $I = 0.001$ mol dm^{-3}, $1/\kappa \approx 10^{-8}$ m, for $I = 0.1$ mol dm^{-3}, $1/\kappa \approx 10^{-9}$ m.

Equation (4.33) has the form

$$W_{el} = AZ^2 \tag{4.35}$$

which on differentiation to obtain the change in electrostatic free energy per mole for a change in Z leads to

$$\Delta G = \Delta G^{o}_{int} + 2AN_{Av}z_i Z \tag{4.36}$$

and thence, by comparison with equation (4.24), to

$$w = \frac{AN_{Av}}{R_G T} \tag{4.37}$$

where R_G is the molar version of the gas constant. Thus the full expression for w in the Debye–Hückel theory is

$$w = \frac{\varepsilon^2}{2DR}\frac{N_{Av}}{R_G T}\left(1 - \frac{\kappa R}{1 + \kappa a}\right) = \frac{\varepsilon^2}{2DRkT}\left(1 - \frac{\kappa R}{1 + \kappa a}\right) \tag{4.38}$$

As κ increases (and the double layer thickness decreases), the term in brackets decreases. As κ is proportional to $I^{1/2}$, w is inversely related to ionic strength, so that the electrostatic effect on proton dissociation decreases as ionic strength increases.

4.6.2 Relationship between w and double layer capacitance

As has already been stated, the treatment outlined above follows that of Tanford (1961) for macromolecules. Tanford preferred to use macromolecular charge as the determinant of electrostatic effects. An alternative, preferred by colloid chemists dealing with mineral surfaces, but equally applicable to macromolecules, is to use potential. The potential at the surface of the particle is given by

$$\psi_s = \frac{2wZkT}{\varepsilon} = \frac{2wZR_GT}{F} \tag{4.39}$$

where F is the Faraday constant (96 487 C mol^{-1}). Thus w is the factor that relates potential to surface charge. (This holds whether or not the Debye–Hückel approximation is used.) In surface chemistry, the electrical double layer associated with a charged particle is traditionally viewed as a condenser, for which the capacitance C (F m^{-2}) connects potential with the surface charge density σ_S (C m^{-2}). Thus

$$\psi_s = \frac{\sigma_s}{C} \tag{4.40}$$

Denoting the molecular weight of the macromolecule by M, and its specific surface area ($m^2\,g^{-1}$) by S, we have

$$C = \frac{\varepsilon^2 N_{Av}}{2wMSkT} = \frac{F^2}{2wMSR_GT} \tag{4.41}$$

Thus the capacitance is inversely proportional to w.

4.6.3 Relationship of *w* to activity coefficients

The expression for w derived above was based on the Debye–Hückel theory, as was the expression in equation (3.22) for the activity coefficients of small ions. Indeed, the full expression for the activity coefficient is

$$\log_e \gamma = -\frac{z^2\varepsilon^2\kappa}{2DkT} \tag{4.42}$$

Now consider the reaction

$$^{Z+1}LH_i = {}^{Z}LH_{(i-1)} + H^+ \tag{4.43}$$

Here, L represents a polyprotic acid with charge $Z + 1$, which is losing a proton by dissociation. The difference between the values of $\log_e K$ for the reaction at ionic strengths I and zero can be obtained from equation (3.24) in terms of activity coefficients

$$\begin{aligned}\log_e K(I) - \log_e K(0) &= \log_e \gamma(LH_i) - \log_e \gamma(LH_{(i-1)}) - \log_e \gamma(H^+)\\ &= \frac{\varepsilon^2\kappa}{2DkT}(-(Z+1)^2 + Z^2 + 1) \\ &= \frac{\varepsilon^2\kappa Z}{DkT}\end{aligned} \tag{4.44}$$

Alternatively, from equation (4.24), we have

$$\log_e K(I) - \log_e K(0) = -2Z(w(I) - w(0)) \tag{4.45}$$

Substituting for w from equation (4.38) gives

$$\log_e K(I) - \log_e K(0) = -2Z \frac{\varepsilon^2}{2DR} \frac{N_{\mathrm{Av}}}{R_{\mathrm{G}}T}\left(1 - \left(\frac{\kappa R}{1 + \kappa a}\right) - 1\right) \tag{4.46}$$

The derivation of the activity coefficient expression includes the proviso that it only applies to low ionic strength, so that $\kappa a \ll 1$, therefore

$$\log_e K(I) - \log_e K(0) = \frac{\varepsilon^2 \kappa Z}{DkT} \tag{4.47}$$

which is the same as equation (4.44). Thus the same expression for the difference in the $\log_e K$ values is obtained from both starting points, demonstrating the equivalence between activity coefficients, as applied to small ions, and the electrostatic terms applied to macroions, as long as the same simplifying assumptions are used.

4.6.4 Magnitude of the electrostatic effect on proton dissociation

To appreciate the influence of electrostatic interactions on proton dissociation, let us compute the term $\exp(2wzZ)$ in equation (4.24). This is the factor by which the intrinsic equilibrium constant is multiplied to obtain the effective 'constant' for a given set of circumstances. Variations in the term are due to variations in Z and w (z is equal to $+1$ for protons). For a given value of Z, $\exp(2wzZ)$ depends upon w, which in turn depends upon particle density and dimensions, and ionic strength. For spheres of density $\rho\,\mathrm{g\,m^{-3}}$, the radius R (m) is related to molecular weight M by

$$R = \left(\frac{3M}{4\pi\rho N_{\mathrm{Av}}}\right)^{\frac{1}{3}} \tag{4.48}$$

and surface charge density σ_{S} ($\mathrm{C\,m^{-2}}$) is related to Z ($\mathrm{eq\,mol^{-1}}$) by

$$\sigma_{\mathrm{S}} = \frac{Z\varepsilon}{4\pi R^2} \tag{4.49}$$

We can now use equations (4.38), (4.39), (4.48) and (4.49) to compute the variation of $\exp(2wzZ)$ and ψ_{S} with molecular weight for either constant Z or constant σ_{S}. The use of equation (4.38) means that we are applying the Debye–Hückel approximation. The results of calculations for molecular weights typical of humic substances are shown in Table 4.2. The following trends emerge:

(a) The value of w decreases with molecular weight, as a consequence of its dependence on R (equation 4.38).

Table 4.2. Variables associated with electrostatic effects, calculated as described in Section 4.6.4.

M	R (nm)	$-Z$	$-\sigma_S$ (C m^{-2})		w		$-\psi$ (V)		exp(2wzZ)		$-\Delta\log K$	
				$I =$	0.001	0.1	0.001	0.1	0.001	0.1	0.001	0.1
1 000	0.74	5	0.118		0.450	0.305	0.115	0.078	0.011	0.047	1.95	1.33
5 000	1.26	5	0.040		0.251	0.142	0.064	0.036	0.082	0.243	1.09	0.61
10 000	1.58	5	0.025		0.193	0.099	0.050	0.026	0.145	0.370	0.84	0.43
1 000	0.74	4.2	0.100		0.450	0.305	0.098	0.066	0.022	0.075	1.65	1.12
5 000	1.26	12.4	0.100		0.251	0.142	0.159	0.090	0.002	0.030	2.70	1.52
10 000	1.58	31.2	0.100		0.193	0.099	0.195	0.100	0.0005	0.020	3.30	1.70

$\Delta\log K$ is the difference in $\log K$ between the value including the electrostatic contribution and the intrinsic value.
Ionic strengths (I) are in mol dm^{-3}.
The constants used were as follows: $\varepsilon = 1.60 \times 10^{-19}$ C; $N_{Av} = 6.02 \times 10^{23}$ mol^{-1}; $F = 96\,500$ C mol^{-1}; $R_G = 8.31$ J deg^{-1} mol^{-1}; $D = 8.79 \times 10^{-9}$ C^2 N^{-1} m^{-2}; $\kappa = 3.29 \times 10^9\ I^{1/2}$ m^{-1}; $k = 1.38 \times 10^{-23}$ J deg^{-1} molecule^{-1}; $\rho = 10^6$ g m^{-3}; $(a - R) = 0.3$ nm; $T = 298$ K.

(b) If Z is held constant, the surface charge density decreases with increasing molecular weight, as does the surface potential, while the exponential term in equation (4.24) increases (because Z is negative). Consequently, the increment in $\log K$ that can be attributed to electrostatic effects ($\Delta\log K$) becomes smaller in magnitude as molecular weight increases.
(c) If σ_S is held constant, Z increases with molecular weight, as does the potential, but the exponential term now decreases, while the magnitude of the increment in $\log K$ increases.

The calculated values of $\Delta\log K$ range from small (0.43), at low charge density and high ionic strength, to large (3.30), at high charge density and low ionic strength. Thus there is a strong case for taking the electrostatic effect on proton dissociation into account.

Tipping *et al.* (1990) applied the Debye–Hückel theory to hypothetical molecules with properties similar to those of fulvic acids. To explore the electrostatic effect they considered spherical, impermeable, polyprotic acids with chemically identical sites located on the outer surface. The molecules, built of units each of molecular weight 200 and carrying a single carboxylic acid group with $pK_{int} = 3.5$, are referred to as '*n*-mers', depending upon the number of units (n). It was assumed that $Z = -r$, i.e. the molecules developed charge by dissociation and did not bind counterions. Titration curves (Z vs. pH) were calculated, and also values of the 'apparent pK', defined by

$$pK_{app} = pH - \log\left(\frac{\alpha}{1-\alpha}\right) \tag{4.50}$$

where α is the degree of dissociation of the acid (equivalents per mole). Note that for a simple monoprotic acid, pK_{app} is constant, i.e. does not vary with α.

Figure 4.5 shows computed titration curves and plots of pK_{app} vs. α for polyprotic acids. The two expected main trends are apparent, the electrostatic influence increasing with molecular weight, and decreasing with ionic strength. The calculated change in pK_{app} over the range of the titration is more than two log units for the largest molecule, a 12-mer, at low ionic strength. The change in pK_{app} is also reflected in the increasing 'smearing' or 'stretching' of the titration curves going from the 1-mer to the 12-mer. The shapes of the plots are very similar to those obtained from calculations assuming only chemical heterogeneity, with no electrostatic effects (Fig. 4.3). Thus smearing can arise for two reasons, site heterogeneity

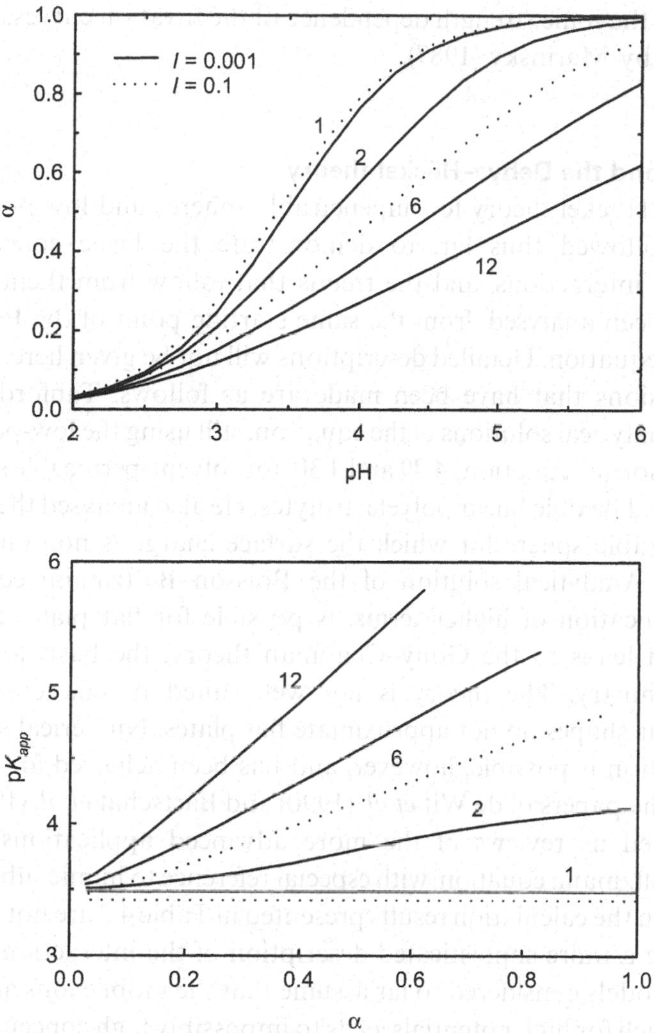

Figure 4.5. Calculated dissociation behaviour of polyprotic acids, with electrostatic effects estimated from the Debye–Hückel theory. The number of acid groups in each molecule (n) is indicated on the curves. The fractional dissociation α is equal to r/n. The calculations were performed assuming that $Z = -r$ (see Section 4.6.4). [Redrawn with permission from Tipping, E., Reddy, M.M. & Hurley, M.A. (1990), Modelling electrostatic and heterogeneity effects on proton dissociation from humic substances, *Environ. Sci. Technol.* **24**, 1700–1705. Copyright 1990 Am. Chem. Soc.]

and electrostatics. The contribution of electrostatics can be gauged by determining the ionic strength dependence of the titration curves, a point emphasised by Marinsky (1987).

4.6.5 Beyond the Debye–Hückel theory

The Debye–Hückel theory for impenetrable spheres and low potentials has been followed thus far, to demonstrate the basic concepts of electrostatic interactions, and the trends that follow from them. Other cases have been analysed, from the same starting point of the Poisson–Boltzmann equation. Detailed descriptions will not be given here, but the main extensions that have been made are as follows. Tanford (1961) described analytical solutions of the equation, still using the low-potential approximation (cf. equations 4.29 and 4.30) for solvent-permeable spheres, long rods, and flexible linear polyelectrolytes. He also analysed the case of an impenetrable sphere for which the surface charge is non-uniformly distributed. Analytical solution of the Poisson–Boltzmann equation, without truncation of higher terms, is possible for flat plates (infinite planes), and leads to the Gouy–Chapman theory, the basis for much surface chemistry. The theory is not well suited to macromolecules because their shapes do not approximate flat plates. Numerical solution of the equation is possible, however, and has been achieved for spheres and rods. The papers of de Wit *et al.* (1990) and Bartschat *et al.* (1992) are recommended as reviews of the more advanced applications of the Poisson–Boltzmann equation, with especial reference to humic substances. The trends in the calculation results presented in Table 4.2 are not affected by adopting a more sophisticated description of the interactions.

All the models considered so far assume that the mobile ions are point charges, which for high potentials leads to impossibly high concentrations of ions near the molecular surface. In surface chemistry, the Stern model has been used to deal with this problem. It postulates specific binding of 'indifferent' ions at surface sites. In the context of organic macromolecules, such binding might be considered outer-sphere complexation, i.e. the binding of a hydrated ion at a specific site (Section 5.1.1). It would give rise to decreased surface charge density and consequently lower surface potential, and smaller electrostatic effects on cation binding. Another approach is the 'counterion condensation' theory (Manning, 1978), which postulates 'territorial binding' of counterions to occur when the charge density on the macromolecule exceeds a critical value.

4.6.6 Binding due to counterion accumulation

A significant aspect of electrostatic interactions is the presence of an excess of mobile counterions in the vicinity of the macroion. If the macroionic charge is negative, as in humic substances, the counterions are cations. Such accumulation, referred to as non-specific binding, contributes to the overall cation binding, together with binding at specific sites. The excess amount of a given ion in the diffuse layer can be calculated by integrating the expressions for potential as a function of distance, combined with the Boltzmann distribution. This is a complicated procedure in systems containing ions of different valence (Bolt, 1982).

4.6.7 Donnan model

An extreme case of a macromolecule being permeable to solvent is where the charge is uniformly distributed within the molecular domain, and the net charge is zero. This is called the Donnan model because it is formally the same as the 'Donnan effect', which accounts for the unequal distribution of charge in macroscopic systems where two solutions are separated by a membrane that is permeable to small (mobile) ions but not to macroions (Fig. 4.6). In the microscale version, the membrane is conceptual. Kinniburgh *et al.* (1996) wrote the electroneutrality condition as

$$\frac{Z}{V_{\mathrm{D}}} + \sum z_j(c_{\mathrm{D},j} - c_j) = 0 \tag{4.51}$$

where Z is the macromolecular charge (eq g^{-1} or eq mol^{-1}), V_{D} the volume of the Donnan solution phase (dm^3 g^{-1} or dm^3 mol^{-1}), z_j the charge of a mobile ion (cation or anion), $c_{\mathrm{D},j}$ its concentration in the Donnan solution phase, and c_j its concentration in the bulk phase. The concentrations in the two phases are related by the Boltzmann law

$$c_{\mathrm{D},j} = c_j \exp\left(-\frac{z_j \varepsilon \psi_{\mathrm{D}}}{kT}\right) = c_j \left[\exp\left(-\frac{\varepsilon \psi_{\mathrm{D}}}{kT}\right)\right]^{z_j} \tag{4.52}$$

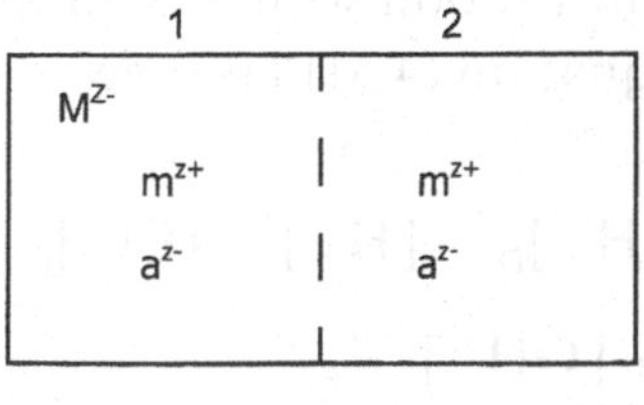

Figure 4.6. Schematic diagram of the Donnan equilibrium. M represents the (anionic) macroion, which cannot pass through the membrane (dashed line). Electrolyte ions are denoted by m (cation) and a (anion). At equilibrium, $[\mathrm{m}]_1 > [\mathrm{m}]_2$ and $[\mathrm{a}]_1 < [\mathrm{a}]_2$.

where ψ_D is the (uniform) Donnan potential. If V_D is known or assumed, and the values of c_j are known, values of $c_{D,j}$ and ψ_D can be calculated. Note that since there must always be at least two types of mobile ion (a cation and an anion), the Boltzmann term, $\exp(-\varepsilon\psi_D/kT)$, can be eliminated from equations like (4.52) to give

$$\left(\frac{c_{D,1}}{c_1}\right)^{1/z_1} = \left(\frac{c_{D,2}}{c_2}\right)^{1/z_2} = \left(\frac{c_{D,3}}{c_3}\right)^{1/z_3} \cdots \tag{4.53}$$

so there is in fact no need to solve for ψ_D. The Boltzmann term does not feature in the macroscopic description, but the same result is obtained.

Binding or dissociation reactions within the Donnan phase are described by conventional equilibrium expressions involving values of $c_{D,j}$. Thus the dissociation of a proton from a weak acid site is given by

$$\mathrm{RAH} = \mathrm{RA^-} + \mathrm{H^+} \qquad K = \frac{[\mathrm{RA^-}]_D[\mathrm{H^+}]_D}{[\mathrm{RAH}]_D} \tag{4.54}$$

where the subscript D indicates concentration expressed in terms of the Donnan phase solution, and $[H^+]_D$ is equivalent to one of the $c_{D,j}$ values. The degree of dissociation, α, is given by

$$\alpha = \frac{K/[\mathrm{H^+}]_D}{1 + K/[\mathrm{H^+}]_D} \tag{4.55}$$

Consider the case where $pK = 4$, $V_D = 0.002\,\mathrm{dm^3\,g^{-1}}$ (typical for humic acids; Kinniburgh *et al.*, 1996), and the macromolecule has 4×10^{-3} mol of proton-dissociating groups per gram (again typical for humic acids). For 50% dissociation, $\alpha = 0.5$ and $[H^+]_D = 10^{-4}\,\mathrm{mol\,dm^{-3}}$. The concentration of dissociated groups expressed in terms of the Donnan phase solution volume then turns out to be $1\,\mathrm{mol\,dm^{-3}}$. This charge has to be neutralised by H^+, together with the accumulated excess mobile ions, in the Donnan phase solution. In a typical titration with NaOH and with NaCl as the background electrolyte, equation (4.51) becomes

$$-\frac{Z}{V_D} = [\mathrm{Na^+}]_D - [\mathrm{Na^+}] + [\mathrm{H^+}]_D - [\mathrm{H^+}] - [\mathrm{Cl^-}]_D + [\mathrm{Cl^-}] - [\mathrm{OH^-}]_D + [\mathrm{OH^-}] \tag{4.56}$$

Under the circumstances in question, $[Na^+] \cong [Cl^-]$, while $[H^+]$, $[Cl^-]_D$, $[OH^-]_D$ and $[OH^-]$ are negligible. Therefore we require

$$[\mathrm{Na^+}]_D = 1 - [\mathrm{H^+}]_D = 0.9999\,\mathrm{mol\,dm^{-3}}$$

From equation (4.53) we have

$$[H^+] = [H^+]_D[Na^+]/[Na^+]_D$$
$$= 1.0001 \times 10^{-4}\,[Na^+] \cong 10^{-4}\,[Na^+]\,\mathrm{mol\,dm^{-3}}$$

Therefore

if $[Na^+] = 0.001\,\mathrm{mol\,dm^{-3}}$ then $[H^+] = 10^{-7}\,\mathrm{mol\,dm^{-3}}$ (pH = 7.01)
if $[Na^+] = 0.01\,\mathrm{mol\,dm^{-3}}$ then $[H^+] = 10^{-6}\,\mathrm{mol\,dm^{-3}}$ (pH = 6.04)
if $[Na^+] = 0.1\,\mathrm{mol\,dm^{-3}}$ then $[H^+] = 10^{-5}\,\mathrm{mol\,dm^{-3}}$ (pH = 5.08)

Thus the apparent pK, computed from α and bulk solution pH, varies by approximately 1 log unit per decadal change in ionic strength. This does not apply at low values of α, where $[H^+]_D$ is significant in the neutralisation of the macromolecular charge. To use the model in cases where the salt dependence does not conform to the decadal relationship, it has to be assumed that the Donnan volume, V_D, varies with ionic strength (see Section 9.5.3).

It will be noted that the Donnan model takes counterion accumulation into account, including the contributions of ions of different charge, much more easily than the integration procedure mentioned in Section 4.6.6.

4.7 Proton dissociation from well-defined polymers

Polymeric compounds can be synthesised which bear acidic groups that are chemically identical. A well-known example is polyacrylic acid, which has the repeating unit $[\text{—}CH(COOH)CH_2\text{—}]_n$. A titration curve of a high molecular weight sample of polyacrylic acid is shown in Fig. 4.7. The curve exhibits the anticipated 'smearing' discussed above, which can be attributed almost entirely to electrostatic effects. The situation is complicated by changes in the conformation, notably expansion, of the macromolecule, which alters the strength of the electrostatic interaction. Polyacrylic acid is a simple polyelectrolyte; other compounds exhibit more complex titration behaviour. For example, the apparent pK of poly-L-glutamic acid increases with α at low α, then remains steady or decreases (depending upon temperature), then increases again at high α, due to a change in conformation from a helix to a random coil (Olander & Holtzer, 1968). Another significant point is that high molecular weight polyelectrolytes undergo polymer–polymer interactions at sufficiently high concentrations, with consequent modification of titration behaviour (Nagasawa *et al.*, 1965). Tanford (1961) and Marinsky (1987) have given detailed accounts of proton dissociation from polyacids.

Proton dissociation from proteins was described by Tanford (1961), who presented a detailed analysis of the titration behaviour of the enzyme ribonuclease. The proton-dissociating groups of amino acid side chains could be identified from the amino acid content. Intrinsic equilibrium constants (equation 4.24) for the groups were also deduced from the titration behaviour by application of the Debye–Hückel theory (equation 4.38). The results from the two approaches were in excellent agreement (Table 4.3). This classic work with proteins was done many decades ago,

Table 4.3. Titratable groups of ribonuclease, determined from the amino acid content and deduced from titration data, analysed by applying the Debye–Hückel correction for electrostatic effects (Tanford & Hauenstein, 1956). [Adapted from Tanford, C. (1961), *Physical Chemistry of Macromolecules*, © Charles Tanford.]

	pK_{int} expected from data for small molecules	No. of sites from amino acid analysis	From titration data	
			pK_{int}	No. of sites
α-COOH	3.75	1 + 4 = 5	4.7	5
Side chain COOH	4.6			
Imidazole	7.0	4	6.5	4
α-NH_2	7.8	1	7.8	1
Phenolic	9.6	6	9.95	6
Side-chain NH_2	10.2	10	10.2	10
Guanidyl	>12	4	4	4

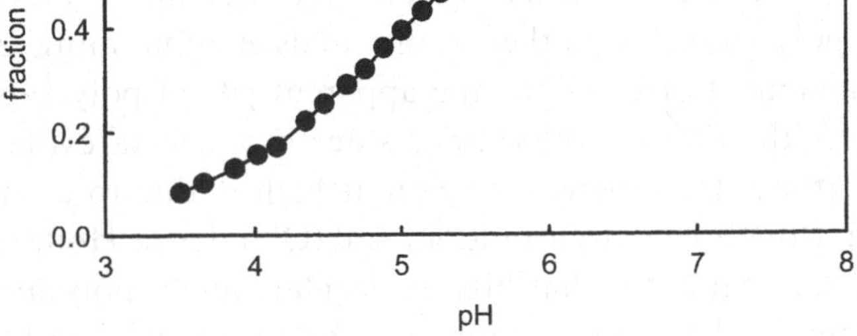

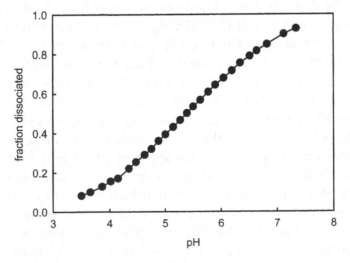

Figure 4.7. Titration of polyacrylic acid at 15 °C, ionic strength 0.1 mol dm^{-3}, drawn using data published by Nagasawa *et al.* (1965). The polymer concentration was 0.008 29 mol dm^{-3} (COOH groups), the molecular weight was 150 000.

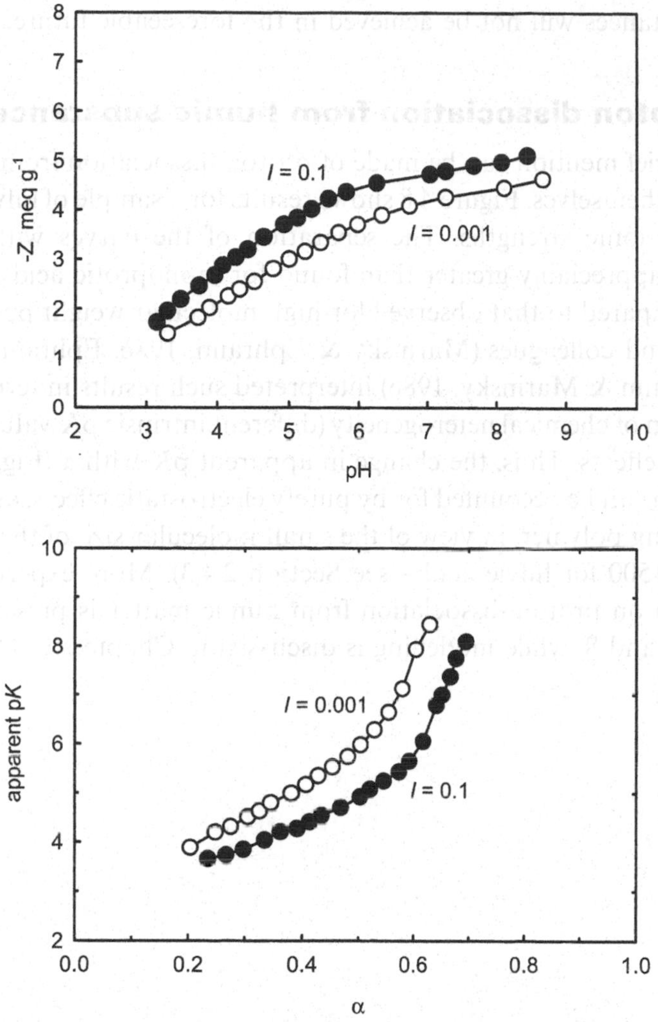

Figure 4.8. Proton dissociation from a sample of fulvic acid. Fractions dissociated were estimated from the experimental data by assuming a total site content of 7.35×10^{-3} mol g^{-1}, obtained from mathematical modelling (see Chapters 9 and 10). The data are also plotted in terms of the apparent pK as a function of fraction dissociated (α), showing the effects of both chemical heterogeneity and electrostatic interactions on proton dissociation from humic substances. [The upper panel is redrawn from Cabaniss, S.E. (1991), Carboxylic acid content of a fulvic acid determined by potentiometry and aqueous Fourier transform infrared spectrometry. *Anal. Chim. Acta* **255**, 23–30, with permission from Elsevier Science.]

but most researchers would agree that such precise assignment of sites in humic substances will not be achieved in the foreseeable future.

4.8 Proton dissociation from humic substances

Finally a brief mention can be made of proton dissociation from humic substances themselves. Figure 4.8 shows results for a sample of fulvic acid at different ionic strengths. The separation of the curves with ionic strength is appreciably greater than found for monoprotic acids, but is modest compared to that observed for high molecular weight polymers. Marinsky and colleagues (Marinsky & Ephraim, 1986; Ephraim *et al.*, 1986; Ephraim & Marinsky, 1986) interpreted such results in terms of a combination of chemical heterogeneity (different intrinsic pK values) and Coulombic effects. Thus, the change in apparent pK with α (Fig. 4.8) is greater than can be accounted for by purely electrostatic effects, as found in a repeating polymer, in view of the small molecular size of the humic sample (*c.* 1500 for fulvic acid – see Section 2.4.3). More experimental information on proton dissociation from humic matter is presented in Chapters 7 and 8, while modelling is discussed in Chapters 9–11.

5

Metal–ligand interactions

The interactions of metals with other solutes – usually termed ligands – are of widespread importance, in industry, biochemistry and medicine, as well as environmental science. Introductory accounts in relation to natural waters have been given by Stumm & Morgan (1996) and Langmuir (1997), while McBride (1994) focuses on soil environments. The intention here is to lay the foundations for considering the interactions of metal cations with humic substances.

5.1 Coordination

Cations in aqueous solution are not independent entities. They interact with molecules or ions containing free pairs of electrons (ligands), to achieve a completed electronic outer shell, a process known as coordination. In the simplest case, the ligands are water molecules. The number of ligands in nearest-neighbour positions is called the coordination number, for which the commonest values are two, four and six. Surrounding the inner coordination sphere is an outer sphere, comprising water molecules with a more ordered structure than the bulk solvent (Fig. 5.1).

5.1.1 Complex formation

When a cation combines with a ligand other than water, the resulting entity is referred to as a complex. Outer-sphere complexes are formed when the ligand is situated in the outer coordination sphere, forming no direct bonds to the metal ion (Fig. 5.1) and held by the relatively weak forces of electrostatic attraction and hydrogen bonding. The formation of inner-sphere complexes involves the displacement of one or more water molecules from the inner coordination sphere; a direct metal–ligand bond is formed by the sharing of the ligand's free electron pair between the metal and the ligand. Some authors refer to outer-sphere complexes as ion pairs, reserving the term complex for inner-sphere complexes only. When writing equations for complex formation, water molecules are usually omitted. Thus the reaction of the nickel (II) ion with sulphate could be written

$$Ni(H_2O)_6^{2+} + SO_4^{2-} = Ni(H_2O)_4SO_4 + 2H_2O \qquad (5.1)$$

but it is more usual simply to write

$$Ni^{2+} + SO_4^{2-} = NiSO_4 \qquad (5.2)$$

5.1.2 Stability constants

The equilibrium constant for the reaction between a metal and one or more ligands is referred to as the stability constant. For one-to-one reactions, the symbol K is conventional. Thus, omitting charges for clarity

$$M + L = ML \qquad K_1 = \frac{[ML]}{[M][L]} \qquad (5.3)$$

$$ML + L = ML_2 \qquad K_2 = \frac{[ML_2]}{[ML][L]} \qquad (5.4)$$

$$ML_{i-1} + L = ML_i \qquad K_i = \frac{[ML_i]}{[ML_{i-1}][L]} \qquad (5.5)$$

If the reactions are written as overall combinations, the symbol β is often used

$$M + iL = ML_i \qquad \beta_i = \frac{[ML_i]}{[M][L]^i} \qquad (5.6)$$

The greater the stability constant, the greater is the strength of bonding and the more stable is the complex.

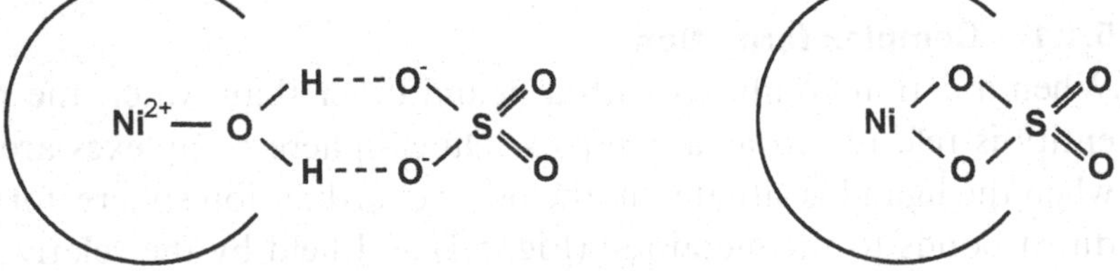

Figure 5.1. Types of coordination, illustrated schematically for Ni^{2+} and SO_4^{2-}. Outer-sphere coordination is shown on the left, and inner-sphere on the right. The part-circle indicates the location of the inner sphere of coordinated water. Simplified from Martell & Hancock (1996).

5.1.3 Trends in reactivity

The stimulating review by Martell & Hancock (1996) of the factors governing the stabilities of metal–ligand complexes has been drawn upon in producing the following introductory account. The weakest complexes are those involving outer-sphere bonding. Their stability constants can be estimated on the basis of electrostatic interactions by the Fuoss (1958) equation. For ion pairs with an opposite charge of one, log K at $I = 0$ is in the range 0–1, increasing to 1.5–2.4 for an opposite charge of two, and to 2.8–4.0 for an opposite charge of three (Stumm & Morgan, 1996).

Figure 5.2 shows trends in the stability constants of inner-sphere complexes formed by the interactions of Ag^+, Pb^{2+} and In^{3+} with halides. It can be seen that Ag^+ strongly prefers I^- to F^-, whereas the opposite is true for In^{3+}, and Pb^{2+} exhibits almost no preference. Such phenomena led to classification schemes for metals and ligands, the most prominent of which is the HSAB (Hard and Soft Acids and Bases) scheme of Pearson (1967). In this context, the acids are Lewis acids, which can accept and share a pair of electrons from a Lewis base, i.e. can form a coordinate bond. The softest cations and ligands are highly polarisable, which means that their electron sheaths are readily deformed by an electric field. They include the cations Ag^+, Cd^{2+}, Hg^{2+} and ligands containing S. Hard acids and bases are not polarisable, and include Na^+, Mg^{2+}, Al^{3+} and Th^{4+}, and ligands containing O atoms, and especially F. In general, hard acids (cations) prefer to associate with hard bases while

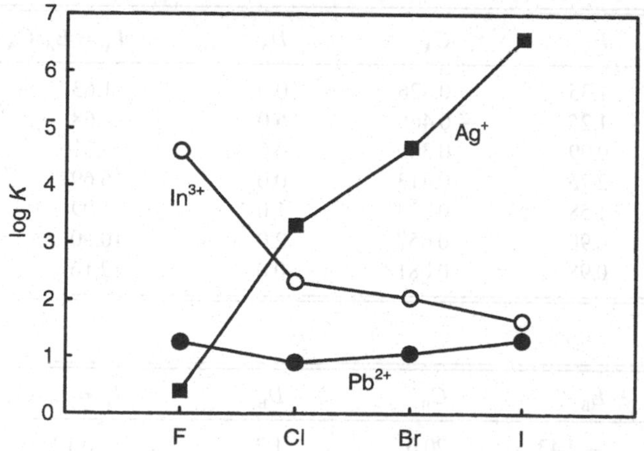

Figure 5.2. Stability constants (log K) for the formation of halide complexes. In the classification of Pearson (1967), the metal ions become harder in the sequence $Ag^+ < Pb^{2+} < In^{3+}$, while the ligands become softer in the sequence $F^- < Cl^- < Br^- < I^-$. Plotted from data assembled by Martell & Hancock (1996).

soft acids prefer soft bases. In bonds involving hard ligand donor atoms, the shared electron pair is generally closer to the ligand donor atom, and the bond has ionic character.

A quantitative approach to complex formation based on the HSAB principles is the equation of Hancock & Marsicano (1980)

$$\log K = E_A E_B + C_A C_B - D_A D_B \tag{5.7}$$

Here, A and B refer to the acid (metal) and base (ligand), E measures the tendency to form electrostatic (ionic) bonds, C the tendency to form covalent bonds, and D corrects for the steric hindrance to solvation on complex formation. The values of D are strongly related to the size of the acid or base. Table 5.1 shows values of E, C and D for some environmentally important metal cations and some simple ligands. The ratio of E/C, designated by I_A or I_B, is a quantitative measure of hardness or softness; the greater is I, the harder is the acid or base. Table 5.2 shows stability constants for complexes of a number of metals with simple ligands containing donor atoms that are present in humic substances. The data in Table 5.1, together with equation (5.7), help to interpret the stability constants. For example, complexes of Hg^{2+} are more stable than those of Ca^{2+} for all ligands because both E_A and C_A for Hg^{2+} are greater than those for Ca^{2+}, while D_A for both metals is zero. The difference is

Table 5.1. Parameters for equation (5.7). The Lewis acids (metal cations) and bases (ligands) are placed in order of increasing hardness (I_A or I_B).

Lewis acids	E_A	C_A	D_A	$I_A = E_A/C_A$
Hg^{2+}	1.35	0.826	0.0	1.63
Cu^{2+}	1.25	0.466	6.0	2.68
Cd^{2+}	0.99	0.300	0.6	3.31
Pb^{2+}	2.76	0.413	0.0	6.69
Mn^{2+}	1.58	0.223	1.0	7.09
Al^{3+}	6.90	0.657	2.0	10.50
Ca^{2+}	0.98	0.081	0.0	12.16

Lewis bases	E_B	C_B	D_B	$I_B = E_B/C_B$
I^-	– 2.43	20.0	1.7	– 0.122
$HO\text{-}(CH_2)_2\text{-}S^-$	– 3.74	38.8	0.9	– 0.096
NH_3	– 1.08	12.34	0.0	– 0.088
OH^-	0.0	14.00	0.0	0.0
CH_3COO^-	0.0	4.76	0.0	0.0
F^-	1.00	0.0	0.0	∞

especially great for the soft ligand β-mercaptoethanol, because the C_A value for Hg^{2+} and the C_B value for the ligand are both large.

As will be seen in Chapter 8, the most abundant cation-binding groups in humic substances are carboxylate (—COO^-), and these are the least discriminatory among metals (Table 5.2). Next come phenolate groups, which are intermediate, in terms of discrimination among metals, between carboxylate and OH^-. Less abundant are N-containing groups, which favour the softer metals, as shown by the log *K* values for ammonia in Table 5.2. The most discriminatory, and least abundant, are groups containing sulphur, which very strongly favour soft metal cations such as Hg^{2+} and Ag^+. These trends have implications for binding site strength, and also for competition among metals. For example, Ca^{2+}, which is abundant in natural waters, may compete effectively with Cd^{2+} at carboxylate sites, but hardly at all at sites with S as the donor atom.

5.1.4 Chelates

Chelating ligands have more than one grouping that can form a bond with the metal ion. The resulting complexes are called chelates. The denticity is the number of bonds in the complex. Thus when a single donor atom takes part, the ligand is referred to as monodentate, when two atoms are involved the ligand is bidentate, three atoms tridentate, and so on. Some examples of chelating ligands are shown in Fig. 5.3. The equilibrium constants shown in Table 5.3 for the formation of some bidentate chelating ligands demonstrate that greater stabilities can generally be achieved compared to monodentate complexes involving the same donor groups (Table 5.2); the more bonds that are formed, the more stable is the complex. For example, the well-known chelator ethylenedinitrilotetraacetate

Table 5.2. Stability constants (log *K*) for the formation of 1:1 complexes with simple ligands.

	CH_3COO^-	OH^-	NH_3	$OH\text{-}(CH_2)_2\text{-}S^-$
Hg^{2+}	6.3	10.6	8.8	*27.0*
Cu^{2+}	2.2	6.3	4.0	*8.1*
Cd^{2+}	1.9	3.9	2.6	6.1
Pb^{2+}	2.7	6.3	1.6	6.6
Mn^{2+}	1.4	3.4	1.0	*1.8*
Al^{3+}	*3.1*	9.0	0.8	*− 2.1*
Ca^{2+}	1.2	1.3	− 0.2	*− 0.5*

The values in italics were calculated from equation (5.7).
$OH\text{-}(CH_2)_2\text{-}SH$ (β-mercaptoethanol) binds through the S atom.

(EDTA) (Fig. 5.3), which can form up to six bonds with the metal ion, binds the metals of Table 5.3 with log *K* values in the range 12–24.

Martell & Hancock (1996) discuss in detail the factors governing the formation and stability of chelates, utilising the large literature on metal interactions with ligands of known structure. A central theme is the structure of the chelate, specifically the relative positions of the ligand's bonding groups and the metal ion. For small bidentate chelating ligands,

Table 5.3. Equilibrium constants (log *K* values) for the formation of metal complexes with bidentate ligands (EN, ethylenediamine).

	Oxalate	Phthalate	EN
Hg^{2+}	*10.5*	*5.7*	8.3
Cu^{2+}	6.2	4.0	10.5
Cd^{2+}	3.9	*3.3*	5.4
Pb^{2+}	4.9	*3.6*	5.0
Mn^{2+}	4.0	*3.4*	2.7
Al^{3+}	*7.3*	*4.5*	—
Ca^{2+}	3.0	2.5	0.1

The values refer to zero ionic strength; those in italics were converted, using activity coefficients from the Davies equation, from values for ionic strengths between 0.1 and 1. Values are from the NIST database (Smith *et al.*, 1993).

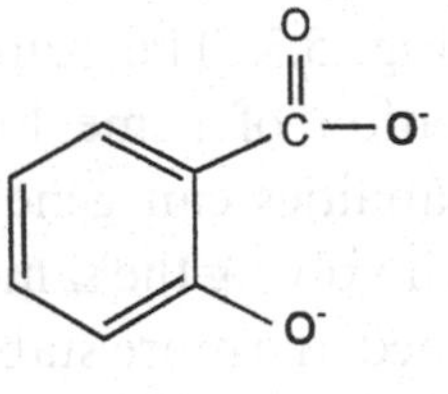

salicylate

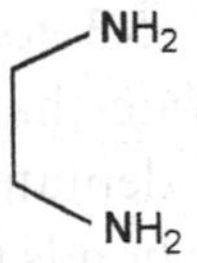

ethylenediamine (EN)

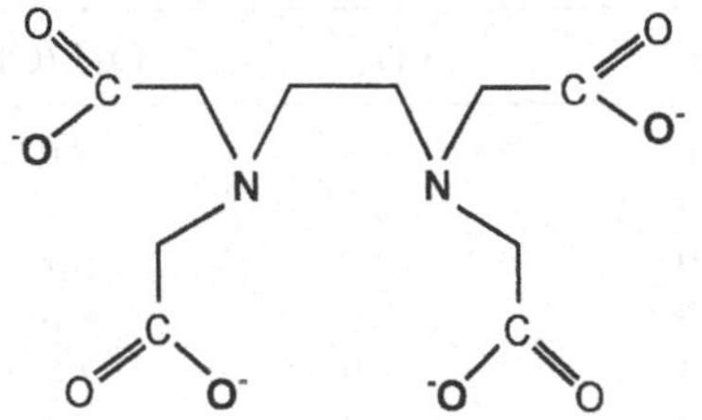

ethylenedinitrilotetraacetate (EDTA)

Figure 5.3. Examples of chelating ligands. The bold atoms are potential donors to metal ions.

the size of the ring formed when the metal is bound is an important factor determining the stability of the complex. Thus, the formation of a four-membered ring, as in carbonate complexes, is favoured by large metal ions with high coordination numbers, which allow the L—M—L bond angles to be small enough to avoid steric strain. For many metals, a five-membered ring is more stable than the equivalent six-membered one, principally because there is less steric strain. The difference in stability depends on the size of the metal ion, so that selectivity toward metal ions of a ligand forming six-membered rings is greater than that of the corresponding ligand forming five-membered rings. An illustration is provided by the data of Table 5.4, which show the greater discrimination among metals by trimethylenediaminetetraacetate (TMDTA) compared to EDTA. For chelate rings with more than six members, stabilities tend to decrease, due to a lower probability of attachment of the second donor atom, an entropic effect. However, this may not apply if the ligand is 'preorganised', in other words if the donor atoms in the free ligand are positioned relative to each other similarly to their positions in the chelate. Preorganisation is greatest in rigid structures. In proteins, the donor groups for metal binding may be separated widely on the polypeptide chain, but due to the specific conformation of the macromolecule they may be held rigidly in favourable multidentate binding positions (see also Section 5.5.2).

There is little information about the relative positions of chelating donor atoms in humic substances, and so it is not known how the factors discussed here may govern cation–humic interactions. Humic substances probably form chelates with a variety of ring sizes, and the ligand groups probably vary in their extents of preorganisation. As a result, heterogeneity in binding strengths, and differences in selectivity towards metals among sites, must arise. Figure 5.4 shows how different kinds of bidentate binding

Table 5.4. Stabilities of chelates with five- and six-membered rings.

Metal ion	Ionic radius, nm	log K EDTA	log K TMDTA	Δlog K
Cu^{2+}	0.057	18.70	18.82	− 0.12
Cd^{2+}	0.095	16.36	13.83	2.53
Pb^{2+}	0.118	17.88	13.70	4.18

Ethylenedinitrilotetraacetate (EDTA) and trimethylenediaminetetraacetate (TMDTA) form five- and six-membered chelates respectively.
Based on data presented by Martell & Hancock (1996).

sites might be formed in humic substances. The ideas could be extended to sites of greater denticity. Some sites may exhibit different denticities toward different metal cations. Experimental information on the natures of humic binding sites is reviewed in Chapter 8.

5.1.5 Mixed complexes

Usually the term 'mixed complex' refers to the interaction of a metal ion with more than one ligand. A simple example is the complex $Al(OH)F^+$, which forms in some acidic natural waters. There are few published data for such complexes. Sharma & Schubert (1969) discussed the estimation of stability constants of mixed complexes from data for the corresponding simple complexes, taking into account statistical effects. Macromolecular ligands with many binding sites can complex several metals simultaneously. This applies especially to humic substances in the natural environment.

5.1.6 Kinetics of metal ion complexation

The following account is taken mainly from Brezonik (1994). The forward reaction, and corresponding rate equation are

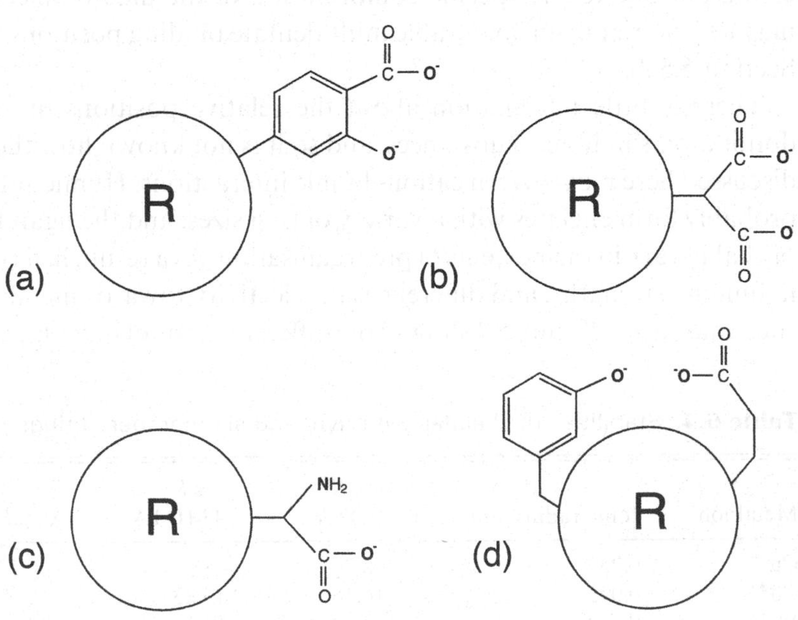

Figure 5.4. Plausible ways in which bidentate sites could form in humic substances (schematic). R represents the rest of the humic molecule (not to scale). In structures (a)–(c), the sites are formed by donor atoms on adjacent groups. In (d) they are formed by groups spaced further apart in the primary structure, brought together by conformational factors.

$$M + L = ML \qquad \frac{d[ML]}{dt} = k_f[M][L] \qquad (5.8)$$

The formation of outer-sphere complexes is diffusion controlled, and readily reversible. As noted in Section 5.1.3, the equilibrium constants for 1:1 outer-sphere complexation are in the range 1–10 $dm^3\,mol^{-1}$. The formation of an inner-sphere complex with a monodentate ligand then requires the displacement of a coordinated water molecule by the ligand, which is known as anation. Anation takes place by an essentially simultaneous interchange, rather than by the ligand entering the coordination sphere before the water leaves (associative pathway), or by the water leaving before the ligand enters (dissociative pathway). The breaking of the water bond is a key part of the process, as evidenced by the correlation between the rates of anation and the water exchange rates of metal ions. The latter vary over a very wide range, from *c.* 10^{-7} to $10^9\,s^{-1}$, although only Rh^{3+}, Cr^{3+}, Ru^{2+} and Co^{3+} have values less than 1 s^{-1}. Rate constants for many monodentate anation reactions are in the range 10^3–10^9 $dm^3\,mol^{-1}\,s^{-1}$. The reaction times can be gauged by calculating times for the reaction to go to 50% completion (half-lives). Table 5.5 shows some examples.

Table 5.5. Half-lives ($t_{1/2}$) in seconds for forward and reverse reactions.

Forward second-order reactions (k_f in $dm^3\,mol^{-1}\,s^{-1}$)

$[L]_0$ ($mol\,dm^{-3}$)	$[M]_0$ ($mol\,dm^{-3}$)	$t_{1/2}$ for $k_f = 10^3$	$t_{1/2}$ for $k_f = 10^6$	$t_{1/2}$ for $k_f = 10^9$
10^{-3}	$< 10^{-4}$	0.693	6.93×10^{-4}	6.93×10^{-7}
10^{-5}	10^{-5}	100	10^{-1}	10^{-4}
10^{-3}	10^{-3}	1	10^{-3}	10^{-6}

Reverse first-order reactions (k_b in s^{-1})

k_b	$t_{1/2}$
10^{-6}	6.93×10^5
10^{-3}	6.93×10^2
1	0.693
10^3	6.93×10^{-4}

For the purposes of the examples, the reactions are considered to be irreversible, i.e. going to completion. If they do not go to completion, the kinetic calculations are somewhat more complicated, but the characteristic times are similar.
Note that 6.93×10^5 s is approximately eight days.
The starting concentrations of ligand and metal are indicated by $[L]_0$ and $[M]_0$ respectively.

Dissociation of monodentate complexes is a hydrolysis reaction, involving the displacement of the ligand by H_2O. The rate expression is

$$ML \rightarrow M + L \qquad \frac{d[ML]}{dt} = -k_b[ML] \tag{5.9}$$

The rate constants are inversely proportional to the stability constant of the complex. For most metals of interest in natural waters, dissociation rate constants for monodentate complexes are in the range 10^3 to $10^4\ s^{-1}$, although those for Al^{3+} and Fe^{3+} are appreciably slower (10 and $100\ s^{-1}$ respectively). These values mean that the reactions take place rapidly. Much smaller values of k_b can apply to chelates; for example a complex with $K = 10^{10}\ dm^3\ mol^{-1}$ and $k_f = 10^6\ dm^3\ mol^{-1}\ s^{-1}$ has $k_b = 10^{-4}\ s^{-1}$. Thus very stable complexes may dissociate very slowly. Table 5.5 shows some half-lives for different dissociation rate constants.

5.2 Chemical equilibria involving metal ions, protons and simple weak acid ligands

Here we build on the discussion of acid dissociation (Chapter 4), extending the analysis to include metal ions along with protons and ligands. By manipulation of the expressions that describe the equilibrium relationships of metals, protons and ligands, the concentrations of the different species can be calculated. Because many ligands, including humic substances, bind protons as well as metals, the following treatment is for ligands that are weak acids. By 'simple' is meant that the ligand molecules have only a single site (not necessarily monodentate) for metal binding.

5.2.1 Monoprotic ligands

Here we shall consider reactions taking place in a solution of a monoprotic ligand, LH, to which have been added known concentrations of a metal salt, MX_z (X^- is monovalent) and either strong acid (e.g. HCl), or strong base (e.g. NaOH), or both. The total concentrations of the reactants are denoted by T_M, T_A, T_B and T_L. The solution may also contain a background electrolyte (e.g. NaCl), but the concentrations of the component ions will be assumed to be equal, and not to interact with the metal of interest or the ligand; therefore they can be neglected from the algebraic expressions developed. By examining theoretical relationships among the solution species, the implications of the equilibrium laws can be appreciated. This is a useful basis for understanding the more complex reactions involving humic substances. In the following, charges are omitted for clarity. The metal is a cation of charge + 1 or greater. The ligand may be anionic (e.g. $RCOO^-$) or neutral (e.g. RNH_2).

As seen in Section 5.1.2, the 1:1 binding of the metal ion to the ligand takes place according to the reaction

$$M + L = ML \tag{5.10}$$

The equilibrium (or stability) constant, in terms of concentrations, is given by

$$K_{ML} = \frac{[ML]}{[M][L]} \tag{5.11}$$

As seen in Chapter 4, the species L^- is also able to react with another solution species, H^+, according to the reaction

$$H + L = HL \tag{5.12}$$

The proton-binding reaction is usually written in the reverse direction, as a dissociation, for which the equilibrium constant is

$$K_{HL} = \frac{[H][L]}{[HL]} \tag{5.13}$$

The mass balances for L and M can be written

$$T_L = [L] + [HL] + [ML] \tag{5.14}$$

$$T_M = [M] + [ML] \tag{5.15}$$

The charge balance expression depends upon the charge on the ligand. If LH is a neutral species (i.e. L is an anion), the balance is given by

$$z[M] + (z - 1)[ML] + [H] + T_B - [L] - [OH] - zT_M - T_A = 0 \tag{5.16}$$

The term zT_M accounts for the concentration of X, added with the metal. Note that the concentration of the neutral species HL does not appear in equation (5.16). An alternative expression for neutral L (LH being a cation) can readily be constructed. Finally, there is the equilibrium relationship

$$K_w = [H][OH] \tag{5.17}$$

The set of six equations (5.11), (5.13), (5.14), (5.15), (5.16) and (5.17) can be solved to obtain the concentrations of the six reactants M, LH, L, LM, H and OH. As with equilibria involving only the weak acid and protons

(Section 3.6.3), this can be done by a series of substitutions, in simple cases, or numerically, by continued improvement of approximate values. The results of such calculations show how the reactions are predicted to proceed under different conditions, assuming that equilibrium is reached.

At different constant pH values, [ML] depends on [M] as shown in Fig. 5.5. The curves have the same basic shape, given by the following relationship, which can be derived from equations (5.11), (5.13), (5.14) and (5.15)

$$v = \frac{[\mathrm{ML}]}{T_\mathrm{L}} = \frac{K_{\mathrm{ML}}[\mathrm{M}]}{1 + K_{\mathrm{ML}}[\mathrm{M}] + ([\mathrm{H}]/K_{\mathrm{HL}})} \tag{5.18}$$

This equation includes the definition of the useful variable v, which here is the average amount of metal bound per mole of ligand. (When working with humic substances, v is better expressed in moles per gram.) It is an average because at the microscopic level at any instant, the ligand will either have a metal atom bound or it will not, i.e. a site can either be filled or empty. Equation (5.18) means that at a given pH (constant [H]), v is a unique function of [M]. In other words, it does not depend upon the total ligand concentration. Equation (5.18) can also be written

$$v = \frac{[\mathrm{ML}]}{T_\mathrm{L}} = \frac{\dfrac{K_{\mathrm{ML}}}{1 + ([\mathrm{H}]/K_{\mathrm{HL}})}[\mathrm{M}]}{1 + \dfrac{K_{\mathrm{ML}}}{1 + ([\mathrm{H}]/K_{\mathrm{HL}})}[\mathrm{M}]} = \frac{K_{\mathrm{M,H}}[\mathrm{M}]}{1 + K_{\mathrm{M,H}}[\mathrm{M}]} \tag{5.19}$$

where $K_{\mathrm{M,H}}$ is the apparent constant at a given value of [H]. Thus as long as the pH is constant the binding of the metal follows the simple equation of a 1:1 chemical reaction. Equation (5.19) has the same mathematical form as that derived by Langmuir for adsorption onto a surface, and is sometimes referred to as the Langmuir equation. An equation of the same form also appears in the Michaelis–Menten model of enzyme kinetics. The main features are (a) an essentially linear relationship (Henry's Law behaviour) between v and [M] at values of [M] sufficiently small for the denominator to be close to unity, and (b) a plateau maximum value of v (unity) at sufficiently high [M] for the numerator and denominator to be close to $K_{\mathrm{M,H}}$[M]. The effect of changing pH is only on the value of $K_{\mathrm{M,H}}$; the maximum amount of binding is independent of pH, although of course greater values of [M] will be required for it to be attained at high [H] (low pH). As the pH passes through the pK of the ligand, the dependence upon pH decreases, the deprotonated form of the ligand becoming dominant.

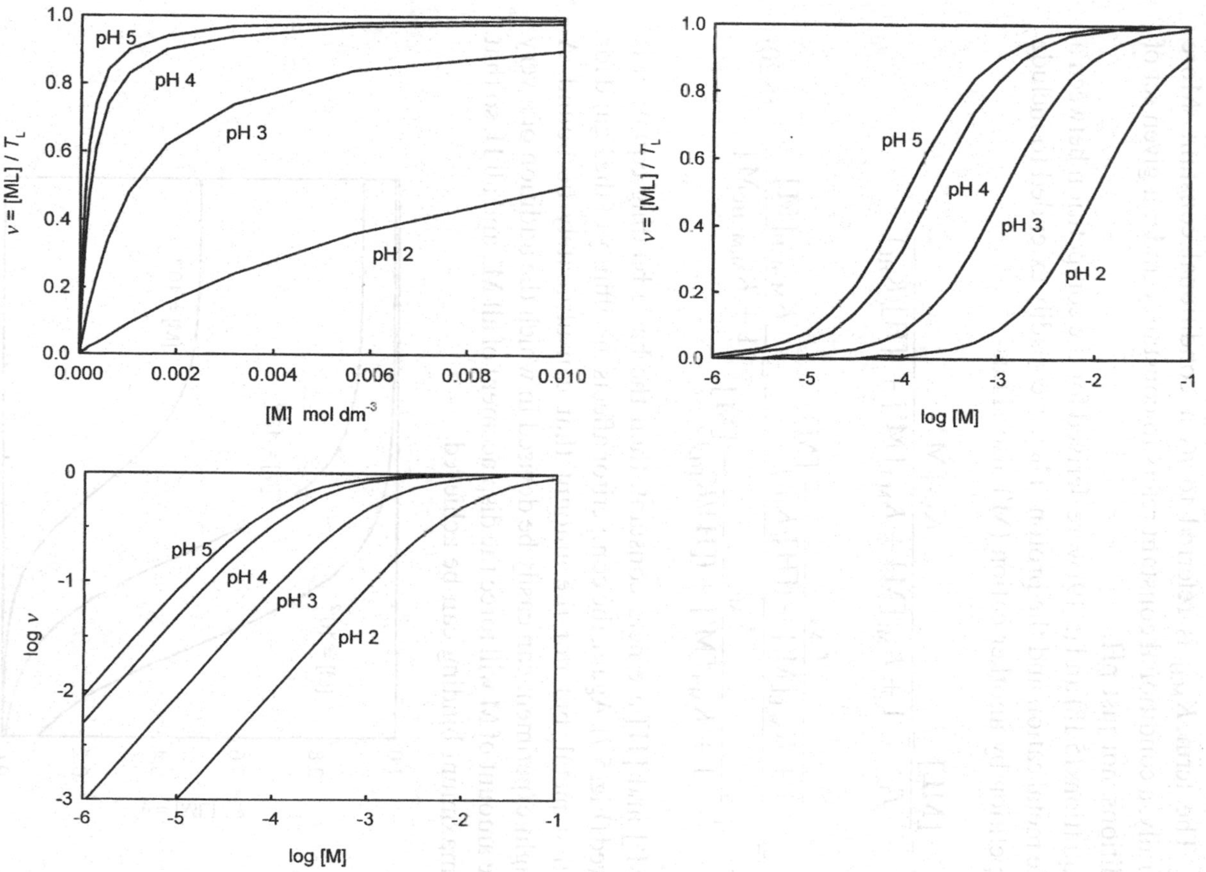

Figure 5.5. Application of equation (5.18) to show the dependence of ν on [M] at different pH values, for a monoprotic ligand, calculated for $\log K_{ML} = 4$, $\log K_{LH} = -4$. Concentrations are in mol dm^{-3}.

Figure 5.6 shows how, at constant [M], metal binding increases with pH. As pH is increased, the competition of H with M for the ligand decreases, and so more metal binding occurs at a given free metal concentration, [M]. The term $K_{M,H}$ is referred to as a conditional constant. More generally, a conditional constant refers to constancy under a given set of conditions, not just pH.

Equations (5.18) and (5.19) were derived for the competition between a single metal cation and the proton. They are readily extended to include competition by another cation (M′) also. Thus

$$\nu = \frac{[ML]}{T_L} = \frac{K_{ML}[M]}{1 + K_{ML}[M] + K_{M'L}[M'] + ([H]/K_{HL})}$$

$$= \frac{\dfrac{K_{ML}}{1 + K_{M'L}[M'] + ([H]/K_{HL})}[M]}{1 + \dfrac{K_{ML}}{1 + K_{M'L}[M'] + ([H]/K_{HL})}[M]} = \frac{K_{M,M',H}[M]}{1 + K_{M,M',H}[M]} \qquad (5.20)$$

If [M′] and [H] are held constant, then the basic binding equation is obeyed (Fig. 5.7). Again, the competitor affects the affinity of the ligand for the first metal, but not the amount that can ultimately be bound. A thought experiment can easily be devised in which the addition of a very large amount of M will force the displacement of all M′, and all H, so that the maximum binding can be achieved.

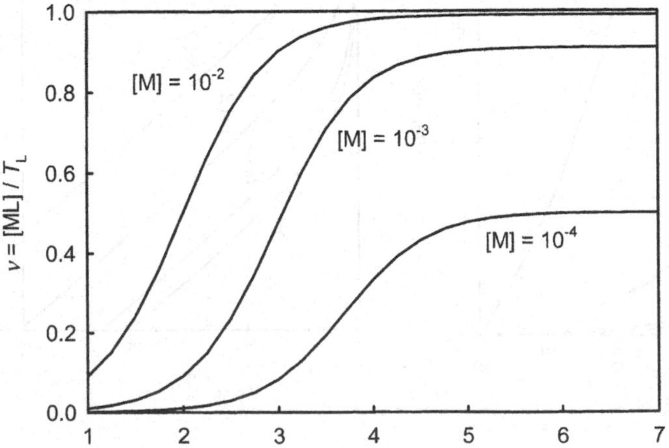

Figure 5.6. Application of equation (5.18) to show the dependence of ν on pH at different [M], for a monoprotic ligand, calculated for log $K_{ML} = 4$, log $K_{LH} = -4$. Concentrations are in mol dm^{-3}.

5.2.2 Interaction of a metal ion with more than one ligand atom or group

It was mentioned in Sections 5.1.2 and 5.1.4 that metal cations can bind to more than one ligand atom or group. This can occur in two ways, by interaction with more than one monodentate ligand, or by interaction with a multidentate ligand. Although the resulting complexes are similar in that two or more metal–ligand bonds are involved, the equilibria are significantly different, as can be seen by comparing the following reactions involving the monodentate ligand L

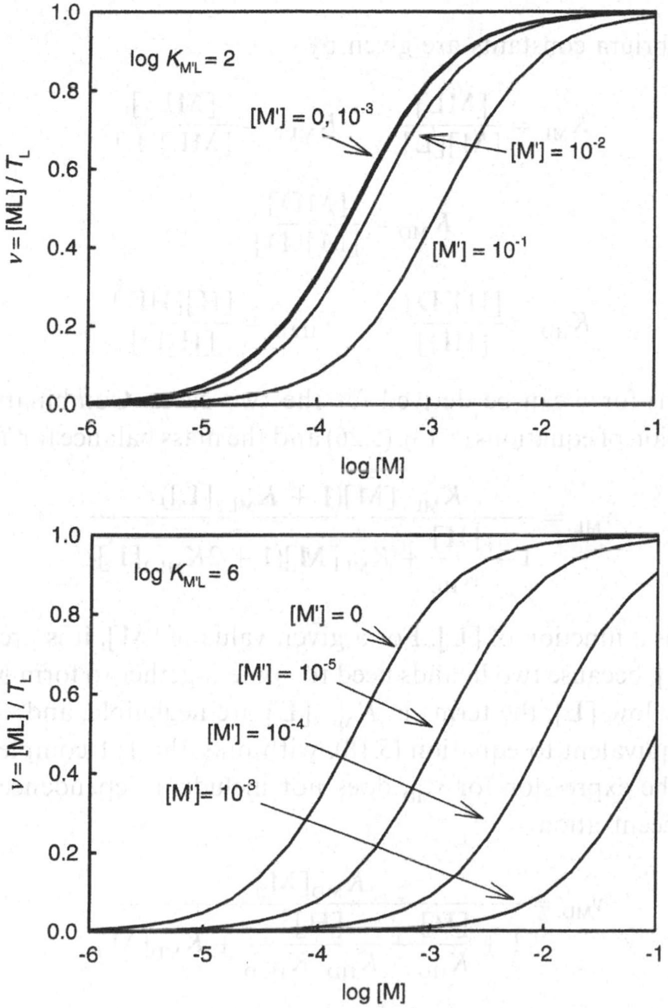

Figure 5.7. Applications of equation (5.20) to show how ν depends on [M] for different concentrations of a competing metal, M', at pH 4. The value of log K_{ML} (first metal) is 4, log K_{HL} = – 4. Concentrations are in mol dm^{-3}.

$$M + L = ML \tag{5.21}$$

$$ML + L = ML_2 \tag{5.22}$$

with the formation of a chelate by the reaction of M with a bidentate ligand D

$$M + D = MD \tag{5.23}$$

The latter also binds protons

$$D + H = HD \tag{5.24}$$

$$HD + H = H_2D \tag{5.25}$$

The equilibrium constants are given by

$$K_{ML} = \frac{[ML]}{[M][L]} \qquad K_{ML2} = \frac{[ML_2]}{[ML][L]} \tag{5.26}$$

$$K_{MD} = \frac{[MD]}{[M][D]} \tag{5.27}$$

$$K_{HD} = \frac{[H][D]}{[HD]} \qquad K_{H2D} = \frac{[H][HD]}{[H_2D]} \tag{5.28}$$

Expressions for v can be derived for the two cases. Combination and manipulation of equations (5.13), (5.26) and the mass balance for T_L, gives

$$v_{ML} = \frac{K_{ML}\ [M](1 + K_{ML2}[L])}{1 + \dfrac{[H]}{K_{HL}} + K_{ML}[M](1 + 2K_{ML2}[L])} \tag{5.29}$$

Thus v_{ML} is a function of [L]. For a given value of [M], it is greater for greater [L], because two ligands need to come together to form ML_2. At sufficiently low [L], the terms in K_{ML2}[L] are negligible, and equation (5.29) is equivalent to equation (5.18), with only the 1:1 complex being formed. The expression for v_{MD} does not include a dependence on the ligand concentration

$$v_{MD} = \frac{K_{MD}[M]}{1 + \dfrac{[H]}{K_{HD}} + \dfrac{[H]^2}{K_{HD}\ K_{H2D}} + K_{MD}[M]} \tag{5.30}$$

The differences in metal binding are illustrated by the plots in Fig. 5.8. It can be seen that a bidentate ligand with similar bonding strength binds significant amounts of metal at appreciably lower concentrations than is possible with a 1:2 complex. This is a demonstration of the 'chelate effect', which is due to differences in the entropy changes of the two reactions (see Martell & Hancock, 1996). The results can be extended to more than two donor atoms (e.g. three monodentate ligands and one tridentate ligand), and the outcome is the same in principle. Both types of reaction have been proposed for humic substances, as will be discussed in Chapter 9. A key

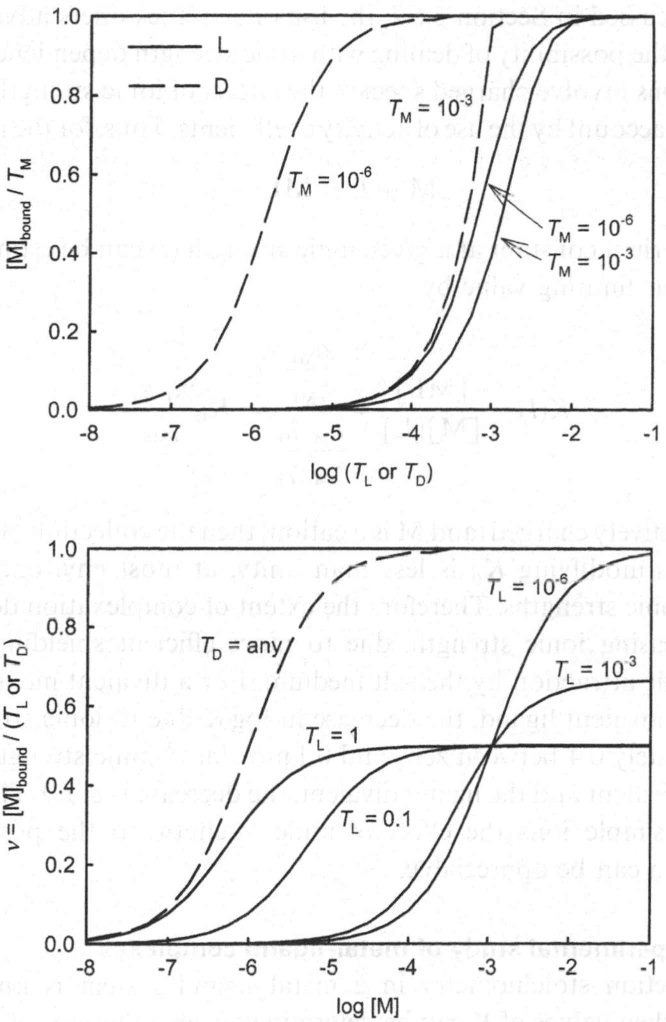

Figure 5.8. Comparison of the binding of a metal M by a bidentate chelating ligand (D) and by a monodentate ligand (L) able to form a 1:2 complex, calculated from equations (5.29) and (5.30). The values of the equilibrium constants are: $\log K_{MD} = 6$, $\log K_{HD} = -3$, $\log K_{H2D} = -5$, $\log K_{ML} = 3$, $\log K_{ML2} = 3$, $K_{HL} = -4$. The pH is 5. Concentrations are in mol dm^{-3}.

result of the analysis is that if the binding of several separate ligands occurs, then metal binding expressed in terms of v will depend upon the total ligand concentration. On the other hand, for multidentate binding to a single ligand there is no concentration dependence.

5.2.3 Effects of ionic strength

So far, the equilibria have been presented in terms of species concentrations. But as discussed in Section 3.6.2, the use of activities offers advantages, especially the possibility of dealing with ionic strength dependence. Since the reactions involve charged species, the effects of ionic strength can be taken into account by the use of activity coefficients. Thus, for the reaction

$$\mathrm{M} + \mathrm{L} = \mathrm{ML} \tag{5.31}$$

the equilibrium constant at a given ionic strength (I) can be estimated in terms of the limiting value by

$$K(I) = \frac{[\mathrm{ML}]}{[\mathrm{M}][\mathrm{L}]} = \frac{\dfrac{a_{\mathrm{ML}}}{\gamma_{\mathrm{ML}}}}{\dfrac{a_{\mathrm{M}}}{\gamma_{\mathrm{M}}}\dfrac{a_{\mathrm{L}}}{\gamma_{\mathrm{L}}}} = K_0 \frac{\gamma_{\mathrm{M}}\,\gamma_{\mathrm{L}}}{\gamma_{\mathrm{ML}}} \tag{5.32}$$

If L is negatively charged (and M is a cation) then the collection of activity coefficients modifying K_0 is less than unity, at most environmentally relevant ionic strengths. Therefore the extent of complexation decreases with increasing ionic strength, due to more efficient shielding of the electrostatic attraction by the salt medium. For a divalent metal cation and a monovalent ligand, the decrease in $\log K$ due to ionic strength is approximately 0.4 between zero and 0.1 mol dm^{-3} ionic strength; if the metal is trivalent and the ligand divalent, the decrease is *c.* 1.0. Thus even for fairly simple ions the effect of ionic strength on the position of equilibrium can be appreciable.

5.2.4 Experimental study of metal–ligand complexes

If the reaction stoichiometry in a metal–ligand system is known or assumed, then values of K can be determined from equations of the type presented above, given experimental measurements of suitable solution variables. Martell & Hancock (1996) present a brief review of methods, and trace the history of this field of research. The most direct approach is to measure concentrations of complexed metal and/or uncomplexed metal. Another method is to monitor pH as acid or base is added to a metal–ligand mixture. If stoichiometries are not known, then for a ligand with a known structural formula, a small number of alternative reaction schemes can be constructed, the best being the one that gives the smallest sum of squared residuals when predictions are compared with observations. The study by Öhman & Sjöberg (1981) of the Al–gallic acid–OH^- system is an example of how stoichiometries, as well as equilibrium constants, can be extracted from experimental data. For this system, which provides a simple model for Al interactions with humic matter, eight complexes were deduced, and their equilibrium constants determined; Al hydrolysis also had to be taken into account. Figure 5.9 shows titration results and the calculated curves, demonstrating how good the results can be when high-precision methods are applied to a definable system.

As well as the determination of amounts of binding, other methods are available to obtain information about the complexes of metals with defined ligands. Various spectroscopic methods provide information about structure, and if the complexes can be crystallised, X-ray structural determinations can also be made.

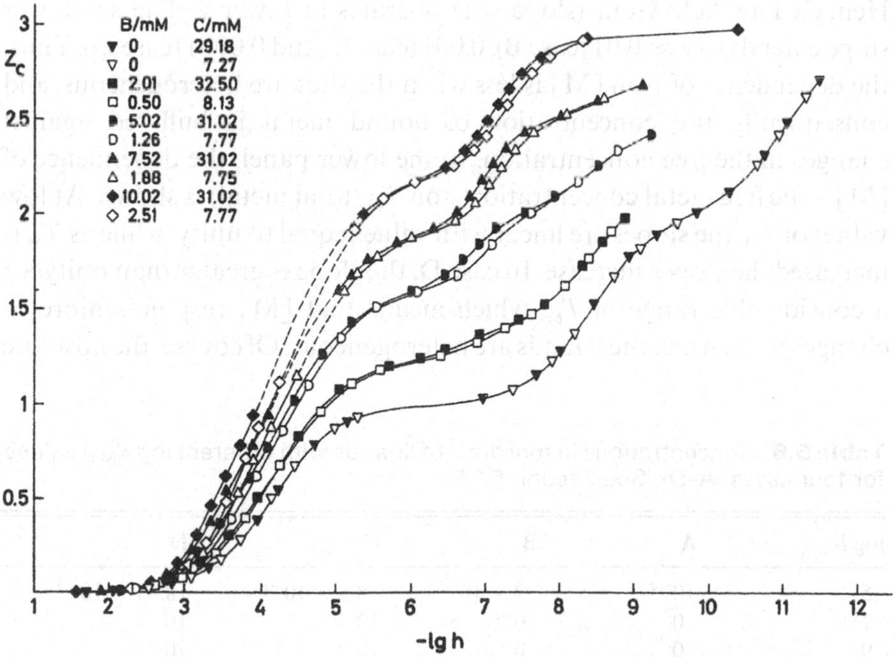

Figure 5.9. Titration data for solutions containing aluminium (B) and gallic acid (C). The *x*-axis is $-\log[H^+]$, the *y*-axis is the net charge per mole of gallic acid. The lines are model fits (see Section 5.2.4). [Reproduced from Öhman & Sjöberg (1981), with permission from *Acta Chemica Scandinavica* and the authors.]

Methods used for defined ligands can often be applied to humic matter, but deriving hard factual information from the results is problematic, because of heterogeneity and the poor definition of the ligands. Methods for studying proton and metal interactions with humic matter are reviewed in Chapter 6.

5.2.5 Metal binding in a mixture of ligands

For many environmental entities, including humic substances, there is a distribution of binding sites. There are a large number of weak sites, and smaller numbers of stronger sites. To explore the implications of such heterogeneity for metal binding at constant pH, consider the ligand conditions of Table 5.6. Four classes of monodentate ligand sites are assumed, with $\log K_{M,H}$ values (equation 5.19) increasing from 5 to 11. In case A there are no ligands with $\log K_{ML} > 5$. In case D, there are ligands with all $\log K_{ML}$ values. Cases B and C are intermediate. The total ligand concentration in each case is the same, 10^{-4} mol dm^{-3}.

Figure 5.10 (upper panel) shows how v depends upon [M]. In the simplest case (A), the plot shows a slope of less than 1 at $v > 0.1$, whereas Henry's Law behaviour (slope = 1) operates at lower v. The shallower slope extends to $v \approx 0.01$ (case B), 0.001 (case C) and 0.0001 (case D). Thus, the dependence of v on [M] is less when the sites are heterogeneous, and consequently the concentration of bound metal is buffered against changes in the free concentration. In the lower panel, the dependence of [M] – the free metal concentration – on T_M (total metal) is shown. At low values of T_M, the slopes are linear with values equal to unity, while as T_M is increased the slopes increase. In case D, the slope is greater than unity for a considerable range of T_M, which means that [M] responds more to changes in T_M when the ligands are heterogeneous. Of course, the absolute

Table 5.6. Concentrations in mol dm^{-3} of ligands with different $\log K_{M,H}$ values, for four cases (A–D). See Section 5.2.5.

$\log K_{M,H}$	A	B	C	D
5	10^{-4}	9×10^{-5}	8.9×10^{-5}	8.89×10^{-5}
7	0	10^{-5}	10^{-5}	10^{-5}
9	0	0	10^{-6}	10^{-6}
11	0	0	0	10^{-7}

In each case the total ligand concentration is 10^{-4} mol dm^{-3}.
The values are used to construct Fig. 5.10.

values of [M] are always lower when there are stronger ligands present.

Modelling of metal interactions with humic substances has been performed on the assumption that the humic binding sites behave like a collection of simple ligands (see Chapter 9). Such an approach allows humic binding site heterogeneity to be represented, but inevitably ignores the influences on binding that result from the multi-site nature of humic matter.

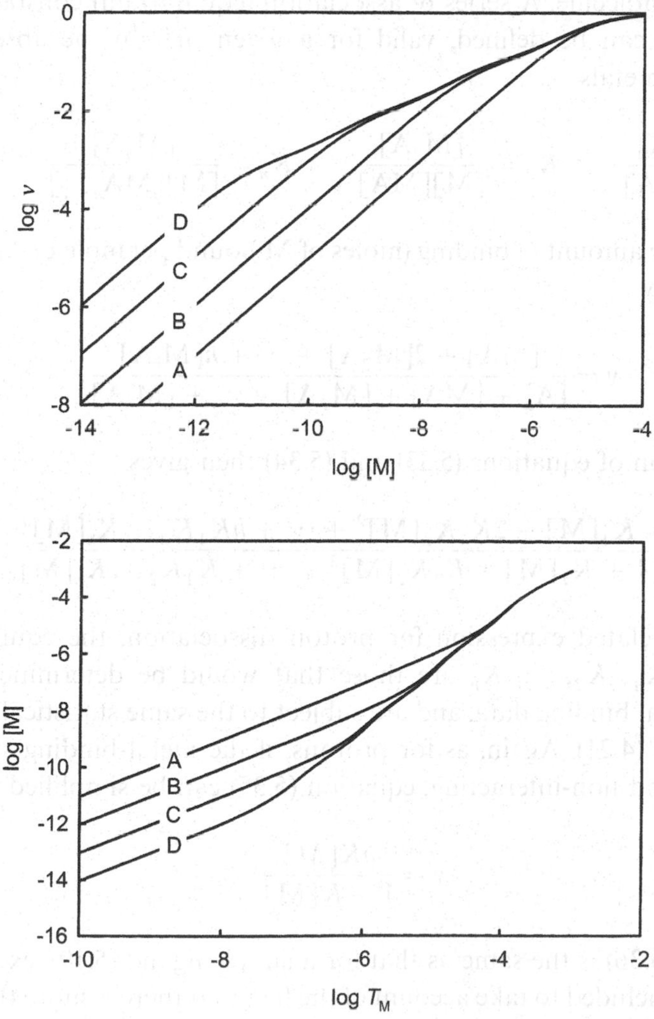

Figure 5.10. Influence of ligand heterogeneity on metal binding. The heterogeneity increases from A to D (see Section 5.2.5 and Table 5.6).

5.3 Multisite ligands

So far we have discussed ligands with a single site, which may be mono- or multidentate. Now we move on to consider ligands with more than one site, which include biologically important macromolecules, (proteins, nucleic acids, polysaccharides), synthetic polymers, and humic substances. To a large extent, the analysis is the same as that presented for proton dissociation in Section 4.5, with the main exception that we shall deal with association, not dissociation, reactions.

Consider the binding of a metal ion (M) to a macromolecule, A, with n binding sites, and without formation of complexes involving more than one macromolecule. A series of association (equilibrium) constants K_1, K_2, ..., K_n can be defined, valid for a given pH. In the absence of competing metals

$$K_1 = \frac{[\mathrm{MA}]}{[\mathrm{M}][\mathrm{A}]} \qquad K_2 = \frac{[\mathrm{M_2A}]}{[\mathrm{M}][\mathrm{MA}]} \qquad K_n = \frac{[\mathrm{M}_n\mathrm{A}]}{[\mathrm{M}][\mathrm{MA}_{n-1}]} \tag{5.33}$$

The average amount of binding (moles of M bound per mole of A) can be expressed by

$$\nu = \frac{[\mathrm{MA}] + 2[\mathrm{M_2A}] + \cdots + n[\mathrm{M}_n\mathrm{A}]}{[\mathrm{A}] + [\mathrm{MA}] + [\mathrm{M_2A}] + \cdots + [\mathrm{M}_n\mathrm{A}]} \tag{5.34}$$

Combination of equations (5.33) and (5.34) then gives

$$\nu = \frac{K_1[\mathrm{M}] + 2K_1K_2[\mathrm{M}]^2 + \cdots + nK_1K_2\ldots K_n[\mathrm{M}]^n}{1 + K_1[\mathrm{M}] + K_1K_2[\mathrm{M}]^2 + \cdots + K_1K_2\ldots K_n[\mathrm{M}]^n} \tag{5.35}$$

As in the related expression for proton dissociation, the equilibrium constants K_1, K_2, ..., K_n are those that would be determined from experimental binding data, and are subject to the same statistical factors of equation (4.21). Again, as for protons, if the metal-binding sites are identical, and non-interacting, equation (5.35) can be simplified, giving

$$\nu = \frac{nK[M]}{1 + K[\mathrm{M}]} \tag{5.36}$$

Equation (5.36) is the same as that for a simple ligand (5.19), except the factor n is included to take account of the fact that there is more than one site per (macro)molecule. Similarly, if there are two classes of identical non-interacting sites, A and B, we have

$$\nu = \nu_A + \nu_B = \frac{n_A K_A[\mathrm{M}]}{1 + K_A[\mathrm{M}]} + \frac{n_B K_B[\mathrm{M}]}{1 + K_B[\mathrm{M}]} \tag{5.37}$$

These expressions can be extended to describe competition between metal ions and protons or other metal cations, and the formation of multidentate sites, following the algebra of Section 5.2.

5.4 Electrostatic interactions

Electrostatic effects on metal binding by macroions can be described on the same basis as for protons. The required expression for the equilibrium association constant at a given value of Z is

$$K = K_{\mathrm{int}} \exp(-2wzZ) \tag{5.38}$$

The charge on the macroion is computed by summing the charges at all the sites. For example, in the simple case of one class of monodentate sites binding both protons and a divalent metal cation, the net average charge is given by

$$Z = -(n - \nu)\alpha + \nu \tag{5.39}$$

where α is the fractional dissociation of the metal-free sites (in the absence of metal, $\alpha = r/n$). The first term is the charge due to metal-free sites, and the second is the charge due to the bound divalent metal, the net value of which is $(+2 - 1) = +1$ for each occupied site. The calculation of charge becomes more complicated when multidentate sites are present, and when several metals of different charge can bind. As in the proton-only case, the value of Z depends on the amount of binding, which depends upon the value of Z, and so iteration is required to solve equilibrium problems.

If the charge on the metal cation is greater than one, then the electrostatic term will be correspondingly greater than for protons, i.e. there will be a greater electrostatic effect on binding. However, this may be offset by the tendency of Z to be less negative when metal is bound.

Metal cations accumulate as counterions in the vicinity of a negatively charged macroion, thereby contributing to the overall binding (cf. Sections 4.6.6 and 4.6.7).

5.5 Results with well-defined macromolecules

Studies of the binding of metal ions by synthetic polyelectrolytes and biological macromolecules provide examples that may be useful in understanding cation binding by humic substances. Generally it can be anticipated that such systems should provide results that can be interpreted more precisely than those for humic matter, because the macromolecules are better defined and less heterogeneous.

5.5.1 Synthetic polyelectrolytes

Mandel & Leyte (1964a,b) carried out acid–base titration studies of polymethacrylic acid, which has the repeating unit [—C(CH_3)(COOH)—CH_2—], in the presence of Mg, Co, Ni, Cu, Zn and Cd. They concluded that the metals formed bidentate complexes, a phenomenon which also occurs with humic substances (Section 7.2). Figure 5.11 shows values of the apparent association constant for metal binding by polymethacrylic acid as a function of the degree of proton dissociation (α) from the polymer, determined from electrochemical measurements. The variation of the apparent constant is ascribed to electrostatic effects, as discussed in Section 5.4. The values of log *K* when α, and hence the electrostatic effect, are small are similar to those for simple dicarboxylic acids, again suggesting the formation of bidentate complexes.

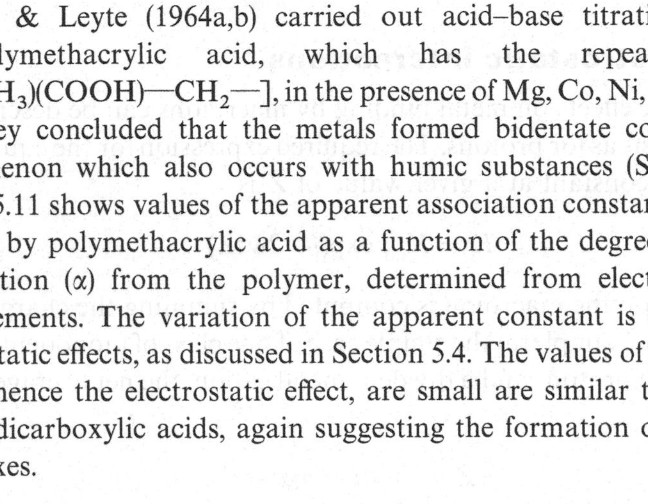

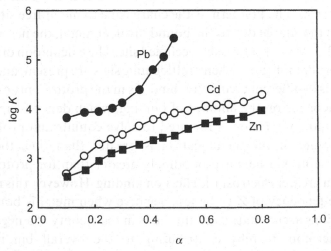

Figure 5.11. Variation of the apparent association constant for metal binding by polymethacrylic acid as a function of the degree of proton dissociation (α). The polymer molecular weight was 26 000, the ionic strength 0.05 mol dm^{-3}. [Redrawn from Cleven, R.F.M.J., de Jong, H.G. & van Leeuven, H.P. (1986), Pulse polarography of metal/polyelectrolyte complexes and operation of the mean diffusion coefficient, *J. Electroanal. Chem.* **202**, 57–68, with permission from Elsevier Science.]

Pomogailo & Wöhrle (1996) presented a review of the interactions of metal cations with synthetic polymers. They argued that a major difference, other than electrostatic effects, between macromolecular and small ligands, is the ability of flexible macromolecules to undergo substantial changes in conformation, thereby allowing favourable binding site geometries to occur. They also proposed the formation of complexes in which the cation interacts with more than one polymer molecule, so that a high denticity can be achieved, for example tetracoordination.

5.5.2 Results with proteins

There are many enzymes in which the metal plays an essential rôle in catalytic activity (Scrutton, 1973). Metals involved in such enzymes include Mg, Ca, Mn, Fe, Co, Ni and Zn. Non-enzymic proteins that bind metals directly include transferrin, which transports iron in the blood, and ceruloplasmin, which binds Cu. In such examples, there is usually a single, highly specific, site for the metal, and therefore the interactions do not have much direct relevance to metal interactions with humic substances. Somewhat more relevant are metal–protein interactions that are unrelated to biological activity *in vivo*. Such interactions have been studied because of the use of metals in protein isolation, studies of denaturation and enzyme inhibition, and as model systems that yield basic information about the interactions (Breslow, 1973). Most of the work reviewed by Breslow referred to Cu and Zn. By carrying out conventional binding studies, together with spectroscopy and X-ray crystallography, and with knowledge of the amino acid sequences of the proteins, it was possible not only to characterise metal binding in terms of numbers of binding sites and their associated equilibrium constants, but also to determine the exact groupings making up the binding sites. A striking general outcome of the studies is the dominance of nitrogen-containing ligand groups in metal binding, which comes about because of the high affinities of N atoms for the metals, and the high nitrogen content of proteins (*c*. 15% by weight). Work with metmyoglobin (see Breslow, 1973 for primary references) showed that (a) the strongest binding sites for Cu and Zn involve groups far apart on the polypeptide chain, (b) the two metals exhibit competitive binding, but bind at different, adjacent, sites, and (c) the binding of metals at ν values greater than 1 mole mole^{-1} (but not at 1 or less) brings about substantial changes in the protein tertiary structure. Whether similar phenomena occur in humic substances is not known – indeed, because of the heterogeneity of humic matter, it is difficult to envisage how they might be investigated – but the findings for the protein demonstrate that cation–macromolecule interactions can be both subtle and complex, a fact that should be borne in mind when attempting to interpret cation–humic binding results.

6

Methods for measuring cation binding by humic substances

Here a review is presented of methods for the quantification of cation binding, mainly at equilibrium. The principle of each method is explained, and its range of application and drawbacks indicated. The focus is on measurement with isolated humic samples, although some of the methods can be applied to field samples. Techniques for the determination of speciation in natural environmental samples – which will invariably involve humic matter – are reviewed in books edited by Tessier & Turner (1995) and Ure & Davidson (1995).

6.1 The humic sample

Ideally, the sample of humic matter to be investigated should be unaltered from its state in the natural environment, except that it should be freed of other chemical components. However, currently available isolation procedures are such that the introduction of artefacts cannot be ruled out (Section 2.2). Samples low in ash should be used, because residual mineral matter (oxides and aluminosilicates) may interact with both protons and metal cations and thereby distort the results.

For acid–base titrations, the humic sample will ideally initially be free of inorganic cations (except for H^+) and anions, a state that can be achieved by passage of the material through a column of hydrogen-saturated cation exchanger, or by exhaustive dialysis. Alternatively, base cations and strong acid anions associated with the sample can be determined analytically, and taken into account when analysing the titration data (Section 6.2.2). However, this is satisfactory only if the base cations are weakly binding, which in practice means Na^+ and K^+.

If trace level binding of metals is to be investigated, then efforts should be made to remove metals present in the isolated sample, otherwise high-energy sites present in low abundance may be masked. For example, Milne *et al.* (1995) performed an extraction with the strong metal complexing agent ethylenedinitrilotetraacetate (EDTA) to minimise the metal content of the peat humic acid on which they worked. Residual metal is less of a problem when binding to the more prevalent low-energy sites is the main interest.

The source and method of isolation of the humic material should be reported, so that as data sets build up it may be possible to discern differences among samples from different environments. With the availability of reference humic materials from the International Humic Substances Society, it is now possible, and desirable, to make comparative measurements, and to use the reference materials for quality control purposes.

Comprehensive proton- and metal-binding studies are preferably performed on a single sample of humic matter, thereby avoiding the need to extrapolate results obtained for one sample to try to explain the behaviour of another. Historically, such comprehensive studies have not been made, and it is somewhat unusual to find metal-binding data that refer to a sample for which proton binding has also been measured.

6.2 Determination of proton binding by potentiometry

Determination of the pH dependence of the extent of proton dissociation (or binding) provides fundamental information about the cation-binding properties of any material containing weak acid groups, and humic matter is no exception. The literature contains many examples of such titrations. Here, the experimental technique is outlined, and the initial interpretation of results described.

6.2.1 Titration methodology

Concentrations or activities of free protons are usually determined with a glass electrode, the operational principles of which are given by Bates (1973). Titrations are performed with a set-up as shown schematically in Fig. 6.1. The experiment provides values of pH for known additions of strong acid or strong base. The commonest experimental approach is to add known amounts of NaOH or KOH to an initially acid solution or suspension of the humic material, containing known concentrations of neutral salt (typically KNO_3 or NaCl) to control ionic strength. Titration of an initially basic solution with acid is also possible.

In carrying out the titration, accurate and precise measurements are of course desirable. In principle, precision is greatest if the concentration of

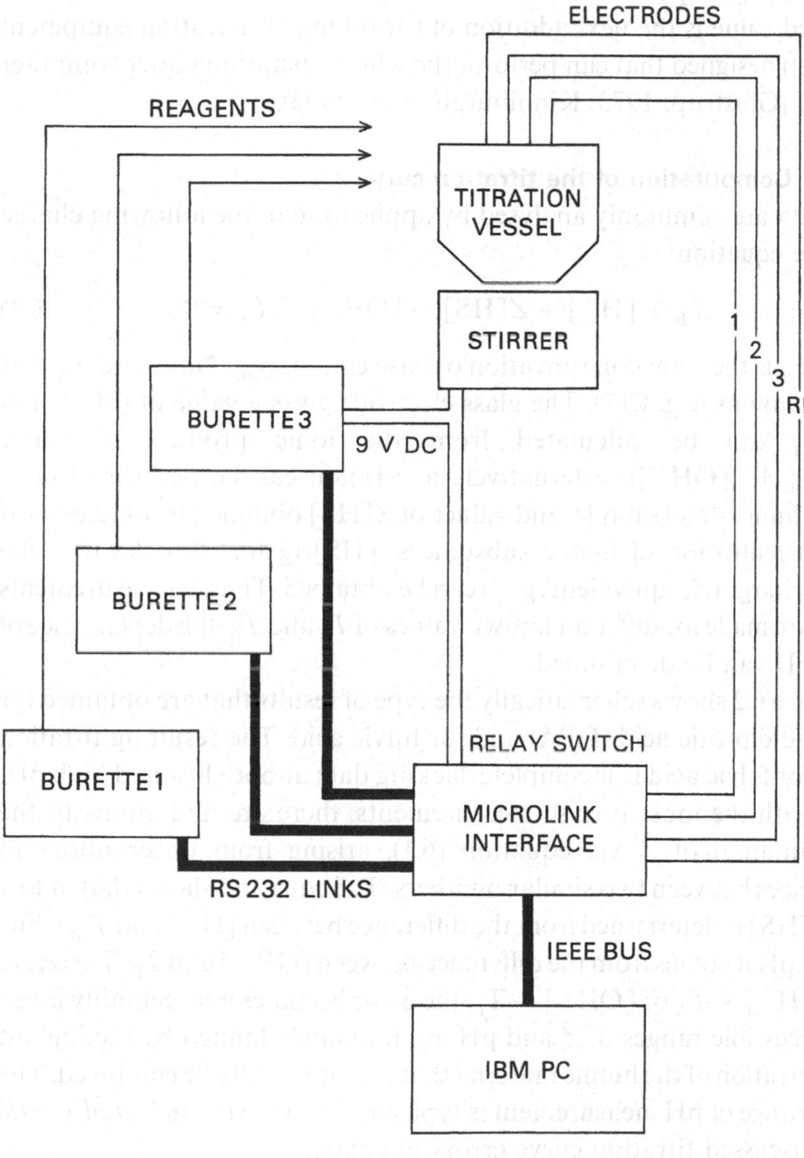

Figure 6.1. Schematic showing communications between hardware devices in the computer-controlled automatic titrator designed by Kinniburgh *et al.* (1995). [Reproduced by permission of the Soil Science Society of America. IPR/18-1C British Geological Survey. ©NERC. All rights reserved.]

humic matter is high, e.g. 1 g dm^{-3}, but then other problems may arise, for example the need to use high concentrations of base, and aggregation effects that slow the attainment of equilibrium. Most authors attempt to ensure that equilibrium has been attained, by setting a criterion for the rate of change of pH with time, so that only when the rate is lower than the specified value is the next addition of titrant made. Titration equipment has been designed that can perform the whole operation under computer control (Ginstrup, 1973; Kinniburgh *et al.*, 1995).

6.2.2 Computation of the titration curve

The data are commonly analysed by application of the following charge balance equation

$$T_B + [H^+] + Z[HS] - [OH^-] - T_A = 0 \qquad (6.1)$$

Here, T_B is the total concentration of base cations (e.g. Na^+) and T_A that of acid anions (e.g. Cl^-). The glass electrode gives a value of $[H^+]$, and $[OH^-]$ can be calculated from the ionic product of water ($K_W = [H^+][OH^-]$). Alternatively a titration can be performed on a blank (humic-free) sample, and values of Z[HS] obtained by difference. If the concentration of humic substances, [HS] (g dm^{-3}), is known, the humic charge (Z, equivalents g^{-1}) can be obtained. Thus, if measurements of pH are made for different known values of T_B and T_A, the dependence of Z on pH can be determined.

Figure 6.2 shows schematically the type of results that are obtained for a simple diprotic acid (DPA) and for fulvic acid. The resulting titration curve for fulvic acid is incomplete, lacking data at both low and high pH. Even with the most precise measurements, there are limitations to the determination of Z via equation (6.1), arising from uncertainties in differences between two similar numbers. The equation shows that at low pH, Z[HS] is determined from the difference between $[H^+]$ and T_A, while at high pH it comes from the difference between $[OH^-]$ and T_B. Therefore when $[H^+] \sim T_A$ or $[OH^-] \sim T_B$, the error becomes unacceptably large. The accessible ranges of Z and pH are ultimately limited by the highest concentration of the humic material that can practically be employed. The useful range of pH measurement is typically 3 to 11. Marshall *et al.* (1995) have discussed titration curve errors in detail.

6.2.3 Interpretation of the titration curve

The titration curve of Fig. 6.2(c) shows a low slope at pH 7–8, which reflects principally the completion of the dissociation of humic carboxylic

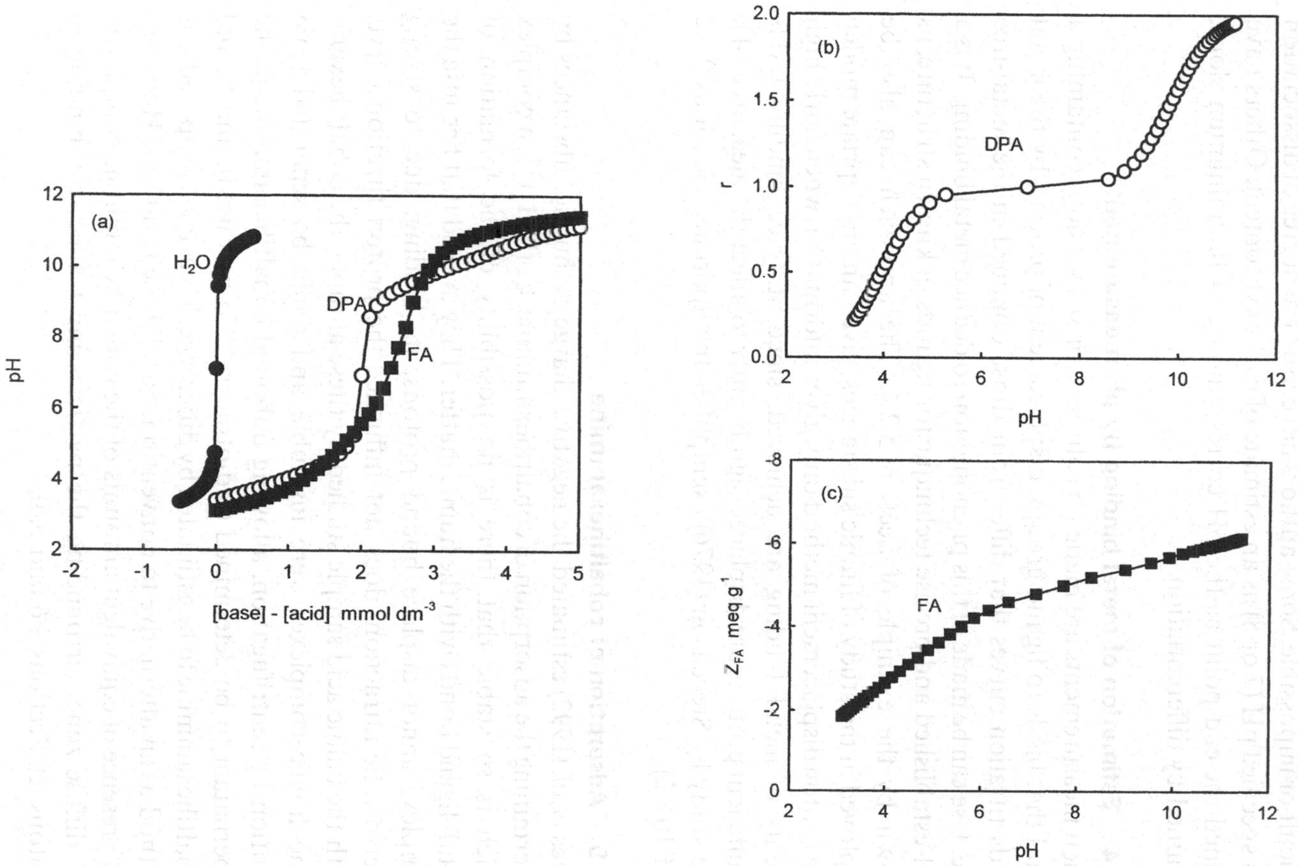

Figure 6.2. Titrations of a diprotic acid (DPA) with pK values of 4 and 10, and fulvic acid. Panel (a) idealised raw results from titrations; (b) converted results for DPA; (c) converted results for fulvic acid.

acid groups. The lack of a defined titration end-point arises from site heterogeneity, and makes the precise determination of site contents difficult or impossible. Some authors have used the degree of dissociation at a specified pH (7 or 8) as an estimate of carboxyl content. Others have defined the end-point as the pH corresponding to the minimum slope, obtained by differentiation.

6.2.4 Estimation of metal binding by pH measurement

When measurements are made on solutions or suspensions containing a metal that binds to ligand groups, displacement of protons by the metal yields titration curves that differ from those obtained in the metal-free case. Use can be made of this phenomenon to deduce metal binding. It is a well-established and precise technique for ligands of known structure, as shown by the example of Section 5.2.4. The approach can also be employed in the study of humic substances, given an appropriate model. The proton displacement method only gives information when fairly high extents of metal binding are achieved, since at trace binding levels insufficient protons are displaced significantly to affect the measured pH. The study by Stevenson (1976) exemplifies the approach (Sections 9.3.2 and 10.8.2).

6.2.5 Adsorption of cobaltihexammine

Maes *et al.* (1992) estimated the negative charge on humic substances by determining the adsorption of cobaltihexammine, $Co(NH_3)_6^{3+}$, a complex which is so stable that there is no possibility of the formation of metal–ligand bonds with the humic matter. They argued that because the complex cannot displace bound protons, its binding, due to strong electrostatic attraction, does not influence the proton titration curve. With the humic acid sample studied by these authors, the cobaltihexammine–humic complexes were insoluble and could be separated from solution by centrifugation, allowing unbound cobaltihexammine in the supernatant to be determined (radiotracer ^{60}Co was used), and bound cobaltihexammine to be estimated by difference. Maes *et al.* proposed this method as an alternative to conventional acid–base titrations. However, the presence of equivalent amounts of the cobaltihexammine complex in the diffuse zone surrounding the particles must affect the binding of protons (cf. Sections 4.6 and 5.4).

6.3 Analytical determination of acid group contents

This subject has been reviewed by Schnitzer & Khan (1972), Perdue (1985) and Stevenson (1994), and those references should be consulted for detailed information on experimental procedures and data interpretation.

6.3.1 Total acidity by barium exchange

A widely used indirect titration method for determining total acidity is the barium hydroxide method (Schnitzer & Khan, 1972). A known excess amount of $Ba(OH)_2$ is added to the humic sample, the aim being to precipitate the humic matter quantitatively as barium complexes. Following removal of the precipitate by centrifugation or filtration, the remaining solution is titrated to neutral pH with acid, to determine the excess Ba. The humic acidity is then found by difference. The method works well for humic acids, but Davis (1982) showed that it fails when incomplete precipitation occurs, since then humic matter in the supernatant is protonated in the titration step, and the total acidity is underestimated. This is especially likely for the more soluble types of humic matter (fulvic acids and aquatic samples). Perdue (1985) suggested the use of ultrafiltration to increase the amount of humic material removed.

6.3.2 Carboxylic acidity by calcium exchange

This is similar in principle to the $Ba(OH)_2$ method, but the reagent is calcium acetate at pH 6–7. The calcium is supposed to react quantitatively with only carboxyl groups, again to form insoluble complexes. After reaction, the excess acetate is determined by titration with base to pH 9.8. The technique has the same problem as the barium hydroxide method in that a complete separation of the Ca–humic substance has to be achieved to make the back-titration results meaningful. Perdue (1985) presented a critical discussion of the method, explaining that it implies a definition of carboxyl content in terms of the pH of the reaction mixture.

6.3.3 Other techniques

Stevenson (1994) reviewed various methods for determining functional groups in humic matter, including proton-dissociating groups. The total humic acidity has been estimated by determining the $—OCH_3$ content following methylation, and by reaction with diborane and with lithium aluminium hydride. Carboxyl group contents have been estimated by methylation/saponification and by an iodometric method, aromatic carboxylic acids by quinoline decarboxylation. The difference between total acidity and carboxyl content provides an estimate of phenolic content. Hydroxyl contents have also been determined by methylation.

Additional information can be obtained from spectroscopy, chiefly nuclear magnetic resonance and infra-red.

6.3.4 Metal-binding sites in relation to proton-binding sites

The determination of acidity, i.e. the total content of proton-dissociating groups in a sample, provides a guide to the content of sites able to bind metal cations. Although metals can coordinate to atoms that do not bind protons (e.g. carbonyl and ester oxygen atoms), the highest observable extents of metal binding are less than the numbers of proton-binding sites. One probable reason is the presence of multidentate sites for metal cations. It should be borne in mind, however, that non-specific binding occurs due to the accumulation of counterions in the vicinity of a charged macromolecule (see Section 5.4). In principle, this can result in the binding of the same amount of counterion (e.g. Na^+) as there are dissociated groups. Total acidity is useful when considering the interactions of humic matter with major cations such as H^+, Na^+, Ca^{2+}, Mg^{2+} and Al^{3+}, which can be present at high concentrations in natural systems. However, it provides little insight into binding sites for cations present at trace levels, for example heavy metals and radionuclides. Binding of such cations may take place principally at rare, high-energy sites which make a negligible contribution to the total acidity. The sites may be multidentate and involve N and S atoms. Binding data and spectroscopic studies strongly suggest the presence of such sites in humic matter (Chapters 7 and 8). In fact, there appear to be sites with a continuous range of affinities. Thus it can be appreciated that analytical estimation of the contents of metal-binding sites may be difficult or impossible. Presently, the most useful approach is to estimate the number of different sites by model fitting, using analytical data on proton-dissociating sites as a model constraint. There is also some possibility of using molecular modelling techniques to estimate the statistical likelihood of certain sites occurring in a humic population, i.e. to estimate the binding site content from knowledge, or supposition, about chemical groupings (see Chapter 11).

6.4 Direct measurement of equilibrium metal binding – principles

6.4.1 Data requirements

The most basic requirement in the measurement of metal binding is to obtain pairs of concentrations of bound and unbound (free) metal for different experimental conditions (total metal and humic matter concentrations, pH, ionic strength, temperature, etc.). Binding is usually expressed in terms of amount of bound metal per amount of humic matter; we shall use the symbol ν to denote moles of metal bound per gram of humic matter. If other solution conditions are maintained constant, then at equilibrium ν is expected to be a unique function of the free metal concentration, unless aggregation of the humic matter affects metal binding, or the metal can bind to more than one humic molecule (cf. Section 5.2.2). Because of these last possibilities, measurements should ideally be made at different humic concentrations.

Most commonly, ν is determined as a function of free metal concentration, [M], at different (constant) values of pH, ionic strength, and temperature. Plots of ν vs. [M] generated from such data are often referred to as binding isotherms. Another widely used approach is to determine ν and [M] as functions of pH at constant total metal and humic concentrations. Competition effects are usually investigated by maintaining a constant total concentration of competing metal. Data so obtained are convenient for plotting and for the graphical analysis methods that were used in the past. Nowadays however, numerical analysis methods and non-linear fitting are straightforward, and the acquisition of 'well-behaved' data is not so necessary.

Often, only one of the (ν, [M]) pair is determined, and the value of the other calculated from the metal mass balance. For such methods, Fish & Morel (1985) identified three principal classes of measurement technique. Methods in Class I determine free ion activity (or concentration). Methods in Class II determine total unbound metal, and free ion activities are calculated from knowledge of the unbound speciation. Methods in Class III determine total bound metal. Fish & Morel showed that the propagation of error in a titration of humic matter with metal is different for each class. The main issue, for measurements above the detection limit, is the value of the variable (ν or [M]) that is obtained from the mass balance, assuming that accurate and precise measurements of the directly determined quantity are possible. Errors tend to be largest when this value is the difference between similar numbers. Thus, for Class I and II methods, the greatest error is in high values of ν, estimated from the difference between total and free metal. For Class III methods, the greatest error is in low values of [M], under which conditions a high proportion of total metal is bound. Errors can be minimised by careful choice of the total concentrations of metal and humic matter. In principle, the best data sets are those established with methods from more than one class.

6.4.2 Ranges of measurement

A significant consequence of the heterogeneity of humic binding sites is that to characterise fully the interaction of a metal with a humic sample, measurements need to be made over a wide range of free metal concentrations. Consider three cases as follows:

A a single ligand with $\log K = 6$
B a two-site ligand with $\log K = 5$ and 7
C a three-site ligand with $\log K = 4$, 6 and 8.

In each case, 50% of the sites are occupied at a free metal concentration, [M], of 10^{-6} mol dm^{-3}. However, as shown by the plots in Fig. 6.3, the range of [M] required to cover the range of site occupancy from 10% to 90% increases greatly from 100-fold in case A to nearly 100 000-fold in case C, which is most similar to humic matter. The ideal method for measuring metal binding by humic substances therefore has a large 'analytical window' (Buffle, 1988), permitting binding to be detected over wide ranges of values of v and [M]. However, only a few techniques provide a large window, and as yet no technique does so for all metals.

Data should have relevance to environmental conditions. This is especially difficult for trace elements because by definition they are present at low concentrations, so that the most significant binding may be at the rare high-energy sites. If so, then the measurement of binding to the abundant low-energy sites may not provide data that can be used to interpret or predict metal–humic interactions in the natural environment.

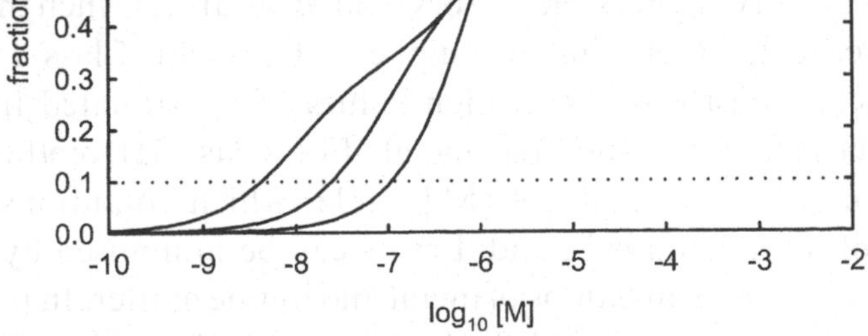

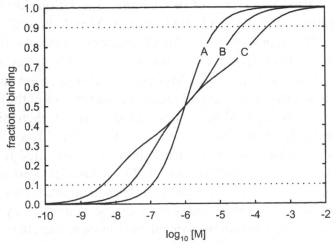

Figure 6.3. Simulated binding as a function of free metal concentration [M] for one-, two- and three-site ligands. See Section 6.4.2 for explanation.

Similarly, it may be required to know how humic matter interacts with a metal within a particular range of pH, which means that measurements for a different pH range are not directly relevant. In general, the available methods do not allow a free choice of experimental conditions, and the limitations of each method must be accepted. Understanding the underlying chemistry of metal–humic interactions is therefore important, in order that interpolation and extrapolation can be made on a sound basis.

6.4.3 Speciation of unbound metal

In the natural environment, metal that is not bound to humic matter may exist in several forms, depending upon the solution composition and other variables. For example, Cu could be present as Cu^{2+}, $CuOH^{+}$, $CuHCO_3^{+}$, $CuCl^{-}$ etc. In most experimental studies, conditions are arranged such that the unbound metal is present only as the simple aquo ion (i.e. Cu^{2+} in the case of copper). This is achieved by keeping the pH sufficiently low to avoid significant hydrolysis and ensuring the absence of other ligands such as HCO_3^{-} or buffer components. By these means the experimental results are made easier to interpret, but the data obtained are inevitably restricted. As a result, little information exists on the possible binding of species other than the aquo ion.

6.5 Separation methods to quantify equilibrium metal binding

A number of methods involve a physical separation of humic-bound metal from unbound metal, or of bound-plus-free metal from free metal. Once the physical separation has been achieved, analytical determination of the metal is required, which may be performed by atomic absorption spectrophotometry, inductively coupled plasma techniques, electrochemistry, use of radiotracers, colorimetry, etc. The range of concentrations that can be investigated depends upon the sensitivity of the analytical method. Free metal concentrations of less than 1 nmol dm^{-3} can be measured with appropriate care and equipment. Very low concentrations can be achieved by the use of radiotracers of known specific activity.

6.5.1 Centrifugation–depletion

This is a standard method for investigating binding by particulate materials. As such it can be used to study humic substances, especially humic acid, under conditions (low pH or high metal binding) where precipitation is essentially complete. Suspensions of the humic material and inorganic components, including the metal(s) of interest, are equilibrated, usually in centrifuge tubes, at known total concentrations and constant temperature. After equilibration, they are centrifuged at sufficiently high force and for sufficiently long to sediment all the humic matter.

The simplest analytical approach is to determine the concentration of metal in the supernatant, and then to subtract it from the known total concentration to determine the amount of bound metal. This is satisfactory as long as it can be assumed that there is no loss of metal to the walls of the centrifuge tubes. If such losses do occur, either they must be quantified, or the total metal concentration in the suspension after equilibration but before centrifugation must be measured. The supernatants have to be monitored for the presence of dissolved humic matter, which tends to vary with solution conditions, especially pH. Such solubilisation restricts the range of conditions over which the method can be applied. An example of the application of the method is the study of Al binding by humic acid carried out by Tipping *et al.* (1988b).

6.5.2 Equilibrium dialysis

A semipermeable membrane is used, with pores small enough to retain the humic material but large enough to allow passage of unbound metal and other low molecular weight solutes. At equilibrium, the distribution of the components is as depicted in Fig. 6.4. The total metal concentration on the humic-free side of the membrane equals the unbound concentration, while in the solution or suspension containing the humic matter, the total is equal to the bound plus the unbound concentrations. The bound concentration is obtained by difference.

Part of a humic sample is often able to diffuse through the membrane, and the extent of passage can vary with pH and ionic strength. Appropriate checks must be made to determine the extent of such diffusion. If it does occur, correction can be attempted by assuming that the diffusing material has the same metal-binding properties as the bulk sample. Then if the humic concentrations are known, the following equations hold

$$T_{M,1} = \nu[HS]_1 + [M] \tag{6.2}$$

$$T_{M,2} = \nu[HS]_2 + [M] \tag{6.3}$$

Here, 1 and 2 refer to the two sides of the membrane. If there is no diffusion, $[HS]_1$ or $[HS]_2$ will be zero. If neither are zero, the equations can be solved for ν and [M]. The most direct way to overcome the problem of humic diffusion through the membrane is to eliminate the diffusing material by carrying out exhaustive dialysis before performing the binding experiments, but then only the higher molecular weight fraction is available for study.

There can be losses of metal by adsorption to the vessel walls and to the membrane. These should not invalidate the results, as long as total metal concentrations on either side of the membrane are determined and equilibrium is truly achieved.

The presence of humic matter bearing an electrical charge on one side of the membrane can give rise to the Donnan effect, the unequal distribution of mobile ions that arises because the non-diffusible humic charge must be compensated to achieve electroneutrality. The compensation is achieved partly by a reduced concentration of co-ions, and partly by an increased concentration of counterions. The microscale version was described in Section 4.6.7 (see also Tanford, 1961). The Donnan effect becomes more significant the higher is the concentration of humic matter and the lower the ionic strength. For example, in an experiment with a humic acid at a concentration of 200 mg dm^{-3}, ionic strength $\sim$ 0.001 mol dm^{-3}, and pH $\sim$ 7, the free concentration of a divalent cation on the non-humic side of the membrane may be only $\sim$ 50% of that on the humic side. The Donnan effect should either be suppressed by

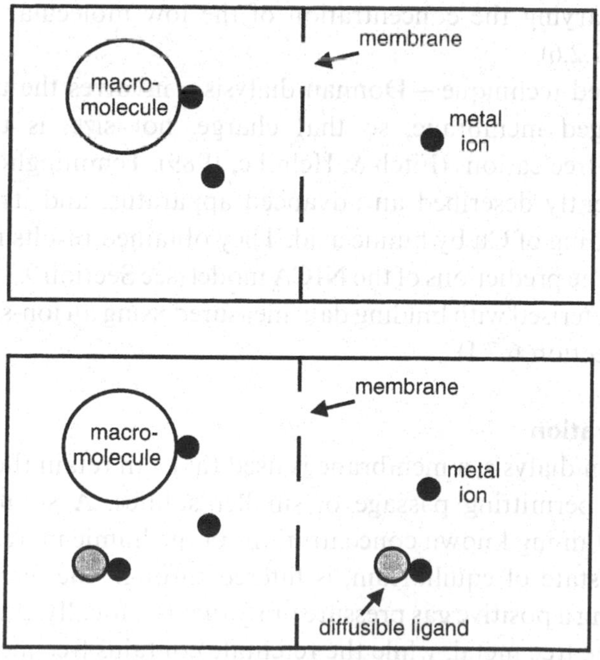

Figure 6.4. Principles of equilibrium dialysis (see Section 6.5.2).

choosing sufficiently high ionic strengths and/or low humic concentrations, or taken into account in data analysis.

Examples of the application of equilibrium dialysis in the study of metal–humic interactions are provided by the studies of Truitt & Weber (1981), Rainville & Weber (1982) and Carlsen *et al.* (1984).

An interesting variant on the dialysis method, known as equilibrium dialysis–ligand exchange (EDLE), was introduced for the study of metal–humic interactions by van Loon *et al.* (1992) and Glaus *et al.* (1995a). It aims to overcome the problem of having to determine low concentrations of metal, in systems where the humic matter binds a large proportion of the total. This is done by introducing a low molecular weight, diffusible, ligand with known binding properties. The diffusible ligand equilibrates across the membrane and binds metal on both sides. The situation at equilibrium is depicted in Fig. 6.4. From the known equilibrium constant(s) for the complexation of the metal by the low molecular weight ligand, [M] can be calculated, while v for the humic matter is obtained from mass balance. A potential problem with the method is the formation of ternary complexes comprising the humic material, the metal and the low molecular weight ligand, but Glaus *et al.* (1995b) showed that this can be avoided by the judicious choice of concentrations. Indeed, they used the method to study mixed complex formation, by varying the concentration of the low molecular weight ligand (Section 7.2.6).

Another related technique – Donnan dialysis – involves the use of a negatively charged membrane, so that charge, not size, is used to discriminate the free cations (Fitch & Helmke, 1989). Temminghoff *et al.* (2000) have recently described an advanced apparatus, and its use to measure the binding of Cu by humic acid. They obtained results in good agreement with the predictions of the NICA model (see Section 9.7), which had been parameterised with binding data measured using an ion-selective electrode (see Section 6.7.1).

6.5.3 Ultrafiltration

As in equilibrium dialysis, a membrane is used that can retain the humic material, while permitting passage of smaller solutes. A solution, or suspension, containing known concentrations of the humic material and the metal, in a state of equilibrium, is filtered through the membrane, either by applying a positive gas pressure or by suction. Ideally, the filtrate contains only the free metal, while the retentate contains free metal plus humic matter and associated bound metal. The amount of metal bound is determined by difference. Only a small part of the mixture is filtered, because otherwise the humic concentration will increase and the extent of metal binding will change. The method can be used in flow-through mode, such that the concentration of trapped humic matter remains constant as new solution is added (Law *et al.*, 1997). Buffle (1988) has discussed the practical application of ultrafiltration in detail, emphasising the need to know the factors that influence passage across the membrane.

Possible problems include passage of humic matter through the membrane, progressive clogging of the membrane, effects of membrane electrical charge, loss of metal to the membrane, and contamination by soluble constituents of the membrane. Nordén *et al.* (1993) measured the binding of Sr^{2+} and Eu^{3+} by fulvic acid using ultrafiltration and Schubert's method (Section 6.6.1), and found that the latter method gave higher binding. The likeliest explanation was passage of small amounts of fulvic acid through the ultrafiltration membrane.

6.5.4 Chromatography

Methods have been designed in which the humic–metal complex passes quickly through a suitable chromatographic column, while unbound metal is either eluted more slowly or is trapped on the column. The amount of bound metal can then be determined by analysis of the eluate. The method of Driscoll (1984) for aluminium binding employs a cation-exchange column, which effects a rapid removal of unbound cations from the solution, while not affecting, or affecting only slightly, the amount bound. Passage through the column must be fast enough to avoid dissociation of the metal–humic complex, but slow enough to ensure that unbound cations are fully removed. Careful establishment of experimental conditions is thus required. The method has mainly been used on field samples, especially in studies of Al speciation in acid waters (Section 13.3.2). Tests can be made to determine whether dissociation of the metal–humic complex is significant (Backes & Tipping, 1987a). Chromatographic separation based on molecular weight has also been used. In the method of Warwick & Hall (1992), the humic compound plus bound metal is separated by high-performance size-exclusion chromatography. As with the cation-exchange column, the method relies on the metal–humic complex not dissociating significantly during elution.

The methods described in the previous paragraph only provide reliable results when the kinetic properties of the system under study are such that the equilibrium is not unacceptably disturbed. In contrast, the chromatographic method of Hummel & Dreyer (1962), originally developed to measure the binding of small molecules by proteins, operates at equilibrium. The chromatography column, usually providing size separation, is eluted with a solution of the metal of interest, at a known concentration (corresponding to [M]), and the humic compound is introduced. As it passes through the column it binds metal, and if the column is long enough equilibrium is reached. The metal concentration in eluted fractions is measured. The resulting profile includes a peak, due to humic-bound metal, and a trough due to metal depleted by humic binding (Fig. 6.5). The amounts of metal bound and depleted are equal, and provide the value of v. The method is time consuming, and there may be problems due to metal sorption to the column material. It was used to study metal–humic interactions by Mantoura & Riley (1975).

6.5.5 Diffusive gradients in thin films (DGT)

In its simplest form DGT (Davison & Zhang, 1994) determines total dissolved concentrations of metals and other solutes. The principle of the method, illustrated in Fig. 6.6, is to expose one face of a hydrogel (the thin film) to the test solution and the other to a binding agent that acts as an essentially infinite sink for the solute. Diffusion of the solute through the hydrogel therefore occurs across a gradient defined by the solute concentration in the test solution. Knowledge about the diffusion coefficient of the solute in the gel, and analytical determination of the amount of solute trapped by the binding agent in a given time, allow the

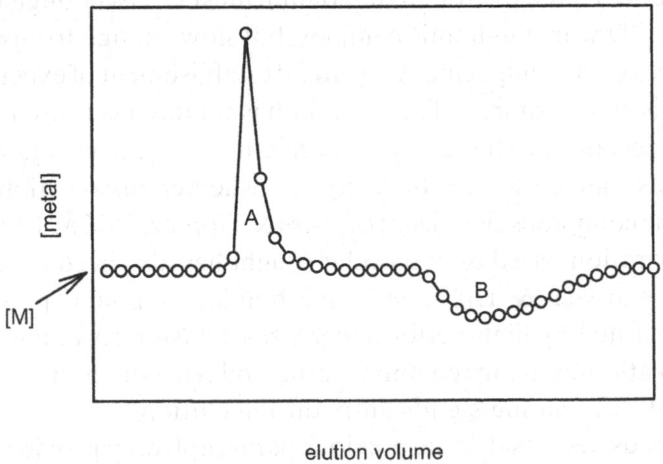

Figure 6.5. Principle of the Hummel–Dreyer chromatographic method for determining metal binding. The humic matter is eluted in A, and carries with it bound metal, in equilibrium with [M], the free metal concentration. The following trough (B) shows the deficit of metal in the column.

solute concentration in the test solution to be determined. The accumulation of metal means that the method can be highly sensitive. Recent developments (Zhang & Davison, 1999) have resulted in the identification of hydrogels in which the diffusion coefficients of humic substances are much smaller than those of the inorganic forms of metals. Thus the method can discriminate, by size, between free metal ions and humic-bound metals, and can therefore be used to determine metal binding by humic substances.

6.6 Competition methods

These rely on the competition between dissolved humic material and a second material, which can be a solid or a separate liquid phase, or a ligand. The relationship between the amount of metal bound to the competitor and the free metal concentration is determined in the absence of humic matter. Then, when the competitor and humic matter are present together, knowledge of the competitor–metal relationship, and determination of the amount of metal bound to the competitor, permit calculation of the free metal concentration. If in addition either the metal bound to humic matter, or the metal bound plus the free metal, can be determined, pairs of v and [M] can be obtained.

6.6.1 Schubert's method

This method, introduced by Schubert (1948), is also known as the ion exchange distribution method (e.g. Nordén *et al.*, 1993). The competitor is a solid-phase cation-exchange resin. In the absence of humic matter, the binding of metal to the solid should ideally follow a simple distribution law

$$D_0 = \frac{v_S}{[M]} \tag{6.4}$$

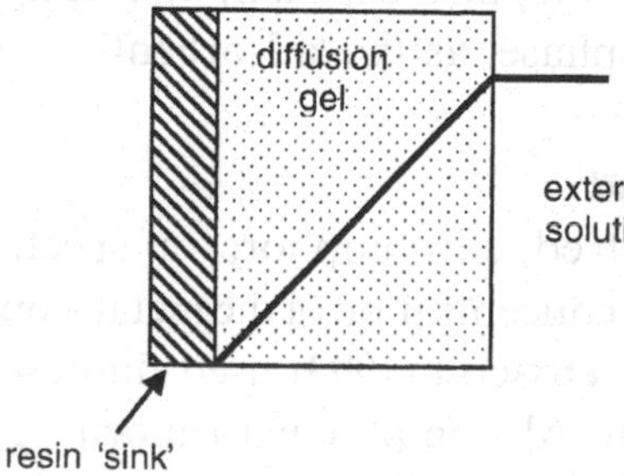

Figure 6.6. Schematic picture of the 'diffusion gradients in thin films' technique (DGT). The bold line indicates the concentration profile of the determinand (see Section 6.5.5).

Here, ν_S is the moles of metal bound per unit mass of the solid. The value of the distribution coefficient D_0 is constant at a given pH, ionic strength, and temperature. If humic matter is present in solution, then a different distribution coefficient is obtained experimentally

$$D = \frac{\nu_S}{[M] + \nu[HS]} \tag{6.5}$$

where ν is moles of metal bound per gram of humic substances and [HS] is the concentration of humic substances in g dm^{-3}. In equation (6.5) the denominator represents the total dissolved metal concentration (free plus humic-bound). The value of [M] is obtained by application of equation (6.4), and ν follows by substitution of [M] into equation (6.5) or from

$$\nu = \frac{\nu_S}{[HS]}\left(\frac{1}{D} - \frac{1}{D_0}\right) \tag{6.6}$$

Ideally, D_0 is constant, but if not then its variation with [M] must be known. Careful manipulation of the experimental conditions is often necessary, to achieve values of D sufficiently different from D_0 to allow precise and accurate determination of ν. All the humic matter must remain in solution, i.e. there must be no significant adsorption at the solid surface or accumulation at the liquid–liquid interface. Adsorption could conceivably occur by the formation of metal 'bridges' between the solid and the humic matter, which would violate the assumption that D_0 is the same in the presence and absence of humic matter.

The method has been used extensively to determine metal binding by humic substances, often in trace level work with radioisotopes. Examples are the studies by Randhawa & Broadbent (1965), Bertha & Choppin (1978), Ephraim *et al.* (1989b), and Higgo *et al.* (1993). Choppin and co-workers (e.g. Nash & Choppin, 1980) also used a lipophilic ligand for the metal, dissolved in an organic phase, as the 'adsorbent'.

6.6.2 Competing dissolved ligand

When the competitor is a dissolved chromophore, a spectroscopic method can be used to determine the concentration of the metal–competitor complex. For example, Browne & Driscoll (1993) used fluorescence to determine the concentration of the Al complex with morin (2,3,4,5,7-pentahydroxyflavone) in their study of the binding of Al by fulvic acid. The Al–morin complex is strongly fluorescent, allowing low concentrations to be detected, and making it possible to arrange the solution conditions so that other equilibria are hardly affected by the presence of morin in the solution. This is highly advantageous in determining Al speciation in natural waters (Browne *et al.*, 1990), but is not necessary for investigating Al interactions with isolated humic matter. A possible complication is the formation of ternary complexes, which may both contribute to the spectroscopic signal and complicate the interpretation of results. A very sensitive version of this method, using electrochemistry to determine the concentration of metal bound by the competing ligand, is described in Section 6.7.4.

6.6.3 Kinetic discrimination

The methods in 6.6.1 and 6.6.2 are equilibrium techniques. Clarke *et al.* (1992) developed a method for aluminium speciation that relies on the rapid formation of the electrically neutral $Al(oxine)_3$ complex (oxine = 8-hydroxyquinoline), followed by its rapid extraction into chloroform and then determination of extracted Al. The procedure is carried out in a flow-injection system. The authors showed that the oxine reacts with Al^{3+}, $AlOH^{2+}$, probably $Al(OH)_2^+$, and Al complexes with sulphate and weak organic ligands, but not with metal bound to fulvic or humic acids. Clarke *et al.* (1995) used the method to measure Al binding by isolated fulvic and humic acids.

6.7 Electrochemical techniques

Two main types of electrochemical techniques have been used to determine metal binding by humic substances. Potentiometric measurements operate at equilibrium, with essentially zero current flowing, while voltammetric methods involve appreciable currents, and are in reality kinetic techniques. In voltammetry, the measured quantity is that of labile metal, which may be simply the aquo ion (e.g. Cu^{2+}, Cd^{2+}), but may also include metal bound in labile complexes. Generally, voltammetric results require more interpretation than potentiometric data to derive equilibrium relationships. Detailed reviews of the theory and practice of these and related methods are given by Buffle (1988) and by Mota & Correia dos Santos (1995). Buffle emphasises the value of applying more than one technique to the system under study, to obtain complementary information, not only about amounts bound, but also about the properties of the metal–humic complexes. The following sections give brief outlines of the principal methods.

6.7.1 Ion-selective electrodes (ISEs)

Electrodes are available that respond to free metal activity, comparable to the glass electrode and protons (Section 6.2). The key feature is a membrane that responds to the external solution by exchanging one of its ions. The exchange creates a variation in electrical charge and consequently a potential difference, which can be measured. For example, Cu^{2+} can be determined with an electrode the membrane of which consists of CuS, while an ISE for Ca^{2+} makes use of a porous organophosphate. The change in potential, for an electrode obeying Nernst's law, is *c.* 59 mV per decadal change in concentration (expressed in equivalents) at room temperature.

Such electrodes have been widely used to determine metal binding by humic substances; they have probably made more contributions to the available data than any other single technique. For humic substances, there has been most interest in the use of ISEs that determine Ca, Cu, Cd and Pb. Under the right circumstances, very low free activities can be determined, and consequently low values of v. The key issue is buffering of the metal ion activity. In experiments with the well-defined ligand ethylenediamine, Avdeef *et al.* (1983) were able to measure activities of Cu^{2+} as low as 10^{-19} mol dm^{-3}. At high concentrations of humic acid, Cu^{2+} activities down to *c.* 10^{-14} mol dm^{-3} were achieved by Benedetti *et al.* (1995). At lower concentrations of humic matter, the detection limit is much higher. Thus it is difficult to carry out experiments under conditions close to those in natural waters, although soil conditions can be met. The main problem with ISEs is from interference, for example the Cu^{2+} electrode is sensitive to Cl^-.

6.7.2 Polarography

Polarography can be used to quantify labile forms of reducible metals. An increasing potential is applied across a pair of electrodes (the cathode is commonly a mercury drop) immersed in the solution of interest, and the current is determined. When the applied potential exceeds the decomposition potential for a particular reducible species, there is a change in the current, which can be related to the amount of the metal species present. The voltage at which the reduction takes place also identifies the metal being reduced. The greatest sensitivity is achieved using pulse techniques. Concentrations of free metal as low as 10^{-12} mol dm^{-3} can be determined (Pinheiro *et al.*, 1996). Complexes of metals with humic substances are not detected, because they diffuse relatively slowly to the mercury drop.

Pinheiro *et al.* (1996) compared potentiometry and polarography for the Cd–humic acid system. Agreement was good at pH < 5, but there was divergence at higher pH, attributed to disaggregation of the humic matter influencing the polarographic results.

6.7.3 Anodic stripping voltammetry (ASV)

This method is most useful for metals that form an amalgam with mercury, e.g. Cu, Zn, Cd, Pb. Labile metal ions are first electro-deposited and reduced on a mercury electrode, then the polarity is reversed and the current corresponding to the reoxidation of the deposited metal is determined at fixed potential. The method is subject to difficulties associated with the reduction of 'non-labile' (i.e. complexed) metal, and the dissociation of metal complexes.

6.7.4 Ligand competition with cathodic stripping voltammetry (CSV)

This method was devised to study metal complexation in marine waters by van den Berg (1984), and has been used for freshwaters by Xue & Sigg (1993, 1994). Xue & Sigg (1999) employed it to determine low-level binding by isolated humic matter. The principle of the method is to use CSV to determine concentrations of the metal complexed by a low molecular weight competing ligand, in a solution containing the humic compound. In the original method, catechol was used as the competing ligand, but others have been used subsequently. Then, with knowledge of the equilibrium constant(s) for complexation of the metal by the competing ligand, the free metal concentration is computed. Metal bound by the humic matter is obtained from the total metal concentration by difference, to provide pairs of ν and [M]. The technique allowed Xue & Sigg (1999) to determine concentrations of Cu^{2+} as low as 10^{-15} mol dm^{-3}.

6.8 Spectroscopic methods

A number of spectroscopic techniques have been used to quantify binding. In general they depend upon a change in a spectroscopic signal, associated either with the metal or the humic matter, that can be related to the extent of binding. Table 6.1 lists some spectroscopic methods, and Fig. 6.7 presents two examples. Spectroscopic methods have the advantage of not perturbing the system under study. The main disadvantages are lack of sensitivity and the rather small ranges of concentrations that can be covered. Furthermore, the signals have to be interpreted, i.e. assumptions have to be made, to extract quantitative data. For example, to estimate

Table 6.1. Some spectroscopic methods used to quantify metal binding by humic substances.

Method	Principle	Examples
Ultraviolet/visible absorption spectroscopy	Absorption band due to metal–humic complex; absorption assumed to be linear with binding	Am/FA & HA (Moulin *et al.*, 1987)
Fluorescence spectroscopy	Humic fluorescence decreased by metal binding; quenching assumed to be linear with binding	Ni, Co, Cu, Pb/FA (Saar & Weber, 1980a) Cu/FA (Ryan & Weber, 1982)
	Metal binding enhances humic fluorescence	Be/FA (Esteves da Silva *et al.*, 1995) Al/FA (Esteves da Silva & Machado, 1996)
	Bound and free metal fluoresce at different wavelengths and/or with different intensities	Eu/FA (Dobbs *et al.*, 1989) Cm/HA (Kim *et al.*, 1991b; Moulin *et al.*, 1992) $U(VI)O_2$ (Czerwinski *et al.*, 1994)
	Total luminescence spectroscopy, multiresponse parameter estimation	Smith & Kramer (1998, 2000)
Electron spin resonance spectroscopy	Bound and free metal exhibit different resonances	Mn/FA (Gamble *et al.*, 1977) V(IV)O/FA (Templeton & Chasteen, 1980)

FA: fulvic acid, HA: humic acid.

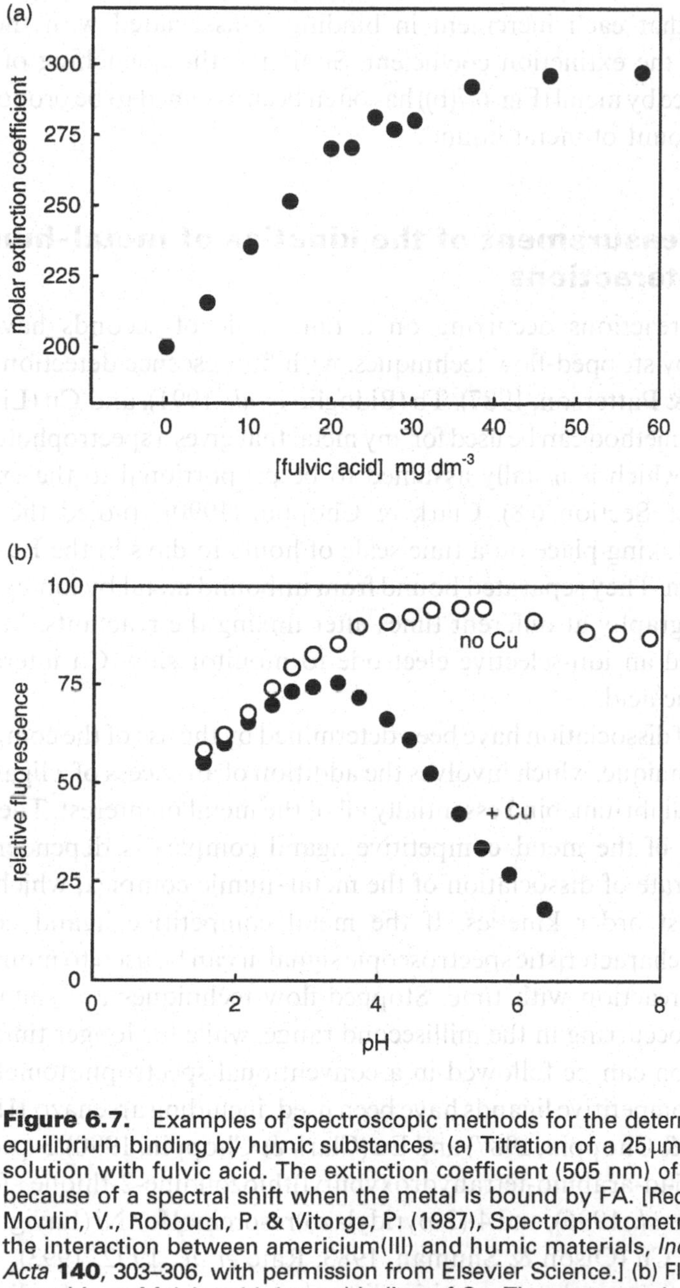

Figure 6.7. Examples of spectroscopic methods for the determination of equilibrium binding by humic substances. (a) Titration of a 25 μmol dm^{-3} Am solution with fulvic acid. The extinction coefficient (505 nm) of Am varies because of a spectral shift when the metal is bound by FA. [Redrawn from Moulin, V., Robouch, P. & Vitorge, P. (1987), Spectrophotometric study of the interaction between americium(III) and humic materials, *Inorg. Chim. Acta* **140**, 303–306, with permission from Elsevier Science.] (b) Fluorescence quenching of fulvic acid due to binding of Cu. The concentrations of fulvic acid and Cu were 32 mg dm^{-3} and 150 μmol dm^{-3} respectively. [Redrawn from Saar, R.A. & Weber, J.H. (1980), Comparison of spectrofluorometry and ion-selective electrode potentiometry for determination of complexes between fulvic acid and heavy-metal ions, *Anal. Chem.* **52**, 2095–2100. Copyright 1980 American Chemical Society.]

amounts of metal bound from the data presented in Fig. 6.7(a), it was assumed that each increment in binding is associated with the same change in the extinction coefficient. Similarly, the quenching of humic fluorescence by metal (Fig. 6.7(b)) has often been assumed to be proportional to the amount of metal bound.

6.9 Measurement of the kinetics of metal–humic interactions

Forward reactions occurring on a time scale of seconds have been followed by stopped-flow techniques, with fluorescence detection, for Al (Plankey & Patterson, 1987), Tb (Bidoglio *et al.*, 1991), and Cu (Lin *et al.*, 1994). The method can be used for any metal that gives a spectrophotometric response, which is usually assumed to be proportional to the extent of binding (cf. Section 6.8). Clark & Choppin (1990) studied the slower reactions taking place on a time scale of hours to days in the Eu–humic acid system. They separated bound from unbound metal by ion-exchange chromatography at different times after mixing the reactants. Ma *et al.* (1999) used an ion-selective electrode to monitor slow Cu interactions with humic acid.

Rates of dissociation have been determined by the use of the competitive ligand technique, which involves the addition of an excess of a ligand that will, at equilibrium, bind essentially all of the metal of interest. The rate of formation of the metal–competitive ligand complex is dependent only upon the rate of dissociation of the metal–humic complex, which obeys pseudo-first order kinetics. If the metal–competitive ligand complex displays a characteristic spectroscopic signal, it can be used to monitor the extent of reaction with time. Stopped-flow techniques are suitable for reactions occurring in the millisecond range, while for longer time scales the reaction can be followed in a conventional spectrophotometer cell. Various competitive ligands have been used, including arsenazo III for Th (Cacheris & Choppin, 1987) and Eu (Clark & Choppin, 1990), 3-propyl-5-hydroxy-5-(d-arabino-tetrahydroxybutyl)thiozolidine-2-thione for Cu (Bonifazi *et al.*, 1996), and 4-(2-pyridylazo)resorcinol for Ni (Lavigne *et al.*, 1987) and Cu (Olson & Shuman, 1983; Rate *et al.*, 1992, 1993).

A variation on the competitive ligand theme is the use of solid-phase materials that adsorb dissociated metal. For example Lu & Chakrabarti (1995) used Chelex resin, with analysis of metal remaining in solution as a function of time. Burba (1994) used a solid in which 1-(2-hydroxyphenylazo)-2-naphthol groups were immobilised on cellulose, and determined metal immobilised on the solid phase, following elution with acid, as a function of time.

Exchange rates of metals bound by humic substances have been followed using different metal isotopes, with separations by size exclusion (Colston *et al.*, 1997; Marx & Heumann, 1999). Österberg *et al.* (1999) used ion-selective electrodes to monitor slow reactions of calcium and copper with humic acid.

7

Quantitative results with isolated humic substances

We now consider the results of experimental studies in which extents of interaction of protons and metal ions with humic substances have been determined. The emphasis is on amounts of binding, rather than the qualitative nature of the complexes, which is dealt with in Chapter 8. The review material is not exhaustive, but key examples are given to cover the present state of knowledge. Some simple interpretations are offered, to prepare the ground for Chapters 9–11, which deal with modelling. The 'isolated' appears in the chapter title because only results with isolated materials are considered; results from field samples are discussed in Chapters 13 and 14.

7.1 Proton dissociation

7.1.1 Titration curves for humic substances

Some results have already been presented in Chapter 4, including comparisons with simple ligands and polyacids. More data are plotted in Fig. 7.1. As has been noted before, the plots of Z vs. pH are rather featureless, with only suggestions of an end-point around neutral pH. Also, there is an appreciable effect of ionic strength, which is greater for humic acid than for fulvic acid, but in both cases less than for high molecular weight polyprotic acids. As mentioned in Chapter 4, the results can be interpreted in terms of a combination of (a) Coulombic effects, which give the ionic-strength dependence and cause some of the 'smearing' of the titration curves, and (b) site heterogeneity. The exact contributions of the two factors are difficult to extract from the experimental data. Instead, modelling can be tried to see which physical representation gives the best agreement with the data (see Chapters 9–11).

Although many titration data sets suggest that proton dissociation is reasonably well described, and understandable, a number of authors have reported observations that are difficult to interpret, or which complicate the simple picture of non-interacting molecules undergoing reversible proton dissociation and binding reactions. Some of the observations are summarised in Table 7.1. Sposito *et al.* (1977) reported a decrease in the number of proton-dissociating groups as the concentration of sewage sludge fulvic acid was increased by two- to four-fold. However such a trend was observed by neither Perdue (1990) for aquatic organic acids (10-fold variation in concentration), nor by Fiol *et al.* (1999) for soil fulvic acids (three-fold variation). To the physico-chemical phenomena listed in Table 7.1 must be added uncertainty due to experimental error, common sources of which are lack of equilibration during titrations, uncertainty in moisture content and inadequate electrode calibration (Santos *et al.*, 1999). There is scope for more detailed study and rationalisation of the various complicating observations. However, the uncertainties do not preclude the use of published data to draw general conclusions about the interactions of protons with humic substances, and this is attempted in the following sections.

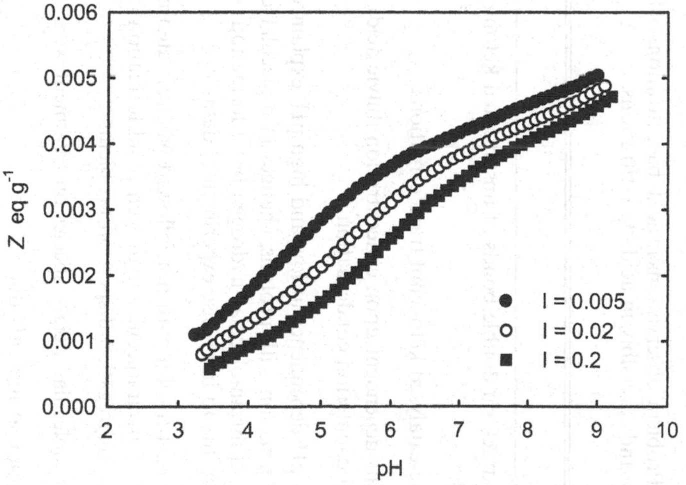

Figure 7.1. Proton dissociation from humic acid. [Redrawn from Tipping, E., Fitch, A. & Stevenson, F.J. (1995b), Proton and copper binding by humic acid: application of a discrete-site/electrostatic ion-binding model, *Eur. J. Soil Sci.* **46**, 95–101, with permission from Blackwell Science.]

Table 7.1. 'Problem' observations in acid–base titrations of humic substances. These results are not consistent with the idea that humic substances behave simply and reversibly in acid–base titrations.

Observation	Reference
Electrophilic attack on double bonds of unsaturated ketonic structures, and associated anion-binding	Stevenson (1994)
Acid- and base-catalysed keto–enol transformations	Stevenson (1994)
Titration results dependent upon concentration (fulvic acid), due to hydrogen-bond formation and counterion condensation	Sposito & Holtzclaw (1977) Sposito *et al.* (1977)
Reversal of Z–pH dependence at low and high pH, explained as	
• ion-pair formation with protons, changes in molecular conformation	Sposito & Holtzclaw (1977)
• the result of intramolecular hydrogen bond formation	Davis & Mott (1981)
• an artefact due to imprecise experimental data	Marshall *et al.* (1995)
Hysteresis – base titration path not followed when titrated back with acid	
• the result of intramolecular hydrogen bond formation	Davis & Mott (1981)
• ascribed to aggregation/dispersion problems	Milne *et al.* (1995)
Release of low molecular weight species on treatment with base	Wilson *et al.* (1988)
Base consumption at high pH due to	
• oxidation when oxygen is present	Borggaard (1974)
• hydrolysis of esters	Gregor & Powell (1987), Bowles *et al.* (1989)
• lack of equilibrium	Marshall *et al.* (1995)

7.1.2 Comparison of results for different samples of fulvic and humic acid

There are only a few studies in which the same authors have used the same technique to compare the acid–base properties of humic samples from different sources. Patterson *et al.* (1992) studied fulvic acids isolated from the surface horizons of four contrasting forest soil environments in the United States (Table 7.2), and drew attention to differences among the samples. However, the acid–base properties of the samples are really quite similar. Ephraim *et al.* (1995) reported the COOH contents of 11 aquatic fulvic acids to be in the range 3.72–9.86 mmol g^{-1}, although eight of the samples had values in the much narrower range of 4.60–5.46 mmol g^{-1}. Fiol *et al.* (1999) found soil fulvic acid samples isolated from three depths in the A horizon of an umbric regosol to exhibit almost identical titration behaviour.

Ephraim *et al.* (1995) demonstrated the complexity of the ionic strength dependence of proton dissociation, showing that ionic strength effects vary not only with molecular weight, but also with polydispersity (i.e. the range of molecular weights within a sample). In a subsequent study on a size-fractionated fulvic acid, Ephraim *et al.* (1996) found the ionic strength effect on proton dissociation to increase with the molecular weight. Powell & Fenton (1996) extracted humic acids from a peat sample at pH 6.0, 9.0 and 12.0. They obtained fractions of increasing molecular size, but decreasing (from 5.2 to 4.3 mmol g^{-1}) content of COOH groups. Fukushima *et al.* (1996) obtained similar results.

Comparisons of titration curves obtained for different samples by different authors can be made for data at $I = 0.1$ mol dm^{-3}. This ionic strength is common to most data sets, and is sufficiently high that Coulombic effects are small, so that comparisons are mainly in terms of site densities and chemical heterogeneity. Table 7.3 summarises values of

Table 7.2. Acid properties of forest soil fulvic acids (Patterson *et al.*, 1992).

Source	Total RCOOH (mmol gC^{-1})	pK_H
Northern hardwood, spodosol	8.9	3.96
Northern coniferous, spodosol	8.3	4.14
Southern hardwood, ultisol	10.4	3.82
Southern coniferous, ultisol	9.8	3.96

The total carboxylic acid content (RCOOH) was estimated by titration to pH 7. The pK_H values refer to a monoprotic single-site model.

Table 7.3. Values of Z (mmol g^{-1}) for different fulvic and humic acid samples, at pH 4, 7 and 9, $I = 0.1$ mol dm^{-3}.

$-Z$ at pH 4	$-Z$ at pH 7	$-Z$ at pH 9	$Z_{pH\ 4}/Z_{pH\ 7}$	$Z_{pH\ 9}/Z_{pH\ 7}$	Reference
Fulvic acids					
3.40	5.64	6.35	0.60	1.24	Dempsey (1981)
3.21	5.31	5.87	0.60	1.11	Dempsey (1981)
2.90	5.51	5.92	0.53	1.07	Paxeus & Wedborg (1985)
2.74	4.66	5.25	0.59	1.13	Perdue *et al.* (1984)
2.14	3.80	4.28	0.56	1.13	Plechanov *et al.* (1983)
2.75	4.70	5.23	0.59	1.11	Cabaniss (1991)
3.11	5.38	6.10	0.58	1.13	Lead *et al.* (1994)
Humic acids					
1.27	3.29	3.99	0.39	1.21	Posner (1964)
1.76	4.17	5.04	0.42	1.21	Fitch *et al.* (1986)
1.35	3.27	4.00	0.41	1.22	Lead *et al.* (1994)
1.31	2.96	3.78	0.44	1.28	Milne *et al.* (1995)

Z for different fulvic and humic acid samples, at pH 4, 7 and 9. Fulvic acids have more groups dissociating at pH 7 than do humic acids, the average values being *c.* 5 and 3.5 meq g^{-1} respectively. The results for different samples hardly overlap, and so there is a clear distinction between the two types of humic matter. Another comparison can be made by taking the ratios of *Z* at pH 4 and at pH 9 to that at pH 7. There are consistent differences: compared to pH 7, humic acid is dissociated less than fulvic acid at pH 4 and more at pH 9. Thus, on average, fulvic and humic acids differ in terms of site densities and their ranges of (apparent) pK_H values.

Christensen *et al.* (1998) compared titration properties of groundwater fulvic acids by normalisation to a common reference point, the charge at pH 7. This procedure allows comparison of pK_H distributions, independently of differences in absolute site densities, and was used to produce the plots of Fig. 7.2, for various fulvic acid and humic acid samples. There are remarkable similarities in the normalised titration curves for each kind of humic matter. The plots also emphasise the differences between the two types of humic matter with regard to the pK_H distributions. The similarities in the normalised plots suggest that the main differences among different samples of a given humic material (fulvic or humic acid) are in site density, but not in pK_H distribution.

The foregoing discussion suggests that there may be sufficient similarity amongst samples of fulvic acid for it to be considered the same in all or most environments, at least in terms of proton dissociation. The same may also be true of humic acid, although there are fewer data to support that contention.

7.1.3 Hydrophilic acids

Compared to studies of fulvic and humic acids, determinations of the proton-dissociating properties of hydrophilic acids are sparse. David & Vance (1991) isolated hydrophobic and hydrophilic acids (cf. Section 2.2.1) from five ponds in central Maine and determined 'exchange acidity', essentially an estimate of groups dissociating at pH below neutrality. The hydrophilic acids had a mean exchange acidity of 13.7 mmol gC^{-1}, while the value for the hydrophobic acids was 11.3 mmol gC^{-1}. The average pK_H for the hydrophilic acids was 4.35, that for the hydrophobic acids was 4.26. Dai *et al.* (1996) studied soluble organic substances in spruce-fir forest floor leachates. The hydrophobic acid samples had a mean exchange acidity of 12.7 mmol gC^{-1} and an average pK_H of 4.10. The corresponding values for the hydrophilic acids were 19.8 mmol gC^{-1} and 3.70 respectively. In these studies the hydrophobic acids are probably equivalent to fulvic acid. Therefore the results suggest that hydrophilic acids have greater contents of proton-dissociating groups, by 20–60%, than fulvic acids. More research is required to establish whether there are consistent differences in pK_H values.

7.1.4 Effects of temperature and enthalpy changes

There does not seem to be any direct information about the effects of temperature on the acid–base behaviour of humic substances. However there have been two studies in which the enthalpies of reaction have been

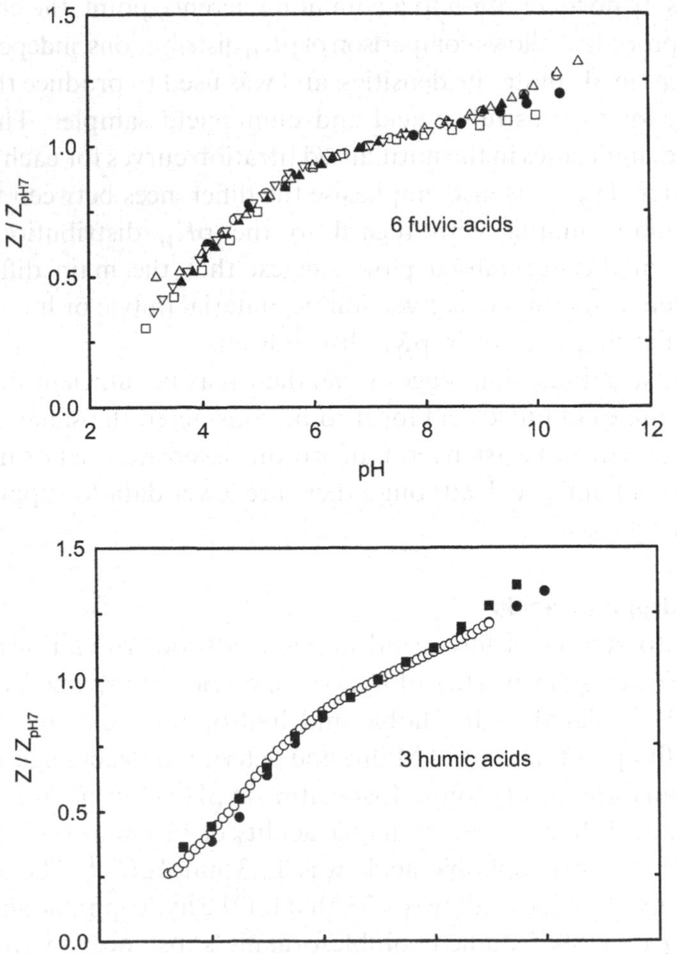

Figure 7.2. Normalised proton dissociation plots for fulvic and humic acids. Fulvic acid data are from Dempsey (1981), Paxeus & Wedborg (1985), Perdue *et al.* (1984), Plechanov *et al.* (1983) and Cabaniss (1991). Humic acid data are from Posner (1964), Tipping *et al.* (1995b) and Milne *et al.* (1995).

measured. Perdue (1978) used titration calorimetry and found that the carboxylic acid groups of humic acid dissociated with an average enthalpy change (ΔH) of $+3.8\ (\pm 3.4)\ \text{kJ mol}^{-1}$, while for phenolic groups ΔH was $-1.7\ (\pm 9.2)\ \text{kJ mol}^{-1}$. Machesky (1993) studied fulvic and humic acids by the same technique and obtained values of ΔH in the range -6 to $+5\ \text{kJ mol}^{-1}$ for dissociation at pH < 7, and values of $+20$ to $+36\ \text{kJ mol}^{-1}$ for pH > 9. The values of ΔH at low pH are as expected from results for simple carboxylic acids. Machesky's results at high pH are consistent with those for simple phenols, but Perdue's are appreciably different.

7.1.5 Conductimetric data

Figure 7.3 shows the results obtained by van den Hoop *et al.* (1990) from conductimetric experiments in which a humic acid sample was neutralised with NaOH and $Ca(OH)_2$. The experiments were performed without background electrolyte. The observed conductivity is due to contributions from protons, metal cations originating from the added base, hydroxyl ions and the humic 'polyion'. The authors investigated how the ions are distributed in this low ionic strength system. They divided the plots into three regions (A, B and C). For the titration with NaOH, the fall in conductivity (K) in region A is mainly due to neutralisation of the highly mobile free H^+ ions, while in regions B and C it increases due to the

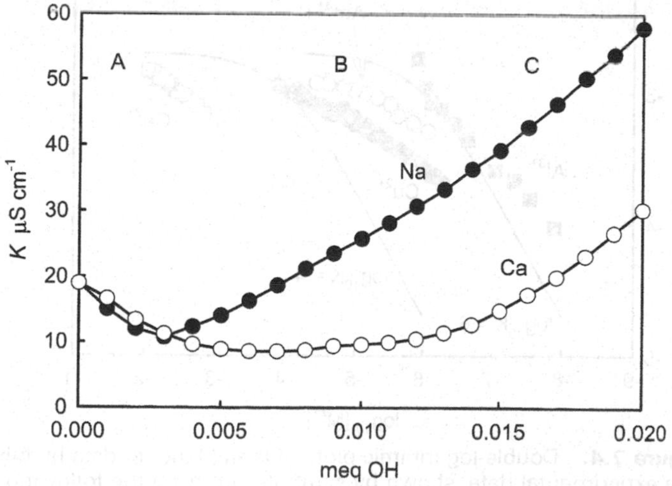

Figure 7.3. Results of conductimetric titrations of humic acid ($\sim 0.1\ \text{g dm}^{-3}$) with NaOH and $Ca(OH)_2$. The plots show every second experimental point. Regions A, B and C are discussed in Section 7.1.5. [Redrawn from van den Hoop, M.A.G.T., van Leeuwen, H.P. & Cleven, R.F.M.J. (1990), Study of the polyelectrolyte properties of humic acids by conductimetric titration, *Anal. Chim. Acta* **232**, 141–148, with permission from Elsevier Science.]

increasing negative charge on the HA and the increasing concentration of free metal ions. In regions B and C there appears to be little binding of Na^+ by the humic acid; i.e. the cations are not held by the macromolecule as it moves in the electric field. This contrasts with the behaviour of the linear polyelectrolyte polymethacrylate, for which the Na^+ curve flattens, due to 'counterion condensation', occurring when the macroionic charge exceeds a critical value (Manning, 1978). The results in Fig. 7.3 show that significant binding of Ca^{2+} does take place in the B region, so that K does not increase as base is added. Van den Hoop *et al.* (1990) calculated that when 70% or more of the humic protons were dissociated, approximately 70% of the total Ca^{2+} was immobilised (bound). They concluded that humate ions behave differently to polymethacrylate, due to differences in the density and distribution of charge. The results for Li^+ and K^+ were similar to those for Na^+, while Ba^{2+} behaved similarly to Ca^{2+}.

7.2 Equilibrium binding of metal ions

7.2.1 Binding as a function of free metal concentration at fixed pH

Figure 7.4 shows some examples of metal binding by fulvic acid as a function of unbound metal concentration. The data are typical of many for humic substances in that they cover a modest range of free metal

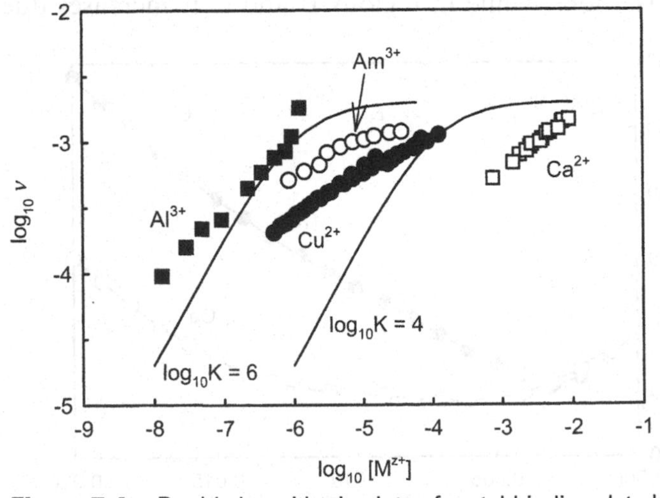

Figure 7.4. Double-logarithmic plots of metal binding data by fulvic acids. The experimental data, shown by symbols, are from the following sources: Al (pH 5) Browne & Driscoll (1993), Ca (pH 9) Dempsey (1981), Cu (pH 5) Bresnahan *et al.* (1978), Am (pH 6) Moulin *et al.* (1987). The lines represent theoretical results for a ligand with a single site at an abundance of 2 mmol g^{-1}, and with log K values as shown. Note that the theoretical slopes tend to unity at low free metal concentrations.

concentrations, and rarely more than one order of magnitude in v. The experimental data yield plots that differ consistently from those for single-site ligands, in that they have shallower slopes. Exactly how to explain the isotherm shapes will be considered in connection with modelling, but generally the shallower slopes provide evidence that the binding sites in humic substances are heterogeneous.

None of the isotherms in Fig. 7.4 reaches a maximum, plateau, value, but humic substances can clearly bind in excess of 10^{-3} mol of metal per gram, thus approaching the content of proton-dissociating sites in fulvic acid, which is typically 8×10^{-3} mol g^{-1} (see Section 7.1). Therefore it is reasonable to conclude that the proton-dissociating sites are responsible for metal binding. Such an idea is, of course, expected on the basis of knowledge about the metal-binding properties of weak acids discussed in Chapter 5. It also follows that at least some of the sites are expected to be common to several metals.

The studies of Benedetti *et al.* (1995) with a peat humic acid have provided some of the most wide-ranging data sets so far, and have been highly influential in model development (Chapters 9 and 10). These authors used ion-selective electrodes (for Ca, Cu and Cd) at concentrations of humic acid (*c.* 1 g dm^{-3}) sufficiently high for nearly all of the metal to be bound, thus providing strong buffering of the free metal ions and thereby allowing low free concentrations to be measured. Results for Cu and Cd are shown in Fig. 7.5. In the case of Cu, the slopes of the plots are approximately 0.3–0.4, much lower than the value of 1 that would be expected at low binding for a simple ligand (as plotted in Fig. 7.4). This is further strong evidence for binding site heterogeneity, and the point is underlined by the plots in Fig. 7.6, which show how the concentration of Cd^{2+} varies with total cadmium for a 1 g dm^{-3} suspension of humic acid. The two lines in the diagram show the expected binding for two hypothetical homogeneous compounds with the same number of binding sites as the real humic acid. The first hypothetical material has one type of binding site with an affinity for Cd the same as that of the weakest sites of the real material. The second has a single type of site with an affinity equal to the strongest sites detected in the experiments. The real material has sites with a range of affinities, giving rise to a dependence of $[Cd^{2+}]$ on total Cd concentration that differs appreciably from the two homogeneous extremes.

7.2.2 Metal binding to the abundant weak sites in humic matter

The affinities of different metals for humic matter can be compared in terms of the free concentrations associated with a given value of v. The lower is the free concentration, the greater is the affinity. The comparisons have to be made for the same pH and ionic strength, since binding varies with both of these (see below). Table 7.4 compares results for $v = 5 \times 10^{-4}$ mol g^{-1}, at pH values near to 6 and at an ionic strength of 0.1 mol dm^{-3}. This value of v corresponds to occupation of approximately

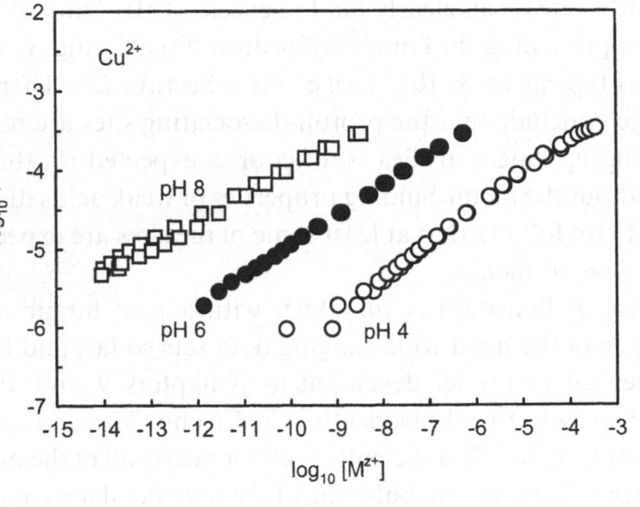

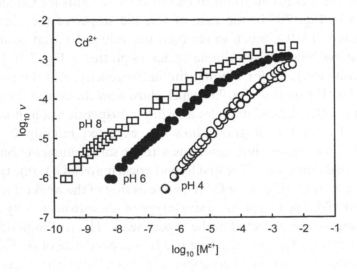

Figure 7.5. The binding of Cu and Cd by humic acid ($I = 0.1$ mol dm^{-3}). [Redrawn from Benedetti, M.F., Milne, C.J., Kinniburgh, D.G., van Riemsdijk, W.H. & Koopal, L.K. (1995), Metal-ion binding to humic substances – application of the nonideal competitive adsorption model, *Environ. Sci. Technol.* **29**, 446–457. Copyright 1995 American Chemical Society.]

10% of the proton-dissociating sites, i.e. it involves mainly the weak abundant sites. From the results in Table 7.4, and bearing in mind the conductimetric study described in Section 7.1.5, the following order of affinities can be drawn up:

Table 7.4. Free cation concentrations for different metals, at $\nu = 5 \times 10^{-4}$ mol g^{-1}, pH 6.0, 0.1 mol dm^{-3} ionic strength.

Metal	$-\log [M^{z+}]$	Reference
Fulvic acid		
Ca^{2+}	2.4, 2.6	Dempsey (1981)
Cu^{2+}	6.2, 5.3	Bresnahan *et al.* (1978); Turner *et al.* (1986)
Cd^{2+}	4.1	Saar & Weber (1979)
Eu^{3+}	5.4	Shin *et al.* (1996)
Pb^{2+}	5.8	Saar & Weber (1980b)
Am^{3+}	5.1	Buckau *et al.* (1992)
Humic acid		
Ca^{2+}	2.5	Benedetti *et al.* (1995)
Cu^{2+}	6.0	Benedetti *et al.* (1995)
Cd^{2+}	4.3	Lee *et al.* (1993)
Eu^{3+}	7.7	Carlsen *et al.* (1984)
Hg^{2+}	11.6	Yin *et al.* (1997)
Am^{3+}	6.5	Kim *et al.* (1991a)

Some interpolations and extrapolations were needed.
Note that the results refer to different samples of humic matter.

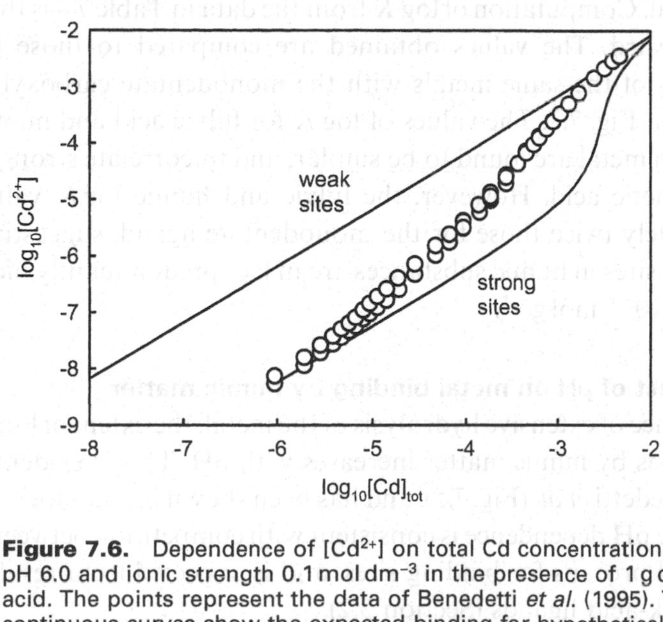

Figure 7.6. Dependence of $[Cd^{2+}]$ on total Cd concentration – $[Cd]_{tot}$ – at pH 6.0 and ionic strength 0.1 mol dm^{-3} in the presence of 1 g dm^{-3} humic acid. The points represent the data of Benedetti *et al.* (1995). The continuous curves show the expected binding for hypothetical compounds each with the same number of binding sites as the real humic acid; see Section 7.2.1.

$$\mathrm{Li^+ \sim Na^+ \sim K^+ < Ca^{2+} \sim Ba^{2+} < Cd^{2+} < Pb^{2+} \sim Cu^{2+} < Am^{3+} < Eu^{3+} < Hg^{2+}}$$

A similar sequence is expected for binding to carboxylate groups, on the basis of equation (5.7), and this fits with the knowledge that carboxylate groups are abundant in humic matter (Chapters 2 and 8).

To take things slightly further, the data for humic substances can be used to estimate apparent equilibrium constants, using a simple model. To construct the simple model, only carboxyl groups are assumed to be involved in binding metals, which is reasonable at pH 6. All groups are supposed to be involved in binding, all are supposed to be identical and monodentate binding is assumed. For the purposes of the modelling we shall assume that the carboxyl contents of fulvic and humic acid are 5 mmol g^{-1} and 3.5 mmol g^{-1} respectively (cf. Section 7.1.2). The apparent equilibrium constant at a particular pH can be written

$$K = \frac{\nu}{\nu_{\mathrm{COO^-}}\,[\mathrm{M}^{z+}]} \tag{7.1}$$

Where $\nu_{\mathrm{COO^-}}$ is the number of moles of free, dissociated, carboxyl groups per gram of humic matter. At pH 6, most of the carboxyl groups will either be coordinated to the metal or dissociated, so that $\nu_{\mathrm{COO^-}}$ is well approximated by the total carboxyl content less the content occupied by bound metal. Computation of log K from the data in Table 7.4 is therefore straightforward. The values obtained are compared to those for the interactions of the same metals with the monodentate carboxyl ligand lactic acid in Fig. 7.7. The values of log K for fulvic acid and humic acid with a given metal are found to be similar, and to correlate strongly with those for lactic acid. However, the fulvic and humic log K values are approximately twice those for the monodentate ligand, suggesting that the binding sites in humic substances are in fact predominantly bidentate at $\nu = 5 \times 10^{-4}$ mol g^{-1}.

7.2.3 Effect of pH on metal binding by humic matter

In the absence of extensive hydrolysis of the metal, the extent of binding of metal cations by humic matter increases with pH. This is evident in the data of Benedetti *et al.* (Fig. 7.5), and has been shown in a number of other studies. The pH dependence is consistent with competition between metal cations and protons for binding sites, and is expected from results with simple weak-acid ligands (Section 5.2).

Just as variations in proton concentration influence metal binding, so does proton dissociation depend upon metal concentration, which is the basis for using pH measurements to follow metal binding (Section 6.2.4). Metal–proton exchange has also been demonstrated by measuring the amount of base required to maintain constant pH when a metal is added to a humic sample. From such data, the ratio of protons displaced per metal bound can be calculated, as long as metal binding is measured at the same time. The determinations require especially precise experimental technique. Tipping *et al.* (1995b) analysed the results of Fitch *et al.* (1986) for Cu(II) binding by a humic acid in the pH range 4–5, and found average ratios of 1.1–1.4. Benedetti *et al.* (1995), also working with humic acid, but at pH 6–8, reported average ratios of *c.* 0.4, 1.5 and 0.6 for Ca, Cu and Cd respectively. The explanation of proton–metal exchange ratios by modelling is considered in Section 10.8.

Gamble (1973) and Bonn & Fish (1993) showed that the binding of alkali metal cations (Li^+, Na^+, K^+) by humic substances depended strongly on pH. Bonn & Fish interpreted their results in terms of charge neutralisation, rather than binding at specific sites.

7.2.4 Competition amongst metals

Since there is good evidence that protons and metal cations compete for binding sites in humic substances, then it follows that different metal cations should compete with one another, and that extents of competition

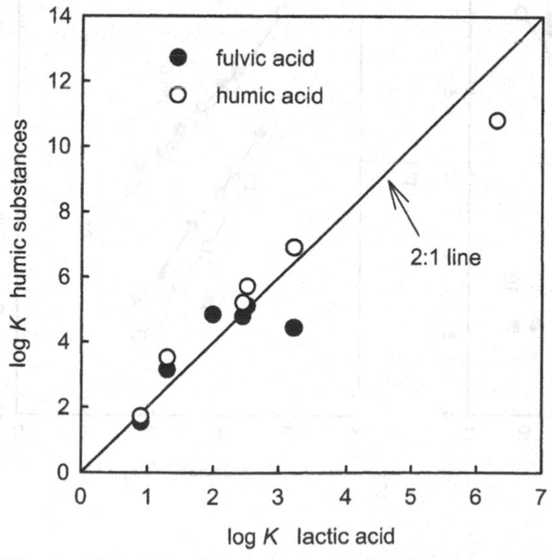

Figure 7.7. Equilibrium constants (log K_{COOH}) for metal binding to humic substances, according to the model of Section 7.2.2, compared with values for lactic acid, taken from Martell & Smith (1977). Each point represents a different metal (see Table 7.4).

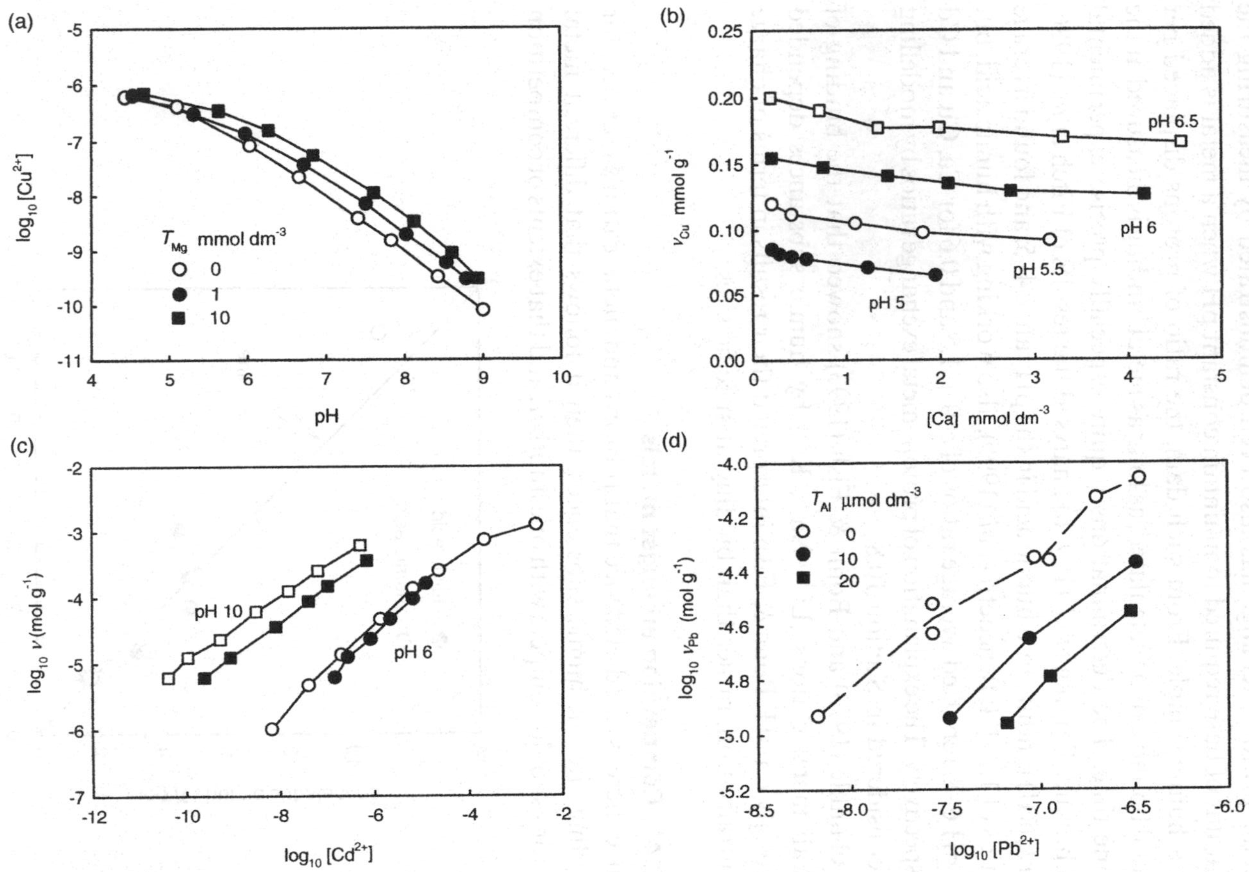
(a)
log10 [Cu2+]
pH
TMg mmol dm-3
0
1
10
(b)
vCu mmol g-1
[Ca] mmol dm-3
pH 6.5
pH 6
pH 5.5
pH 5
(c)
log10 v (mol g-1)
log10 [Cd2+]
pH 10
pH 6
(d)
log10 vPb (mol g-1)
log10 [Pb2+]
TAl µmol dm-3
0
10
20

Figure 7.8. Competition among metals for binding by humic matter. The lines are only for guidance. (a) Competition by Mg for Cu(II) binding by fulvic acid. The results refer to experiments at constant concentrations of fulvic acid ($10\,mg\,dm^{-3}$) and Cu ($1\,\mu mol\,dm^{-3}$). [Redrawn from Cabaniss, S.E. & Shuman, M.S. (1988), Copper binding by dissolved organic matter; I. Suwannee River fulvic acid equilibria, *Geochim. Cosmochim. Acta* **52**, 185–193, with permission from Elsevier Science.] (b) Competition by Ca for Cu(II) binding by aquatic dissolved organic matter, derived from data reported by Buffle *et al.* (1980). (c) Competition by Ca for Cd binding by humic acid. The open symbols refer to results in the absence of Ca, the closed symbols refer to a total Ca concentration of $10^{-3}\,mol\,dm^{-3}$. Only some of the data points are shown. [Redrawn from Kinniburgh, D.G., van Riemsdijk, W.H., Koopal, L.K., Borkovec, M., Benedetti, M.F. & Avena, M.J. (1999), Ion binding to natural organic matter: competition, heterogeneity, stoichiometry and thermodynamic consistency, *Coll. Surf. A* **151**, 147–166, with permission from Elsevier Science. IPR/18-1C British Geological Survey. ©NERC. All rights reserved.] (d) Competition by Al for Pb binding by humic acid, based on data published by Mota *et al.* (1996).

should depend upon relative binding strengths and relative concentrations. Competitive effects have been demonstrated by a number of studies, as shown in Fig. 7.8. It is difficult to decide, simply on the basis of experimental results, whether the competition results are 'expected', since the data have to be interpreted with a model that describes the investigator's concepts of the nature of the binding process. As will be seen in Chapters 9 and 10, recent models can account reasonably well for competition, on the basis of binding-site heterogeneity and electrostatic effects.

7.2.5 Effects of ionic strength

Figure 7.9 shows how metal binding by humic matter decreases with ionic strength. This is a general phenomenon in complexation reactions involving reactants of opposite charge, and reflects the shielding by neutral salt of the electrostatic attraction. The magnitude of the ionic strength effect for humic matter is larger than that found for low molecular weight ligands, due to the higher molecular charge. The modelling of ionic strength dependence is discussed in Chapter 9.

7.2.6 Ternary complexes

The simplest type of ternary complex is that formed when a hydrolysed metal ion binds to humic matter. As pH is raised, metal hydrolysis increases, as well as the affinity of the metal for humic matter. If the hydrolysis products of the metal (MOH, $M(OH)_2$ etc.) do not bind to the humic matter, or bind less strongly, then OH^- acts as a competitive ligand for the metal. Under such circumstances, metal binding by the humic matter will decrease with pH, as hydrolysis becomes more intense. The point is illustrated by the theoretical plots in Fig. 7.10, which show how Al and Cu might bind to fulvic acid. When hydrolysis products of the metal are assumed not to bind, both metals display a maximum in binding, arising from the general increase that goes with increasing pH, balanced against the decrease due to competition by OH^-. The same trend is seen when the first hydrolysis products ($AlOH^{2+}$ or $CuOH^+$) are allowed to bind with the same affinity as the parent aquo-ions (Al^{3+}, Cu^{2+}), but there is a shift in the maximum towards higher pH.

Dierckx *et al.* (1994) provided evidence for mixed complex formation involving humic acid, Eu^{3+}, and several low molecular weight ligands (CO_3^{2-}, oxalate, acetylacetone, iminodiacetic acid and OH^-). Figure 7.11 shows the results of the equilibrium dialysis studies of Glaus *et al.* (1995b) in which the complexation of Co^{2+} by a fulvic acid was measured at different concentrations of oxalate. The apparent equilibrium constant for

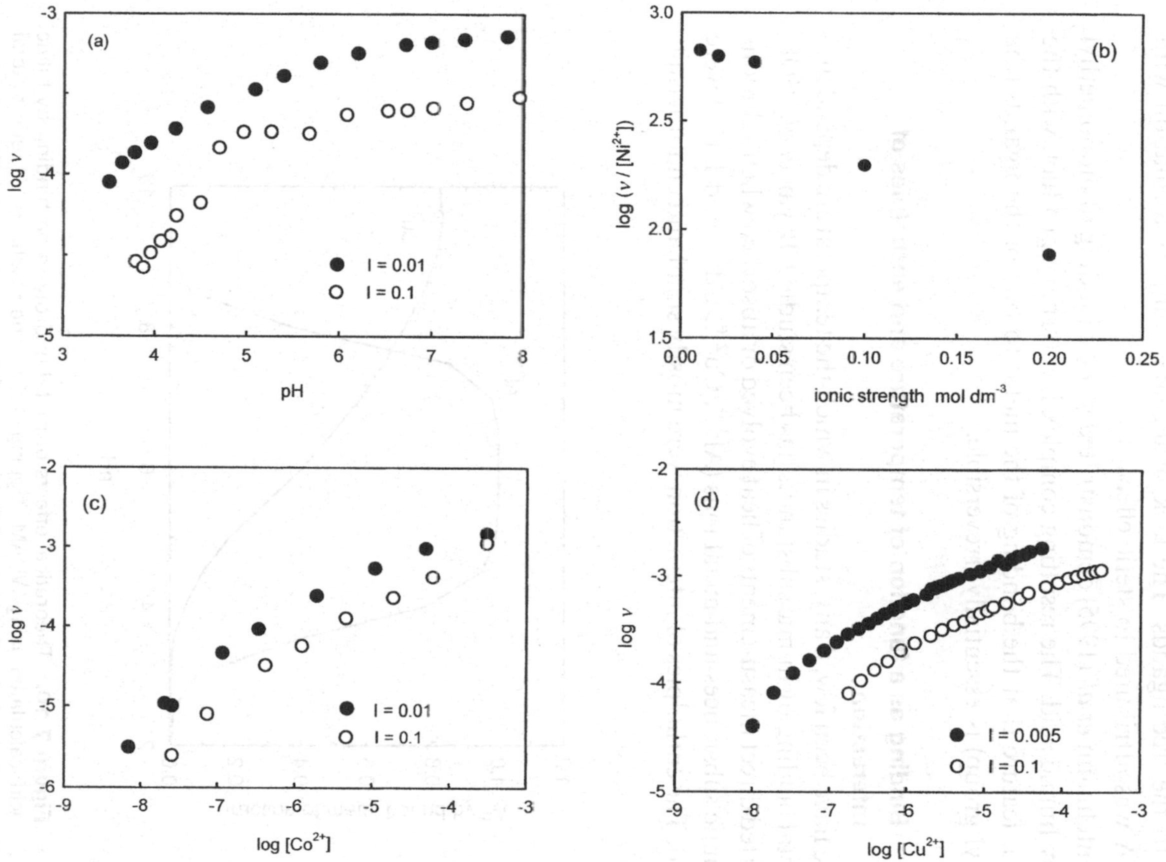

Figure 7.9. Effects of ionic strength upon metal ion binding by humic substances: (a) variation of ν with pH at constant concentrations of Ca and fulvic acid (data of Mathuthu, 1987); (b) binding of Ni by fulvic acid at pH ~7 (data of Higgo *et al.*, 1993); (c) binding of Co by humic acid at pH 6.7 (data of Zachara *et al.*, 1994); (d) binding of Cu by humic acid at pH 4 (data of Fitch *et al.*, 1986).

Co binding to the fulvic acid increases with oxalate concentration, as is expected if mixed complex formation occurs. A similar result was obtained with UO_2^{2+} and oxalate, but not with UO_2^{2+} and ethylenedinitrilotetraacetate (EDTA). Glaus *et al.* found that metal binding to the low molecular weight ligand in the ternary oxalate complexes was weaker than to the free ligands. The lack of ternary complex formation with EDTA was attributed to steric effects.

Hintelmann *et al.* (1995) demonstrated strong binding of monomethyl Hg by humic acid. The resulting complex is a ternary product, with the special feature that the bonding of the metal to one of the ligands (the methyl group) is essentially irreversible.

7.2.7 Binding as a function of temperature and enthalpies of interaction

There have been few, if any, studies in which the temperature dependence of metal binding by humic substances has been studied. Bryan *et al.* (1998) reported direct measurements of heat evolved or absorbed when solutions of humic substances and metal ions (Al^{3+}, Cu^{2+}, Cd^{2+} and La^{3+}) were mixed. The enthalpies of interaction were mostly small and endothermic,

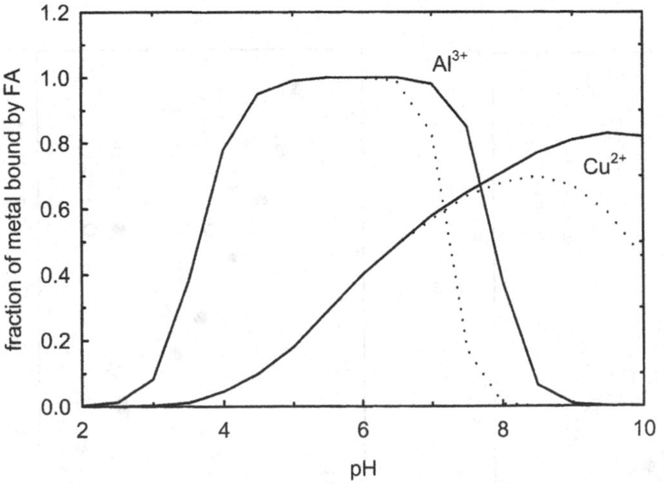

Figure 7.10. Theoretical effects of metal hydrolysis on binding by fulvic acid, calculated using WHAM (Tipping, 1994). The results for Al refer to total concentrations of metal and fulvic acid of 10^{-6} mol dm^{-3} and 10 mg dm^{-3} respectively, those for Cu to 10^{-6} mol dm^{-3} and 1 mg dm^{-3} respectively. The dotted lines were obtained by assuming that only the aquo ions (Al^{3+}, Cu^{2+}) can bind, the full lines by assuming that the first hydrolysis products ($AlOH^{2+}$, $CuOH^{+}$) can bind with the same affinity as the aquo ions.

of the order of 10 kJ per mole of metal bound. Values for acetic acid are similar, but metal-binding reactions with other carboxylic acids are slightly exothermic. The authors argued that the discharge of the diffuse layer that occurs on metal binding makes a significant contribution to the observed enthalpies. In a later paper (Bryan *et al.*, 2000), the components (enthalpy and entropy) of the free energy of metal binding to humic substances were examined further, by comparison with simple ligands, and it was concluded that for both Eu^{3+} and UO_2^{2+} the enthalpies and entropies of binding were positive, implying opposing contributions to the binding energy. The entropy changes were attributed to both dehydration of the metal ion on binding, and to relaxation of the electrical double layer surrounding the humic particles.

7.2.8 Effects of humic concentration

There is no consistent evidence that metal ion binding, expressed in terms of moles of metal bound per gram of humic matter, varies with the concentration of humic matter. Saar & Weber (1979) found that Cd binding decreased with increasing fulvic acid concentration (eight-fold range), but found no concentration dependence with Cu. Fitch *et al.* (1986)

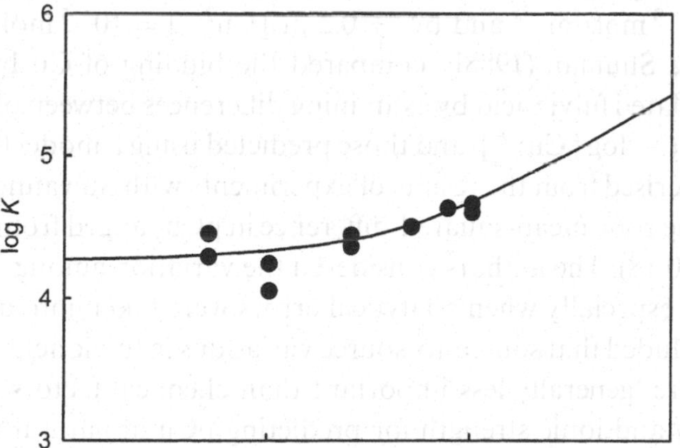

Figure 7.11. Evidence for mixed complex formation in the fulvic acid–Co–oxalate system. The conditional stability constant for Co–fulvic binding increases as the oxalate concentration is raised and more metal is bound in the ternary complex. The solid line is a fit to a model that includes the ternary complex. [Redrawn from Glaus, M.A., Hummel, W. & van Loon, L.R. (1995), Stability of mixed-ligand complexes of metal ions with humic substances and low molecular weight ligands, *Environ. Sci. Technol.* **29**, 2150–2153. Copyright 1995 American Chemical Society.]

reported that binding of Cu increased with humic acid concentration (three-fold range). Perdue & Carreira (1997) found that Cu binding by riverine dissolved organic matter (DOM) was the same at DOM concentrations of 50 and 1000 mg dm^{-3} at pH 4, but at pH 7 there was more binding at the higher concentration. Li *et al.* (1998) determined Cd binding by Suwannee River natural organic matter (NOM), and obtained essentially identical results at 95 and 9500 mgC dm^{-3}.

7.2.9 Metal binding by different humic samples and humic fractions

In the preceding sections, it has been tacitly assumed that all fulvic acid samples bind metals in the same way, and that all humic acids do so as well. There have been few studies in which different samples have been compared directly, by the same method. The most comprehensive is that of McKnight *et al.* (1983), who measured copper binding by aquatic fulvic acids of different origins. The data were fitted to a two-site model, the parameters for which (two site densities and two equilibrium constants per sample) are shown in Table 7.5. The parameters show broad similarities, but also variations. To give an overall impression of variability among samples, the parameter values were used to construct idealised binding plots (Fig. 7.12), which show the values of log v to vary by ± 0.5 at $[Cu^{2+}] = 10^{-8}$ mol dm^{-3} and by ± 0.2 at $[Cu^{2+}] = 10^{-5}$ mol dm^{-3}.

Cabaniss & Shuman (1988b) compared the binding of Cu by seven samples of isolated fulvic acid by examining differences between observed values of pCu ($-\log[Cu^{2+}]$) and those predicted using a model that had been parameterised from the results of experiments with Suwannee River fulvic acid. The root-mean-squared-difference in pCu ranged from 0.10 to 0.31 (average 0.18). The authors considered the variation among samples to be modest, especially when analytical errors were taken into account, and they concluded that source-to-source variations in fulvic acid binding properties were 'generally less important than chemical factors such as pH, alkalinity and ionic strength for predicting pCu in natural waters'.

Comparisons of the binding of Cd by three different humic acids, at high metal loadings, were made by Lee *et al.* (1993). They found variations of up to a factor of two in site densities, but almost no variation in conditional stability constants. Similarly, Kim *et al.* (1991a) reported very similar conditional stability constants for the binding of Am by two different humic acids.

There have been several studies in which humic samples have been fractionated and the binding of metals by the different fractions determined.

Table 7.5. Conditional association equilibrium constants, and site contents ($mmol\,gC^{-1}$) for Cu binding by aquatic fulvic acids (McKnight *et al.*, 1983), obtained from fitting to a two-site model.

Source of sample	n_1	$\log K_1$	n_2	$\log K_2$
Shawsheen River	0.90	6.0	0.21	7.6
Ogeechee River	1.00	6.1	0.22	7.9
Ohio River	0.65	6.6	0.12	8.3
Missouri River	0.56	5.9	0.14	7.6
South Platte River	0.70	6.1	0.11	8.3
Bear River	1.10	6.0	0.26	7.8
Como Creek	1.30	5.9	0.32	8.1
Deer Creek	0.83	5.9	0.11	8.3
Hawaiian River	1.90	5.6	0.65	7.2
Black Lake	0.74	6.3	—	—
Island Lake	0.92	6.4	—	—
Brainard Lake	1.20	5.4	0.38	7.0
Merril Lake	1.30	5.8	0.31	7.6
Suwannee River	1.20	5.9	0.27	7.8
Hawaiian Marsh	1.90	6.0	0.57	8.0
Yuma Canal	0.59	6.0	0.12	8.5
Biscayne Aquifer	1.00	5.6	0.29	7.5

The experiments were done at pH 6.25, in a background electrolyte of $10^{-3}\,mol\,dm^{-3}$ KNO_3.

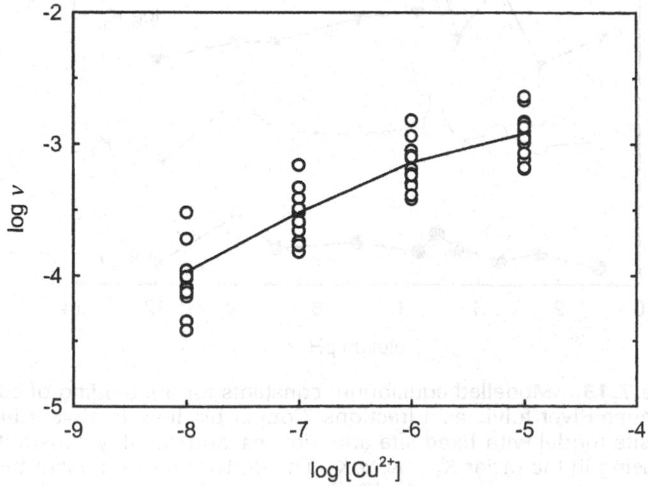

Figure 7.12. Copper binding by different fulvic acid samples at pH 6.25, calculated from the parameter values of McKnight *et al.* (1983) shown in Table 7.5.

Lakshman *et al.* (1996) used ultrafiltration to fractionate a soil fulvic acid into three fractions of differing molecular weight (<500, 500–1000, 1000–10 000). They measured Al binding by a fluorescence technique, and found that the conditional stability constant increased by a factor of 4–5 with molecular weight. Van den Hoop & van Leeuwen (1997) fractionated a sample of humic acid by varying the pH of precipitation, and obtained three fractions of weight-average molecular weight 7000, 9000 and 14 000. They found little difference in the binding of Zn and Cd by the different fractions.

Brown *et al.* (1999) fractionated an aquatic fulvic acid by desorption from XAD-8 resin at different pH values, obtaining 19 fractions in all. The binding of Cu by 10 of the fractions was studied by potentiometry with an ion-selective electrode, and the results were fitted to a three-site model with the same assumed binding site contents for each fraction. The log K values so derived are shown in Fig. 7.13. The binding constant for the high-concentration, low-affinity site (K_3) varied little, whereas that for the low-concentration, high-affinity site (K_1) showed appreciable variation, suggesting that the fractions differ most with respect to the high-affinity sites. There was no simple dependence upon elution pH, indicating a fairly uniform distribution of binding sites among fractions of differing

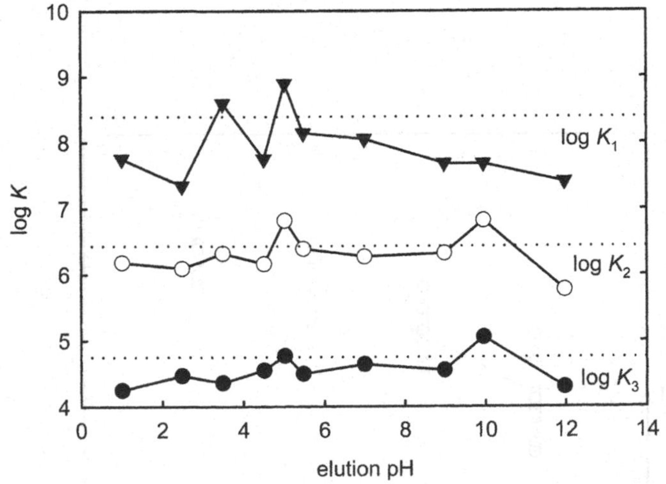

Figure 7.13. Modelled equilibrium constants for the binding of copper by Suwannee River fulvic acid fractions. Copper binding data were fitted to a three-site model with fixed site abundances, and stability constants decreasing in the order $K_1 > K_2 > K_3$. The dotted lines represent the values for the unfractionated sample. [Redrawn from Brown G.K., Cabaniss S.E., MacCarthy P. & Leenheer J.A. (1999), Cu(II) binding by a pH-fractionated fulvic acid, *Anal. Chim. Acta.* **402**, 183–193, with permission from Elsevier Science.]

hydrophobicity. The binding constants did not correlate well with other properties of the fractions, for example contents of different types of carbon (carboxyl, aromatic, heteroaliphatic, ester).

Definite conclusions cannot be drawn from the results summarised above. Variations in binding properties exist, but have been found to be greater in some studies than in others. Differences in cation-binding properties have not yet been related to differences in other chemical properties, nor to the sources of the samples. Relationships may emerge as more measurements are made, but an extensive, systematic programme of research is needed to provide firm answers. In applying binding information obtained in laboratory studies to the field, uncertainty analysis could take into account variation in humic binding properties.

7.3 Kinetics of metal ion binding

A number of studies have been made of the rates of interaction of metal ions with isolated humic substances. Broadly, the experiments can be divided into two types, those investigating association reactions, and those in which the dissociation of metal–humic complexes is followed.

7.3.1 Rates of forward reactions

The forward reactions are almost complete within a few minutes at most. Results from Lin *et al.* (1994) are shown in Fig. 7.14. These authors used a stopped flow technique with fluorescence detection to study Cu complexation with a fulvic acid. They could fit the data by assuming two kinds of site, with rate constants of 10^6 and $5.15 \times 10^3\,\mathrm{mol^{-1}\,s^{-1}}$. Other examples of fast forward reactions are for Eu and UO_2 with humic acid (Choppin & Clarke, 1991), Al with fulvic acid (Plankey & Patterson, 1987), and Tb with fulvic and humic acids (Bidoglio *et al.*, 1991). Ma *et al.* (1999) reported that 99% of added Cu was bound very quickly (within a few minutes) to humic acid at pH 8, but there then followed a slow further binding of the residual free metal, taking place over 24 hours, during which the concentration of Cu^{2+} decreased by an order of magnitude.

7.3.2 Rates of dissociation

Numerous authors have studied pseudo-first order dissociation of metals from complexes with humic substances, using the competitive ligand method (Section 6.9). In general, the overall dissociation is due to contributions from a range of sites. Shuman and colleagues (Olson & Shuman, 1983; Shuman *et al.*, 1983) analysed their data using the ‘affinity spectrum' approach, which assumes a continuous distribution of dissociation rate constants. The experimental data are transformed to obtain a continuous function $H(k,t)$ where k is the dissociation rate constant and t is time. The product $H(k,t)$ d(ln k) is the probability of finding a site with a dissociation rate constant in the region between (ln k) and (ln k + d(ln k)). An example of the output is shown in Fig. 7.15. In most other studies, data analysis has been performed using two or more discrete dissociation rate constants, and fitting data by optimising the rate constants and the amounts of each kinetically characteristic fraction. Rate *et al.* (1992) proposed the use of a lognormal distribution of rate constants, the dissociation behaviour being described by the mean and standard deviation. For present purposes, it is sufficient to consider ranges of values of k. Table 7.6 summarises values for a number of different metals studied under different conditions and using a variety of ligands and solids as 'sinks' for dissociated metal.

In considering the data collected in Table 7.6, the importance of the 'analytical window' has to be borne in mind, i.e. the range of values of k obtained in a given study depends upon the analytical technique and the time over which dissociation is followed. For example, the methodology of Sekaly *et al.* (1998) only allowed the determination of k in the range 0.0002–0.01 s^{-1}. Some authors report detection of almost all the released

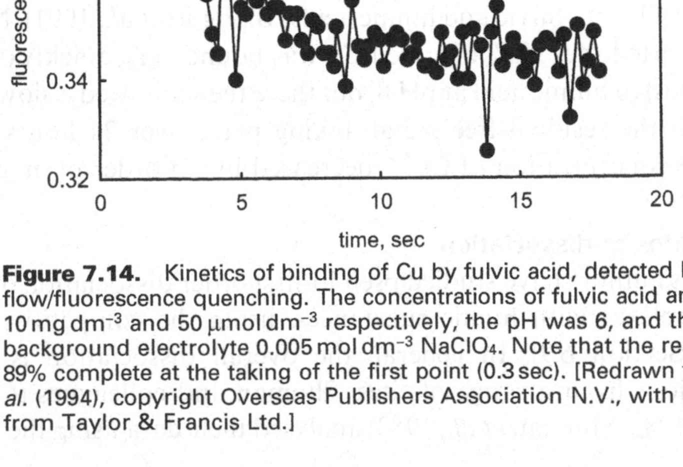

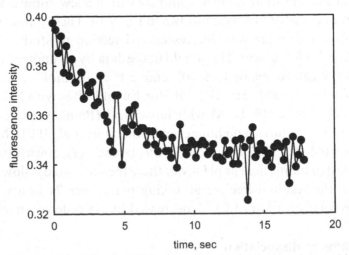

Figure 7.14. Kinetics of binding of Cu by fulvic acid, detected by stopped flow/fluorescence quenching. The concentrations of fulvic acid and Cu were 10 mg dm^{-3} and 50 μmol dm^{-3} respectively, the pH was 6, and the background electrolyte 0.005 mol dm^{-3} $NaClO_4$. Note that the reaction was 89% complete at the taking of the first point (0.3 sec). [Redrawn from Lin *et al.* (1994), copyright Overseas Publishers Association N.V., with permission from Taylor & Francis Ltd.]

Table 7.6. Ranges of kinetic constants.

System	pH	I (mol dm^{-3})	v (mol g^{-1})	Equilibration time	log k (s^{-1}) low	log k (s^{-1}) high	Reference
Al–FA	5.0	$\sim 10^{-5}$	10^{-3}	$\sim$ 24 h	−4.0	−3.3	Sekaly *et al.* (1998)
Ni–FA	4.0–6.4	0.1	10^{-4}–10^{-3}	$\geq$ 24 h	−2.6	−0.8	Lavigne *et al.* (1987)
Ni–FA	7.4	0.002	$\sim 5 \times 10^{-4}$	12–24 h	−2.6	−1.5	Cabaniss (1990)
	7.4	0.03	$\sim 5 \times 10^{-4}$	12–24 h	−2.7	−1.4	Cabaniss (1990)
	8.4	0.002	$\sim 5 \times 10^{-4}$	12–24 h	−3.0	−1.7	Cabaniss (1990)
	8.4	0.1	$\sim 5 \times 10^{-4}$	12–24 h	−2.6	−1.4	Cabaniss (1990)
Ni–FA	5.1	—	$\sim 10^{-5}$–10^{-3}	—	−5.7	−3	Mandal *et al.* (1999)
Cu–DOM	7.5	0.1	$\sim 2 \times 10^{-4}$	—	−1.8	$> +1.2$	Shuman *et al.* (1983)
Cu–DOM	7.5	0.1	2×10^{-4}	12 h	< -1.6	+1.6	Olson & Shuman (1983)
Cu–HA	6.0, 7.0	—	$\sim 10^{-3}$	24 h	−2.1	> -1	Bonifazi *et al.* (1996)
Cu–aqHS	8	—	$\sim 10^{-6}$	0.5 h–18 d	−5.2		Burba (1994)
Cu–HA	6	0.1	1.5×10^{-4}	24 h	−3.1	>0.9	Rate *et al.* (1992)
Zn–FA	~ 5	$\sim 10^{-5}$	$\sim 2 \times 10^{-4}$	$\sim$ 24 h	−3.4	−1.6	Sekaly *et al.* (1998)
Cd–FA	5	$\sim 10^{-5}$	$\sim 10^{-4}$	$\sim$ 24 h	−3.1	−1.6	Sekaly *et al.* (1998)
Sm–HA	4.2	0.1	$\sim 2 \times 10^{-4}$	48 h	−4.6	+0.03	Clark & Choppin (1990)
Pb–FA	5.6	$\sim 10^{-5}$	6×10^{-5}–2×10^{-4}	$\sim$24 h	−3.0	−1.2	Sekaly *et al.* (1998)
Th–HA	4.2–5.9	0.1	2.5×10^{-4}	48 h	−4.5	+0.6	Cacheris & Choppin (1987)
UO_2–HA	4.2	0.1	2.5×10^{-4}	48 h	−4.0	−0.2	Choppin & Clark (1991)

Key: FA fulvic acid, HA humic acid, aqHS aquatic humic substances, DOM dissolved organic matter.

metal, whereas others observe only an intermediate fraction. For example, Olson & Shuman (1983) reported recovery of *c.* 90% of bound Cu, whereas in the experiments of Rate *et al.* (1992) approximately 60% of the metal was released during the mixing time at the start of the experiment, while 10% or more remained bound at the end.

The most noticeable general result is the wide range of dissociation rate constants, which is consistent with the heterogeneity in binding site strength revealed by equilibrium studies (Section 7.2). The value of k tends to be smallest when v is low, as illustrated by the results in Table 7.6 from Burba (1994) and Mandal *et al.* (1999). This is consistent with the finding from equilibrium studies that there are small numbers of strong (slowly dissociating) sites in humic matter. Cabaniss (1990) carried out a systematic study of the dissociation of Ni from a fulvic acid and found that dissociation rates fall with increases in pH, decreases in ionic strength, and decreases in metal–humic ratios. Rate *et al.* (1992) demonstrated that, for Cu and humic acid, rates increase with temperature. Burba (1994) compared the dissociation behaviours of a number of metals with aquatic humic substances, and found the following order of lability

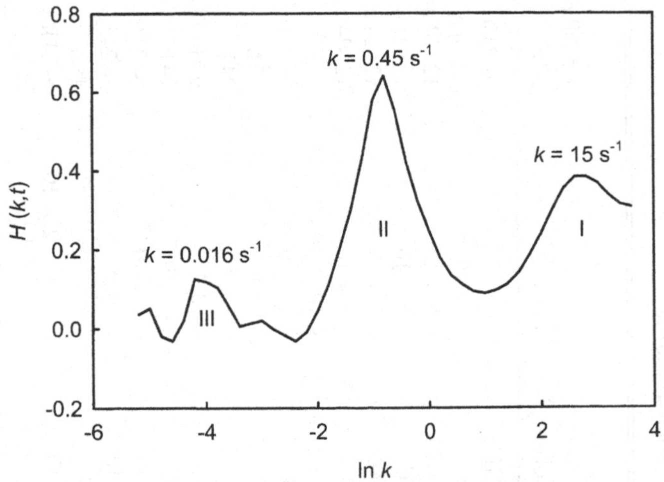

Figure 7.15. Kinetic results for the dissociation of Cu bound to a sample of estuarine dissolved organic matter, redrawn from Shuman *et al.* (1983). The solution was pre-equilibrated at pH 7.5, 21 mg dm^{-3} dissolved organic carbon, 7.1 μmol dm^{-3} Cu, I = 0.1 mol dm^{-3}, and 25°C, giving > 99% binding of the metal. Dissociation was followed by a stopped flow method (up to 30 sec) using an excess of the colorimetric reagent 4-(2-pyridylazo)resorcinol. From the areas under the curve, the contributions of the three main components – I, II and III – were estimated to comprise 9%, 12% and 1% respectively of the total copper. The rest of the metal dissociated too quickly to be observed by the stopped flow technique used in the experiments.

$$Mn > Zn > Co > Pb > Ni > Cu \gg Al > Fe(III).$$

Marx & Heumann (1999) investigated the kinetic stability of the complexes of Cr(III) and Cu(II) using radioactive isotopes to determine rates of metal ion exchange. They found that the Cr(III) complexes were much more kinetically stable, as expected from knowledge about complexes with simpler ligands.

7.3.3 Change of dissociation rates with time after complex formation

A number of authors have noted that the longer is the time between mixing of the reactants and measurement of dissociation, the smaller are the rate constants, or the greater is the fraction of metal dissociating slowly. An illustration comes from the work of Choppin & Clark (1991), in which the extent of removal of Eu from its complex with humic acid by passage through a column of ion-exchange resin was measured after different equilibration times (Fig. 7.16). Similar observations, made by several other methods, were reported for Mn, Fe(III), Ni, Cu, Zn and In(III) by Burba (1994), Cu by Rate *et al.* (1993), Th (Cacheris & Choppin, 1987), and UO_2 (Choppin & Clark, 1991).

Choppin & Clark (1991) explained the ageing effect as follows. When the metal ions are first brought into contact with the humic matter,

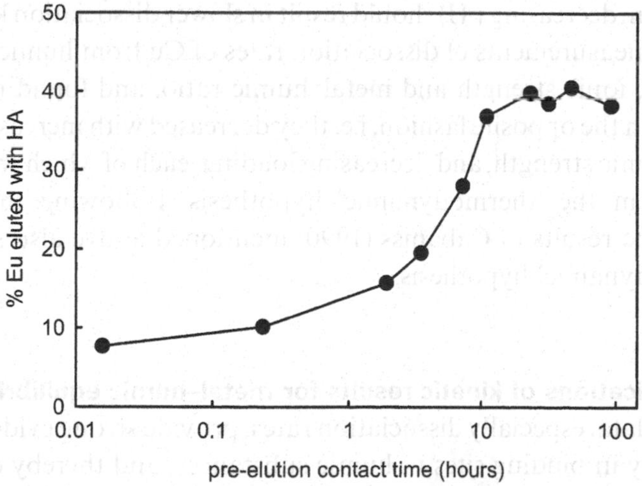

Figure 7.16. Effect of pre-equilibration time on the kinetic lability of Eu complexes with humic acid (HA). Mixtures of Eu (2.5 μmol dm^{-3}) and HA (26 mg dm^{-3}) were prepared at pH 4.2, $I = 0.1$ mol dm^{-3}, and after different times were subjected to cation-exchange chromatography. The higher is the percentage of metal eluting with the humic acid, the less labile is the Eu–HA complex. [Redrawn from Choppin, G.R. & Clark, S.B. (1991), The kinetic interactions of metal ions with humic acids, *Marine Chem.* **36**, 27–38, with permission from Elsevier Science.]

binding takes place rapidly and 'non-specifically' at sites on the outer parts of the humic molecules. Such binding perturbs electrostatic and hydrogen-bonding interactions, and consequently the molecular conformation changes, opening the structure and allowing migration of the metal ions to specific (strong) sites, a process taking place on a time scale of hours or even days. Dissociation is slower from the strong sites, and so the longer the system is left to move towards equilibrium, the more the strong sites contribute to the observed dissociation behaviour. This may be termed the 'internalisation' hypothesis. It would explain not only the ageing effect, but also the observed range of dissociation rates, which would be due to differences in accessibility, essentially a physical phenomenon. An alternative explanation – which might be called the 'thermodynamic' hypothesis – is that all the sites are exposed and unchanging, but that it takes some time for the metal ions to find the strongest ones, simply on the grounds of probability. Thus, the fall in dissociation rate with time would reflect the gradual loading of the strong sites, and the range of dissociation rate constants would be due to site heterogeneity. Mainly in the context of trying to explain the wide range in rate constants, Rate *et al.* (1993) pointed out that if the 'internalisation' theory is correct, then factors which cause the molecular structure of humic substances to contract (increasing ionic strength, degree of metal complexation, decreasing pH) should result in slower dissociation kinetics. They made measurements of dissociation rates of Cu from humic acid at different pH, ionic strength and metal: humic ratio, and found that the rates varied in the opposite fashion, i.e. they decreased with increasing pH, decreasing ionic strength, and decreasing loading, each of which would be expected from the 'thermodynamic' hypothesis. Following the same argument, the results of Cabaniss (1990) mentioned above also support the 'thermodynamic' hypothesis.

7.3.4 Implications of kinetic results for metal–humic equilibria

The kinetic data, especially dissociation rates, provide strong evidence for heterogeneity in binding sites in humic substances, and thereby confirm the conclusions drawn from equilibrium data. However, they also raise some questions about the validity of the 'equilibrium' data. Especially for the stronger binding sites, it appears possible that in studies where titrations of humic matter with metal ions are carried out, true equilibrium may not be achieved, because the transfer to the strongest sites is so slow. Some data assumed to represent equilibrium conditions might therefore tend to underestimate binding affinities.

8

Cation binding sites in humic substances

This chapter summarises information about the chemical groupings responsible for the binding of protons and metal cations by humic substances. The bonding atoms themselves are considered, together with structural arrangements at the molecular level (conformation). Such information complements the quantitative binding data reviewed in Chapter 7, and can help in the formulation of mechanistic models of cation binding by humic matter (Chapters 9–11).

8.1 Proton-dissociating groups

The proton dissociation properties of humic substances (Section 7.1) point very strongly to the presence of carboxylic acid groups in their structures. Direct evidence for the chemical nature of groups dissociating in the acid-to-neutral pH range was provided by Cabaniss (1991), who used Fourier-transform infra-red (FTIR) spectroscopy to show that the groups were nearly all carboxylic acids (Fig. 8.1).

Leenheer *et al.* (1995a, b) carried out detailed studies to investigate the chemical natures of carboxylic acid groups in Suwannee River fulvic acid. They used chemical modifications specific to individual groupings, titration data, model compound studies, NMR, and theoretical arguments, to estimate the contributions of different groups. They pointed out that a significant number of the groups are quite strongly acidic. For the fulvic acid sample in question, they estimated that one-third of the carboxyl groups had pK values of 3.0 or less. The main aim of their work was to account for such acidity. Table 8.1 summarises their findings. It was deduced that the strong carboxyl acidity is associated with polycarboxylic α-ether and α-ester structures. It follows that a substantial proportion of

Table 8.1. Carboxyl group structures in Suwannee River fulvic acid, from Leenheer *et al.* (1995a,b).

Acid group structure	Content ($mmol\,g^{-1}$)	% of total COOH
S and N acids	0.18	3.0
Oxalate half-esters	≤ 0.02	≤ 0.3
Substituted malonic acids	≤ 0.02	≤ 0.3
Keto acids (pK ≤ 3.0)	0.20	3.3
Aromatic and olefinic acids (pK ≤ 3.0)	0.46	7.7
Aromatic and olefinic acids (pK ≥ 3.0)	0.86	14.3
α-Ether and α-ester acids (pK ≤ 3.0)	1.28	21.3
Carboxylic acids associated with α-ether and α-ester acids		
if one additional acid	1.28	21.3
if two additional acids	2.56	42.6
Total acids accounted for	4.26–5.58	71–93

The total content of COOH groups was 6.0 $mmol\,g^{-1}$.

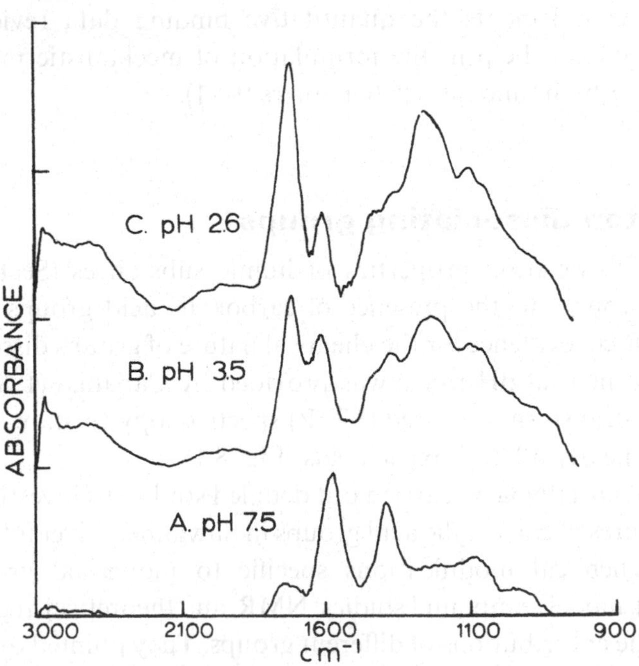

Figure 8.1. FTIR spectra of Lake Drummond fulvic acid in aqueous solution. The peaks at 1250 and 1710 cm^{-1} at low pH are due to —COOH groups. Those at 1390 and 1590 cm^{-1} at higher pH are due to —COO^-. [Reproduced from Cabaniss, S.E. (1991), Carboxylic acid content of a fulvic acid determined by potentiometry and aqueous Fourier transform infrared spectrometry, *Anal. Chim. Acta* **255**, 23–30, with permission from Elsevier Science.]

the carboxyl groups must be quite localised within the fulvic acid structure. There are obvious implications for the formation of multidentate binding sites for metals (see below), and also for the amphiphilic properties of the compounds (cf. Section 3.1.2). Leenheer *et al.* (1995b) suggested some possible structures of Suwannee River fulvic acid consistent with the chemical natures of the COOH groups (Fig. 8.2).

There is relatively little information available on the weak acid groups in humic substances. The pK values suggest a significant contribution of phenolic groups, but alcohols and β-dicarbonyl acids may also contribute (Perdue, 1985).

8.2 Binding sites for metals – information from binding studies

The simplest kind of binding of metal cations by humic substances is the weak electrostatic attraction of counterions. The dissociation of protons, and their incomplete replacement by coordinated metal ions, mean that humic matter nearly always carries a negative charge, to which counterions (cations) will be attracted. Coulombic attraction is sufficient to explain the binding of ions such as Na^+ and K^+, which display very weak coordinative binding (Gamble, 1973; Bonn & Fish, 1993). Stronger binding involves specific complexation reactions.

As was discussed in Section 7.2, equilibrium binding data provide some insight into the natures of complexation sites for metals. Weak sites present in large amounts are due mostly to carboxyl groups, and also to weaker acids. The binding of metal ions by these sites is generally stronger than would be expected for monodentate binding. Therefore, we deduce the existence of sites that involve at least two donor atoms. Since there is no strong evidence that metal binding by humic substances is dependent upon the concentration of humic substances, the multiple donor atoms must be on the same molecule. It follows that chelation (Section 5.1.4) plays a rôle in the binding of metal ions at the abundant weak binding sites.

Equilibrium binding data show the presence in humic matter of small numbers of sites with affinities for metal cations much greater than those of the abundant sites (Section 7.2). From modelling (Section 10.3) it can be deduced that not all metals are affected equally by such strong sites, and the results may be explained, somewhat speculatively, by the presence in the strong sites of soft donor atoms such as N. This would explain why, at low levels, Cu^{2+} can be much more strongly bound than Eu^{3+}, even though at higher concentrations they have similar affinities. Studies of

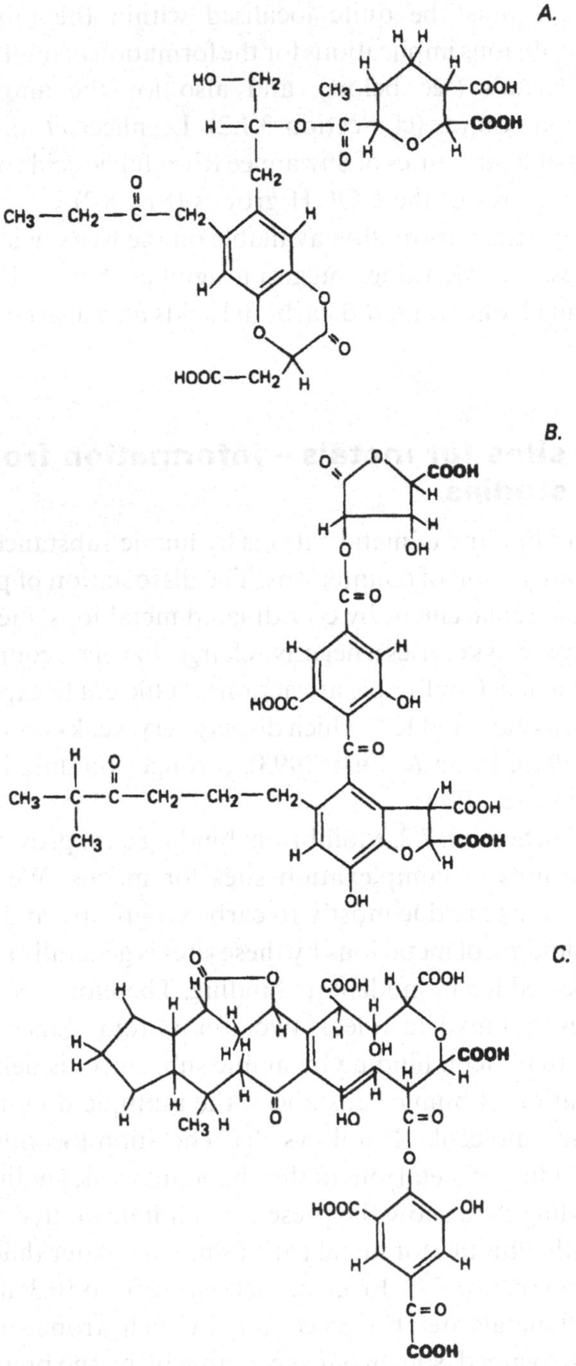

Figure 8.2. Structural models of fulvic acid. Carboxyl groups in bold have p*K* values of less than 3.0. [Reproduced from Leenheer, J.A., Wershaw, R.L. & Reddy, M.M. (1995), Strong-acid, carboxyl-group structures in fulvic acid from the Suwannee River, Georgia. 2. Major structures, *Environ. Sci. Technol.* **29**, 399–405. Copyright 1995 American Chemical Society.]

Table 8.2. Results from electron spin resonance studies of metal complexes with humic acid (HA) and fulvic acid (FA).

Metal	Complex	Result	Reference
VO^{2+}	HA	V has approximately axial symmetry. Complexation by either O donor groups or mixed O and N.	Goodman & Cheshire (1975)
	HA	V bound by O atoms in inner sphere complexes.	McBride (1978)
Mn(II)	HA	Outer sphere complexes at low pH. Some inner sphere complexation at pH > 5.	McBride (1978, 1982)
	FA*	High-spin Mn hexahydrate, outer sphere complexes.	Cheshire *et al.* (1977)
Fe(III)	FA*	Strong binding of high-spin Fe(III) ions in tetrahedral or octahedral sites, inner sphere complexes. Evidence for binding at weaker – 'surface' – sites.	Senesi (1990)
	FA	Sites of different strength, depending upon loading.	Senesi et al. (1977)
Cu(II)	HA*	Interaction with the N groups of porphyrin structures.	Goodman & Cheshire (1973)
	HA	Inner sphere complexation, binding by O atoms (high loadings).	McBride (1978)
	HA	At least one Cu^{2+} H_2O is displaced by a carboxylate O atom. No evidence for N donors.	Bloom & McBride (1979)
	HA	Cu forms two equatorial bonds with O donor atoms from HA in *cis* positions.	Boyd *et al.* (1981, 1983)
	HA	Cu is bound in a square-planar arrangement, to two aliphatic N and two carboxylate groups.	Lakatos *et al.* (1977)
	HA	Fe(III) can displace Cu from strong binding sites.	Senesi *et al.* (1986)
	FA	Inner sphere complexes, square-planar coordination. Bonding either all by O atoms, or mixed O and N.	Senesi (1990)

The asterisk (*) denotes complexes of metal present in samples isolated from the field.
Otherwise, the complexes were prepared by adding metal in the laboratory.

metal dissociation kinetics also reveal a range of binding sites (Section 7.3.2). Fast-dissociating metal may be in outer sphere complexes, while slower dissociation is from a variety of inner sphere complexes.

8.3 Information from spectroscopy

The use of spectroscopic methods to determine extents of binding is described in Chapter 6. Here we are concerned with their use to obtain information about the atoms and molecular structures that constitute humic binding sites. It should be borne in mind firstly that a spectroscopic method may not respond to all the metal in the complex, and secondly that the spectroscopic response may vary with the amount of metal bound per g of humic matter, i.e. it will depend upon loading. Therefore, different techniques can give different results, depending upon the method and the experimental conditions. For reference, the maximum value of v (mol bound per g humic matter) for metals is between 10^{-3} and 5×10^{-3}, while the abundance of 'strong' sites is of the order of 10^{-5} mol g^{-1} or less. Therefore if the metal loading is high – often a necessary condition to obtain a spectroscopic signal – only the weaker sites may be probed.

8.3.1 Electron spin resonance (ESR)

ESR was one of the earliest spectroscopic methods used to obtain direct insight into humic binding environments. It relies on the absorption of microwave radiation by unpaired electrons placed in a magnetic field, and is therefore suitable for metals such as Cu(II), Fe(III), V(IV) and Mn(II). ESR spectra are sensitive to the chemical environment of the absorbing electron, and therefore provide information about the molecular structure (Senesi, 1990). Information on humic substances has mostly been obtained by comparing results with those for well-defined ligands. Results from ESR studies are summarised in Table 8.2. Under most conditions examined, the main coordination is with O atoms, but there is also evidence for the participation of N atoms, especially for Cu.

8.3.2 Nuclear magnetic resonance (NMR)

Deczky & Langford (1978) studied the effects of paramagnetic cations (Mn^{2+}, Fe^{3+}, Cu^{2+}) on the relaxation in the proton magnetic resonance signal of solvent water, at acid pH, and at high metal–fulvic acid ratios. They concluded that Mn^{2+} formed outer sphere complexes, while inner sphere complexes were formed by Fe^{3+} and Cu^{2+}.

Chung *et al.* (1996) and Li *et al.* (1998) used ^{113}Cd NMR to study metal complexation by fulvic acids. The free Cd^{2+} ion produces a single sharp peak in the NMR spectrum, the chemical shift of which depends upon its bonding to different donor atoms. In the study of Chung *et al.* (1996), loadings of 10^{-4} mol g^{-1} or more were used, and the data were interpreted to mean that ligand groups other than carboxylates were involved in binding, hydroxyl groups and neutral N-donor groups being suggested as possibilities. In the study of Li *et al.* (1998), the experimental conditions were manipulated to maximise the possibility of observing binding by N or S donor atoms, by working at high ratios of fulvic acid to Cd. High concentrations of both Cd (> 0.001 mol dm^{-3}) and fulvic acid (DOC up to 10 g dm^{-3}) were required to allow a good NMR signal to be observed. At acidic pH, Cd exchanged rapidly between bound and free forms, while at higher pH exchange was slower. Under nearly all conditions of pH examined (3.6 to 9.0) and all Cd loadings (ν ranged from 5×10^{-5} to 3×10^{-4} mol g^{-1}) Cd was primarily coordinated by O donor atoms, probably in carboxyl groups. There was some evidence of coordination by N donor atoms at high pH, but no evidence of coordination by S donor atoms, even though the Cd loading was significantly less than the S content of the fulvic acid.

8.3.3 Fluorescence

According to Cabaniss (1992), 'fluorescence spectroscopy offers a unique perspective on metal–DOM binding because it observes the DOM ligand directly'. He used synchronous quenching spectra – the difference between synchronous scans of ligand fluorescence before and after metal addition – to investigate binding site heterogeneity in fulvic acid, and competition between metals (Mg, Al, Fe(III), Mn(II), Co, Ni, Cu, Pb). The values of ν were of the order of 10^{-4} mol g^{-1}. Three distinct types of quenching spectra, typified by Mg, Al and Cu, were observed (Fig. 8.3). Those for Mn(II), Co, Ni, Cu and Pb were similar in shape to one another at both pH 5.0 and 7.5, suggesting that all the metals bind to the same sites. The results for Mg were markedly different, fluorescence being enhanced at pH 5 and 7.5, perhaps due to conformational change in the fulvic acid on Mg binding. The quenching spectrum of Fe(III) at pH 5 was similar to that of Cu, suggesting common binding sites, but at pH 7.5 hydrolysis of iron restricted its binding to the fulvic acid. Aluminium showed fluorescence enhancement at pH 5 but quenching at pH 7.5, although the spectrum was different to that of the other metals at the higher pH. Principal components analysis demonstrated that multiple fluorescent binding sites for divalent transition metals, and Pb, are present in the fulvic acid at pH 7.5, but there

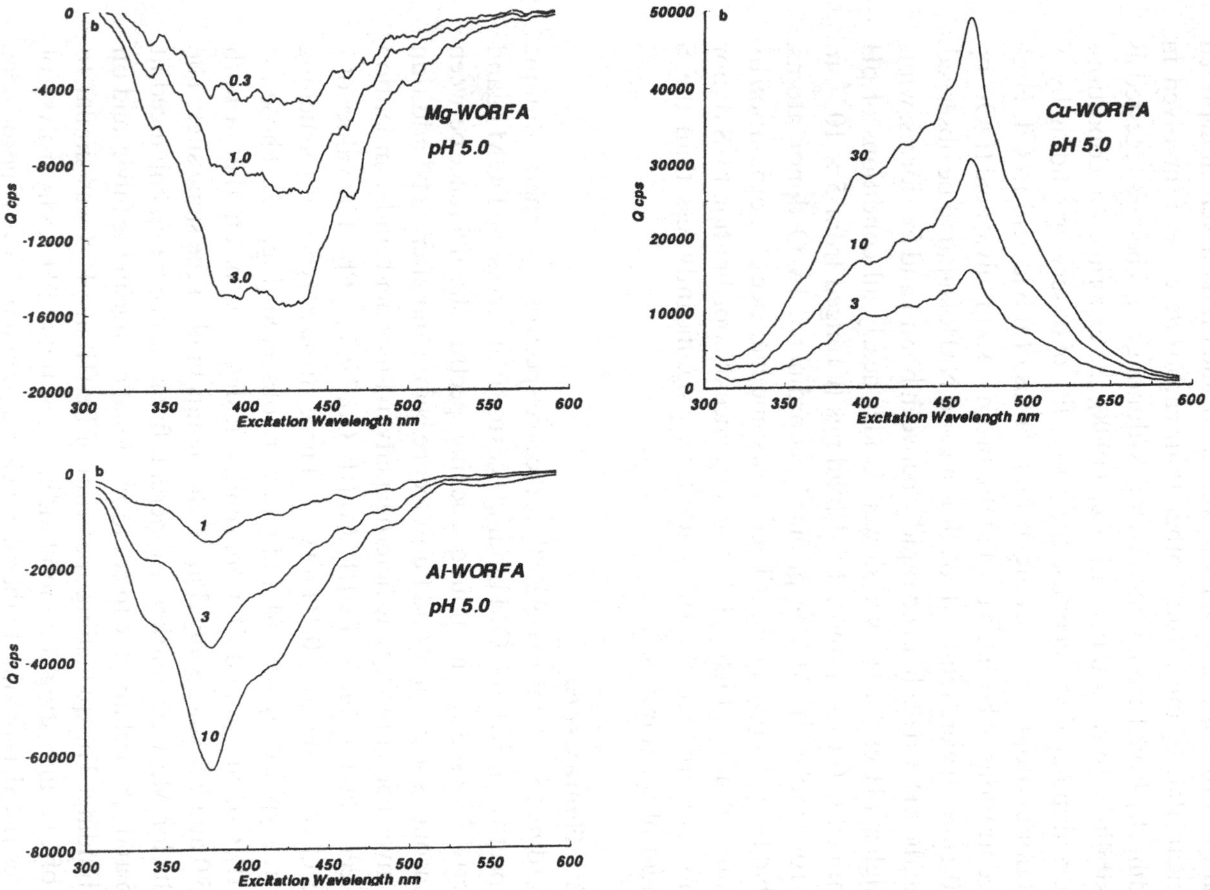

Figure 8.3. Fluorescence quenching synchronous spectra of metal–fulvic acid complexes. The *y*-axis is the difference in fluorescence between a solution of White Oak Run fulvic acid (5 mg dm^{-3} DOC) and a solution having the same fulvic acid concentration but with metal added at the concentration indicated (mmol dm^{-3} for Mg, μmol dm^{-3} for Al and Cu). Positive values indicate that the fluorescence of the fulvic acid is quenched by the bound metal, negative values that the fluorescence is enhanced. [Reproduced from Cabaniss, S.E. (1992), Synchronous fluorescence spectra of metal–fulvic acid complexes, *Environ. Sci. Technol.* **26**, 1133–1139. Copyright 1992 American Chemical Society.]

was no evidence for multiple sites at pH 5.0. Results from experiments involving more than one metal showed that Al, but not Mg, competes with Cu for binding, and that Al displaces bound Mg.

Thomason *et al.* (1996) investigated the binding of Eu by aquatic humic substances using lanthanide ion probe spectroscopy. The fluorescence lifetime of Eu bound by the humic matter depends upon the number of water molecules in the inner coordination sphere of the metal, which can be determined by comparing fluorescence decay in H_2O and D_2O solutions. The number of ligand donor atoms can then be estimated by difference. In addition, the position of a fluorescence emission band associated with a hypersensitive electronic transition ($^5D_0 \rightarrow {}^7F_2$) depends upon the charge of the Eu atom in the Eu–humic complex. The results, which refer to a pH of *c.* 3.5, showed that at the lowest Eu loadings ($\nu \sim 3 \times 10^{-4}$ mol g^{-1}) approximately four humic ligand atoms were involved in binding the metal. At higher loadings, fewer ligand atoms were involved. The authors envisaged a series of binding sites in the series mono-, bi-, tri-, tetra-dentate, probably due to carboxylate groups. The change in charge of the Eu ion was in accord with these conclusions.

A different use of fluorescence to examine metal binding was developed by Green *et al.* (1992), who used fluorescence probes that do not form coordinative bonds, but may be attracted to the humic molecules by Coulombic forces. Quenching of fluorescence was much greater with a cationic nitroxide probe than with a neutral one. Quenching by the cationic probe was sensitive to ionic strength, pH, and humic matter molecular weight, confirming the rôle of Coulombic attraction.

Lakshman *et al.* (1996) investigated complexes of Al with fulvic acid fractions of different molecular weight using fluorescence polarisation spectroscopy. They deduced that binding of the metal to the lowest molecular weight fraction caused no change in the conformation of the humic material, whereas there were conformational changes in higher molecular weight fractions.

8.3.4 X-ray and γ-ray techniques

Extended X-ray absorption fine structure (EXAFS) spectroscopy depends upon the ejection of electrons from metal atoms by X-rays. Interference between electrons ejected directly from the metal and those backscattered from neighbouring atoms provides information on their identities and positions. Xia, Bleam and colleagues (Xia *et al.*, 1997a,b; 1999) used X-ray absorption spectroscopy to characterise trace metal complexation by humic matter. Metal complexes were prepared at high loadings, and freeze-dried before analysis. Some typical spectra are shown in Fig. 8.4. The main results are summarised in Table 8.3. Information was obtained for atoms in two shells around the metal atom. The presence of O in the first shell and C in the second would be consistent with C—O—M bonds, which led the authors to conclude that all the metals were bound by inner sphere coordination. Coordination by O atoms was prevalent, with the exception of Zn bound by soil humic substances, and Hg bound by humic acid, where thiol groups were involved.

A complementary technique to EXAFS is XANES (X-ray absorption near-edge structure), which was used by Frenkel *et al.* (2000) to investigate the binding of copper by aquatic humic substances. The results, for solutions at pH 4, showed that, at Cu:humic ratios less than 2×10^{-4} mol g^{-1}, nitrogen atoms were mainly responsible for binding the metal, whereas at higher ratios carboxyl groups became dominant.

Kupsch *et al.* (1996) used time-differential perturbed angular correlation of γ-rays to study the interaction of humic acids with the short-lived isotopes ^{111m}Cd and ^{199m}Hg, under acid conditions. The results indicated two-fold coordination of Hg at low loadings, but higher coordination at higher loadings. The authors tentatively suggested this to indicate a

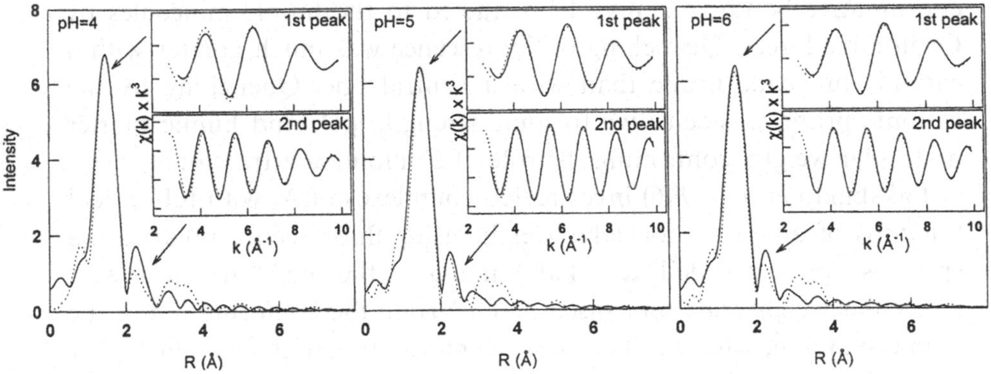

Figure 8.4. EXAFS spectra of Pb complexes with humic acid. The main panels show the radial structure function (RSF), in which the locations of peaks represent distances between the absorber atom (Pb) and successive shells of neighbouring atoms, while the amplitudes of the peaks depend on the number and identities of the backscattering atoms. The insets show Fourier transforms of peaks in the RSF. Structural information is obtained by computing the functions for model structures, and finding the best agreement with observations. The dotted lines are experimental data, and the full lines simulations. See Section 8.3.4 and Table 8.3. [Reproduced from Xia, K., Bleam, W. & Helmke, P.A. (1997), Studies of the nature of Cu^{2+} and Pb^{2+} binding sites in soil humic substances using X-ray absorption spectroscopy, *Geochim. Cosmochim. Acta* **61**, 2211–2221, with permission from Elsevier Science.]

change from coordination by two S atoms to coordination by four N atoms. No specific binding of Cd was detected.

8.3.5 Infra-red (IR) spectroscopy

Byler *et al.* (1987) determined IR spectra of a fulvic acid and of the same material with added Fe(III), at a ratio of *c.* 10^{-3} mol g^{-1}. More COO^- groups were evident in the presence of the metal, indicating the formation of Fe(III)—OOC bonds. Alberts & Filip (1998) used FTIR to study the properties of metal complexes of estuarine humic and fulvic acids, formed at high metal loadings. By comparing the spectra with those for the protonated forms of the humic substances, they obtained information about the functional groups involved in metal binding. The results for Cu

Table 8.3. Information on metal–ligand bonding in humic substances obtained from X-ray spectroscopy by Xia *et al.* (1997a,b; 1999).

Metal	HS	pH	Shell	Type	No.	BD
Co	FA,HA,CHS	4	1st	O_M	6	2.04
			2nd	C	1	2.91
Ni	FA,HA,CHS	4	1st	O_M	6	2.09
			2nd	C	2	2.93
Cu	FA,HA	4	1st	O_L	4	1.94
			1st	O_W	2	1.98
			2nd	C	4	3.06
	CHS	4,5,6	1st	O_L	4	1.94
			1st	O_W	2	2.01
			2nd	C	4/5	3.13
Zn	FA,HA	4	1st	O_M	6	2.10
			2nd	C	1	2.89
	CHS	4	1st	O_W	4	2.13
			1st	S	2	2.33
			2nd	C	2	3.29
Pb	CHS	4,5,6	1st	O_M	4	2.38
			2nd	C	2	3.26
Hg	HA		1st	O_L	1	2.02
			1st	S	1	2.38
			2nd	C	1	2.78
			2nd	S	1	2.93

Key: CHS soil humic substances extracted with Chelex resin; O_W oxygen atom in water; O_L ligand oxygen atom; O_M oxygen atoms in water and ligand; BD bond distance in Å.

Where results are given for more than one type of humic substance (HS), or pH, the bond distances are means of values covering small ranges.

with salt marsh sediment humic acid are shown in Fig. 8.5. The main differences between the two spectra are (a) the disappearance of the peak at 1221 cm^{-1}, which is consistent with the formation of Cu—OOC bonds, and is also seen in fulvic acid, and (b) the appearance of peaks at 1539, 1265 and 1230 cm^{-1}, due to Cu bonding to N atoms, which are not observed for fulvic acid. Alberts & Filip (1998) concluded that Cu may be bound preferentially by oxygen-containing functional groups in fulvic acid, while N-containing groups play a greater rôle in humic acid. They also concluded that there are significant differences in the electronic structures of the complexes with different metals, suggesting that different metals bind at different sites. In particular, Mg, Al, Ca, Mn(II), Fe(III), Ag(I) and Hg(II) appear to bind similarly, predominantly through carboxyl groups, but differently from Cu. Both Au(I) and Pb appear to bind through carboxyl and alcoholic moieties.

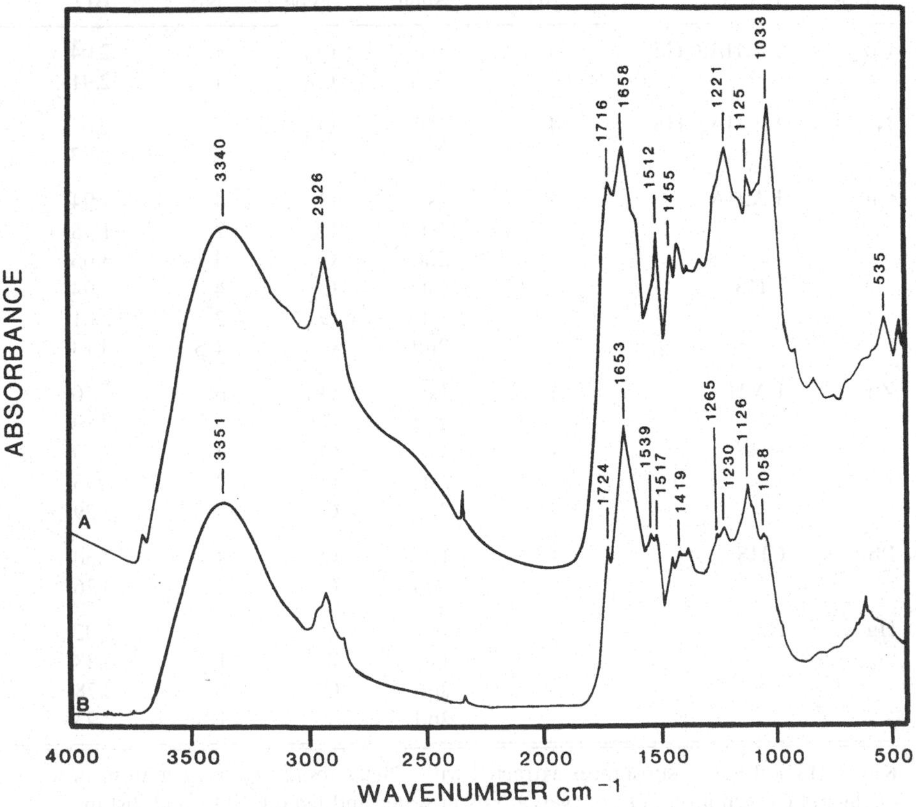

Figure 8.5. FTIR spectra of humic acids isolated from salt marsh sediments, either (A) protonated or (B) complexed with Cu(II). [Reproduced from Alberts & Filip (1998), with permission from Selper Ltd.]

8.3.6 Combined spectroscopic methods

Leenheer *et al.* (1998) used ^{13}C NMR, ^{1}H NMR and FTIR to characterise the molecular structure of a fraction of Suwannee River fulvic acid that had a high affinity for Ca^{2+}. The characterisation data suggested that carboxyl groups were clustered in short-chain aliphatic dibasic structures. A model structure for the Ca complex was proposed in which the metal was coordinated by five fulvic oxygen atoms, and a water molecule. An outer sphere complex was suggested for Cd^{2+}. The authors cautioned that the proposed structures might not be representative of all fulvic acids.

8.4 Viscometry

Viscometric measurements do not yield information about binding sites *per se*, but they can provide insight into the effects of cation binding on humic conformation. Thus, Ghosh & Schnitzer (1980) showed that humic molecules become more rigid as metal binding occurs, which they suggested is due to the bridging of neighbouring segments of the macromolecules.

8.5 Summary

The review of information about the qualitative aspects of metal–humic interactions presented above has not been critical. The findings, and the interpretations given by the researchers who obtained them, have simply been assembled. Although most of the results are consistent with what is known about the chemical properties of humic substances, conclusions drawn from the different studies are not fully in agreement. The inconsistencies may arise from different experimental conditions, different humic samples, or misinterpretations, or partial interpretations, of data obtained for these exceedingly complex and heterogeneous materials. It is difficult at present to establish a full picture of the natures of the interactions. However, some general conclusions can be drawn. On the basis of the results summarised here, and bearing in mind the quantitative binding data summarised in Chapter 7, the binding sites in humic matter for metal cations fall into at least three categories, as follows.

(1) *Non-specific sites*, binding at which is due to electrostatic attraction of the cations by negatively charged humic matter. There will be little selectivity here in terms of chemical differences among cations, but cations of higher charge will be preferentially bound. The number of sites will be equivalent to the net charge of the

humic matter, taking cation valence into account. Typically, the amounts bound will be of the order of 10^{-3} mol g^{-1}. In the natural environment, the sites will be occupied mainly by major cations such as Na^+, Mg^{2+}, K^+ and Ca^{2+}, and also H^+ and cationic Al species under acid conditions.

(2) *Abundant weak sites* formed by oxygen-containing ligands, chiefly carboxyl groups, also phenolic and other weak acid groups. Metal binding at these sites parallels binding to simple organic acids with oxygen functionalities. One or more ligand atoms may be involved in binding. For some cations, inner sphere species may form, for others outer sphere complexation may predominate. In the natural environment, cations occupying these sites to the greatest extents will be alkaline earth cations, cationic Al species, and protons, but a significant fraction will be unoccupied, conferring a net negative charge on the humic matter. Site heterogeneity occurs due to differences in chemical characteristics and denticity.

(3) *Stronger less abundant sites* formed by combinations of O, N and S. These sites could exhibit a wide range of affinities and relative abundances, and the category could be divided further. For example, there could be tridentate and tetradentate sites formed by O groups, and S-containing sites for Hg. The participation of N in metal binding is evidenced by several of the spectroscopic studies. These sites differentiate among metals more than the weaker abundant sites.

There is spectroscopic evidence for competition between metals, with suggestions that it is more complicated than expected for the straightforward sharing of binding sites. Some of the studies provide evidence for conformational changes associated with cation binding.

9

Parameterised models of cation–humic interactions

There have been many attempts to model mathematically the reactions of cations with humic substances. The models developed to date, nearly all of which have been equilibrium models, can be classified as parameterised and predictive. The former operate by data fitting, i.e. by using experimental data to obtain parameter values, while the latter use independent knowledge, or assumptions, to predict the interactions. Parameterised models predominate, with only a few attempts at a predictive approach (see Chapter 11). In this chapter, the concepts behind various parameterised models are presented, while Chapter 10 discusses applications of the most advanced models. Earlier reviews of parameterised models include those by Buffle (1984, 1988), Dzombak *et al.* (1986), Sposito (1986), and Stevenson (1994).

9.1 Overview and philosophy

9.1.1 Purposes of equilibrium modelling

In studies of well-defined ligand–proton–metal systems, modelling is often used to obtain information about the nature of the complexes. The underlying idea is that the correctly chosen stoichiometry will, when incorporated into an equilibrium model, lead to agreement with the observations. The Al–gallic acid–OH system was presented as an example in Section 5.2.4. Humic substances are much less amenable to such an analysis because stoichiometries are difficult to determine in these poorly characterised, heterogeneous systems, and clear-cut choices among possible reaction schemes are not easily made. Nonetheless, a model of cation binding by humic substances that is based on plausible assumptions, consistent with other knowledge about the materials, may be constructed and if it works can help to confirm the underlying assumptions, or at least not rule them out. If a researcher is trying to understand the properties of these compounds, a model provides a way of trying out the state of knowledge, and a means by which experiments can be designed. Thus, the first purpose of modelling is to formalise hypotheses. A second purpose is to encapsulate the substantial body of knowledge about cation–humic interactions, and thereby to permit its practical use, i.e. the prediction of reactions in the natural environment.

9.1.2 Mixture and quasiparticle models

Sposito (1986), reviewing models of metal ion binding by humic substances, distinguished two types of model, 'mixture' and 'quasiparticle'. The first is a conceptual mixture of known, low molecular weight, organic ligands that, acting together, have binding properties similar to those of humic substances. The second is based on the assumption that humic substances can be represented by a set of hypothetical particles, having binding properties that coincide with those of the real material. In principle, the mixture approach allows numerous measurements on real humic matter to be circumvented, and the ready incorporation of the complexation properties of small ligands into speciation programs. The quasiparticle approach has been adopted more widely, probably because it is potentially more realistic. It also offers the possibility of connecting ion binding to other processes such as aggregation and adsorption.

9.1.3 Opinions

A number of authors have commented on the process, aims and use of cation–humic modelling, as follows:

Perdue (1985)

> Given a fairly smooth, featureless (*acid–base*) titration curve, almost any mathematical model with several adjustable fitting parameters can be used to empirically fit the data, making it impossible to use goodness-of-fit to determine whether the mathematical model is also a sound chemical model. That judgement must be reached primarily by chemical intuition and secondly by goodness-of-fit.

De Wit (1992)

> In every quasi particle model several arbitrary assumptions have been made which depend on the purpose of the model and on the good taste and scientific background of the scientist. As a consequence all models are on the edge of science and fiction.

Milne *et al.* (1995)

> The combination of chemical heterogeneity and electrostatic effects means that a completely rigorous description of ion binding (*by humic substances*) in complex electrolytes is probably impractical.

Ephraim *et al.* (1986)

> It is quite apparent that only a scheme which accommodates the main sources of complication can lead to the development of a uniform physicochemical description of these systems.

Westall *et al.* (1995)

> The foremost goal is an accurate representation of the experimental data... The set of adjustable parameters should be small and orderly...

Buffle & Altmann (1987)

> For heterogeneous complexants, no *a priori* assumption can be made concerning the nature of the complexation reaction at each site since, owing to their very great structural complexity, any realistic guess would require measurement of too large a number of physicochemical parameters.... A rational normalization of (*the properties of heterogeneous ligands*) can be based on statistical representations, by applying the classical concept of a probability distribution to complexation reactions.

Kinniburgh *et al.* (1999)

> ...a fully mechanistic molecular approach to metal ion binding by humic/fulvic acids is likely to remain Utopian for the forseeable future...

Thus it can be appreciated that the topic has its controversies, its optimists and its pessimists, and therefore its opportunities. In the remainder of the chapter, various models are described and discussed. The aim here is neither to present a catalogue of all models, nor to put them in chronological order, but to use published models to highlight the different aspects that different authors have emphasised, and to discuss their validity and usefulness.

9.2 Models that describe the binding of a single cation

9.2.1 Formation of a 1:1 complex

The 1:1 model is based on the assumption that the binding of a cation can be expressed in terms of an equivalent mass of humic matter, with which a single binding site, or class of non-interacting sites, is associated. The approach has been applied only to the binding of metals, because proton-binding data invariably cover fairly large ranges of $[H^+]$ and humic charge, and so cannot be fitted by a simple one-site model. Binding is usually expressed by the equation

$$v = \frac{nK[M]}{1 + K[M]} \tag{9.1}$$

where n is the content of binding sites, in mol g(humic matter)$^{-1}$, and K is an (apparent) equilibrium constant, with units of dm^3 mol^{-1}. The equilibrium constant is widely referred to as the conditional stability constant, i.e. the constant that applies for the conditions (pH, ionic strength, concentrations of other cations, temperature etc.) of the experiments.

Equation (9.1) was derived in Chapter 5 from the equilibrium expression for the formation of a simple 1:1 complex. Langmuir derived an equation of the same form for gas adsorption, and the equation is often referred to as the Langmuir isotherm, especially in connection with adsorption at solid surfaces. Rearrangement of equation (9.1) gives

$$\frac{v}{[M]} = K(n - v) \tag{9.2}$$

which is known as the Scatchard (1949) equation. It is really no more than a device to permit a linear plot of binding data (v/[M] vs. v), and such a plot is referred to as a Scatchard plot. Figure 9.1 shows ideal plots of equation (9.1) in linear, semi-logarithmic and logarithmic forms, and of equation (9.2).

This model can fit data only if the ranges of v and [M] are small, since it cannot take into account the binding-site heterogeneity that is always present in humic samples. There are many examples in the literature where binding data have been fitted with equation (9.2), or by non-linear fitting of equation (9.1). The parameters (n and K) derived from such fitting are useful only as a means of summarising the data: they do not have much physical meaning. This is illustrated in Fig. 9.2 by the application of equation (9.2) to Cd–humic acid binding data published by Benedetti *et al.* (1995), which cover a wide range of v, from 10^{-6} to 10^{-3} mol g^{-1}. When v/[M] is plotted vs. v for the full data set, the strongly concave curve of Fig. 9.2(a) is obtained, clearly showing that the model is not applicable. Now imagine that the experimental data were more restricted, to ranges of v of one order of magnitude or less. Arbitrarily considering four segments of the data, we obtain the plots shown in panels (b)–(e) of Fig. 9.2. In each case the data provide a reasonable fit to the

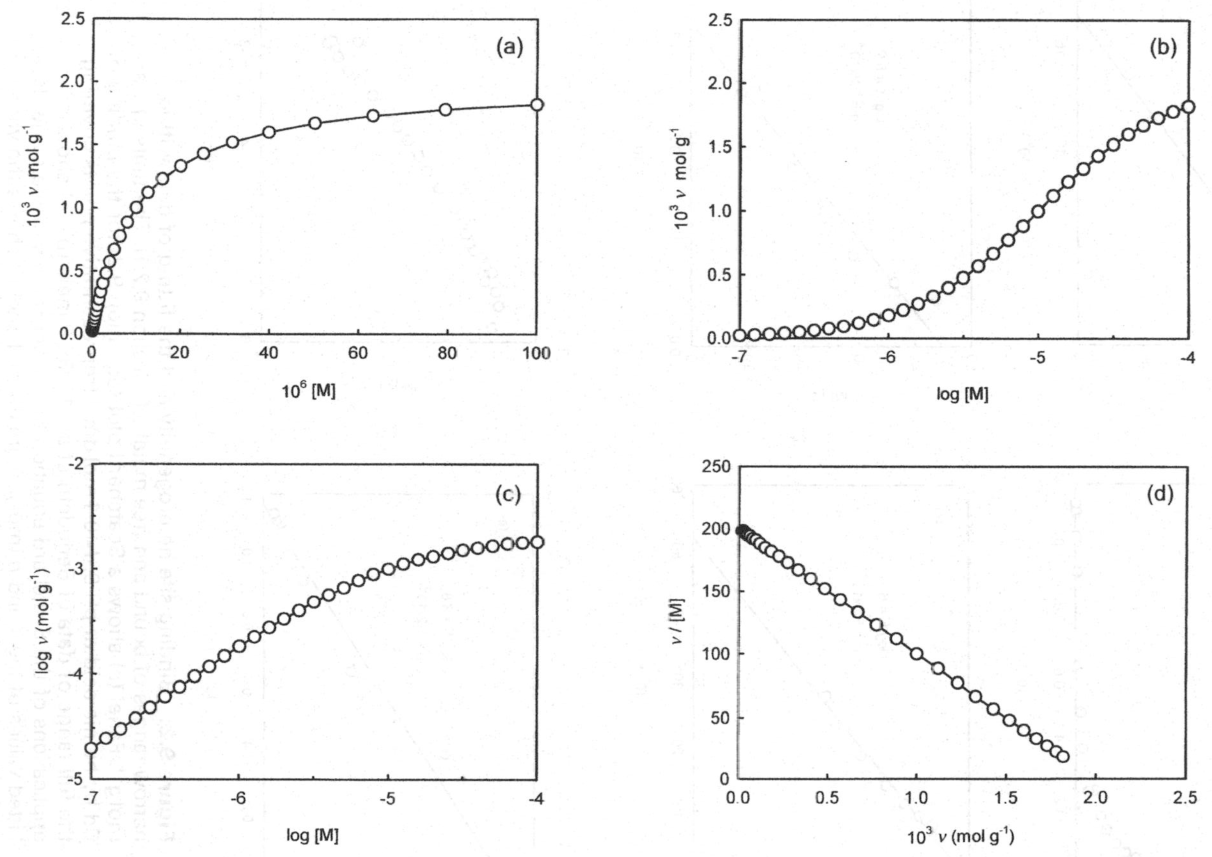

Figure 9.1. Idealised plots for the 'Langmuir' model: (a) linear, (b) log–linear, (c) log–log, (d) Scatchard plot. The example is for a hypothetical material with 2 mmol g^{-1} of identical sites, and an equilibrium constant of 10^5 dm^3 mol^{-1}. The units of [M] are mol dm^{-3}.

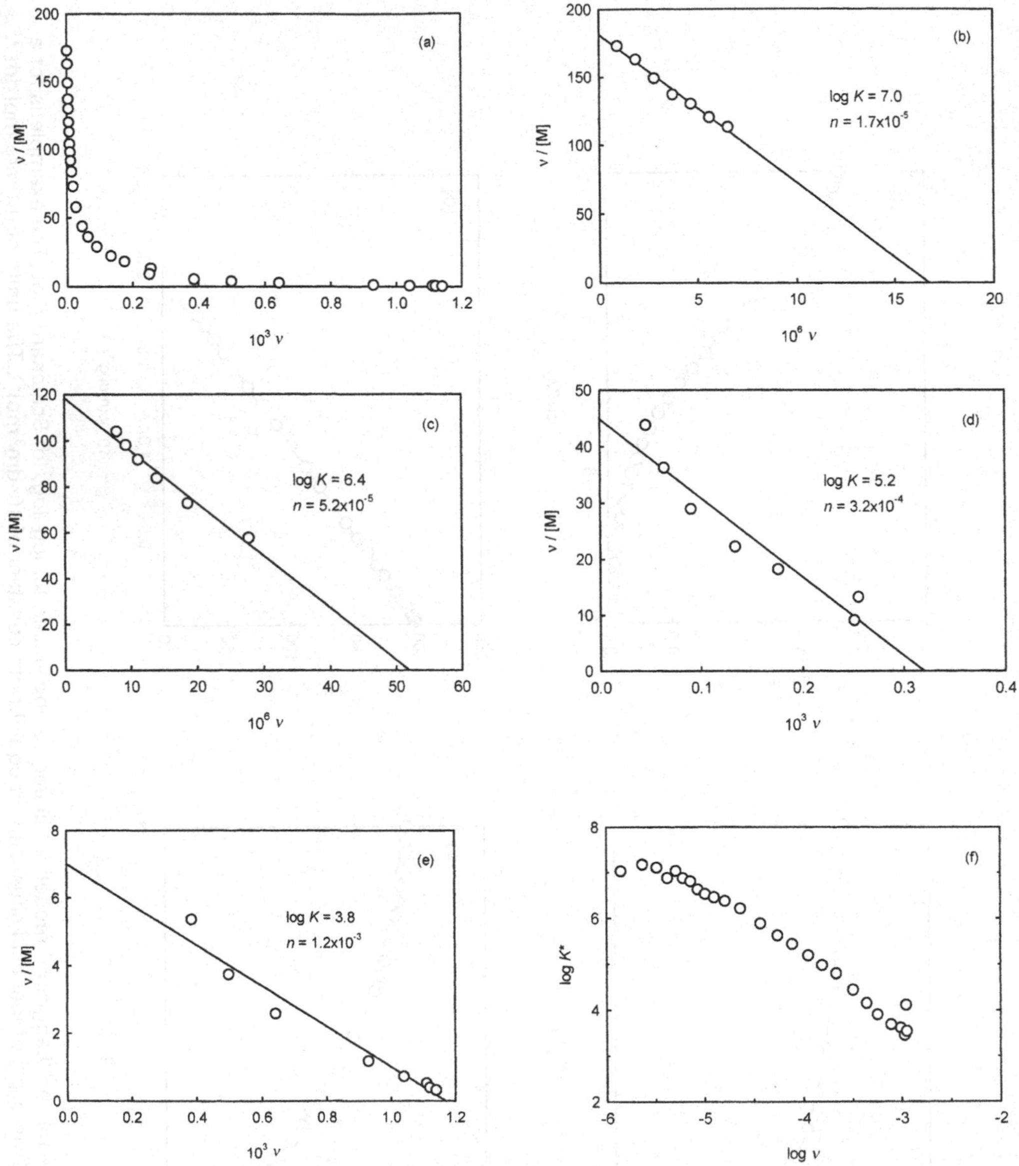

Figure 9.2. Binding site heterogeneity and the fitting of data within narrow ranges of bound and free metal (cf. Section 9.2.1). The units of ν are $mol\,g^{-1}$. Panel (a) shows a Scatchard plot (equation 9.2) for the binding of Cd by humic acid at pH 6, $I = 0.1\,mol\,dm^{-3}$ (representative points covering the full range of data of Benedetti *et al.*, 1995). Panels (b)–(e) show applications of the Scatchard equation to smaller ranges of the same data. Fitted values of log K and n ($mol\,g^{-1}$) are given. Panel (f) shows the variation of the 'local' equilibrium constant (K*) with ν.

equation, with r^2 values in the range 0.92 to 0.99, and so values of n and K can be derived. However, they provide no more than a partial description of the binding properties of the sample, being characteristic only of the range of binding data considered. Thus K is also conditional on the ranges of v and [M] to which the measurements apply. Another way to view the data is to compute values of the 'local' equilibrium constant (K^*), by taking successive pairs of values of v/[M] and v. Figure 9.2(f) shows how K^* progressively falls, as v increases over some three orders of magnitude. A primary aim of more comprehensive models is to account for such variation.

In principle, the 1:1 model could be extended to include competition and ionic strength effects, but the evident limitations of attempting to use a single site to represent humic ligands mean that there is little or no point in doing so.

9.2.2 Multiple discrete sites

The inapplicability of the one-site model demonstrated in the previous section leads logically to models with multiple sites. The commonest format is that of equation (4.23) for proton dissociation or (5.37) for metal ion binding which are extended forms of equation (9.1). The ligand molecule is assumed to carry more than one class of sites, and the 'crossover' terms that arise for a poly-ligand (Section 4.5) are ignored. For proton dissociation we have

$$r = \frac{n_1 K_1/[\mathrm{H}^+]}{1 + (K_1/[\mathrm{H}^+])} + \frac{n_2 K_2/[\mathrm{H}^+]}{1 + (K_2/[\mathrm{H}^+])} + \frac{n_3 K_3/[\mathrm{H}^+]}{1 + (K_3/[\mathrm{H}^+])} + \cdots \quad (9.3)$$

where r is the moles of protons dissociated per gram of humic matter, and n_1, n_2, n_3 etc. are the numbers of sites in each class, per gram of humic matter. The equivalent expression for metal binding, at a given pH, is

$$v = \frac{n_1 K_1/[\mathrm{M}]}{1 + K_1[\mathrm{M}]} + \frac{n_2 K_2[\mathrm{M}]}{1 + K_2[\mathrm{M}]} + \frac{n_3 K_3[\mathrm{M}]}{1 + K_3[\mathrm{M}]} + \cdots \quad (9.4)$$

Sposito (1986) called this the 'Scatchard model' (cf. Section 9.2.1).

An example of the model's application is its use by Paxeus & Wedborg (1985) to describe proton dissociation from fulvic acid, in which they found it necessary to assume six classes of sites, each with a different n and K. Applications to metal binding at constant pH were made by McKnight *et al.* (1983), using two classes of sites, and by Fish *et al.* (1986), using three classes but over a wider range of v and [M] (Fig. 9.3). It should be recognised that a successful fit assuming some number of site classes does not constitute evidence for that number of classes. It simply means that the site heterogeneity can be accounted for within a given range, to some acceptable precision.

9.2.3 Empirical isotherms

The two models described so far have their roots in simple chemical equilibria. Completely empirical equations can also be used to describe binding data. The simplest empirical approach is the Freundlich isotherm which, like the Langmuir isotherm (equation 9.1), was developed to describe adsorption to solid surfaces. The Freundlich equation is

$$\nu = K[\mathrm{M}]^n \tag{9.5}$$

where K and n are empirical constants. The value of n lies between 0 and 1 for 'normal' heterogeneous complexants. Figure 9.4 compares the shapes

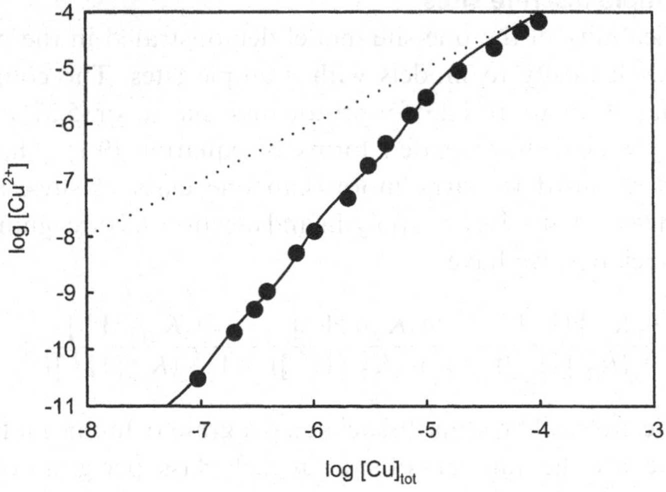

Figure 9.3. The use of a three-site model (see Section 9.2.2) to fit data for Cu binding by fulvic acid (~10 mg dm^{-3}). The points are experimental data, the solid line is the model simulation, and the dotted line indicates the expected result for no binding. The model parameters were three site concentrations ($[L]_1$–$[L]_3$) and three equilibrium constants (K_1–K_3), as follows:

i	1	2	3
$[L]_i$	1.12×10^{-7}	6.76×10^{-7}	6.76×10^{-6}
$\log K_i$	10.83	8.80	6.48

[Redrawn from Fish, W., Dzombak, D.A. & Morel, F.M.M. (1986), Metal–humate interactions. 2. Application and comparison of models, *Environ. Sci. Technol.* **20**, 676–683. Copyright 1986 American Chemical Society.]

of the Freundlich and Langmuir isotherms. One significant feature of the Freundlich isotherm is that a plot of log v vs. log [M] is a straight line with slope n. Another is the lack of a maximum value of v. A number of other empirical isotherms have been devised (see e.g. Buffle, 1988). Until recently, there have been few if any direct applications of empirical isotherms to cation binding by humic substances. However, they have come to the fore in the development of the NICA and NICCA models (Section 9.7).

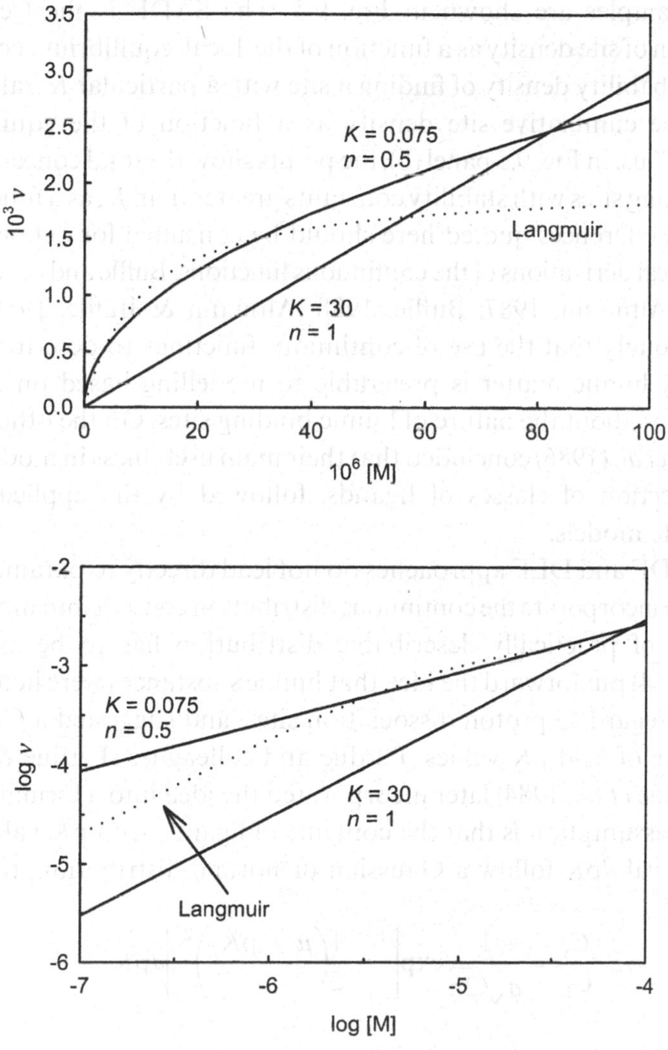

Figure 9.4. Plots of the Freundlich model (equation 9.5), with parameter values of K and n indicated. The Langmuir model plots, shown for comparison, are the same as in Fig. 9.1.

9.2.4 Continuous distribution models

A number of authors have concluded that the heterogeneity of humic substances is sufficiently great for there to be essentially a continuum of binding strengths, and therefore of equilibrium constants. Methods have been devised for transforming experimental data, principally for metals, to obtain insight into the binding strength distributions. Altmann & Buffle (1988) classified them as the Site Affinity Distribution Function (SADF), introduced to humic science by Shuman *et al.* (1983), and the Differential Equilibrium Function (DEF), introduced by Gamble *et al.* (1980). Examples are shown in Fig. 9.5. The SADF is the frequency distribution of site density as a function of the 'local' equilibrium constant, i.e. the probability density of finding a site with a particular K value. The DEF is the cumulative site density as a function of the equilibrium constant. Thus, in Fig. 9.5 panel (b), the points show the total concentration of complexing sites with stability constants greater than K, as a function of $\log K$. The references quoted here should be consulted for details of the mathematical derivations of the continuous functions. Buffle and co-workers (Buffle & Altmann, 1987; Buffle, 1988; Altmann & Buffle, 1988) have argued strongly that the use of continuous functions to describe cation binding by humic matter is preferable to modelling based on *a priori* assumptions about the nature of humic binding sites. On the other hand, Dzombak *et al.* (1986) concluded that their main usefulness in modelling is in the selection of classes of ligands, followed by the application of discrete-site models.

The SADF and DEF approaches do not lead directly to parameterised models. To incorporate the continuous distribution concept into modelling, some sort of practically describable distribution has to be assumed. Posner (1964) put forward the idea that humic substances were heterogeneous with regard to proton dissociation sites, and suggested a Gaussian distribution of acid pK values. Perdue and colleagues (Perdue & Lytle, 1983; Perdue *et al.*, 1984) later incorporated the idea into a formal model. The basic assumption is that the contents of ligands with pK values in a given interval dpK follow a Gaussian or normal distribution, thus

$$\frac{C_i}{C_\mathrm{L}} = \frac{1}{\sigma\sqrt{2\pi}}\exp\left[-\frac{1}{2}\left(\frac{\mu - \mathrm{p}K_i}{\sigma}\right)^2\right]d\mathrm{p}K \qquad (9.6)$$

where C_i and pK_i are respectively the content of sites and the average pK, in the interval dpK, C_L is the total site content, and σ is the standard deviation of the distribution of pK_i values around the mean value of μ.

The degree of proton dissociation is given by

$$r = \frac{1}{(\sigma\sqrt{2\pi})}\int_a^b \frac{K}{K + [H^+]} \exp\left[-\frac{1}{2}\left(\frac{\mu - pK}{\sigma}\right)^2\right] dpK \quad (9.7)$$

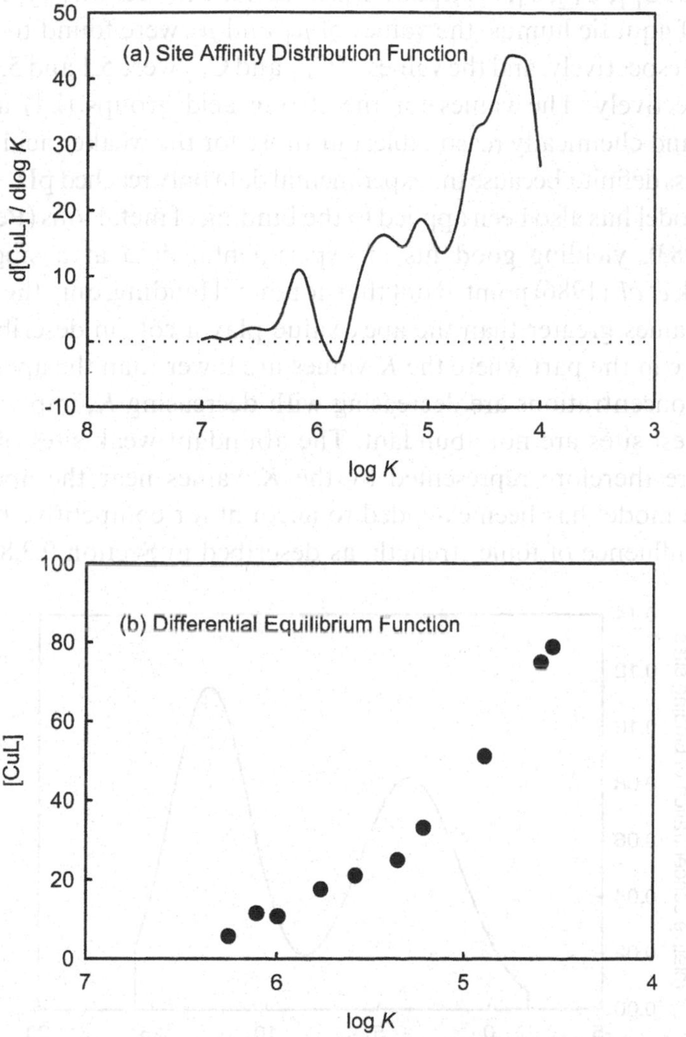

Figure 9.5. Distributional models describing Cu binding by organic matter. Panel (a) refers to aquatic organic matter, panel (b) to fulvic acid. See Section 9.2.4. [Redrawn from Altmann, R.S. & Buffle, J. (1988), The use of differential equilibrium functions for interpretation of metal binding in complex ligand systems: its relation to site occupation and site affinity distributions, *Geochim. Cosmochim. Acta* **52**, 1505–1519, with permission from Elsevier Science. Panel (a) is from Shuman *et al.* (1983), panel (b) from Buffle, J. (1988), *Complexation Reactions in Aquatic Systems: An Analytical Approach*, with permission from Ellis-Horwood Limited, Chichester.]

The distribution, which is infinite in both directions, is truncated for practical application by setting the limits of integration, *a* and *b*, to finite values (Fig. 9.6). For humic substances, two such distributions are required, one characterising carboxylic acid groups, and the other the weaker acid (e.g. phenolic) groups. Thus there are six parameters to be optimised (C_{L1}, C_{L2}, σ_1, σ_2, μ_1 and μ_2). When the model was applied to a sample of aquatic humus, the values of μ_1 and μ_2 were found to be 3.66 and 12.5 respectively, and the values of C_{L1} and C_{L2} were 5.1 and 5.3 mmol g^{-1} respectively. The values for the strong acid groups (L1) are well defined, and chemically reasonable, but those for the weaker acid groups (L2) are less definite, because the experimental data only reached pH $\sim$ 10.75.

The model has also been applied to the binding of metal ions (Perdue & Lytle, 1983), yielding good fits of experimental data at a single pH. Dzombak *et al.* (1986) pointed out that, for metal binding, only the ligands with *K* values greater than the apex value play a rôle in describing the data, since in the part where the *K* values are lower than the apex value, the site concentrations are decreasing with decreasing *K*, implying that the weakest sites are not abundant. The abundant weak sites of humic matter are therefore represented by the *K* values near the apex. The Gaussian model has been extended to account for competitive binding, and the influence of ionic strength, as described in Section 9.3.8.

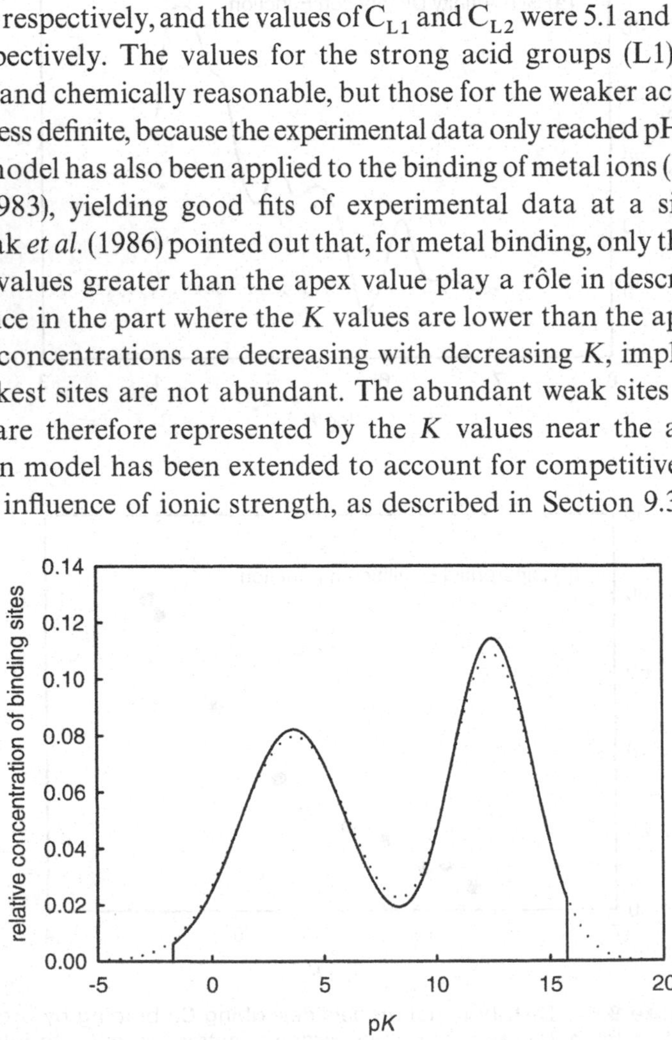

Figure 9.6. Relative concentrations of proton binding sites in aquatic humus as a function of p*K*, according to the Gaussian model. The dotted line shows the unconstrained model, the full line the truncated version, calculated with values of *a* and *b* (equation 9.7) of −1.74 and 15.74 respectively. [Redrawn from Perdue, E.M., Reuter, J.H. & Parrish, R.S. (1984), A statistical model of proton binding by humus, *Geochim. Cosmochim. Acta* **49**, 1257–1263, with permission from Elsevier Science.]

9.2.5 Electrostatic models for proton dissociation

Dempsey & O'Melia (1983) applied a simple electrostatic model to their acid–base titration data for fulvic acid in the acid pH range (up to about pH 7), by means of the equation (rewritten using p notation)

$$pK = pK_{int} - 0.868wr \tag{9.8}$$

Here, K is the apparent equilibrium constant, r the number of dissociated carboxylic acid groups per molecule of fulvic acid, w the electrostatic interaction term (cf. Section 4.6), and K_{int} the intrinsic equilibrium constant for the carboxylic acid groups, which are all assumed to be identical. Thus K_{int} is the value of K for $r = 0$. The experimental data for four fulvic acid fractions yielded linear plots of pK vs. r, with pK_{int} values in the range 1.4–2.0. Dempsey & O'Melia showed that the derived values of w were of the correct magnitude, based on calculations using the Hermans–Overbeek theory for flexible linear polyelectrolytes (Tanford, 1961). However, the dependence of w on ionic strength was smaller than expected. Bearing in mind the expected high degree of chemical heterogeneity in humic carboxyl groups, the authors considered the good fits of the model to be 'remarkable' and the results to be 'more thought-provoking than useful'. In the light of later work (see below), it can be appreciated that the pK_{int} values are too low, and that the good fits were achieved with unreasonably high values of w.

Perdue (1985) pointed out the chemical unreasonableness of 'intrinsic site' models, such as that of Dempsey & O'Melia (1983), which ascribe variations in the apparent acid strength of humic matter solely to overall electrostatic effects, assuming a single class of binding sites. Nonetheless, this work made a significant contribution by highlighting electrostatic effects in humic matter. See Section 9.5 for further consideration of electrostatic effects.

9.3 Simpler models that include competition

The term 'simpler' is somewhat arbitrary here. Most of the examples are attempting to deal with the effect of pH on metal binding, although in principle they can be extended to deal with competition by other metal cations as well as protons.

9.3.1 Mixture models

Sposito (1981, 1986) discussed the use of models in which the binding of protons and metals by humic matter is simply represented by a collection of low molecular weight compounds, the coordination chemistry of which is already known. An example of a collection of simple ligands that could account for proton- and metal-binding data obtained from experiments with a fulvic acid sample is shown in Table 9.1. Morel *et al.* (1975) had previously employed a slightly simpler mixture model (six ligands) to estimate trace metal speciation in a wastewater discharge. According to Sposito (1986), 'numerical simulations ... with mixture models will only have a limited, *ad hoc* relevance to trace metal–humate associations and ... generalized conceptual schemes based on the models should be viewed as only suggestive, at best, in their application to natural waters...'. In particular, he regarded the inability of such models to take into account the polyelectrolyte behaviour (i.e. electrostatic interactions) and conformational lability of humic matter to be serious limitations. Mixture models can of course be used to fit experimental data, but, other than possible computational simplicity, they offer no advantage over models in which binding site densities and equilibrium constants are optimised.

9.3.2 Formation of 1:1 and 1:2 complexes

Models have been formulated based on the assumption that the metal and protons compete at a single site, and that the metal can form both 1:1 and 1:2 complexes. In the model of Buffle *et al.* (1977), the reactions are written

$$M + LH_x = ML + xH^+ \tag{9.9}$$

$$M + 2LH_x = ML_2 + 2xH^+ \tag{9.10}$$

where the value of x may be non-integral. The model was applied to binding data obtained for natural water samples, by optimising the

Table 9.1. Simple ligands making up the mixture model of Mattigod & Sposito (1979).

Ligand	% of total C
Benzenesulphonic acid	9.6
Citric acid	13.0
Maleic acid	19.2
Phthalic acid	19.2
Salicylic acid	9.6
Arginine	5.8
Lysine	7.8
Ornithine	7.8
Valine	7.8

equilibrium constants for reactions (9.9) and (9.10), together with x and the molecular weight of the dissolved organic matter (i.e. of L). The values of x for Cu and Pb were 0.6–0.7 in the pH range 3–7. The optimised molecular weights for different samples were between 500 and 2100, agreeing quite well with values determined by vapour pressure osmometry.

Stevenson (1976) carried out acid–base titrations of humic acids in the presence and absence of metal ions, and analysed the data using concepts from polyelectrolyte and coordination chemistry. Thus he applied the Henderson–Hasselbalch equation to describe the relationship between pH and the degree of proton dissociation of the humic acids (α)

$$pK_{app} = pH - n \log_{10}\left(\frac{\alpha}{1-\alpha}\right) \tag{9.11}$$

Here, K_{app} is the apparent equilibrium constant. Stevenson determined values of K_{app} and n for different ionic strengths, and used them to compute α under different conditions, thereby obtaining estimates of the concentrations of ionised and protonated sites at a given pH. He postulated the existence of MA and MA_2 complexes, formed by the reactions

$$HA + M^{2+} = MA^+ + H^+ \tag{9.12}$$

$$HA + MA^+ = MA_2 + H^+ \tag{9.13}$$

Equilibrium constants for reactions (9.12) and (9.13) were calculated from acid–base titration data obtained for solutions of humic acid with different amounts of added metal (Cu, Cd, Pb). The approach relies on the assumption that protons are displaced from the humic matter by the bound metal ions, which gives rise to a different dependence of pH on added base than in the absence of the metal (Section 6.2.4). Concentrations of MA^+ and MA_2, and thence the equilibrium constants, are obtained by solving the mass and charge balance equations.

Torres & Choppin (1984) introduced a similar model to analyse the binding of Eu and Am by humic acid at trace levels. Again, account is taken of the variation of the apparent proton dissociation constant with the degree of ionisation and the reactions are formulated in terms of MA and MA_2 complexes. Because trace levels of metal are involved, the metal binding has negligible effect on the degree of dissociation.

The formulations of binding sites in this model are unrealistic, firstly because the site contents of humic substances, determined on a mol g^{-1} basis, are too high for each molecule to bear only a single site, as is required by all three formulations. For example, values of ν for metal binding exceeding 10^{-3} mol g^{-1} are commonly achieved, without approaching saturation of sites (Chapter 7). It follows that, even for a molecular weight of 1000 (typical for fulvic acids; Section 2.4.3), the number of sites per molecule must be appreciably greater than one. A humic acid of molecular weight 10 000 must have tens of sites per molecule. Secondly, the size of the molecules, and their significant negative charge, make it difficult to accept that 1:2 (metal:ligand) complexes can readily form, because of steric interference and electrostatic repulsion. Buffle *et al.* (1977) recognised this difficulty, but suggested that fulvic acids were small enough for 1:2 complexes to be possible. Recall that if 1:2 complexes can indeed form, then metal binding will depend upon the concentration of humic matter (Section 5.2.2), which means that the inclusion of such complexes in a model, as a way to generate site heterogeneity, has wider implications.

9.3.3 *N*-site model

Cabaniss & Shuman (1988a) developed a model to describe Cu binding by fulvic acids as a function of [Cu^{2+}] and pH. Five different binding sites were postulated, each of which was characterised by three parameters, an integral proton dependence of Cu binding, a site density and an equilibrium constant. The fitted parameter values are shown in Table 9.2. Ionic strength effects were taken into account by assuming a fixed charge on the fulvic acids and applying the Guntelberg equation (Section 3.6.2) to calculate activity coefficients; an optimised charge of -5 eq mol^{-1} at pH 7 provided good simulations, but results at other pH values were relatively poor. The *N*-site model provided good fits to experimental data for Cu binding, and was also useful predictively (Cabaniss & Shuman, 1988b). While it can describe metal binding as a function of pH, it does not describe proton binding explicitly, and so does not fully take cation–humic interactions into account.

9.3.4 The 'multiple equilibria' model

Wilson & Kinney (1977) devised a model for proton and metal binding based on the analysis by Tanford (1961) of multiple equilibria in proteins. The model assumes a single class of carboxylic acid sites and a single class of phenolic OH sites, and includes a description of electrostatic interactions based on the Debye–Hückel model (cf. Section 4.6). In the model application, Wilson & Kinney assumed that metals competed with protons in the formation of 1:1 complexes at the carboxylic sites, thus achieving a description of metal and proton binding in the acid pH range.

Much later, Tipping and coworkers employed essentially the same model structure (see Section 9.6), but with a more elaborate picture of the binding sites.

9.3.5 Triprotic acid model

Schecher & Driscoll (1995) described a simple triprotic acid model, designed to help with the interpretation of acid–base and aluminium chemistry in acid waters containing dissolved organic matter. The model consists of the following equilibria

$$H_3A = H^+ + H_2A^- \tag{9.14}$$

$$H_2A^- = H^+ + HA^{2-} \tag{9.15}$$

$$HA^{2-} = H^+ + A^{3-} \tag{9.16}$$

$$Al^{3+} + A^{3-} = AlA \tag{9.17}$$

$$Al^{3+} + H^+ + A^{3-} = Al(H)A^+ \tag{9.18}$$

Equilibrium constants have been fitted from field data, but the model would be equally applicable to laboratory data. The structure of the model would not allow a good representation of metal binding over wide ranges of v and [M], being limited in its ability to cope with binding site heterogeneity. However, it has proved valuable in its intended use (see e.g. Driscoll *et al.*, 1994).

9.3.6 Discrete log *K* spectrum model

Westall *et al.* (1995) formulated a model in which proton binding is represented by four acid sites with fixed pK values, the abundance of each site being an adjustable parameter. The equations governing the dissociation of protons at different ionic strengths are

Table 9.2. Reactions and optimised parameter values for the *N*-site model of Cabaniss & Shuman (1988a). The equilibrium constants refer to the general reaction $Cu^{2+} + LH_x = CuL + xH^+$.

Site no.	Proton dependence (x)	Site density (mol gC^{-1})	log K
1	0	5.0×10^{-3}	3.90
2	1	1.9×10^{-4}	1.494
3	1	1.1×10^{-3}	– 0.364
4	2	1.5×10^{-4}	– 7.483
5	2	9.6×10^{-3}	– 10.05

$$HL_i = L_i^- + H^+ \qquad K_a(i) = \frac{[L_i^-][H^+]}{[HL_i]} \tag{9.19}$$

$$HL_i + Na^+ = NaL_i + H^+ \qquad {}^*K_{Na}(i) = \frac{[NaL_i][H^+]}{[HL_i][Na^+]} \tag{9.20}$$

The novelty of this approach is to include Na^+/H^+ exchange as a process occurring at specific sites, rather than to describe the variation in K_a with salt concentration as a non-specific phenomenon (see e.g. Section 4.6, also Section 9.5). In the fitting exercise, a single equilibrium constant for Na^+ binding could be used, for the reaction

$$L_i^- + Na^+ = NaL_i \qquad K_{Na} = \frac{[NaL_i]}{[L_i^-][Na^+]} \tag{9.21}$$

The model was fitted to acid–base titration data for a humic acid, at two ionic strengths, and found to give satisfactory fits. It was then used to fit binding data obtained for Co^{2+}, by finding the best values of the equilibrium constants for the four reactions

$$L_i^- + Co^{2+} = CoL_i^+ \qquad K_{Co}(i) = \frac{[CoL_i^+]}{[L_i^-][Co^{2+}]} \tag{9.22}$$

Again, satisfactory fits to the experimental data were obtained. Table 9.3 summarises the parameter values. The derived equilibrium constants are not realistic on the basis of comparison with simple ligands. For example the complexation of Na^+ by a carboxyl group would have a value of $\log K$ in the range 0–1, and for Co^{2+} the range would be 1–2.

In effect, the model represents humic matter in terms of four monoprotic ligands, which can bind protons and metal cations, including monovalent cations like Na^+. For protons, the parameter values are quite reasonable, but the equilibrium constants for Na^+ and Co^{2+} are appreciably greater

Table 9.3. Parameter values for the discrete log K model of Westall *et al.* (1995). The values of log K_a were fixed, the other values were optimised by fitting experimental data.

Site	$\log K_a$	$\log K_{Na}$	$\log K_{Co}$	T_{HL} (mmol g^{-1} HA)
1	− 4	1.71	—	2.9
2	− 6	1.71	5.38	1.3
3	− 8	1.71	6.38	0.9
4	− 10	1.71	—	1.2

than would be expected for simple monoprotic ligands. As the authors themselves point out, the values for Co are similar to those for bidentate complexes. Thus the success of the model in fitting data does not mean that the underlying assumptions are correct. However, Westall *et al.* (1995) argue that their approach is as physically correct as many more complex models. In particular they do not favour the use of an electrostatic sub-model, because it 'would require information on size, shape and charge distribution'. They contend that their structured approach is well suited to practical environmental applications.

Robertson & Leckie (1999) have shown that the discrete log K spectrum model as presently formulated is unable to fit data covering wide ranges of ν and [M], because there are insufficient different classes of binding sites to account for the observed site heterogeneity.

9.3.7 The Charge Neutralisation Model

This model, developed by Kim and coworkers (e.g. Buckau *et al.*, 1992; Czerwinski *et al.*, 1994) is used to interpret experimental data for the binding of actinides by humic substances. It may also be used to evaluate methods for nuclear waste management (Kim & Czerwinski, 1996; Choppin & Labonne-Wall, 1997). The first premise of the model is that the number of humic ligand groups making up a binding site is equal to the charge of the cation. The second is that the concentration of ligand groups available for cation binding depends upon the degree of proton dissociation. Adapting the equations presented for the interaction of Am^{3+} with a fulvic acid (Buckau *et al.*, 1992) to the general case, we have

$$\beta = \frac{[\mathrm{MFA}(z)]}{[\mathrm{M}^{z+}][\mathrm{FA}(z)]_\mathrm{f}} \tag{9.23}$$

Here, square brackets indicate concentrations in mol dm^{-3}, $[\mathrm{M}^{z+}]$ is the concentration of metal, $[\mathrm{FA}(z)]_\mathrm{f}$ is the concentration of free sites on the fulvic acid, with denticity z, $[\mathrm{MFA}(z)]$ is the concentration of fulvic-bound metal and β is the complexation constant. The loading capacity for the metal, LC, is defined as

$$LC = \frac{z[\mathrm{MFA}(z)]^*}{C_\mathrm{H}} \tag{9.24}$$

where z is the charge on the cation (three in this case), $[\mathrm{MFA}(z)]^*$ is the maximum concentration of metal ion that can be bound at the pH in question, and C_H (mol dm^{-3}) is the proton exchange capacity, i.e. the total concentration of proton-dissociating sites. The concentration of free fulvic sites for Am^{3+} is then given by

$$[FA(z)]_f = [FA(z)]_t LC - [MFA(z)] \qquad (9.25)$$

where $[FA(z)]_t$ is the total concentration of metal-binding sites, i.e. C_H/z. Equations (9.23)–(9.25) can be manipulated to yield the following expression for the concentration of bound metal

$$[MFA(z)] = \frac{[FA(z)]_t LC\,\beta[M^{z+}]}{1 + \beta[M^{z+}]} \qquad (9.26)$$

This equation is the same as equation (9.1) with $n = [FA(z)]_t\ LC$, and $K = \beta$. Thus the model is equivalent to assuming that binding takes place by the formation of a 1:1 complex (as in Section 9.2.1), but with the capacity defined by a combination of the content of proton-dissociating groups and the loading capacity. The variation of LC with pH then provides the pH dependence of binding. The Charge Neutralisation Model can be criticised on the following grounds:

(1) The assumption that all the binding of Am^{3+} takes place at tridentate sites is unrealistic. It may be possible at low loadings, but the model assumes that the sites could fill to full capacity at higher pH values (where LC approaches 1). It is very difficult to envisage the proton-dissociating sites of humic matter being able to arrange themselves into so many tridentate sites.

(2) The development of the model does not follow from the Law of Mass Action. Referring to Section 5.2.1 it can be seen that the effect of pH on the binding of a metal to a ligand is effectively to modify the equilibrium constant, but not the capacity. Thus, if the concentration of metal ion is made large enough, the Law of Mass Action will predict that all the protons can be displaced, so that LC will not depend upon pH, but β will. By writing equations (9.25) and (9.26), the authors are assuming that metal added to a humic substance cannot displace protons from their binding sites, i.e. it can bind only to sites already free of protons, the proton-bound sites staying as they are. Clearly, if two cations are competing for the same ligand, then in principle enough of one (e.g. Am^{3+}) can be added to displace nearly all of the other (H^+).

(3) For the reasons explained in Section 9.2.1, the model can only apply to a narrow range of free concentrations and amounts bound, because it is based on a single class of binding sites. Thus, it

does not account for site heterogeneity. To date, parameter values have been derived from data at high and narrow ranges of metal loading, which cannot reliably be extrapolated to estimate the very low loadings which, we hope, apply to actinide–humic interactions in the environment.

9.3.8 The Competitive Gaussian Model

The Gaussian model of Perdue and colleagues (Section 9.2.4) applies to the binding of a single cation. It was extended by Carreira and colleagues (Dobbs *et al.*, 1989; Susetyo *et al.*, 1990, 1991) to describe competition among cations for binding. In the non-competitive version for protons, a class of binding sites was described with the parameters μ_H (mean $\log K$ value), the standard deviation of $\log K$ values (σ_H), and the total concentration of proton-binding sites (C_H). In the Competitive Gaussian Model, the same sites are assumed to bind metals, so that $C_M = C_H = C_L$. Also, $\sigma_M = \sigma_H = \sigma$ (width of the Gaussian distribution). Binding is assumed to be monodentate for metals, and the ratio of K_M to K_H (or the difference between μ_M and μ_H) is taken to be the same for the whole site distribution. In other words, the more strongly a given site binds protons, the more strongly it binds any metal cation. The model described thus far is equivalent to a collection of monoprotic ligands, each able to bind protons and metal cations, with variations in $\log K$ described by a single Gaussian distribution for each class of sites. An additional feature is the application of the Davies equation to compute activity coefficients, and thereby take into account the effects of ionic strength.

The model has been used successfully to fit data for the binding of Eu at high metal loadings, at pH 6 or below, obtained by fluorescence spectroscopy (Dobbs *et al.*, 1989; Susetyo *et al.*, 1990, 1991). Only the carboxyl groups were considered in the modelling exercise. The optimal value of σ was 1.7, and a humic charge of $-2.8\ \mathrm{eq\ mol^{-1}}$ was required to simulate the ionic strength dependence of binding. Values of μ_M for other metals were estimated from the extents of their competitive effects towards Eu. The full set of μ values is shown in Table 9.4. The parameter set was the first to cover a wide range of metals, obtained by fitting experimental data to a unifying model. Perdue & Carreira (1997) found that the model cannot fit Cu data that cover wider ranges of pH, v and $[Cu^{2+}]$ than those for Eu, probably because it does not take into account weaker acid sites.

9.4 The site heterogeneity/polyelectrolyte models of Marinsky and colleagues

In 1986, Marinsky and his colleagues (Marinsky & Ephraim, 1986; Ephraim & Marinsky, 1986; Ephraim *et al.*, 1986) published a series of papers in which they described for the first time how binding site heterogeneity, electrostatic ('polyelectrolyte') effects, and competition among cations could all be taken into account in describing cation–humic interactions. One could reasonably argue that such knowledge had been available for some time prior to 1986 (see above). It can also fairly be said that the ideas had been developed many years earlier in connection with proton- and metal-binding sites in proteins (Tanford, 1961). However, they had not been stated so clearly in connection with humic substances. Moreover, the papers were published at a time when there was growing interest in the development of models to compute cation–humic binding in various environmental situations. Thus the work of Marinsky had a major influence.

Marinsky and coworkers based their analysis of the polyelectrolyte properties of humic substances on experience with well-defined synthetic repeating polymers. Two possibilities were suggested. Firstly, the humic molecules were considered to be gels, with acid dissociating groups dispersed throughout their interior, and counterions able to penetrate the entire volume. Thus the humic molecule is regarded as a separate phase,

Table 9.4. Values of μ_H or μ_M for the Competitive Gaussian Model, applied to Suwannee River dissolved organic matter, from Susetyo *et al.* (1991).

Cation	μ_H or μ_M
H^+	3.87
Be^{2+}	3.5
Mg^{2+}	< 2.0
Al^{3+}	5.2
Ca^{2+}	2.9
Cr^{3+}	5.6
Fe^{3+}	7.7
$FeOH^{2+}$	5.5
Ni^{2+}	3.3
Cu^{2+}	4.9
Zn^{2+}	3.5
Cd^{2+}	3.3
Ba^{2+}	3.1
Eu^{3+}	6.4
Pb^{2+}	5.2

the zone surrounding it making no contribution to its interactions with species in the bulk solution (Donnan model: Section 4.6.7). The alternative model was a salt (and proton) impermeable molecule, with counterions being attracted to (and co-ions repelled from) the molecular surface according to Boltzmann statistics (Section 4.6).

Ephraim *et al.* (1986) showed that if the ionic strength is made high enough (≥ 1 mol dm^{-3}), the electrostatic influence on proton dissociation from fulvic acids becomes negligible, so that the remaining variation in apparent pK (equation 4.50) is due only to binding site heterogeneity. They interpreted the experimental results for proton dissociation in the absence and presence of metal cations (Cu^{2+}, Eu^{3+}) in terms of known or probable entities in the fulvic samples, and by comparison with information about simple ligands. They proposed four principal sites to be responsible for proton dissociation in the pH range 3–6.5, adjusting their abundances to match the experimental data (Table 9.5). Chemical arguments based on knowledge about simpler ligands were also used to interpret and thereby model metal binding. For example, eight binding sites for Cu in a fulvic acid were assigned. Four were monodentate sites coinciding with the proton-dissociating sites of Table 9.5. Four were bidentate, an aminocarboxylic acid site, a salicylic acid site, a dihydroxy site, and a hydroxycarbonyl site (Ephraim & Marinsky, 1986). The site assignments provided reasonable simulations of the observed data. Further assignments of proton and metal ion binding sites have been made in subsequent publications (Marinsky *et al.*, 1995, 1999; Mathuthu & Ephraim, 1995; Mathuthu *et al.*, 1995).

The 'chemical argument' modelling approach leads to feasible site assignments, but can be criticised on the grounds (a) that it oversimplifies the great complexity of sites in humic substances, (b) that it is subjective and unconstrained, and (c) that it can only be applied to the abundant sites. Furthermore, to analyse all data sets in the required detail is impractical. Therefore, while the approach of Marinsky and colleagues is

Table 9.5. Proton-dissociating sites and their abundances in Armadale fulvic acid, deduced by Ephraim *et al.* (1986). See Section 9.4.

Nature	pK	Fraction of total
Strongly acid COOH group	1.8	0.245
Salicylic acid-type COOH	3.2	0.304
Weakly acidic COOH	4.2	0.224
Acidic alcoholic OH (enol) group	5.7	0.227

useful for considering the nature of the chemical interactions involved in cation binding by humic matter, it does not provide a readily applicable, general means of fitting experimental data.

9.5 Modelling electrostatic effects in humic substances

Recent models have acknowledged the need to take into account electrostatic effects on binding, which are clearly demonstrated by the salt dependence of proton dissociation and metal binding (Chapter 7). The magnitudes of electrostatic effects were discussed in Sections 4.6 and 5.4, where it was shown that the effective value of K might be changed by up to four orders of magnitude ($\log K$ could change by four) compared to the intrinsic constant. Furthermore, the electrostatic effect contributes to apparent heterogeneity in binding at constant ionic strength, since as cation binding to the humic molecules increases, the net charge decreases and the electrostatic effect diminishes. If a model without electrostatics is employed, the 'polyelectrolyte' effect is accounted for by parameters purporting to describe other aspects of the interactions, notably site heterogeneity. Therefore to achieve a reasonable description of the chemical heterogeneity of humic substances, the electrostatic interactions need to be recognised.

9.5.1 Modelling based on the Poisson–Boltzmann (PB) equation

The simplest model is the Debye–Hückel theory (Section 4.6). The applications by Wilson & Kinney (1977) and Dempsey & O'Melia (1983) have been described in Sections 9.3.4 and 9.2.5 respectively. Tipping *et al.* (1990) used the model to predict the influence of ionic strength on proton dissociation from the molecular dimensions of Suwannee River fulvic acid, assuming the material to consist of homogeneous impenetrable spheres, and achieved good results. They also explored the effect of molecular size heterogeneity on average behaviour and found that the weight-average molecular weight is preferable to the number-average value for estimating the dimensions of the average fulvic acid molecule. Other applications based on the Debye–Hückel theory have been through the use of activity coefficients (Sections 9.3.3 and 9.3.8).

The most comprehensive study in which the electrostatic properties of humic substances have been evaluated is that of de Wit *et al.* (1993a), who analysed data for 10 different samples using numerical solutions of the PB equation for impermeable rigid spheres and cylinders. They fitted the data by finding the value of the particle radius that gave the best merging of titration curves at different ionic strengths, assuming a particle density of 1.0 g cm^{-3}. Neither the spherical nor the cylindrical model proved to be superior. An example of the merging of curves is shown in Fig. 9.7. At charges of 0.005 eq g^{-1}, a typical value at pH 7 for fulvic acids, the calculated potentials for spheres were *c.* 70 mV at $I = 0.1$ mol dm^{-3}, and *c.* 130 mV at $I = 0.001$ mol dm^{-3}. For humic acids, the charge at pH 7 is typically 0.003 eq g^{-1}, and the corresponding potentials are *c.* 100 and 150 mV. These values are appreciably higher than the potential at which the Debye–Hückel approximation (Section 4.6.1) breaks down, which therefore justifies the application of the more correct theory.

The results for different humic samples obtained by de Wit *et al.* (1993a) are summarised in Table 9.6. The derived values of molecular weight for fulvic acids and aquatic humic substances range from 500 to 7000, in reasonable agreement with published values obtained by direct methods (Section 2.4.3). The estimated molecular weights for the two humic acids, 4360 and 14700, are at the lower end of the range of published values. In their discussion of the results, de Wit *et al.* pointed out that their analysis

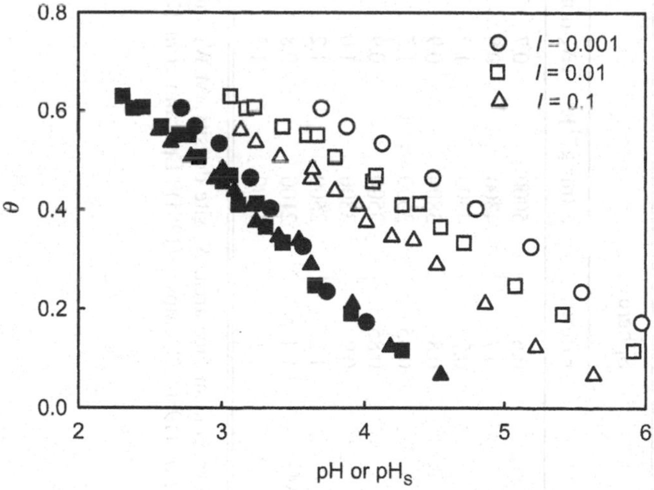

Figure 9.7. Example of the application of the 'master curve' approach to the analysis of proton binding by aquatic humic substances. The variable θ is the degree of occupation by protons of the carboxyl groups (data of Tipping *et al.*, 1988b). The open symbols are values of θ plotted against bulk solution pH. The solid symbols are the same values of θ, plotted against the pH at the molecular surface (pH$_S$), obtained by application of an electrostatic model for a sphere. [Redrawn from de Wit, J.C.M., van Riemsdijk, W.H. & Koopal, L.K. (1993), Proton binding to humic substances. 1. Electrostatic effects, *Environ. Sci. Technol.* **27**, 2005–2014. Copyright 1993 American Chemical Society.]

Table 9.6. Analysis of electrostatic properties of humic substances by de Wit *et al.* (1993a). See Section 9.5.1 for explanation.

	Sphere				Cylinder		
Sample	r (nm)	S ($m^2\,g^{-1}$)	N_S (nm^{-2})	M.Wt	r (nm)	S ($m^2\,g^{-1}$)	N_S (nm^{-2})
Suwannee River FA[a]	0.6	5000	0.7	545	0.19	10 500	0.3
Swedish FA[a]	0.7	4300	0.8	865	0.25	8000	0.4
FA#3[b]	0.7	4300	1.2	865	0.3	6600	0.8
Armadale Bh FA[a]	0.8	3800	0.9	1290	0.28	7100	0.5
FA#1[b]	0.85	3500	1.5	1550	0.4	5100	1.0
Bersbo FA[c]	0.85	3500	0.9	1540	0.31	6500	0.4
Lochard Forest HS[d]	0.9	3300	1.0	1840	0.32	6300	0.5
Mosedale Beck HA[d]	1.2	2500	1.2	4360	0.5	4000	0.7
Penwhirn Reservoir HS-A[d]	1.4	2100	0.8	6920	0.6	3300	0.5
Humic acid[e]	1.8	1700	1.3	14 700	1.0	2000	1.1

Abbreviations: r radius, S specific surface area, N_S site density, M.Wt. molecular weight, FA fulvic acid, HA humic acid, HS humic substances.
Data sources: [a]Ephraim *et al.* (1986); [b]Dempsey (1981); [c]Ephraim *et al.* (1989a); [d]Tipping *et al.* (1988b); [e]Marinsky *et al.* (1982).

assumed the molecules to be homogeneous with respect to size, whereas humic substances are known to be heterogeneous. They suggested that although a given sample may contain a wide range of molecular weights, the range in radius, which for spheres is proportional to the cube root of molecular weight for constant density, would be much smaller. For cylinders, it could be envisaged that the molecules within a given sample differ in length but not radius, so that size heterogeneity might not influence the results.

Other authors have performed the same kind of analysis, but have drawn different conclusions. Barak & Chen (1992) assumed cylindrical geometry and calculated that the radii of two humic acids varied with the degree of dissociation (α), from approximately 1 nm at $\alpha = 0.1$ to approximately 0.25 nm at $\alpha = 0.9$. They attributed this to the uncoiling of a flexible linear polyelectrolyte. Robertson & Leckie (1999) attempted to model the ionic strength dependence of proton dissociation from a soil humic acid, but did not consider their fits to be satisfactory, using either impermeable spheres or cylinders.

Bartschat *et al.* (1992) presented a penetrating review of the application of the PB equation to humic substances, with emphasis on the need to take size heterogeneity into account. They analysed data that showed a much larger electrostatic effect on Cu^{2+} binding than on H^+ binding, even allowing for the higher charge on the metal ion. They explained the difference between Cu^{2+} and H^+ in terms of the greater electrostatic effect in larger molecules, by using a model based on a small molecule (radius 0.77 nm, molecular weight 711) and a large one (1.5 nm, 5260), both bearing two classes of copper binding site and a class of acidic sites able to bind protons but not Cu^{2+}. It should be noted that the purpose of the study was not to obtain high-precision fits of experimental data (which would require more elaborate site assignments) but to show the significance of electrostatic effects, combined with size heterogeneity, on cation binding.

9.5.2 Empirical model of Tipping *et al.*

When applying the Poisson–Boltzmann equation to humic substances, a model must be chosen (impenetrable sphere, rod, penetrable sphere etc.). If information is available on molecular size and size distribution then electrostatic effects can be predicted. If it is not, then one or more adjustable parameters are needed to fit experimental data. As we saw in the previous section, de Wit *et al.* (1993a) assumed one of two models, and adjusted the molecular radius. An alternative approach was adopted by Tipping and coworkers in which the basic PB approach was adopted (i.e. the term $\exp(-2wzZ)$ was used in modifying intrinsic equilibrium constants) but w was obtained from empirical expressions containing adjustable parameters. Tipping *et al.* (1988b) used the following two empirical expressions

$$w = P \log I \tag{9.27}$$

$$w = P(\log I)\exp(Q|Z|) \tag{9.28}$$

where I is ionic strength, $|Z|$ is the modulus of Z, and P and Q are adjustable parameters. It was suggested that the term $\exp(Q|Z|)$ accounted for changes in the molecular size of the humic matter due to variations in electrostatic repulsion as Z varied. In later work however, the use of this term was abandoned, and the simpler equation (9.27) is used in Humic Ion Binding Models V and VI (see Section 9.6). Equation (9.27) gives a somewhat stronger ionic strength dependence of w than the PB equation.

The model approximates counterion accumulation with a Donnan-type sub-model (Section 4.6.7), in which co-ions are assumed to be completely absent (Tipping & Hurley, 1988, 1992). The size of the zone in which accumulation takes place is determined by the molecular radius (R), obtained from particle density and assumed molecular weight, together with a diffuse layer thickness, which is set to $1/\kappa$, as defined by equation (4.34). Thus the humic molecule plus its diffuse layer is regarded as an electrically neutral sphere of radius $(R + 1/\kappa)$. The values of R are taken to be 0.8 nm for fulvic acid and 1.72 nm for humic acid (corresponding to molecular weights of 1500 and 15 000 respectively and a density, including solvent, of 1.2 g cm^{-3}). The use of the Donnan model provides a rapid calculation procedure for estimating counterion accumulation without abandoning the notion that concentrations increase through the diffuse layer up to the molecular surface.

9.5.3 The Donnan model

The Donnan model, as described in Section 4.6.7, has been used by the 'Wageningen–Wallingford' group of workers (Benedetti *et al.*, 1995; Kinniburgh *et al.*, 1996, 1999), with an empirical equation to describe the dependence of the Donnan volume (V_D; equation 4.51) on ionic strength

$$\log V_D = a + b \log I \tag{9.29}$$

A value for V_D of 0.1 dm^3 kg^{-1} (or 0.1 cm^3 g^{-1}) at very high ionic strength (10 mol dm^{-3}) was estimated for both humic and fulvic acids, and with this value fixed, equation (9.29) becomes

$$\log V_D = -1 + b(\log I - 1) \tag{9.30}$$

From fitting proton dissociation data at different ionic strengths, it is found that *b* is typically −0.3 to −0.5 for humic acids and −0.7 to −0.9 for fulvic acids. The implications of equation (9.30) and its parameter values are discussed in the next section. In its 'purest' form, the Donnan model pictures the diffuse zone as being contained completely within the hydrodynamic volume of the particle, but Kinniburgh *et al.* (1999) point out that at low ionic strengths, the diffuse zone volumes are too large to account for the volume of hydrated fulvic acid particles alone, and so some of the diffuse zone must partly be placed in the region surrounding the particle.

9.5.4 Remarks on electrostatic modelling

The applications of models based on impermeable or partly permeable molecules (Section 9.5.1) do not yield unambiguous results concerning the shapes and flexibility of humic substances. One reason is that one or more parameters can be adjusted to provide an acceptable fit, a second is that knowledge about the physical properties of humic substances is insufficient to decide whether the optimised parameter values are in accord with reality. The problem is compounded by the heterogeneity of humic substances, with respect to both size and chemical properties. However the models described above do provide useful quantitative descriptions of electrostatic effects, or to be more precise they describe the influence of ionic strength on proton dissociation. We shall see in Chapter 10 that the parameterisation of electrostatic interactions on this basis can also provide reasonable descriptions of the influence of ionic strength upon metal binding.

Implicit in the electrostatic models described above is the notion that the ions not bound at specific sites are mobile. Thus no account is taken of the specific binding of the (monovalent) 'indifferent' cations such as Na^+, K^+ and NH_4^+. Although such binding may be weak, it is in fact likely to occur, by analogy with simple ion-pair interactions in which outer sphere complexes are formed (Section 5.1). Thus a more complete picture would be obtained by including such binding, which would be equivalent to the Stern layer in classical surface chemistry. The model of Westall *et al.* (Section 9.3.6) goes to the other extreme, attributing all monovalent ion binding to interactions at specific sites.

As we saw in Section 4.6, Coulombic effects depend upon molecular size. Physical measurements suggest that humic molecules expand and contract as net charge and ionic strength vary. For example, Avena *et al.* (1999) reported that humic acid intrinsic viscosity (proportional to hydrodynamic volume) could change by a factor of up to six on going from low to high degrees of proton dissociation, corresponding to a 1.8-fold change in radius. The electrostatic sub-model should, in principle, take such an expansion into account, but the possibility has been included only in the abandoned empirical model of equation (9.28). No strong evidence has yet been produced that changes in molecular size exert a significant effect upon electrostatic contributions to cation binding.

Finally, let us consider the volumes of diffuse layers or zones in comparison to the total volume of the system or suspension. Values computed from the impenetrable sphere and Donnan models (Table 9.7) show that the volumes required by the former are substantially larger, especially at low ionic strength. For example, in a 10 mg dm^{-3} solution of fulvic acid at an ionic strength of 0.001 mol dm^{-3}, which might be typical of a natural surface water, the impenetrable sphere model predicts that about 2% of the total volume of water would be in the diffuse zone, whereas the Donnan model predicts only about 0.2%. At such a low concentration of humic matter, the difference hardly affects the bulk solution. However, at higher concentrations significant differences – and difficulties – do emerge. For example, in a 1 g dm^{-3} solution of fulvic acid at $I = 0.001$ mol dm^{-3}, the impenetrable sphere model predicts a diffuse zone volume of 1.9 dm^3 per dm^3 of solution, a physical impossibility. The same problem arises for the Donnan model at a concentration of c. 10 g dm^{-3}. The paradox arises because the models are based on analyses and parameter fitting that refer to dilute systems, in which the diffuse zones are unconstrained. At high concentrations, diffuse zone overlap or compression take place, and additional modelling assumptions have to be made.

Table 9.7. Diffuse layer volumes (cm^3 g^{-1}) for the impenetrable sphere and Donnan models.

	Impenetrable sphere		Donnan	
I	0.001	0.1	0.001	0.1
Fulvic acid	1900	8.3	160	4.0
Humic acid	240	2.4	4.0	0.6

Dimensions and molecular weights given in Section 9.5.2 were used to calculate the values for impenetrable spheres.
Donnan volumes were calculated from equation (9.30) with $b = -0.4$ for humic acid and -0.8 for fulvic acid.
Ionic strength (I) is in mol dm^{-3}.

9.6 Humic Ion-Binding Models V and VI

Several discrete site/electrostatic models have been developed by Tipping and coworkers (Backes & Tipping, 1987b; Tipping *et al.*, 1988b; Tipping & Hurley, 1988, 1992; Tipping, 1998b). Initially the aim was to calculate proton, aluminium, and base cation binding only (Models II–IV), but Models V and VI take all cations into account, in principle. The models use a structured formulation of discrete, chemically plausible, binding sites for protons, to allow the creation of regular arrays of bidentate (and tridentate in Model VI) binding sites for metals. Electrostatic interactions are dealt with as described above (Section 9.5.2). In applying the Donnan sub-model for counterion accumulation, each counterion can be assigned a selectivity coefficient (K_{sel}), so that accumulation can be made to depend on more than just the counterion charge. For example, Ca^{2+} can be favoured over Mg^{2+}.

9.6.1 Model V

Fulvic and humic acids are assumed to be rigid spheres of uniform size, with the dimensions and molecular weights given in Section 9.5.2. Ion-binding groups are positioned randomly on the molecular surfaces. Proton dissociation is represented by postulating eight groups with different acid strengths. The dissociation reaction can be written generally as

$$(\mathrm{HumH})^{Z} = (\mathrm{Hum})^{Z-1} + \mathrm{H}^{+} \tag{9.31}$$

where Hum represents the humic molecule, and Z is the net charge. The reactions are characterised by intrinsic equilibrium constants, the negative logarithms of which are denoted by $\mathrm{p}K_1$–$\mathrm{p}K_8$. The four most strongly acid groups (groups 1–4) are referred to as type A groups, and consist mainly of carboxylic acids, while the remaining four groups (type B) represent weaker acids, such as phenolic acids. The eight $\mathrm{p}K_i$ values are expressed in terms of four constants ($\mathrm{p}K_A$, $\mathrm{p}K_B$, $\Delta\mathrm{p}K_A$ and $\Delta\mathrm{p}K_B$) as follows

$$\text{for } i = 1\text{–}4 \qquad \mathrm{p}K_i = \mathrm{p}K_A + \frac{(2i-5)}{6}\Delta\mathrm{p}K_A \tag{9.32}$$

$$\text{for } i = 5\text{–}8 \qquad \mathrm{p}K_i = \mathrm{p}K_B + \frac{(2i-13)}{6}\Delta\mathrm{p}K_B \tag{9.33}$$

Thus the values of $\mathrm{p}K_A$ and $\mathrm{p}K_B$ are the average $\mathrm{p}K$ values of the two types of group, while $\Delta\mathrm{p}K_A$ and $\Delta\mathrm{p}K_B$ are measures of the spread of the individual $\mathrm{p}K_i$ values around the means. Positive values of $\Delta\mathrm{p}K_A$ and

ΔpK_B mean that the pK_i values increase as i increases, i.e. the groups become progressively weaker acids. Each type A group is assigned an abundance of $n_A/4$ mol g^{-1} humic matter, and each type B group an abundance of $n_A/8$ mol g^{-1}. Thus, within a type, each group is present in equal amounts, and there are half as many type B groups as type A groups. The imposed regularity of the groups facilitates the formulation of multidentate binding sites for metals.

Metal ions, and their first hydrolysis products, compete with each other, and with protons, for the type A and type B groups. Monodentate binding at type A sites is formulated as metal–proton exchanges

$$(\text{HumAH})^Z + \text{M}^{z+} = (\text{HumAM})^{Z+z-1} + \text{H}^+ \qquad (9.34)$$

$$(\text{HumAH})^Z + \text{MOH}^{(z-1)+} = (\text{HumAMOH})^{Z+z-2} + \text{H}^+ \qquad (9.35)$$

and analogous reactions occur at the type B sites. The (intrinsic) equilibrium constant for the type A sites is given by

$$K_{\text{MHA}} = \frac{[\text{HumAM}][\text{H}^+]}{[\text{HumAH}][\text{M}]} = \frac{[\text{HumAMOH}][\text{H}^+]}{[\text{HumAH}][\text{MOH}]} \qquad (9.36)$$

and there is an analogous constant for the type B sites. Thus, the equilibrium constant for metal ion binding at a deprotonated type A site i (the conventional association reaction) is given by

$$K_{\text{MA},i} = \frac{[\text{HumAM}]}{[\text{HumA}][\text{M}]} = \frac{[\text{HumAMOH}]}{[\text{HumA}][\text{MOH}]} = \frac{K_{\text{MHA}}}{K_i} \qquad (9.37)$$

where K_i is the proton dissociation constant of site i. Therefore we have

$$\log K_{\text{MA},i} = pK_i - pK_{\text{MHA}} \qquad (9.38)$$

and the strength of binding of metal ions to the four type A sites parallels the strength of proton binding to those sites. Again, the expressions are analogous for the type B sites. When fitting data with the model, the values of pK_{MHA} and pK_{MHB} are the optimised quantities for each metal ion.

Bidentate sites are generated by combining pairs of proton-binding sites, and can be A–A or A–B sites, for example

$$\left(\text{Hum}\begin{matrix}\text{AH}\\ \text{BH}\end{matrix}\right)^Z + \text{M}^{z+} = \left(\text{Hum}\begin{matrix}\text{A}\\ \text{B}\end{matrix}\text{M}\right)^{Z+z-2} + 2\text{H}^+ \qquad (9.39)$$

The equilibrium constants are the products of the values for the individual sites

$$K_{\text{MH}_2\text{A}_2} = (K_{\text{MHA}})^2 \qquad K_{\text{MH}_2\text{AB}} = K_{\text{MHA}}\,K_{\text{MHB}} \qquad (9.40)$$

The pairing of proton-binding sites to form bidentate sites is arranged so that all sites are equally represented, and all bidentate sites are present in equal abundance, but only 12 of the possible 36 combinations are used, to avoid unnecessary computation (Table 9.8). Only proton-dissociating sites that are sufficiently close together are able to form bidentate sites. Proximities are estimated statistically by assuming the sites to be randomly positioned on the surface of the sphere (Tipping & Hurley, 1992). Pairs of groups form bidentate sites if they are less than or equal to 0.45 nm apart. Sites that are close enough to pair up take part only in bidentate reactions with metal cations, i.e. they cannot form monodentate complexes. The fraction of proton-dissociating sites able to form pairs is denoted by f_{pr}.

The parameter values required by Model V are shown in Table 9.9. The maximum number that can be optimised is six for proton dissociation and three for each metal cation. Three of the parameters (radius, molecular weight and proximity factor) are fixed from the start, from literature information and *a priori* assumptions. In practice it has been found that reasonable results can be obtained by establishing a relationship between the two parameters for metal binding, pK_{MHA} and pK_{MHB} (see Chapter 10), and for dilute systems all values of K_{sel} can be set to unity.

9.6.2 Model VI

Model VI was developed from Model V firstly to obtain a wider range of binding affinities for metal ions, and secondly to relax the parallel

Table 9.8. Monodentate (M) and bidentate (B) binding sites for metal cations in Humic Ion-Binding Model V.

Site	Proton-binding sites	Abundance	Site	Proton-binding sites	Abundance
M1	1	$n_A(1-f_{pr})/4$	B1	1,2	$n_A f_{pr}/16$
M2	2	$n_A(1-f_{pr})/4$	B2	1,4	$n_A f_{pr}/16$
M3	3	$n_A(1-f_{pr})/4$	B3	1,6	$n_A f_{pr}/16$
M4	4	$n_A(1-f_{pr})/4$	B4	1,8	$n_A f_{pr}/16$
M5	5	$n_A(1-f_{pr})/8$	B5	2,3	$n_A f_{pr}/16$
M6	6	$n_A(1-f_{pr})/8$	B6	2,5	$n_A f_{pr}/16$
M7	7	$n_A(1-f_{pr})/8$	B7	2,7	$n_A f_{pr}/16$
M8	8	$n_A(1-f_{pr})/8$	B8	3,4	$n_A f_{pr}/16$
			B9	3,6	$n_A f_{pr}/16$
			B10	3,8	$n_A f_{pr}/16$
			B11	4,5	$n_A f_{pr}/16$
			B12	4,7	$n_A f_{pr}/16$

relationship between metal and proton binding. With regard to proton-dissociating groups, the assumed molecular dimensions, and the electrostatic sub-model, Models V and VI are identical. They differ only in the way in which metal-binding reactions are formulated. Instead of using metal–proton exchanges, equilibrium constants for the binding of metal ions at deprotonated sites are employed. Monodentate binding takes place according to the general reaction

$$\mathrm{Hum}^{Z} + \mathrm{M}^{z} = (\mathrm{HumM})^{Z+z} \tag{9.41}$$

The equilibrium constants are given by

$$\text{for sites 1–4} \quad \log K(i) = \log K_{\mathrm{MA}} + \frac{(2i-5)}{6}\Delta LK_{\mathrm{A1}} \tag{9.42}$$

$$\text{for sites 5–8} \quad \log K(i) = \log K_{\mathrm{MB}} + \frac{(2i-13)}{6}\Delta LK_{\mathrm{B1}} \tag{9.43}$$

where ΔLK_{A1} and ΔLK_{B1} are constants estimated from data fitting. Thus the values of $\log K(i)$ are evenly spaced around the mean of $\log K_{\mathrm{MA}}$ but the spacing is not necessarily the same as that for protons, which is the

Table 9.9. Summary of parameters in Humic Ion-Binding Model V (see Section 9.6.1).

Parameter	Description	How found
n_{A}	Abundance of type A sites (mol g^{-1})	Fitted
n_{B}	Abundance of type B sites (mol g^{-1})	$= 0.5 \times n_{\mathrm{A}}$
$\mathrm{p}K_{\mathrm{A}}$	Intrinsic proton dissociation constant for type A sites	Fitted
$\mathrm{p}K_{\mathrm{B}}$	Intrinsic proton dissociation constant for type B sites	Fitted
$\Delta\mathrm{p}K_{\mathrm{A}}$	Distribution term that modifies $\mathrm{p}K_{\mathrm{A}}$	Fitted
$\Delta\mathrm{p}K_{\mathrm{B}}$	Distribution term that modifies $\mathrm{p}K_{\mathrm{B}}$	Fitted
$\mathrm{p}K_{\mathrm{MHA}}$	Intrinsic equilibrium constant for metal binding at type A sites	Fitted
$\mathrm{p}K_{\mathrm{MHB}}$	Intrinsic equilibrium constant for metal binding at type B sites	Fitted, or correlated with $\mathrm{p}K_{\mathrm{MHA}}$
P	Electrostatic parameter	Fitted
K_{sel}	Selectivity coefficient for counterion accumulation	Fitted, or set to unity
f_{pr}	Fraction of proton sites that can make bidentate sites	Calculated from geometry
M	Molecular weight	Estimated from literature
r	Molecular radius	Estimated from literature

case in Model V, i.e. in general ΔLK_1 is not equal to $-\Delta pK_A$. The same applies to the type B sites.

In Model VI, metal cations and their first hydrolysis products may bind at both bidentate and tridentate sites. For a bidentate site comprising single sites j and k, the association constant $K(j,k)$ is given by

$$\log K(j,k) = \log K(j) + \log K(k) + x\Delta LK_2 \tag{9.44}$$

where ΔLK_2 is an adjustable parameter. The value of x is zero for 90.1% of the sites, 1 for 9%, and 2 for 0.9%. This generates heterogeneity, with a small number of strong sites, a larger number of moderate ones, and a majority of weak ones. Analogous rules are used for the tridentate sites, made up of single sites l, m and n

$$\log K(l,m,n) = \log K(l) + \log K(m) + \log K(n) + y\Delta LK_2 \tag{9.45}$$

Here, y is 0, 1.5 or 3 for the 90.1%, 9% and 0.9% abundances. Each metal ion has a characteristic ΔLK_2.

The fractions of single sites that contribute to the bidentate or tridentate sites are determined statistically, as in Model V, by assuming the proton sites to be randomly positioned on the surfaces of the (spherical) molecules. The fraction of sites forming bidentate sites is denoted by f_{prB}, that for tridentates is f_{prT}. For fulvic acid, $f_{prB} = 0.42$, $f_{prT} = 0.03$, while for humic acid, the corresponding values are 0.50 and 0.065.

In Model VI, if all combinations of proton sites are allowed, 36 different bidentate sites and 120 different tridentate sites can form, for each of which there are three binding strengths, if ΔLK_2 is non-zero. Together with the eight monodentate sites, there could be 476 different sites in all. The site abundances depend on their composition, in terms of type A and type B sites, the latter being present at half the total abundance of the former. For example, tridentate sites comprising three type A proton sites are eight times more abundant than those comprising three type B sites. To avoid over-complication, and to speed computations, a sub-set of sites is used in the model, while maintaining the relative proportions of the contributing proton sites. Bidentate or tridentate sites are allowed to consist only of different proton sites, and only 24 representative combinations are adopted. The 24 allowed combinations lead to 72 different sites, because of the heterogeneity terms (equations 9.44 and 9.45). With the addition of the eight monodentate sites, there are 80 different sites in all. Table 9.10 shows the combinations. For humic acid the commonest (monodentate type A sites) have an abundance of c. $4 \times 10^{-4}\,\mathrm{mol\,g^{-1}}$, the rarest (tridentate sites consisting of three type B proton sites) *c.* 9×10^{-9} mol g^{-1}. The second figure corresponds approximately to 1 site per 7500 molecules for humic acid of molecular weight 15 000; for fulvic acid of molecular weight 1500, the corresponding figure is 1 in 10^5.

In Model VI, the maximum number of parameters that can be optimised to describe proton dissociation is six, exactly as in Model V. For each metal cation (plus its first hydrolysis product), up to six might be optimised (K_{MA}, K_{MB}, ΔLK_{A1}, ΔLK_{B1}, ΔLK_2, K_{sel}), although the number can be reduced substantially by taking into account data for many metals (Chapter 10).

9.7 The NICA and NICCA models

The work of de Wit *et al.* (1993a) was mentioned in connection with electrostatic modelling in Section 9.5.1. The approach taken by these authors was to 'correct' for electrostatic effects, and then model the resulting binding curve, taking account of binding site heterogeneity. For

Table 9.10. Monodentate (M), bidentate (B) and tridentate (T) binding sites for metal cations in Humic Ion-Binding Model VI.

Site	Proton-binding sites	Abundance	Site	Proton-binding sites	Abundance
M1	1	$(1 - f_{prB} - f_{prT})\, n_A/4$	T1	1,2,3	$n_A\, f_{prT}/27$
M2	2	$(1 - f_{prB} - f_{prT})\, n_A/4$	T2	1,2,4	$n_A\, f_{prT}/27$
M3	3	$(1 - f_{prB} - f_{prT})\, n_A/4$	T3	1,3,4	$n_A\, f_{prT}/27$
M4	4	$(1 - f_{prB} - f_{prT})\, n_A/4$	T4	2,3,4	$n_A\, f_{prT}/27$
M5	5	$(1 - f_{prB} - f_{prT})\, n_A/8$	T5	1,2,5	$n_A\, f_{prT}/18$
M6	6	$(1 - f_{prB} - f_{prT})\, n_A/8$	T6	3,4,6	$n_A\, f_{prT}/18$
M7	7	$(1 - f_{prB} - f_{prT})\, n_A/8$	T7	1,3,7	$n_A\, f_{prT}/18$
M8	8	$(1 - f_{prB} - f_{prT})\, n_A/8$	T8	2,4,8	$n_A\, f_{prT}/18$
			T9	1,5,6	$n_A\, f_{prT}/36$
B1	1,2	$n_A\, f_{prB}/6$	T10	2,7,8	$n_A\, f_{prT}/36$
B2	3,4	$n_A\, f_{prB}/6$	T11	3,5,7	$n_A\, f_{prT}/36$
B3	1,5	$n_A\, f_{prB}/12$	T12	4,6,8	$n_A\, f_{prT}/36$
B4	2,6	$n_A\, f_{prB}/12$	T13	5,6,7	$n_A\, f_{prT}/216$
B5	3,7	$n_A\, f_{prB}/12$	T14	5,6,8	$n_A\, f_{prT}/216$
B6	4,8	$n_A\, f_{prB}/12$	T15	5,7,8	$n_A\, f_{prT}/216$
B7	5,6	$n_A\, f_{prB}/24$	T16	6,7,8	$n_A\, f_{prT}/216$
B8	7,8	$n_A\, f_{prB}/24$			

Note that each type of bidentate or tridentate site is sub-divided into weak sites (90.1%), moderate sites (9%) and strong sites (0.9%), as defined by equations (9.44) and (9.45).

example, de Wit *et al.* (1993b) used the Langmuir–Freundlich equation

$$\theta = \frac{(Kc)^m}{1 + (Kc)^m} \tag{9.46}$$

where θ is the fractional occupancy of sites and m is a parameter that characterises the site heterogeneity. Figure 9.8 illustrates the performance of the equation for different values of m. The slope of the log–log binding curve can readily be adjusted to simulate binding site heterogeneity, while adjusting K allows the overall binding strength to be varied. The equation can be generalised to include multiple binding entities (protons or metal ions)

$$\theta_i = \frac{K_i c_i}{\sum K_i c_i} \cdot \frac{\left(\sum K_i c_i\right)^m}{1 + \left(\sum K_i c_i\right)^m} \tag{9.47}$$

When applied to a number of binding ions, the model requires that the heterogeneity term is the same for each, which limits the ability of the model to fit data, especially at low occupancies. A partial solution was to add an extra set of high-affinity sites, but inevitably this required additional adjustable parameters. The problem was solved by Koopal *et al.* (1994) who introduced the NICA (Non-Ideal Competitive Adsorption) equation

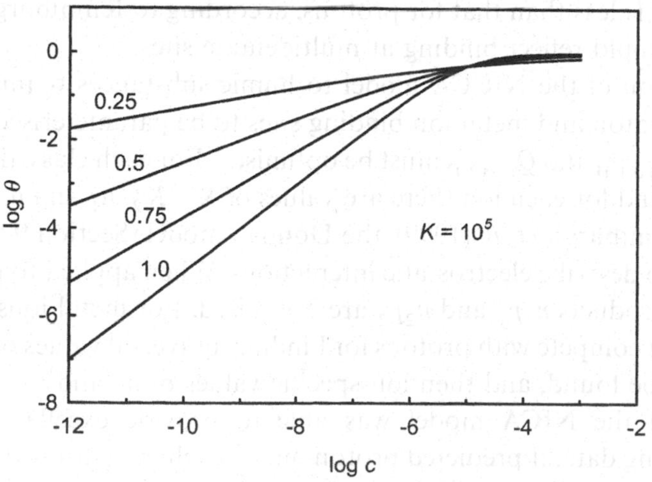

Figure 9.8. Plots of the Langmuir–Freundlich equation (equation 9.46) for different values of *m*.

$$\theta_i = \frac{(K_i c_i)^{n_i}}{\sum (K_i c_i)^{n_i}} \cdot \frac{\left[\sum (K_i c_i)^{n_i}\right]^p}{1 + \left[\sum (K_i c_i)^{n_i}\right]^p} \tag{9.48}$$

Thus in the NICA equation there are two exponents, one of which (n_i) applies to an individual ion (H^+ or metal cation), while p is the same for all ions. To obtain the amount of each ion bound (Q_i – e.g. in mol g^{-1}), θ_i is combined with the content of sites, Q_{max}, to give

$$Q_i = \theta_i \, Q_{max} \tag{9.49}$$

where the species i is assumed to react with one site. Kinniburgh *et al.* (1999) explained that the combination of equations (9.48) and (9.49) is thermodynamically inconsistent unless all values of n_i are equal. They therefore introduced the improved expression

$$Q_i = \theta_i \, n_i \, Q_{max} \tag{9.50}$$

They found it preferable to scale according to (n_i/n_H) where n_H is the value of n for protons, thus

$$Q_i = \theta_i \frac{n_i}{n_H} Q_{max,H} \tag{9.51}$$

where $Q_{max,H}$ is the maximum binding capacity for protons. The combination of equations (9.48) and (9.51) constitutes the 'consistent NICA' model (NICCA). Note that when (n_i/n_H) is less than one, the total site density for the species i is less than that for protons; according to Kinniburgh *et al.* (1999) this could reflect binding at multidentate sites.

Application of the NICCA model to humic substances requires two classes of proton and metal ion binding sites to be parameterised. Thus, values of $Q_{max1,H}$ and $Q_{max2,H}$ must be optimised. For each class, there is a value of p, and for each ion there are values of K_1, K_2, n_1 and n_2. In the work of Kinniburgh *et al.* (1999), the Donnan model (Section 9.5.3) was employed to describe electrostatic interactions. When applied to protons alone, the products $n_1 p_1$ and $n_2 p_2$ are optimised. For metal ions, which always must compete with protons for binding, universal values of p_1 and p_2 have to be found, and then ion-specific values of n_1 and n_2.

Although the NICA model was able to provide excellent fits of metal-binding data, it predicted proton–metal exchange ratios that were appreciably lower than those observed. As will be seen in Chapter 10, this failing was overcome by the NICCA model. A previous attempt to resolve the problem was the CONICA model (van Riemsdijk *et al.*, 1996), a variant of the NICA model in which bidentate binding sites for metals were included. The CONICA equations are more complicated than either the NICA or NICCA equations, and will not be presented here. In the judgement of Kinniburgh *et al.* (1999), the CONICA model provides 'a better (*than NICA*), but still not satisfactory' description of proton–metal exchange ratios.

9.8 Summary

This chapter has reviewed most of the types of parameterised model that have been used to describe cation binding by humic substances. There will no doubt be further developments, especially as new data become available, but clearly models can be formulated to take into account proton and metal binding, competition, and variations in ionic strength, and in these respects they are suitable to predict environmental phenomena involving ion binding by humic substances. The two most advanced models currently available are Model VI (Section 9.6.2) and the NICCA model (Section 9.7), although their immediate forerunners (Model V and NICA) have already proved very useful. In addition, simpler, less general, models may be appropriate for systems where there are only a few significant components, for example the triprotic model of Schecher & Driscoll (Section 9.3.5).

Model V and the NICA model have been compared by other authors (Jones & Bryan, 1998; Perdue, 1998), while Model VI and the NICCA model have yet to come under such scrutiny. In Chapter 10, the abilities of the models to fit experimental data are considered. The main difference between the modelling approaches is that Models V and VI are based on conventional chemical reactions, described by the Law of Mass Action, while the NICA and NICCA models employ mathematical formulations that are somewhat removed from such reactions. As a result, the parameters of Models V and VI are more interpretable in terms of 'normal' chemistry (see Chapter 10). On the other hand, NIC(C)A is more of a mathematical description of heterogeneity. It is also extraordinarily elegant. The two modelling approaches are similar in that they both recognise the need to take into account electrostatic effects and site heterogeneity, while keeping the number of adjustable parameters to a minimum, but without oversimplifying the complex systems they seek to describe.

10

Applications of comprehensive parameterised models

The aims here are (a) to examine whether or not parameterised models can fit experimental data from laboratory experiments with isolated humic substances, (b) to discuss the meanings of the parameter values, and (c) to explore some of the predictions that the models can make. There is little point in considering applications of all the models discussed in Chapter 9. The simpler ones can fit data over small concentration ranges, or at a single pH, but their applicability is restricted. Therefore we shall confine our discussion to the models that might be able to fit most or all of the available data, and which can therefore be termed comprehensive models. Thus we shall just be examining the performances of Models V and VI and the NIC(C)A models.

10.1 Interactions with protons

Simulation of the dissociation of protons from humic substances is the first task for any model that purports to describe cation binding comprehensively. Figures 10.1 and 10.2 illustrate applications of Model V/VI and the NICCA–Donnan model to proton-binding data at different ionic strengths. Note that Models V and VI are identical in their descriptions of proton binding, and the same applies to the NICA and NICCA models. It can be seen that the models work well, reproducing the shapes of the curves and the ionic strength dependence. Generally, the NIC(C)A model provides superior fits, with smaller and more random residuals. This is because Model VI is based on a limited number of discrete sites, insufficient to produce full smoothing, even when the extra parameter n_B is used in the fitting.

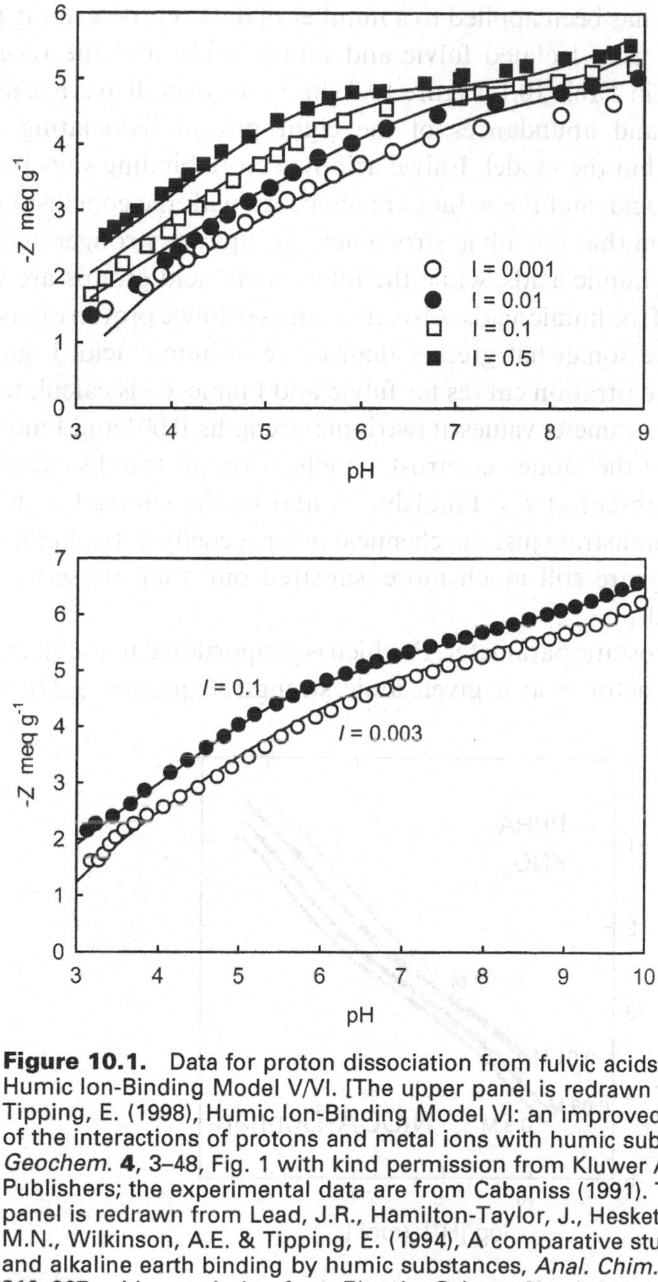

Figure 10.1. Data for proton dissociation from fulvic acids, fitted with Humic Ion-Binding Model V/VI. [The upper panel is redrawn from Tipping, E. (1998), Humic Ion-Binding Model VI: an improved description of the interactions of protons and metal ions with humic substances, *Aq. Geochem.* **4**, 3–48, Fig. 1 with kind permission from Kluwer Academic Publishers; the experimental data are from Cabaniss (1991). The lower panel is redrawn from Lead, J.R., Hamilton-Taylor, J., Hesketh, N., Jones, M.N., Wilkinson, A.E. & Tipping, E. (1994), A comparative study of proton and alkaline earth binding by humic substances, *Anal. Chim. Acta* **294**, 319–327, with permission from Elsevier Science.]

10.1.1 Parameter values in Model V/VI

Model V/VI has been applied to a number of data sets describing proton dissociation from isolated fulvic and humic acids, and the results are summarised in Table 10.1. Figure 10.3 shows the overall average intrinsic pK values and abundances of the eight proton-dissociating groups assumed within the model. Fulvic acid has more binding sites per gram than humic acid, and the values are in accord with the conclusions from Section 7.1, in that the fulvic strong acid groups are stronger acids than those of the humic acids, while the fulvic weak acid groups are weaker than those of the humic acids. Also, the ranges of fulvic proton dissociation constants are somewhat greater than those of humic acid. Figure 10.4 compares the titration curves for fulvic and humic acids calculated from the average parameter values at two ionic strengths, 0.001 and 1 mol dm^{-3}. According to the model, electrostatic effects on proton dissociation are completely absent at $I = 1\,\text{mol}\,\text{dm}^{-3}$, and so the curves for this ionic strength demonstrate just the chemical heterogeneity in the binding sites. Clearly, they are still much more 'smeared out' than those for simple diprotic acids.

The electrostatic parameter, P, which is proportional to the electrostatic interaction factor w at a given ionic strength (equation 9.27), is more

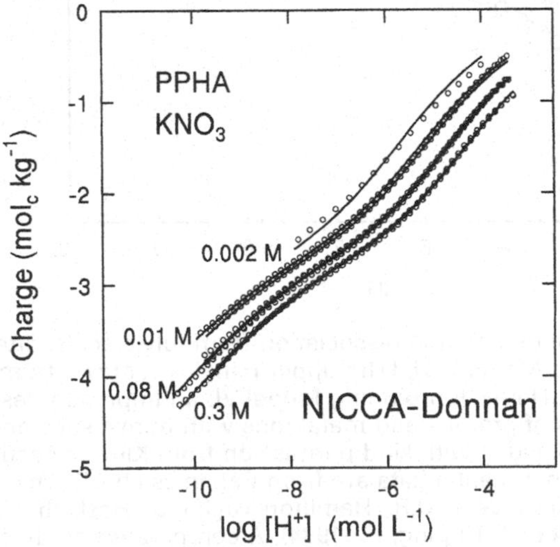

Figure 10.2. Proton dissociation from purified peat humic acid fitted with the NIC(C)A–Donnan model. [From Kinniburgh, D.G., van Riemsdijk, W.H., Koopal, L.K., Borkovec, M., Benedetti, M.F. & Avena, M.J. (1999), Ion binding to natural organic matter: competition, heterogeneity, stoichiometry and thermodynamic consistency, *Coll. Surf. A* **151**, 147–166, with permission from Elsevier Science. IPR/18-1C British Geological Survey. ©NERC. All rights reserved.]

negative for humic acid. This is anticipated for larger molecules in the impenetrable sphere model, when w is expressed on a mass, rather than molar, basis.

10.1.2 Parameter values in the NIC(C)A model

Parameter values for the NIC(C)A model are presently only available for one data set (Table 10.2). Note that only the product, m, of the two heterogeneity parameters n and p can be derived from data for a single

Table 10.1. Proton-binding parameters for Model V/VI, applied to different published data sets. [Reproduced from Tipping, E. (1998), Humic Ion-Binding Model VI: an improved description of the interactions of protons and metal ions with humic substances. *Aq. Geochem.* **4**, 3–48, Table V, with kind permission from Kluwer Academic Publishers.]

Code	n_A (mmol g^{-1})	pK_A	pK_B	ΔpK_A	ΔpK_A	$-P$
FH-01	5.7	3.1	9.9	3.9	3.9	99
FH-02	5.1	3.1	9.9	3.1	6.1	107
FH-03	4.3	—	—	—	—	73
FH-04	4.1	—	—	—	—	148
FH-05	5.0	3.5	10.3	2.9	8.5	73
FH-06	4.8	3.2	9.9	4.4	3.7	—
FH-07	3.3	3.3	8.1	2.5	4.9	—
FH-08	5.6	—	—	—		117
FH-09	4.9	3.1	9.3	3.2	2.7	186
FH-10	4.9	3.2	8.4	3.0	4.1	116
Mean	4.8	3.2	9.4	3.3	4.9	115
HH-01	3.5	3.8	—	2.5	—	480
HH-02	3.4	4.0	8.3	2.1	3.0	—
HH-03	3.4	4.3	—	0.1	—	380
HH-04	2.9	3.9	—	1.7	—	380
HH-05	3.3	4.0	8.8	2.5	3.9	250
HH-06	4.3	4.0	8.9	1.9	4.6	239
HH-07	2.5	4.3	—	1.1	—	435
HH-08	3.4	4.2	8.9	3.4	2.7	171
HH-09	3.2	4.0	8.9	3.2	4.0	303
Mean	3.3	4.1	8.8	2.1	3.6	330

Data sources: FH-01 Dempsey (1981); FH-02 Dempsey (1981); FH-03 Ephraim (1986); FH-04 Ephraim (1986); FH-05 Paxeus & Wedborg (1985); FH-06 Perdue *et al.* (1984); FH-07 Plechanov *et al.* (1983); FH-08 Tipping *et al.* (1988b); FH-09 Cabaniss (1991); FH-10 Lead *et al.* (1994); HH-01 Marinsky *et al.* (1982); HH-02 Posner (1964); HH-03 Stevenson (1976); HH-04 Tipping *et al.* (1988b); HH-05 van Dijk (1959); HH-06 Fitch *et al.* (1986); HH-07 Lead *et al.* (1994); HH-08 Lead *et al.* (1994); HH-09 Milne *et al.* (1995).

Table 10.2. Proton-binding parameters for the NIC(C)A–Donnan model optimised for purified peat humic acid, obtained by Kinniburgh *et al.* (1999). See Section 10.1.2.

	Carboxylic/low affinity sites	Phenolic/high affinity sites
Site density, $Q_{max,H}$ (mol kg^{-1})	2.30	4.32
Heterogeneity	$m_{H1} = (n_{H1} \times p_1) = 0.547$	$m_{H2} = (n_{H2} \times p_2) = 0.251$
log K_H	2.89	8.83

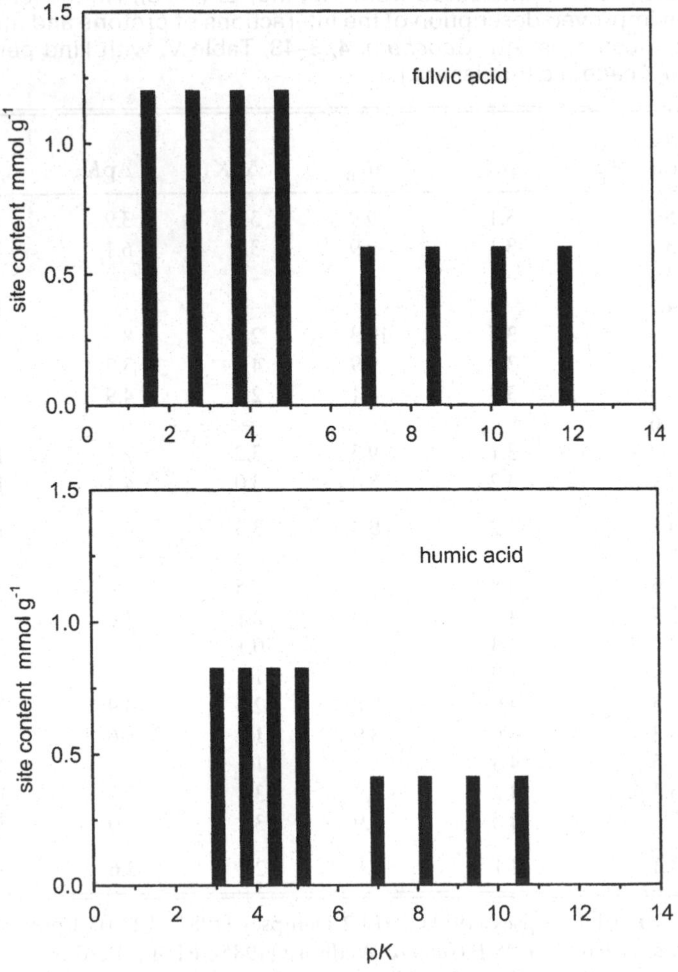

Figure 10.3. Abundances and pK values of proton-dissociating sites in Model V/VI. The values are based on the overall mean parameter sets for fulvic and humic acids (Table 10.1).

cation (i.e. H^+); the separation of n and p requires data for two or more cations (see Section 10.4).

The $\log K_H$ values in the NICCA model do not refer to simple equilibrium reactions, because of the heterogeneity parameters. Thus, equations (9.46)–(9.48) show that only if m is unity do the reactions correspond to the conventional (mass action) binding process. For example, the value of $\log K_H$ (= 2.89) for the 'low affinity' sites cannot be interpreted as some sort of average or representative binding constant for

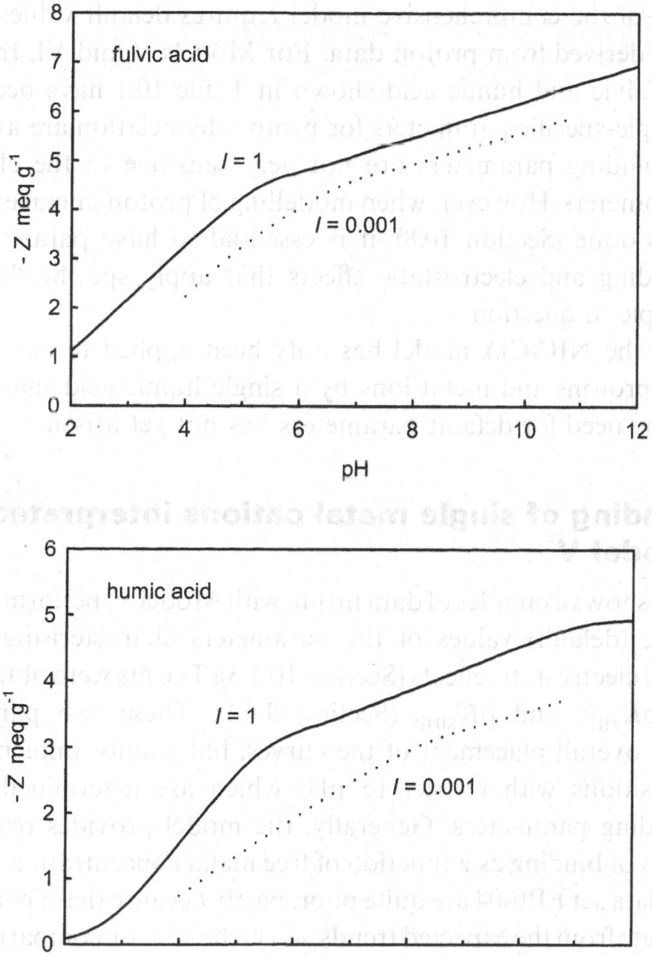

Figure 10.4. Proton dissociation from fulvic acid (FA) and humic acid (HA), calculated using overall mean parameters from Model V/VI, at ionic strengths of 0.001 and 1 mol dm^{-3}.

the reaction $RCOO^- + H^+ = RCOOH$ (where R represents the main part of the humic molecule).

10.1.3 Proton-binding parameters and the modelling of metal binding

To use a comprehensive model to simulate metal binding, parameters for both proton binding and electrostatic effects have to be established. Ideally, they would refer to the same humic sample used for the metal-binding determinations. However, this is not always possible, because many published studies on metals refer to humic samples for which proton dissociation data are not available. In such cases the application of the comprehensive model requires default values for the parameters derived from proton data. For Models V and VI, the mean values for fulvic and humic acid shown in Table 10.1 have been used. Where sample-specific parameters for proton dissociation are available, the metal-binding parameters are not very sensitive to the choice of proton parameters. However, when modelling of proton–metal exchange reactions is done (Section 10.8), it is essential to have parameters for proton binding and electrostatic effects that apply specifically to the humic sample in question.

To date, the NIC(C)A model has only been applied to data for the binding of protons and metal ions by a single humic acid sample, and therefore the need for default parameters has not yet arisen.

10.2 Binding of single metal cations interpreted with Model V

Figure 10.5 shows examples of data fitting with Model V, performed using the average (default) values of the parameters characterising proton binding and electrostatic effects (Section 10.1.3). The fits were obtained by adjusting pK_{MHA} and pK_{MHB} (Section 9.6.1). These two parameters control the overall placement of the curves, but cannot influence their relative positions with respect to pH, which are determined by the proton-binding parameters. Generally, the model provides reasonable descriptions of binding as a function of free metal concentration and pH. The fits of data set FPb-04 are quite poor, partly because the experimental values deviate from the expected trends, as can be seen by comparing them with the points in the other panels.

Tipping (1993) established the following linear relationships between pK_{MHA} and pK_{MHB} for a range of metals

$$\text{fulvic acids} \qquad pK_{MHB} = 3.96pK_{MHA} \qquad (10.1)$$

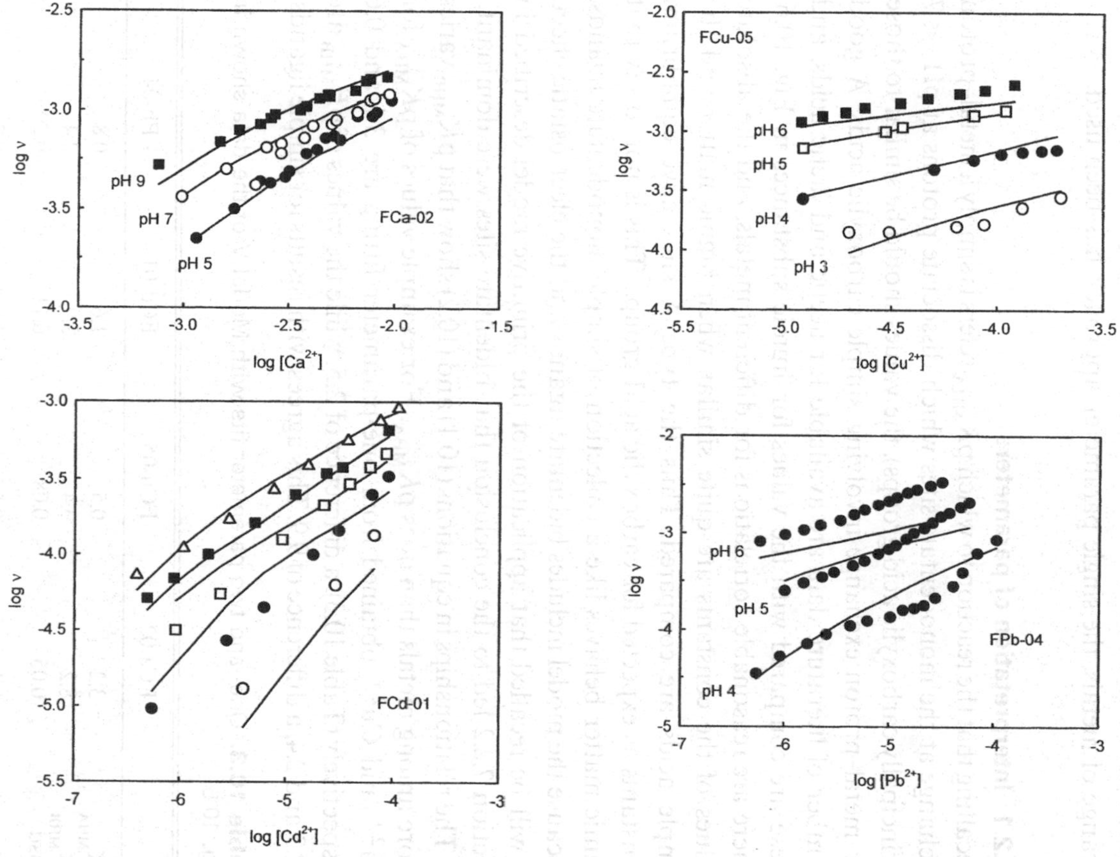

Figure 10.5. Metal binding by fulvic acids. The points are experimental data, the lines Model V fits. The data sources are: FCa-02 Dempsey (1981), FCu-05 Saar & Weber (1980a), FCd-01 Saar & Weber 1979, FPb-04 Saar & Weber (1980b). The ionic strength was $0.1\,mol\,dm^{-3}$ in each case.

$$\text{humic acids} \qquad pK_{MHB} = 3.0pK_{MHA} - 3.0 \qquad (10.2)$$

These enable the model to be applied to small data sets, because only a single parameter (pK_{MHA}) has to be estimated. In principle, a value of pK_{MHA} can be obtained from a single measurement. In some cases, but not all, the fits are poorer when a single parameter is adjusted, as illustrated by Table 10.3. To obtain a self-consistent set of average parameter values for a range of metals, the single parameter approach has been used.

10.2.1 Interpretation of parameters

Recalling that the reaction to which pK_{MHA} refers is simply a metal–proton exchange at the monodentate sites which dissociate protons at pH < 7 (principally carboxylic acid groups), the values should be similar to those for metal–proton exchange involving simple carboxylic acids. A good number of literature data are available for acetic and lactic acids, and these are compared with the values for humic substances in Fig. 10.6. There are reasonable correlations for different metals, and the absolute values of the constants are quite similar when humic matter and the simple acids are compared. Thus the type A sites have equilibrium constants as expected for carboxylic acid groups. This is not to say that humic matter behaves like a collection of simple monodentate ligands, because the model includes bidentate ligands and the electrostatic effect. It will be recalled that application of the primitive model described in Section 7.2.2 led to the conclusion that bidentate sites were dominant.

The relationships in equations (10.1) and (10.2) show that pK_{MHB} varies more among metals than does pK_{MHA}. For example values of pK_{MHA} for Ca^{2+} and Cu^{2+} obtained from one-parameter fitting are 3.1 and 0.6 respectively (Table 10.3), a difference of 2.5, while the values of pK_{MHB} are 12.3 and 2.4, a difference of 9.9. This agrees with results for simple ligands,

Table 10.3. One- and two-parameter fits with Model V of the data shown in Fig. 10.5.

	FCa-02	FCu-05	FCd-01	FPb-04
pK_{MHA}	3.2	0.5	1.6	0.8
pK_{MHB}	8.2	3.4	5.9	2.4
rmsd	0.05	0.08	0.11	0.20
pK_{MHA}	3.1	0.6	1.6	0.7
rmsd	0.05	0.11	0.11	0.20

The rmsd (root mean squared deviation) provides an overall measure of fit: it is equal to $\sqrt{(\log \nu_{calc} - \log \nu_{obs})^2/n}$ where n is the number of data points.

showing correlations between equilibrium constants for ligands with phenolic and carboxyl groups, but a bigger range of constants for the former (Martell & Hancock, 1996). Thus, both the type A sites (mainly carboxylic acids) and the type B sites (mainly phenolic acids) have the binding properties anticipated from those of simple compounds, and in this sense the model is chemically reasonable.

10.2.2 The failure of Model V

The successes of Model V can be attributed to its adequate formulation of the cation-binding reactions at the major sites of humic substances, in essence those sites present in sufficient numbers to determine proton binding within the pH range 3–11. Binding of metal cations is consistent with the model's picture of a mixture of monodentate and bidentate sites, made up of the proton-binding sites, in accord with the chemistry of simple weak acids. However, as shown in Fig. 10.7, Model V failed when confronted with the Cu and Cd binding data of Benedetti *et al.* (1995) described in Section 7.2.1. The key point about these data is that they extend to low values of free metal concentration and *v*. Because Model V did not allow smaller numbers of stronger sites, it could not deal with binding at low loadings. Thus, Model V had proved useful in bringing together many data sets, and showing that consistency and therefore predictability was possible, but ultimately it was flawed. Consequently, Model VI was formulated (Section 9.6.2).

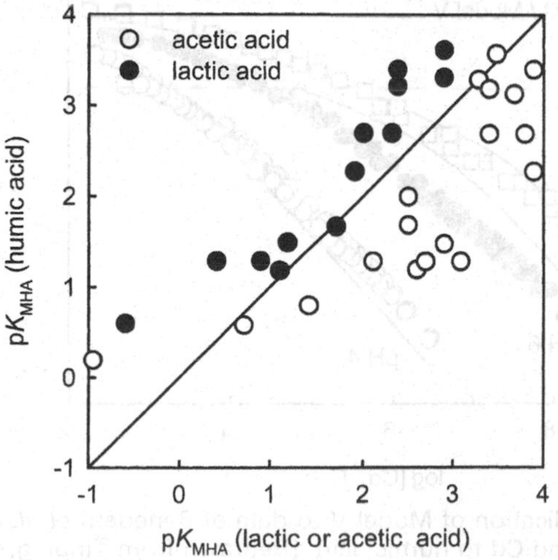

Figure 10.6. Comparison of pK_{MHA} values derived for humic acid in Model V with those for acetic and lactic acids. Each point represents a different metal cation. The 1:1 line is shown.

10.3 Binding of single metal cations interpreted with Model VI

Figure 10.8 shows the Cu and Cd binding data of Benedetti *et al.* (1995) fitted using Model VI, with default parameters for proton binding and electrostatic effects, and with adjustment of four parameters (K_{MA}, K_{MB}, ΔKL_1 and ΔKL_2). Model VI can fit wide-ranging data better than Model V because it has four adjustable parameters (or six if ΔLK_{A1} and ΔLK_{B1} are allowed to differ, and K_{sel} values are adjusted) instead of the two

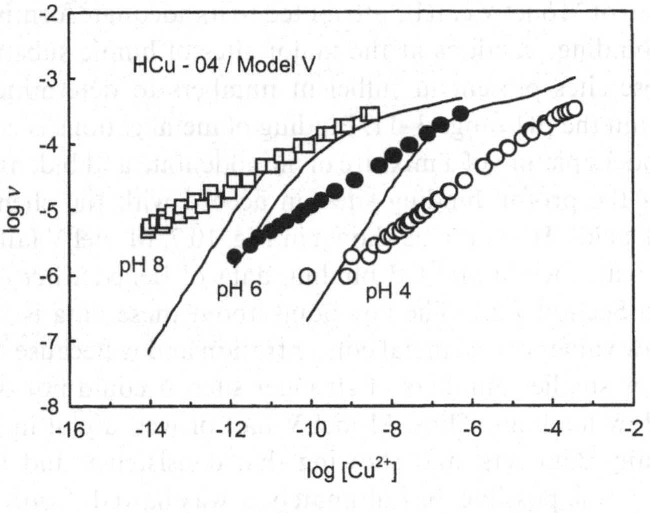

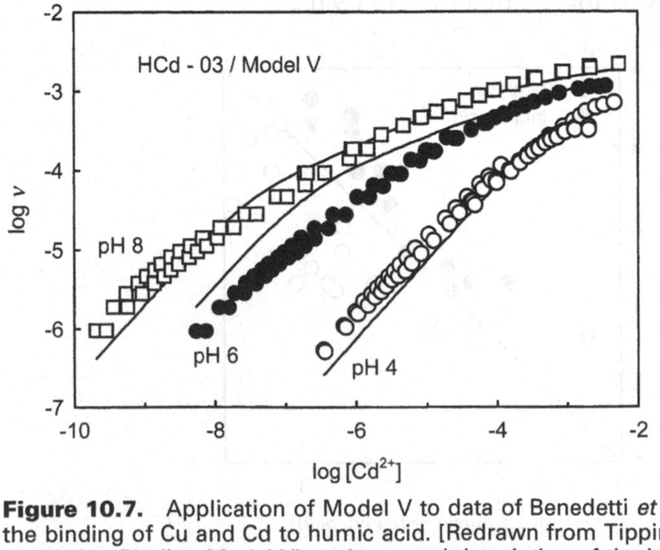

Figure 10.7. Application of Model V to data of Benedetti *et al.* (1995) for the binding of Cu and Cd to humic acid. [Redrawn from Tipping, E. (1998), Humic Ion-Binding Model VI: an improved description of the interactions of protons and metal ions with humic substances, *Aq. Geochem.* **4**, 3–48, Fig. 2, with kind permission from Kluwer Academic Publishers.]

available in Model V. The extra parameters allow the small numbers of strong binding sites to be represented. Table 10.4 shows the optimised parameter values for the two data sets of Fig. 10.8.

10.3.1 Reducing the number of adjustable parameters in Model VI

Although all four parameter values could be estimated from the data in Fig. 10.8, most other available data sets are too small to allow them all to

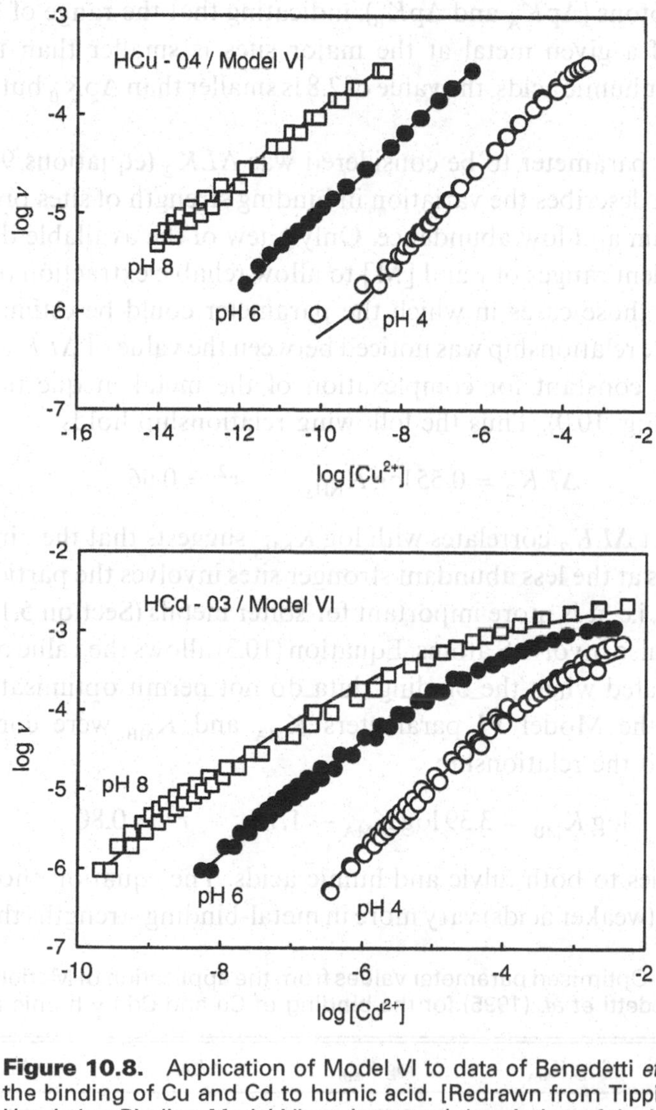

Figure 10.8. Application of Model VI to data of Benedetti *et al.* (1995) for the binding of Cu and Cd to humic acid. [Redrawn from Tipping, E. (1998), Humic Ion-Binding Model VI: an improved description of the interactions of protons and metal ions with humic substances, *Aq. Geochem.* **4**, 3–48, Fig. 2, with kind permission from Kluwer Academic Publishers.]

be found unambiguously. Therefore ways were sought to reduce the number of adjustable parameters in the model, while preserving its structure (Tipping, 1998b). Firstly, it was decided to set the parameters ΔLK_{A1} and ΔLK_{B1} (equations 9.42 and 9.43) to the same value for each metal, called ΔLK_1. Then analysis of data sets for which fitting was sensitive to the value of ΔLK_1 showed that it was possible to set it to the same value (2.8) for all metals. Comparison with Table 10.1 shows that, for fulvic acids, this value is smaller in magnitude than the corresponding ones for protons (ΔpK_A and ΔpK_B), indicating that the range of binding strengths of a given metal at the major sites is smaller than that for protons. For humic acids, the value of 2.8 is smaller than ΔpK_B but greater than ΔpK_A.

The next parameter to be considered was ΔLK_2 (equations 9.44 and 9.45), which describes the variation in binding strength of sites present in high, medium and low abundance. Only a few of the available data sets cover sufficient ranges of ν and [M] to allow reliable extraction of ΔLK_2 values. For those cases in which the parameter could be estimated, an approximate relationship was noticed between the value of ΔLK_2 and the equilibrium constant for complexation of the metal in question with ammonia (Fig. 10.9). Thus the following relationship holds

$$\Delta LK_2 = 0.55 \log K_{NH_3} \qquad r^2 = 0.66 \qquad (10.3)$$

The fact that ΔLK_2 correlates with $\log K_{NH_3}$ suggests that the binding of some metals at the less abundant stronger sites involves the participation of N atoms, i.e. it is more important for softer metals (Section 5.1.3). The sites might also involve S atoms. Equation (10.3) allows the value of ΔLK_2 to be estimated when the binding data do not permit optimisation.

Finally, the Model VI parameters K_{MA} and K_{MB} were correlated, according to the relationship

$$\log K_{MB} = 3.39 \log K_{MA} - 1.15 \qquad r^2 = 0.80 \qquad (10.4)$$

which applies to both fulvic and humic acids. The equation shows that type B sites (weaker acids) vary more in metal-binding strengths than type

Table 10.4. Optimised parameter values from the application of Model VI to the data of Benedetti *et al.* (1995) for the binding of Cu and Cd by humic acid.

Data set	$\log K_{MA}$	$\log K_{MB}$	ΔLK_1	ΔLK_2
HCu-04	1.96	5.68	1.40	2.24
HCd-03	1.66	3.27	2.29	0.92

A sites (carboxyl groups), as concluded from Model V (Section 10.2.1), and as expected from results with simple ligands.

By these procedures, the number of adjustable parameters for each metal in Model VI can be varied, depending upon the available data. For a very complete data set, six parameters (K_{MA}, K_{MB}, ΔLK_{A1}, ΔLK_{B1}, ΔLK_2 and K_{sel}) can be optimised, while a sparse set can be fitted with a single parameter (K_{MA}).

10.3.2 Linear free energy relationships (LFERs)

Figure 10.10 shows that the values of the principal equilibrium constant in Model VI, K_{MA}, for different metals with humic acid, correlate with the values for fulvic acid, and that values for the two types of humic matter correlate with those for lactic acid (LA). The following LFERs apply

$$\log K_{MA}(\mathrm{FA}) = 0.64 \log K_{MA}(\mathrm{HA}) + 0.79 \qquad r^2 = 0.84 \qquad (10.5)$$

$$\log K_{MA}(\mathrm{FA}) = 0.56 \log K_{MA}(\mathrm{LA}) + 0.82 \qquad r^2 = 0.79 \qquad (10.6)$$

$$\log K_{MA}(\mathrm{HA}) = 0.66 \log K_{MA}(\mathrm{LA}) + 0.36 \qquad r^2 = 0.81 \qquad (10.7)$$

As with Model V, the good correlations with lactic acid values, and the similar values of the constants, support the chemical basis of the model, for the abundant weak sites. The LFERs can be used to estimate K_{MA} values when experimental data are absent.

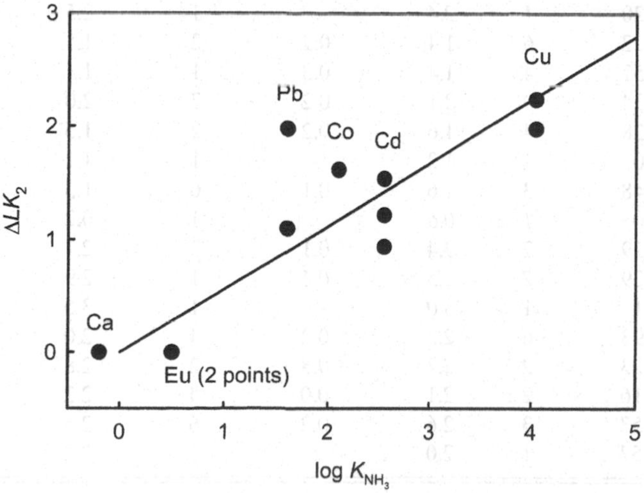

Figure 10.9. Relationship between the Model VI parameter ΔLK_2 and the equilibrium constant for metal–NH_3 complexation.

10.3.3 Model VI parameter values and their influence

Table 10.5 lists the default parameter values obtained by the application of Model VI to data for metal binding by fulvic and humic acids. The values of ΔLK_2 were obtained from equation (10.3), and the values of log K_{MA} by fitting, or from LFERs. Values of log K_{MB} to use in calculations would be obtained from equation (10.4). The data base is reasonably large, although there remains a need to obtain more direct binding data for several environmentally important metals, notably Fe(III), Ag(I) and Hg(II).

Illustrations of how the parameter values influence binding are provided by Fig. 10.11, which shows the variation of $v/[M]$ with v for Ca, Co and Cu calculated using the default values from Table 10.5. The term

Table 10.5. Default values of the constants ΔLK_2 and log K_{MA} in Model VI. [Adapted from Tipping, E. (1998), Humic Ion-Binding Model VI: an improved description of the interactions of protons and metal ions with humic substances. *Aq. Geochem.* **4**, 3–48, Table X, with kind permission from Kluwer Academic Publishers.]

		Fulvic acid			Humic acid		
Metal	ΔLK_2	n	log K_{MA}	sd	n	log K_{MA}	sd
Mg	0.12	2	1.1	0.3	2	0.7	0.4
Al	0.46	2	2.5	0.1	2	2.6	0.2
Ca	0.0	9	1.3	0.3	5	0.7	0.3
V(IV)O	1.74	1	2.4	—	*1*	*2.5*	—
Cr(III)	1.97	*1*	*2.2*	—	1	2.2	—
Mn	0.58	1	1.7	—	1	0.6	—
Fe(II)	0.81	*1*	*1.6*	—	1	1.3	—
Fe(III)	2.20	*1*	*2.4*	—	1	2.5	—
Co	1.22	6	1.4	0.2	2	1.1	0.1
Ni	1.57	4	1.4	0.3	1	1.1	—
Cu	2.34	9	2.1	0.2	7	2.0	0.2
Zn	1.28	4	1.6	0.2	2	1.5	0.2
Sr	0.0	1	1.2	—	1	1.1	—
Cd	1.48	3	1.6	0.1	6	1.3	0.1
Ba	0.0	*1*	*0.6*	—	1	− 0.2	—
Eu	0.29	7	2.4	0.1	3	2.1	0.2
Dy	0.29	*2*	*2.5*	*0.1*	1	2.9	—
Hg	5.1	1	*3.0*	—	1	3.5	—
Pb	0.93	6	2.2	0.2	4	2.0	0.2
Th	0.23	2	*2.7*	*0.3*	2	2.8	0.0
$U(VI)O_2$	1.16	2	2.1	0.0	4	2.2	0.3
Am	1.57	3	2.6	0.2	6	2.5	0.3
Cm	1.57	1	2.0	—	2	2.2	0.4

Values in italics are derived from linear free energy relationships.
The number of estimates in each case is denoted by n, sd is standard deviation.
Source: Values from Tipping (1998b) and Lead *et al.* (1998).

$v/[M]$ can be regarded as a 'local' equilibrium constant, applying to the particular value of v, so the plots demonstrate how binding strength varies with site occupancy. In terms of generating sites with high affinity, both ΔLK_2 and the presence of the tridentate sites are important. The three metals (Ca, Co and Cu) differ greatly in their strengths of binding to humic matter. Thus Ca, for which ΔLK_2 is zero, binds less strongly than Co at low values of v, and since the values of log K_{MA} for the two metals are similar, this is due to the relatively high value of ΔLK_2 for Co. The binding

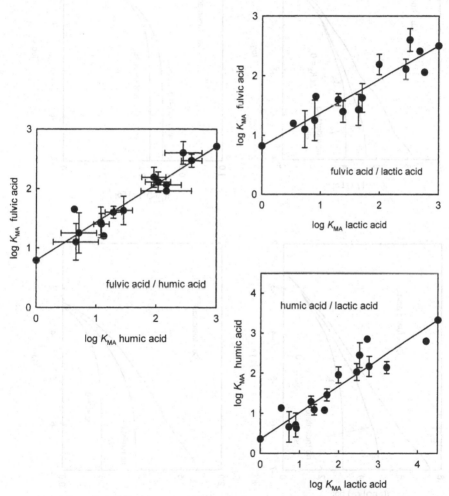

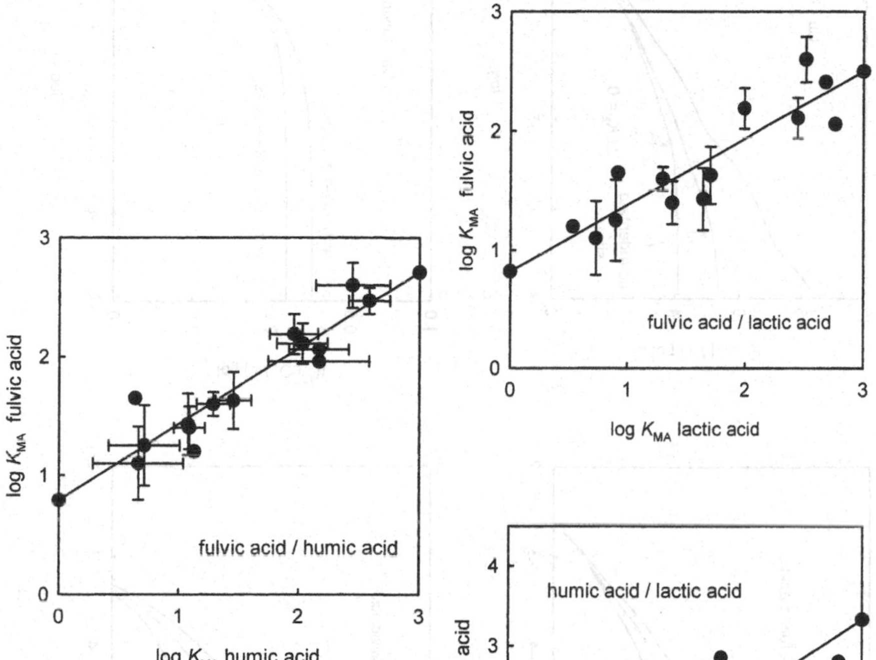

Figure 10.10. Linear free energy relationships derived from values of log K_{MA} for fulvic acid, humic acid and lactic acid. Each point represents a different metal. The error bars represent one standard deviation, and the lines are regression fits. [Redrawn from Tipping, E. (1998), Humic Ion-Binding Model VI: an improved description of the interactions of protons and metal ions with humic substances, *Aq. Geochem.* **4**, 3–48, Fig. 6, with kind permission from Kluwer Academic Publishers.]

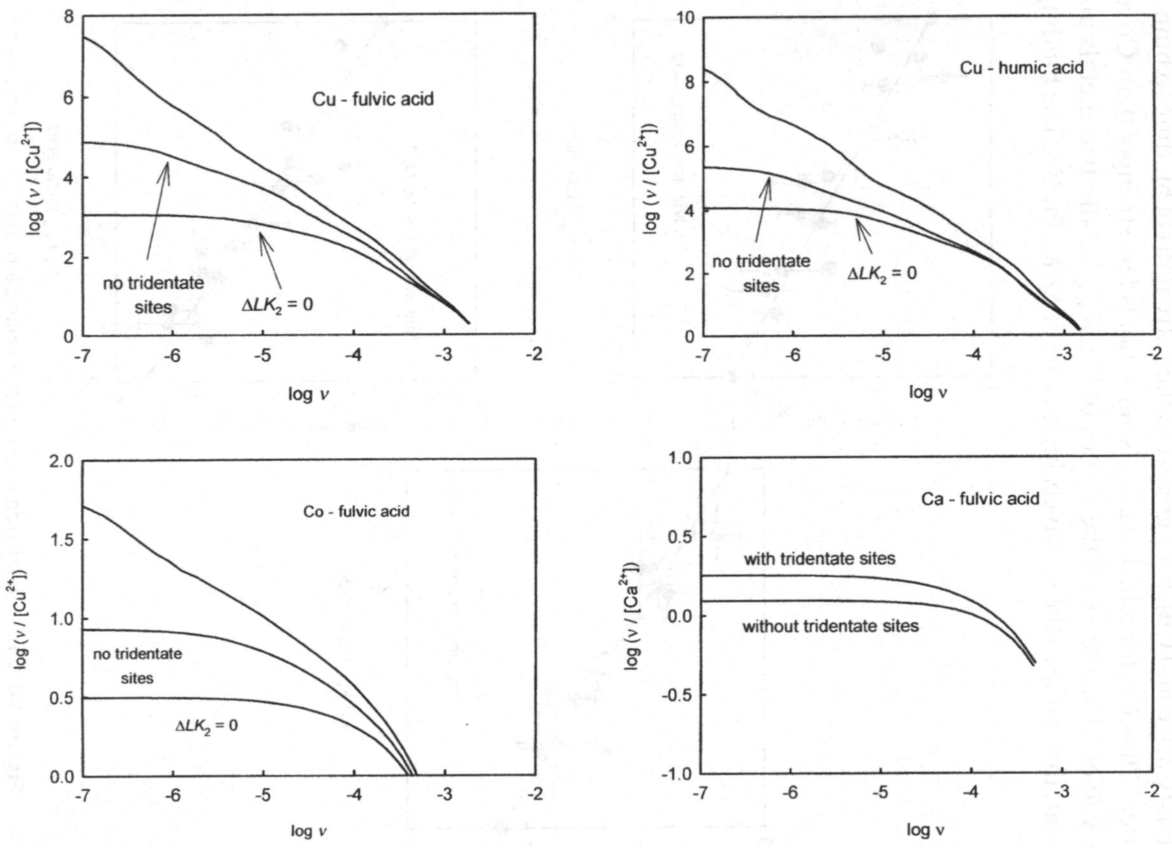

Figure 10.11. Binding of metals by humic substances computed with Model VI using default parameters (Table 10.5). The unlabelled line in each panel is for the full model. The other lines show the effect of omitting tridentate sites, but with ΔLK_2 kept at the default value, and of setting ΔLK_2 to zero, while maintaining the tridentate sites. All plots refer to pH 6, $I = 0.1$ mol dm^{-3}.

of Cu at low v is very much stronger than that of Co, due to its greater values of both $\log K_{MA}$ and ΔLK_2, which means that the model firstly makes stronger abundant sites for Cu, and then increases their affinity to a greater degree in generating the strong sites. It can therefore be appreciated that ΔLK_2 plays a major rôle in determining the strength of trace level binding of metals by humic substances.

Table 10.5 shows that very strong binding by the rare sites is expected for some metals (Cr(III), Fe(III), Cu, Hg), but not for others (e.g. alkaline earths, Al, Eu, Th). Since the strength of binding at the rare sites is based on a rather sketchy correlation with ammonia complexation (equation 10.3), it follows that there is a major research need to investigate strong sites for more metals. If they really are dependent upon nitrogen-containing moieties then the possibility arises of differences among humic samples from different origins, because N contents are variable (Table 2.4). If so, similarities in metal binding among different humic samples might be restricted to the weaker abundant sites.

10.3.4 Contributions of different binding sites

Figure 10.12 gives some insight into how the Model VI binding sites fill as the free metal concentration is increased. At $\log [Cu^{2+}] = -15$, six sites have fractional occupancies (θ) greater than 0.1, and they are ranked in order of θ in Table 10.6, which also shows the top six sites in terms of v. All the sites are strong tridentate sites. Because the sites with the highest affinity (i.e. highest θ) are not necessarily as abundant as those of lower affinity, the sites in each list are not in the same order, although they are the same six in this case. The type B proton-binding sites (i.e. 5, 6, 7 and 8)

Table 10.6. Strong metal-binding sites for Cu in Model VI, ranked by values of θ and v_S/v (cf. Table 9.10 and Fig. 10.12) for $\log [Cu] = -15$.

Site	θ	Proton-binding sites	Abundance	Site	v_S/v	Proton-binding sites	Abundance
T13	0.714	5,6,7	6×10^{-9}	T9	0.408	1,5,6	3.6×10^{-8}
T9	0.583	1,5,6	3.6×10^{-8}	T11	0.181	3,5,7	3.6×10^{-8}
T11	0.259	3,5,7	3.6×10^{-8}	T5	0.165	1,2,5	7.2×10^{-8}
T14	0.225	5,6,8	6×10^{-9}	T6	0.097	3,4,6	7.2×10^{-8}
T5	0.118	1,2,5	7.2×10^{-8}	T13	0.083	5,6,7	6×10^{-9}
T6	0.070	3,4,6	7.2×10^{-8}	T14	0.026	5,6,8	6×10^{-9}

All the sites are from the strong class, i.e. comprising 0.9% of the total (Section 9.6.2). Abundances are given in mol g^{-1}.

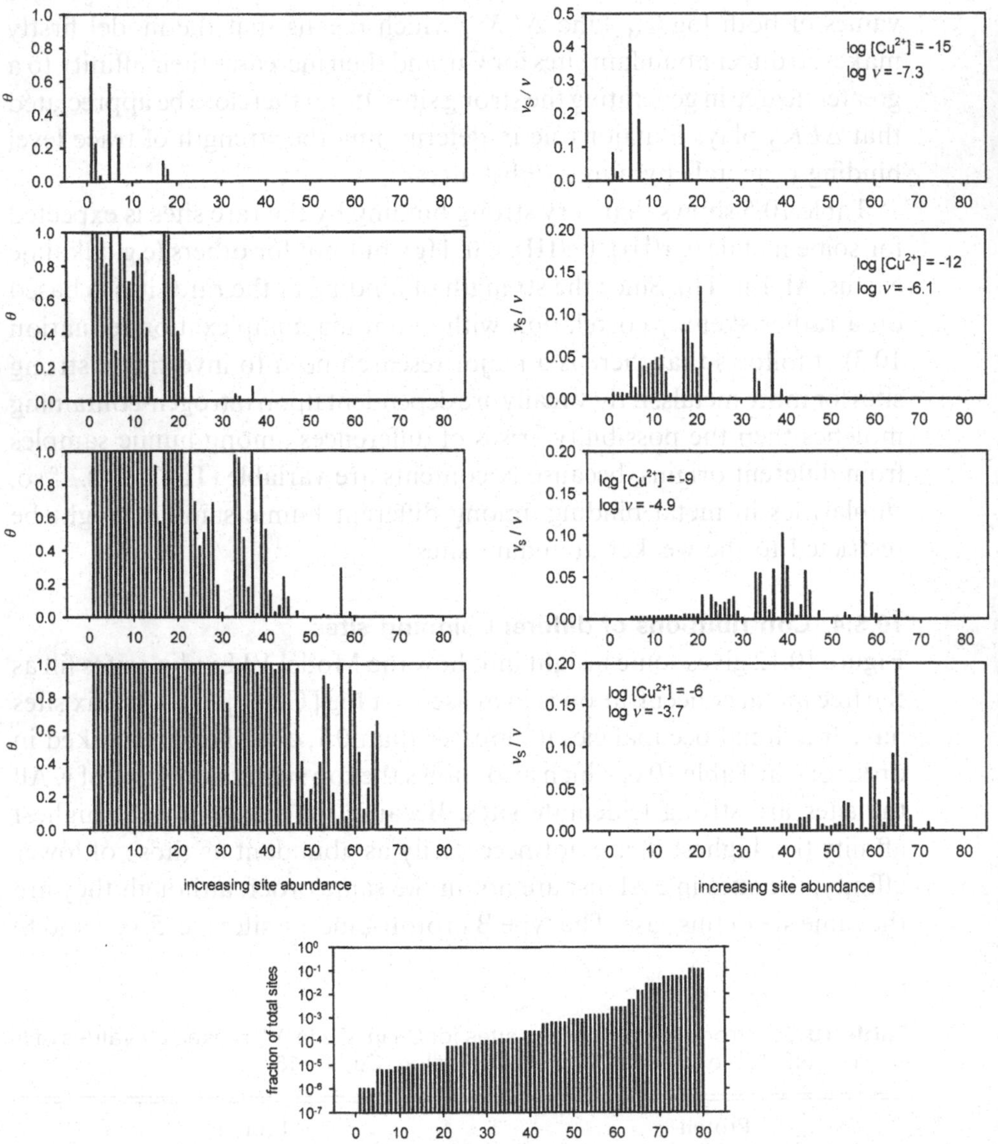

Figure 10.12. Occupancy of individual binding sites in Model VI, exemplified by Cu binding to fulvic acid at pH 6, $I = 0.1$ mol dm^{-3}. Left-hand panels show fractional site occupancy (θ), right-hand panels show values of ν for individual sites (ν_S), relative to the overall value of ν (mol g^{-1}). The *x*-axis shows binding sites ranked in order of abundance (cf. Table 9.10). The bottom centre panel shows the frequency distribution of site abundances.

are important in forming the strong metal-binding sites. They outnumber the type A sites by two to one in both lists, although in the humic material as a whole the type A sites are twice as abundant as the type B. The order is not the same for each condition, because the different combinations respond differently to pH, ionic strength, competition etc.

At higher log $[Cu^{2+}]$, more sites become filled, and the contribution of the weaker sites to the total binding becomes increasingly significant. At log $[Cu^{2+}] = -6$, the abundant (weak) bidentate sites are dominant, and under such conditions Model VI is operating similarly to Model V.

10.3.5 Uncertainty in default values of log K_{MA}

The default values of log K_{MA} derived for Model VI (Table 10.5) are simple averages of values obtained from different data sets. Variations in log K_{MA} among data sets may be due to errors in the data, and to errors in the model, arising from its application to data covering different experimental conditions. However, the main reason for the variation is probably differences among the samples of humic substances used to obtain the data. In turn, such differences may arise from differences in isolation procedures, or they may reflect variability in the natural materials. Whatever the reason(s), it is informative to consider errors in the prediction of binding that might arise from the use of the model with its default parameter set. Inspection of Table 10.5 shows standard deviations of 0.1 to 0.4 in log K_{MA} for different metals. Let us therefore examine the effect on the prediction of binding of adding or subtracting 0.3 to log K_{MA}, varying log K_{MB} also, using equation (10.4). The plots in Fig. 10.13 show that predicted values of log ν for Ca, Co and Cu can vary by up to 1.1 from values predicted by the default parameters, although overall the difference is more like 0.5. The differences are greatest at low loadings. They are greater than the range in log K_{MA} because of the formation of multidentate sites. When using the model to estimate cation–humic interactions in the environment, such uncertainty in prediction should be borne in mind, and there is a clear need to compare model predictions with measurements made on natural samples (see Chapters 13 and 14).

10.4 Application of the NICCA model

Kinniburgh *et al.* (1999) applied the NICCA–Donnan model to their own data for the binding of several metals by a sample of 'purified peat humic acid' (PPHA). The model fits to the data are excellent, as shown in Fig. 10.14. The parameter values, listed in Table 10.7, are difficult to interpret

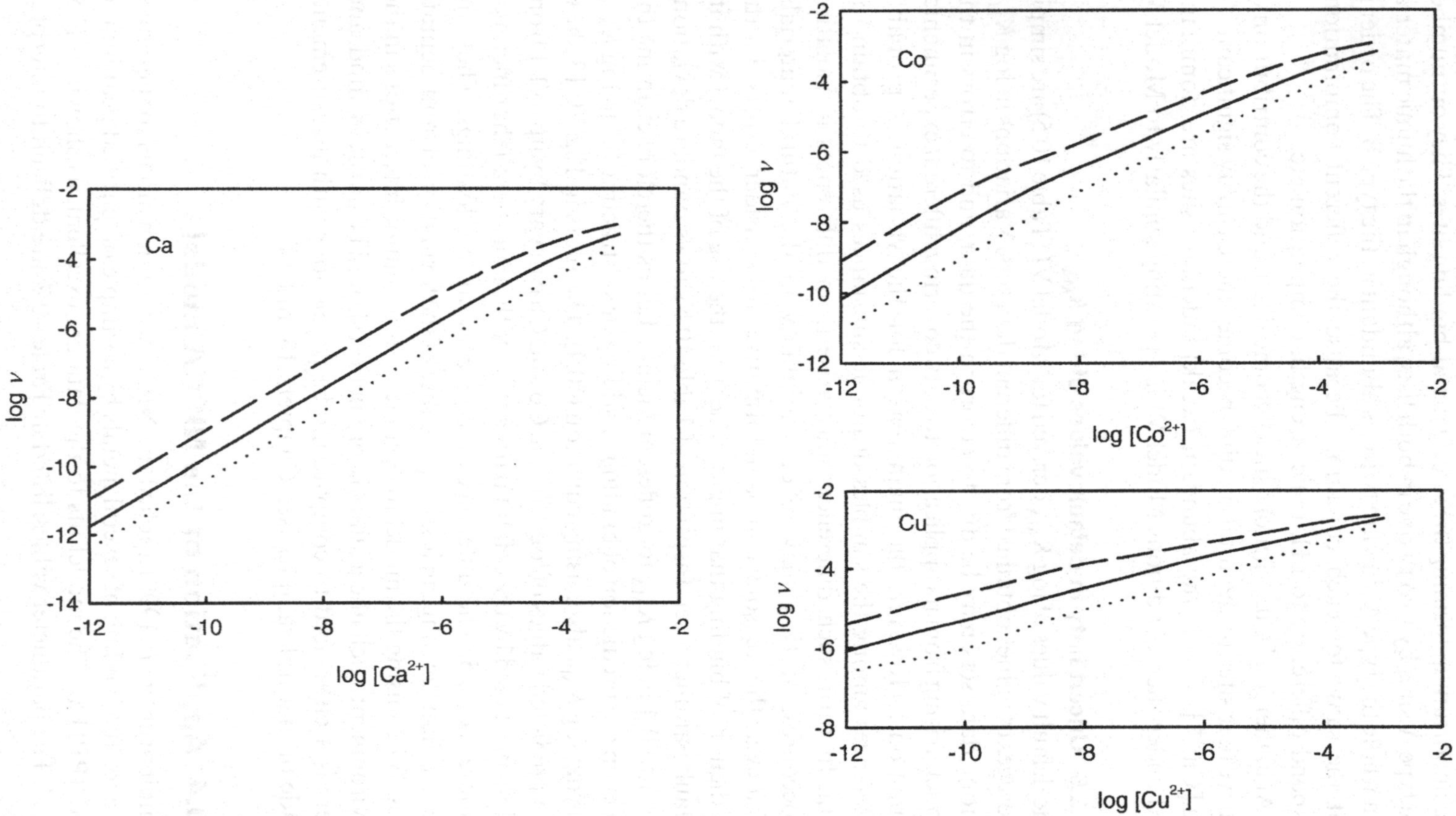

Figure 10.13. Influence of differences in log K_{MA} on the prediction by Model VI of metal binding by fulvic acid at pH 6.0, $I = 0.1$ mol dm^{-3}. In each case the solid line refers to the default value of log K_{MA}, the dashed line to (log $K_{MA} + 0.3$), and the dotted line to (log $K_{MA} - 0.3$).

in terms of simple chemistry, because the binding is governed by a combination of values of log K and the heterogeneity parameters n_1, n_2, p_1 and p_2. Therefore comparisons of log K values for different cations are not meaningful. The values of n show how the binding site strengths vary for the different metals: the smaller is n, the more heterogeneous are the binding strengths. This can be appreciated by considering the plots of $\log(\nu/[M])$ vs. $\log \nu$ for pH 6, $I = 0.1 \text{ mol dm}^{-3}$, shown in Fig. 10.15 (upper panel), which were generated with the parameter values of Table 10.7. According to the NICCA–Donnan model, the site binding strengths

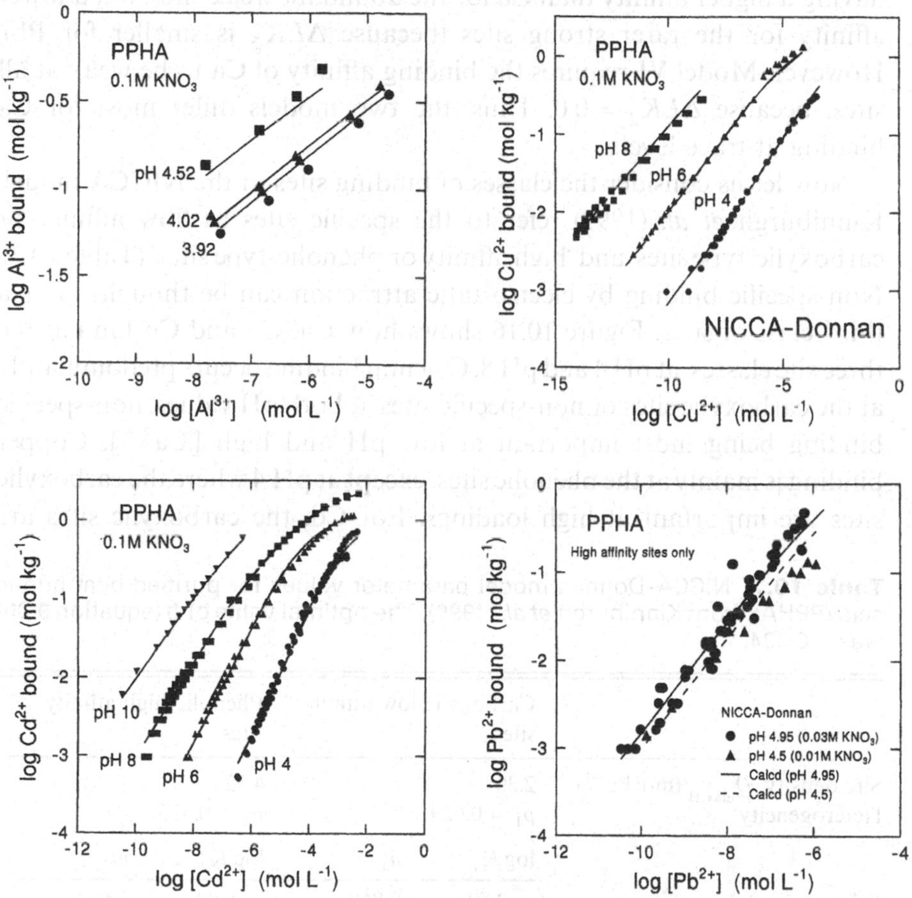

Figure 10.14. Metal binding data for purified peat humic acid, fitted with the NICCA–Donnan model. [From Kinniburgh, D.G., van Riemsdijk, W.H., Koopal, L.K., Borkovec, M., Benedetti, M.F. & Avena, M.J. (1999), Ion binding to natural organic matter: competition, heterogeneity, stoichiometry and thermodynamic consistency, *Coll. Surf. A* **151**, 147–166, with permission from Elsevier Science. IPR/18-1C British Geological Survey. ©NERC. All rights reserved.]

vary more for Ca and Cu than they do for Cd and Pb, which leads to the somewhat paradoxical prediction that at $\log v < -7$ the binding of Ca is stronger than that of Cd, while below $\log v = -9.5$ Ca binding is stronger than that of Pb (not shown in Fig. 10.15). There are no experimental data for these metals at such low values of v, and so the predictions are not testable. Indeed, the data set from which the Ca parameters were extracted only covered a narrow range of $\log v$, -4 to -3. Plots derived from Model VI for the same data are shown in the lower panel of Fig. 10.15. In common with the NICCA–Donnan model, Model VI predicts a 'crossover' for Cu and Pb, which it attributes to Pb having a higher affinity than Cu for the abundant weak sites, but a lower affinity for the rarer strong sites (because ΔLK_2 is smaller for Pb). However, Model VI requires the binding affinity of Ca to be weak at all sites, because $\Delta LK_2 = 0.0$. Thus, the two models differ most for Ca binding at trace levels.

Now let us consider the classes of binding sites in the NICCA model. Kinniburgh *et al.* (1999) refer to the specific sites as 'low affinity or carboxylic-type sites' and 'high affinity or phenolic-type sites' (Table 10.7). Non-specific binding by electrostatic attraction can be thought of as a third class of sites. Figure 10.16 shows how Ca, Cu and Cd bind at the three site classes, at pH 4 and pH 8. Calcium binding occurs predominantly at the carboxylic sites or non-specific sites at both pH values, non-specific binding being most important at low pH and high $[Ca^{2+}]$. Copper binding is mainly at the phenolic sites, except at pH 4 where the carboxylic sites are important at high loadings. For Cd, the carboxylic sites are

Table 10.7. NICCA–Donnan model parameter values for purified peat humic acid (PPHA), from Kinniburgh *et al.* (1999). The optimal value of b (equation 9.30) was -0.334.

	Carboxylic/low affinity sites		Phenolic/high affinity sites	
Site density, $Q_{max,H}$ ($mol\,kg^{-1}$)	2.30		4.32	
Heterogeneity	$p_1 = 0.629$		$p_2 = 0.423$	
	$\log K_1$	n_1	$\log K_2$	n_2
H^+	2.89	0.869	8.83	0.594
Al	—	—	11.49	0.285
Ca	-2.18	0.539	-2.11	0.265
Cu	0.69	0.538	7.41	0.347
Cd	0.00	0.797	2.30	0.498
Pb	—	—	6.26	0.632

dominant at pH 4 and high [Cd^{2+}], while the phenolic sites play a greater rôle at pH 8 and at low [Cd^{2+}]. In their modelling of data for Al and Pb binding, Kinniburgh *et al.* (1999) assumed that only the phenolic sites were involved in metal binding. The importance of the phenolic sites in the NICCA model as applied to PPHA stems from (a) their greater abundance, compared to carboxyl sites, and (b) the need to account for the strong pH dependence of binding displayed by the metals with high affinity (i.e. Al, Cu, Pb).

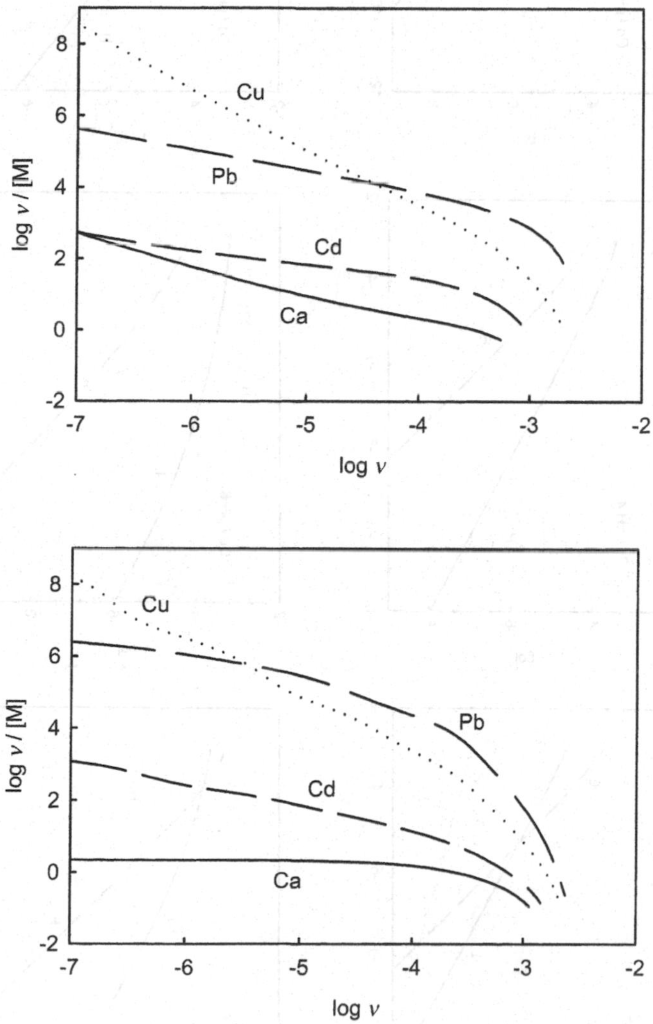

Figure 10.15. Variation of (ν/[M]) with ν, predicted by the NICCA–Donnan Model (upper panel) and by Model VI (lower panel), for the binding of different metals by purified peat humic acid. The value of (ν /[M]) is a measure of the affinity of the humic acid for the metal.

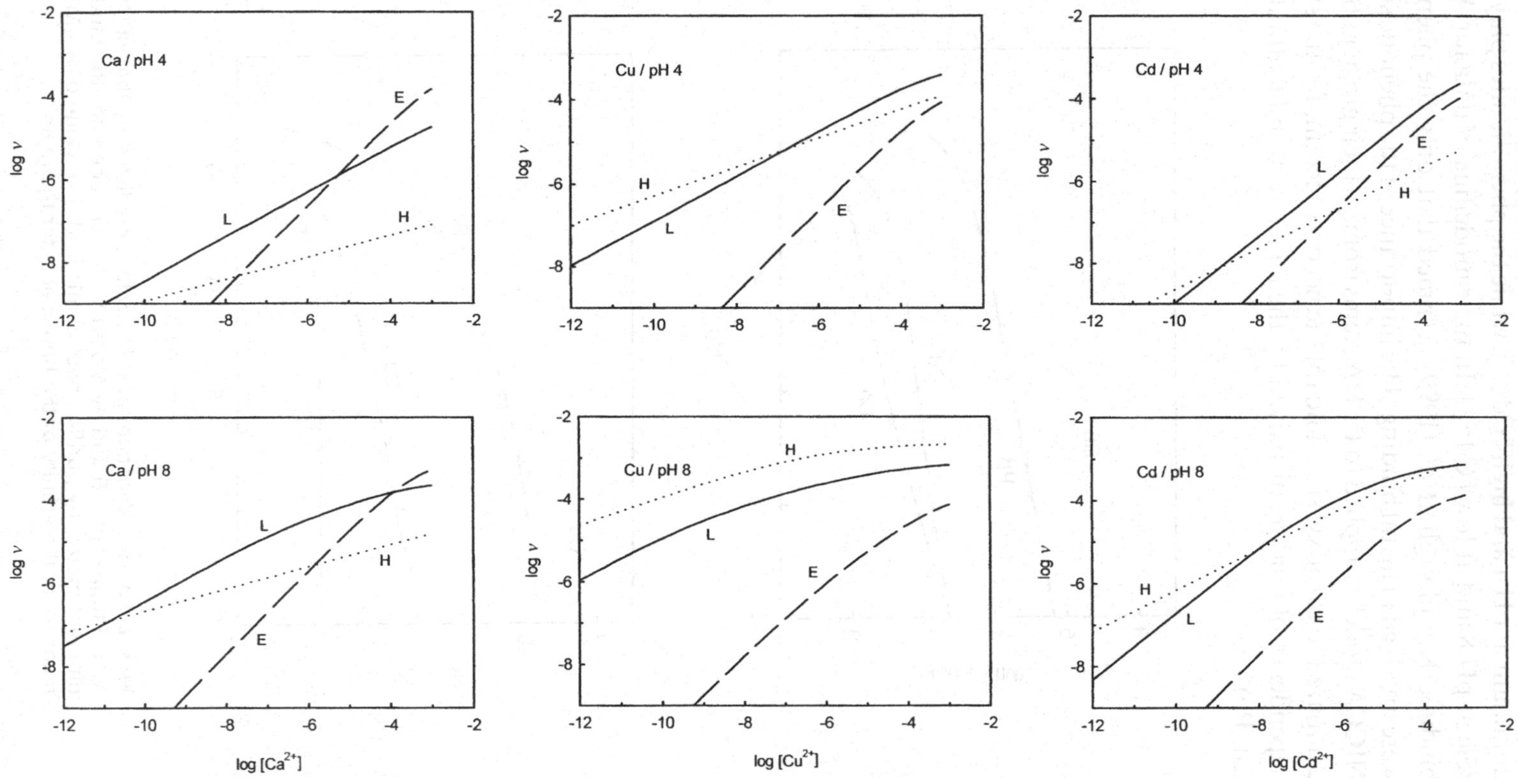

Figure 10.16. Contributions of different sites in the NICCA–Donnan model to the binding of Ca, Cu and Cd by purified peat humic acid, predicted for pH 4 and pH 8, $I = 0.1\ \text{mol dm}^{-3}$. The site classes are H (high affinity/phenolic), L (low affinity/carboxylic) and E (electrostatic).

In their discussion of the application of the NICCA–Donnan model to PPHA, Kinniburgh *et al.* (1999) suggest that the scaling term, i.e. the ratio n_i/n_H in equation (9.51), 'can be interpreted in part at least in terms of multidentate binding and site sharing'. This is reasonable insofar as the ratios force the total number of specific binding sites for a given metal to be fewer than the total number of sites for protons. Furthermore, a tendency towards multidentate binding (greater for lower n_i) might also be expected to give rise to greater heterogeneity in binding affinity (also greater for lower n_i). Kinniburgh *et al.* (1999) conclude that the NICCA–Donnan model is now best described as 'semi-empirical', since it does not provide a rigorous treatment of multidentate binding.

10.5 Metal binding as a function of ionic strength

Both Model V/VI and the NICCA–Donnan model simulate ionic strength effects on metal binding by sub-models that are parameterised only with data for the ionic strength dependence of proton binding. In Model V/VI, the ionic strength effects are predicted from the parameter P (equation 9.27) and from the assumptions about molecular size and the thickness of the diffuse layer. In the NICCA–Donnan model electrostatic effects are determined by the size of the Donnan volume, determined by the parameter b (equation 9.30). Therefore the effects of ionic strength on the binding of metals are predicted rather than fitted, and the faithfulness with which they are reproduced is an independent model test.

10.5.1 Ionic strength effects predicted with Model VI

The examples shown in Fig. 10.17 demonstrate that Model VI provides a reasonable account of the effects of ionic strength. The results suggest that the effect of ionic strength on binding is greatest for weakly binding metals. Figure 10.18 shows predictions of the effects of ionic strength on the binding of Ca, Cu, Cd and Pb by PPHA, made with Model VI. The plots confirm that log v changes with ionic strength in the order Ca > Cd > Cu. The main reason for the difference among the metals is that the type A (mainly carboxylic) sites are more important for Ca binding than for Cu binding, while the reverse is true for the type B (mainly phenolic) sites. The contributions of the two kinds of site to Cd binding are more similar. The ionic strength effect on the type A sites is greater because the sites are more dissociated. Thus, the binding of cations is to more negatively charged sites, and the electrostatic contribution,

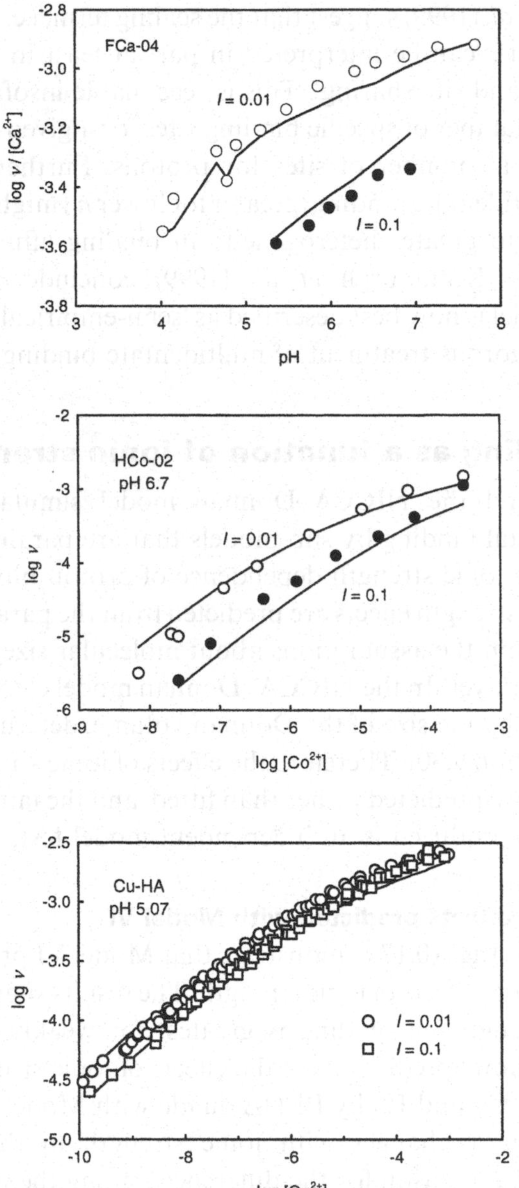

Figure 10.17. Ionic strength dependence of metal binding, observed (points) and calculated with Model VI (lines). The data for Ca are from Marinsky *et al.* (1992), those for Co from Westall *et al.* (1995), and those for Cu from Robertson (1996). [Redrawn from Tipping, E. (1998), Humic Ion-Binding Model VI: an improved description of the interactions of protons and metal ions with humic substances, *Aq. Geochem.* **4**, 3–48, Fig. 7, with kind permission from Kluwer Academic Publishers.]

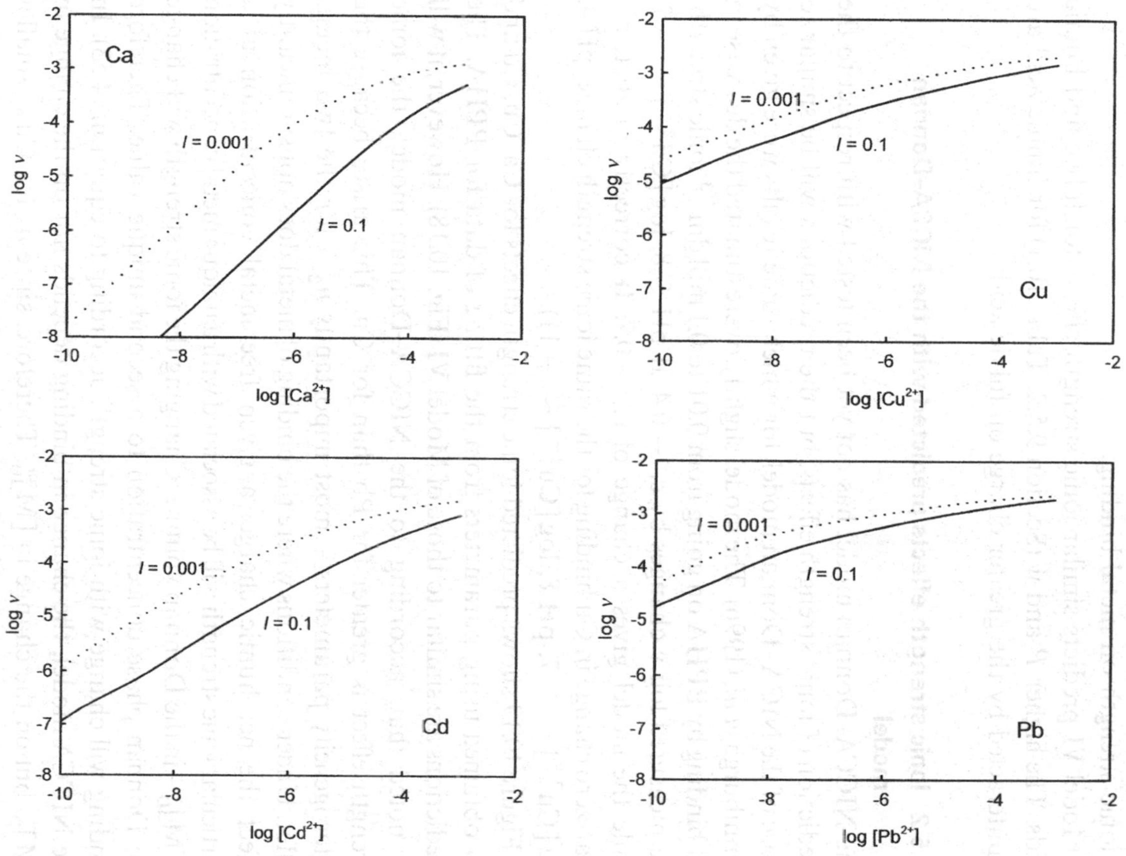

Figure 10.18. Ionic strength dependence of metal binding by purified peat humic acid at pH 6.0, calculated with Model VI.

which depends on the net change in charge at the binding site, is near maximal. The type B sites are significantly protonated at pH values even as high as 10, and so metal binding involves a smaller net change in charge, and therefore a smaller electrostatic effect. The complexity of the sites in Model VI (Table 9.10) precludes a simple description of the effects of ionic strength on metal binding.

Model VI predicts similar ionic strength effects for fulvic and humic acids. The higher *P* and *w* (Section 9.5.2, Table 10.1) for humic acid are counteracted by the greater charge on fulvic acid.

10.5.2 Ionic strength effects predicted with the NICCA–Donnan model

The NICCA–Donnan model has not yet been tested with respect to the prediction of ionic strength effects, but the predictions will be similar to those of the NICA–Donnan model, for which some results were given by Kinniburgh *et al.* (1996). The model slightly overestimated the decrease in Cd binding by PPHA on going from 0.01 to 0.1 mol dm^{-3} ionic strength; the measured log v changes by *c.* -0.4 at pH 6–8, $\log[Cd^{2+}] \sim -7$, while the model gives a change of *c.* -0.7. It correctly predicted a near-zero change in Cu binding for the same ionic strength change (pH 4, $\log[Cu^{2+}] \sim -7$; pH 8, $\log[Cu^{2+}] \sim -11$).

Figure 10.19 shows predicted ionic strength effects for Ca, Cu, Cd and Pb obtained using parameters from the fitting of data for PPHA. The predictions are similar to those of Model VI (Fig. 10.18). However, it will be noted that, according to the NICCA–Donnan model, the ionic strength effect is greater for Pb than for Cu. This arises because the heterogeneity parameters – most importantly n_2 – for the two metals differ. Under conditions where the binding of metal does not significantly affect the net humic charge, a given free metal concentration at a particular ionic strength will be associated with a unique metal concentration – $[M]_D$ – in the Donnan volume. Changing the ionic strength will change the Donnan phase concentration to a second unique value. Therefore binding will change with ionic strength according to equation (4.53). In the NICCA model, the change in binding depends not on the change in $[M]_D$, but on the change in $[M]_D^n$. Therefore, since n_2 for Cu is smaller than the value for Pb (Table 10.7) the change in binding of Cu is less than that for Pb, for a given change in $[M]_D$. Thus the NICCA–Donnan model predicts that sensitivity towards ionic strength increases with n_i.

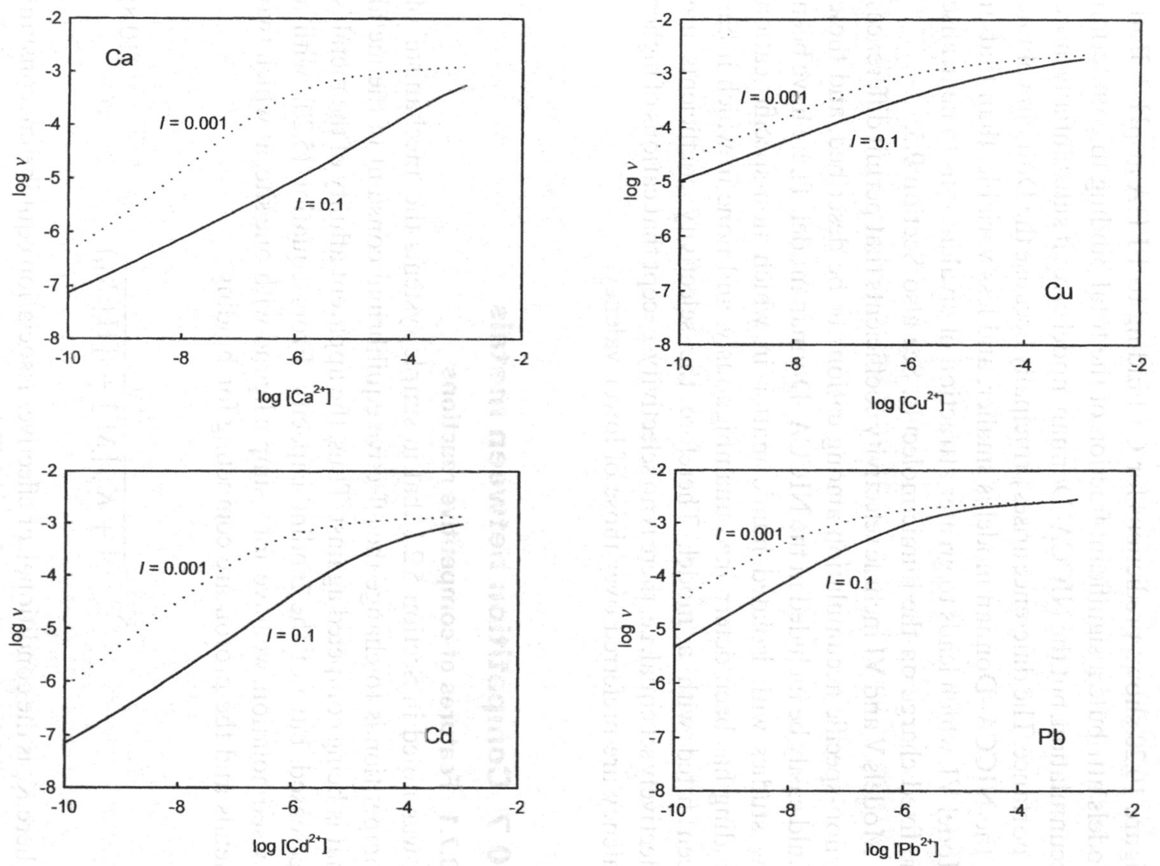

Figure 10.19. Ionic strength dependence of metal binding by purified peat humic acid at pH 6.0, calculated with the NICCA–Donnan model.

10.6 Non-specific binding

Both Model VI and the NICCA–Donnan model include non-specific binding (accumulation of counterions) in their formulations. The contribution of non-specific binding is greatest for metals that bind weakly. Figure 10.20 shows predictions for Ca binding to PPHA at pH 7. Both models attribute a significant fraction of the total binding to counterion accumulation, but the NICCA–Donnan model gives it substantially more importance. The difference arises principally because the Donnan volume in the NICCA–Donnan model is smaller, and less variable, than that in Model VI, which leads to greater attraction of divalent ions to neutralise the fixed charge on the humic molecule. See also Section 9.5.

Models V and VI include selectivity coefficients that permit differences in non-specific accumulation among cations to be described, and these could easily be included in the NICCA–Donnan model. There have been few studies with isolated humic matter in which non-specific cation binding has been determined unambiguously, and none in which it has been fitted with a model. Therefore the selectivity coefficients are effectively set equal, i.e. there is no selectivity, except that cations of higher valency are preferred over those of lower valency.

10.7 Competition between metals

10.7.1 Features of competitive reactions

It was noted in Section 5.2.1 that in simple systems the 'mechanism' of competition is to change the effective equilibrium constant of the metal that is being competed against. Thus, the apparent affinity of the metal is decreased, but not the binding capacity. From equation (5.20), with a revised notation, we have, for a simple ligand with one site, at which two metals and the proton are competing for binding

$$K_c = \frac{K_M}{1 + K_{M'}[M'] + ([H]/K_H)} \qquad (10.8)$$

where K_c is the conditional, or effective, association equilibrium constant for metal M, at given values of [M′], the free concentration of the competing metal, and [H], the free proton concentration. We see that K_c can be diminished by increasing [M′] or [H]. For example, if the terms $K_{M'}[M']$ and $[H]/K_H$ are both equal to 1, then K_c is decreased by a factor of 3, compared to the case where both terms are near zero (low [M′], high

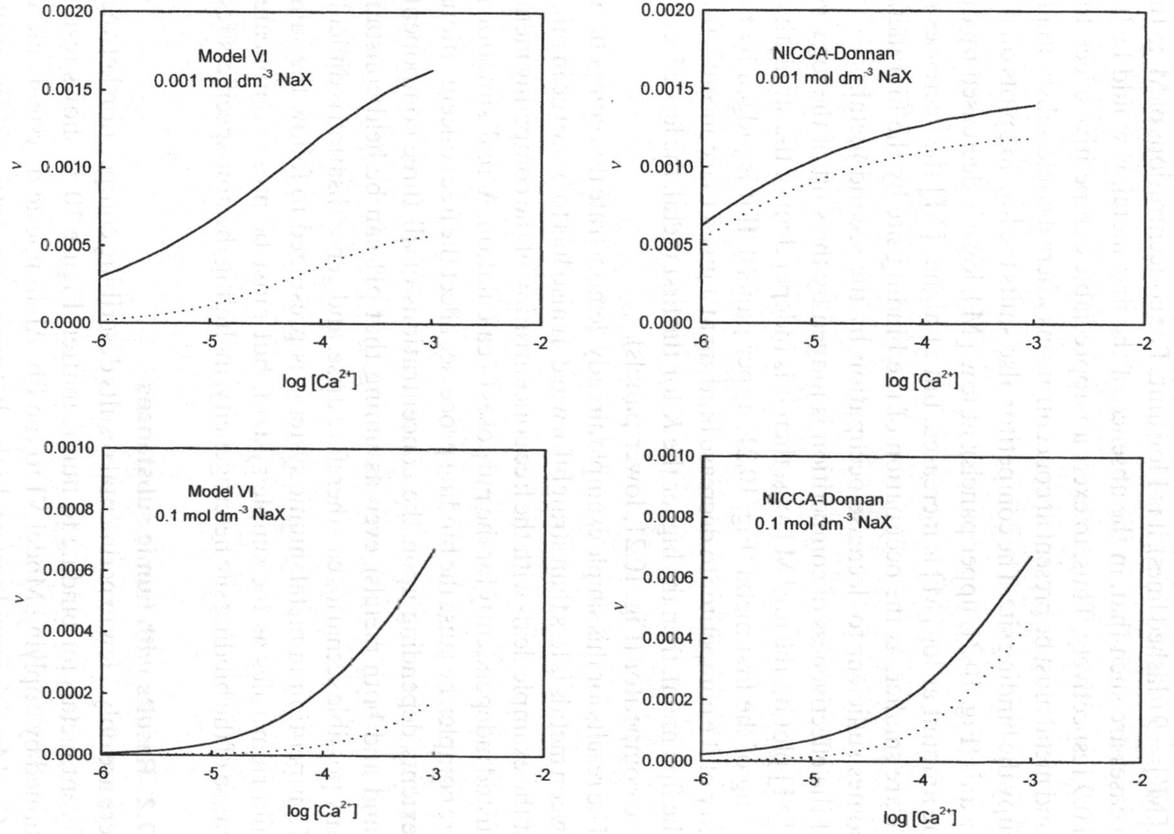

Figure 10.20. Comparison of total (full lines) and non-specific (dotted lines) binding of Ca by purified peat humic acid predicted with Model VI and the NICCA–Donnan model. A 1:1 background electrolyte (NaX) is assumed.

pH). At a given pH, the effect of the competing metal on a plot of log v vs. log [M] is simply to shift it along the log [M] axis (Fig. 10.21, upper panels); if $K_{M'}[M'] = 1$ (dotted lines), the shift is 0.3 log units, while if $K_{M'}[M'] = 9$ (dashed lines), it is 1 log unit. The concentrations of M′ in the two cases are such that, in the absence of the first metal, v' would be 0.5 and 0.9 respectively. Thus, to exert an appreciable competitive effect, the second metal must be present at concentrations where it could substantially occupy the binding site. The competitor also shifts the log–log plots on the log v axis (Fig. 10.21, upper panels); at low [M], log v is decreased by the same amount as log [M] is increased, but at higher [M] the decreases in log v are smaller, as the occupation of the binding site by the first metal becomes sufficient to decrease occupation by the second metal.

If the effectiveness of competition is judged by the shift of the log v vs. log [M] plot on the log [M] axis, then it is independent of the strength of binding of the first metal (Fig. 10.21, upper panels). If it is judged by the ability of the competitor to decrease binding at a given free concentration of the first metal, then the higher the K for the first metal, the less effective is the competitor (Fig. 10.21, lower panels).

The results for this simple example already demonstrate that competition between metals is less than straightforward. It also has to be borne in mind that the example deals with the free concentration of the competing metal, adjusted independently for the purposes of calculation. A 'real' situation is more complex because the binding processes affect the free concentrations, to extents depending upon the concentrations of all three components (ligand and both metals), even assuming that pH can be held constant. Many possible scenarios can therefore arise, and generalisation is difficult.

Competition in metal–humic systems is presumed to follow the same equilibrium rules as the simple system, but must be more complicated because of the binding site heterogeneity, including the non-specific 'sites'.

10.7.2 Results with humic substances

There are only a few experimental results describing competition between different metals for binding to humic matter. Figure 10.22 shows results obtained by applying Model VI to the Pb–Al–humic acid system, and to the Cu–Mg–fulvic acid system. In the first case, optimisation of one model parameter (log K_{MA} for Pb) was necessary, while in the second the log K_{MA} values for both metals had to be adjusted. Note, however, that the optimised values are similar to the default values of Table 10.5. Thus the description of binding as formulated in the model can account for competition effects, but a complete validation of the model with respect to

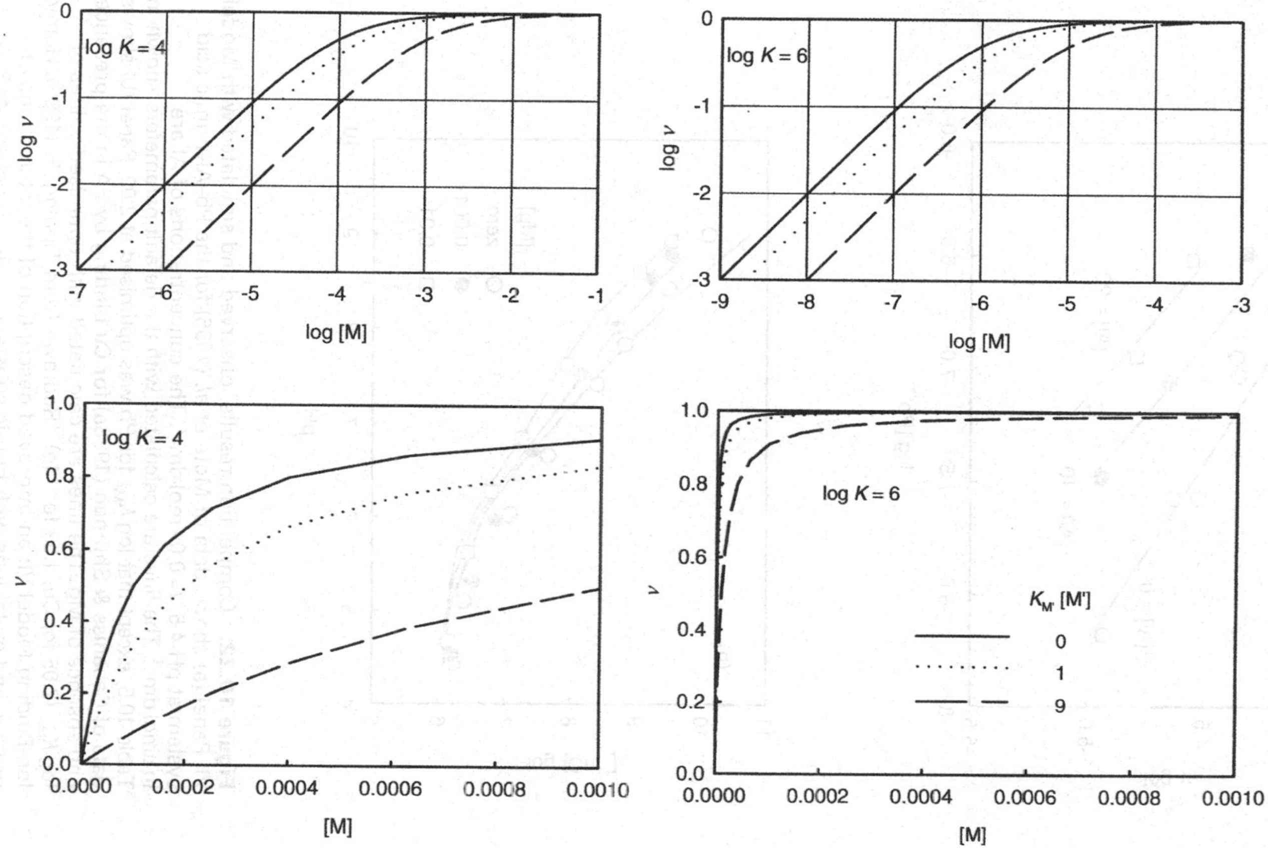

Figure 10.21 Binding of metal M by a simple monodentate ligand, as influenced by metal M'. The ligand binds protons (pK = 4) as well as metals. The plots refer to pH 7. Note the different x-axis scales in the top panels. See Section 10.7.1.

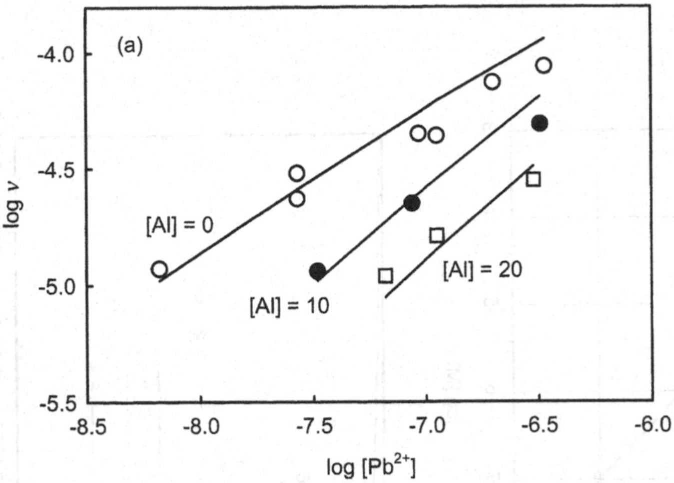

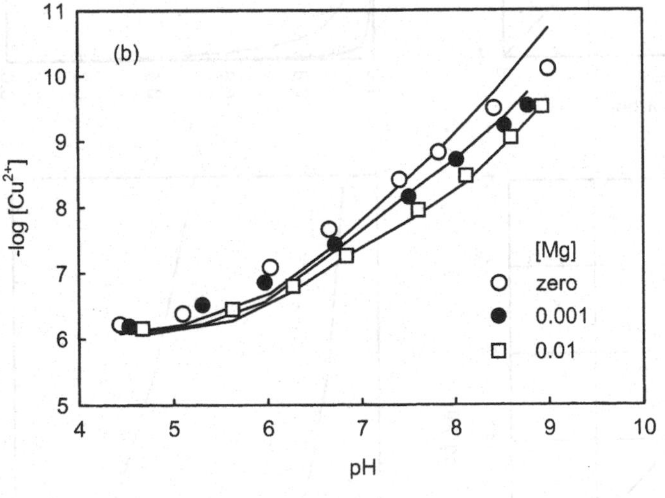

Figure 10.22. Competition results, observed and simulated with Model VI. Panel (a) shows data of Mota *et al.* (1996) for the Pb–Al–humic acid system at pH 4.5, $I = 0.01\ \mathrm{mol\,dm^{-3}}$. The concentrations of Al are in $\mu\mathrm{mol\,dm^{-3}}$. The lines are calculated with the default parameters shown in Table 10.5, except that $\log K_{MA}$ for Pb was optimised at 2.09. Panel (b) shows data of Cabaniss & Shuman (1988a) for Cu binding by FA in the presence and absence of Mg. The lines are calculated with optimised values of $\log K_{MA}$ (2.05 for Cu, 1.09 for Mg). [Redrawn from Tipping, E. (1998), Humic Ion-Binding Model VI: an improved description of the interactions of protons and metal ions with humic substances, *Aq. Geochem.* **4**, 3–48, Figs. 8 and 10, with kind permission from Kluwer Academic Publishers.]

competition has not been achieved because the data available for testing are too few.

The NICCA–Donnan model has also been applied to competition results with success (Fig. 10.23). In the cases shown, parameter values obtained from binding measurements on the single cation systems are

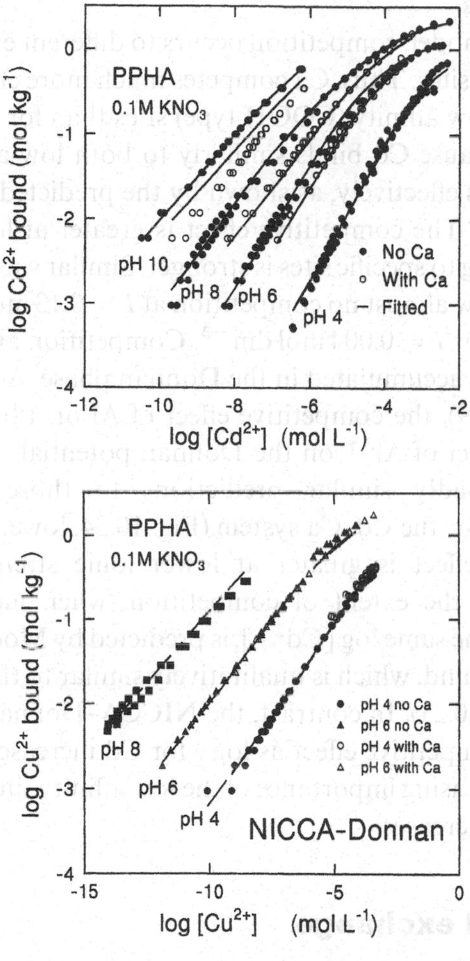

Figure 10.23. Effects of Ca on the binding of Cu and Cd by humic acid, observed and simulated with the NICCA–Donnan model. When Ca was present, its total concentration was 10^{-3} mol dm^{-3}. The simulations of competition effects were obtained using parameter values determined from the results of single-metal experiments. [From Kinniburgh, D.G., van Riemsdijk, W.H., Koopal, L.K., Borkovec, M., Benedetti, M.F. & Avena, M.J. (1999), Ion binding to natural organic matter: competition, heterogeneity, stoichiometry and thermodynamic consistency, *Coll. Surf. A* **151**, 147–166, with permission from Elsevier Science. IPR/18-1C British Geological Survey. ©NERC. All rights reserved.]

able to predict quite well the binding observed when both metals are present. In the Ca–Cu–humic acid system, there is nearly no observed effect of Ca on Cu binding, in agreement with predictions of the NICCA–Donnan model. In terms of model testing, such 'no effect' cases are as important as those where a substantial effect is found.

10.7.3 Model predictions

In the NICCA–Donnan model, competition occurs to different extents at the two classes of binding sites. Thus, Ca competes much more effectively with Cd and Cu for the low affinity (COOH type) sites than for the high affinity (ϕ-OH) sites. Because Cd binds similarly to both low and high affinity sites, Ca competes effectively, as shown by the predicted plots in Fig. 10.24 (upper panels). The competitive effect is greater at low ionic strength, where Ca binding to specific sites is stronger. Similar simulations for the Cu–Ca system show almost no competition at $I = 0.13\,\mathrm{mol\,dm^{-3}}$, and modest competition at $I = 0.004\,\mathrm{mol\,dm^{-3}}$. Competition also takes place among counterions, accumulated in the Donnan phase. According to Kinniburgh *et al.* (1999), the competitive effect of Al on Pb binding arises solely from the effect of Al^{3+} on the Donnan potential.

Model VI gives broadly similar predictions to those of the NICCA–Donnan model for the Cd–Ca system (Fig. 10.24, lower panels). Again, the competitive effect is greater at lower ionic strength. An interesting point is that the extent of competition, when judged by comparing values of ν at the same $\log\,[Cd^{2+}]$, is predicted by Model VI to decrease as more Cd is bound, which is qualitatively similar to the trends for a simple ligand (Fig. 10.21). In contrast, the NICCA–Donnan model predicts an increasing competitive effect as $\log \nu$ for Cd increases, which occurs because of the increasing importance of the low affinity sites for Cd binding, as $\log\,[Cd^{2+}]$ increases.

10.8 Proton–metal exchange

10.8.1 Release of protons due to metal binding

When metals bind to humic matter, they tend to displace protons from the binding sites, although this cannot happen if the sites are fully dissociated at the pH in question. For example, a site consisting of one or more carboxyl groups at pH 8 would be almost completely in the dissociated form, and metal binding would not cause any proton release. If the site comprised one or more phenolic OH groups then displacement of protons would occur. Proton release causes a pH decrease unless base is added to compensate. Experiments in which base was added to maintain constant

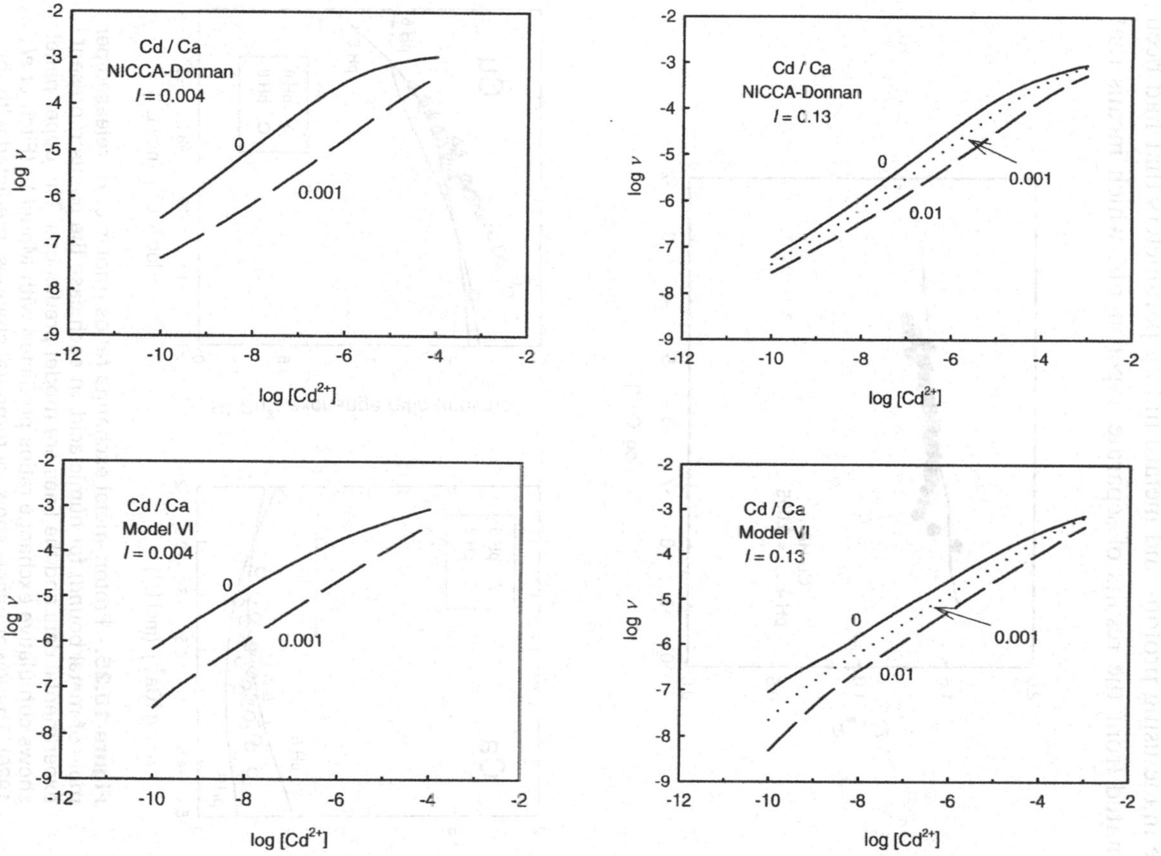

Figure 10.24. Competition by Ca for Cd binding by humic acids, predicted with the NICCA–Donnan model (upper panels) and Model VI (lower panels). The figures next to the individual curves are the (constant) free concentrations of Ca^{2+} in mol dm^{-3}.

pH have been performed, and the data obtained provide information about proton–metal exchange (Section 7.2.3).

Figure 10.25 shows examples of experimental results, and the predictions of Model VI and the NICCA–Donnan model. In each case, the predictions were made using proton- and metal-binding parameters that had been estimated from the results of separate experiments, which means that

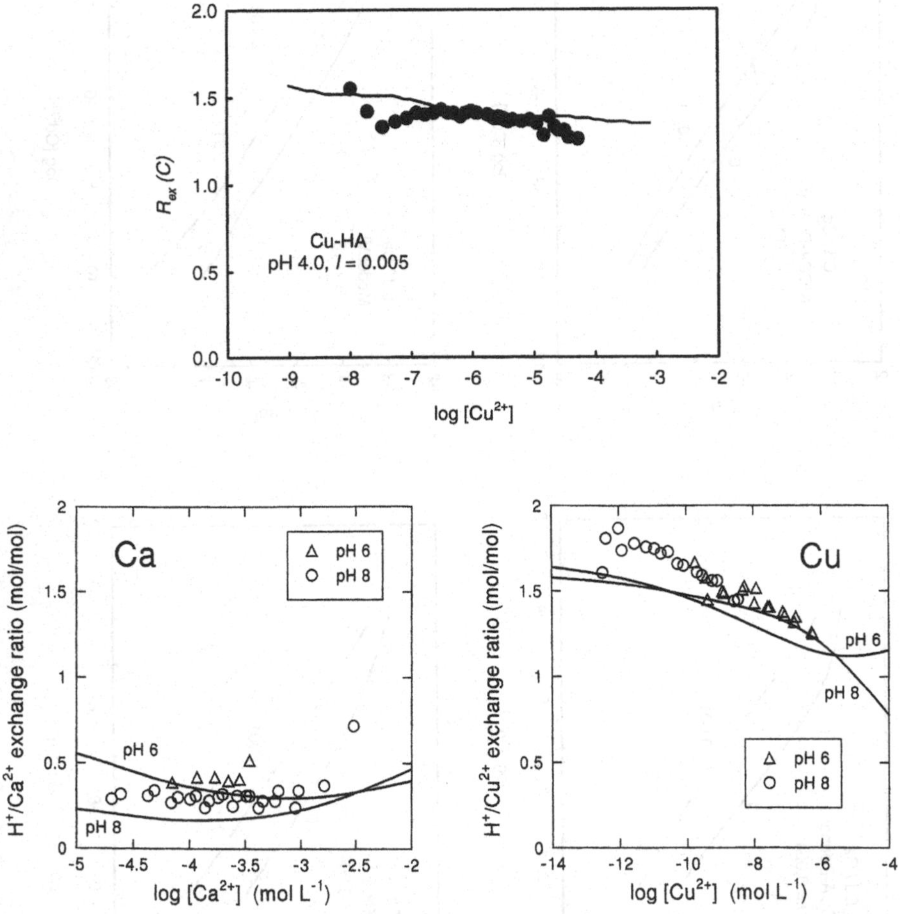

Figure 10.25. Proton–metal exchange ratios (moles of H^+ released per mole of metal bound) for humic acid. In each case the points represent experimental data and the lines are model predictions. The upper panel shows cumulative exchange ratios predicted with Model VI (Fitch *et al.*, 1986). The lower panels show incremental changes, predicted with the NICCA–Donnan model. [The lower panels are reproduced from Kinniburgh, D.G., van Riemsdijk, W.H., Koopal, L.K., Borkovec, M., Benedetti, M.F. & Avena, M.J. (1999), Ion binding to natural organic matter: competition, heterogeneity, stoichiometry and thermodynamic consistency, *Coll. Surf. A* **151**, 147–166, with permission from Elsevier Science. IPR/18-1C British Geological Survey. ©NERC. All rights reserved.]

comparison between observation and prediction is an independent model test.

In Models V and VI, the proton–metal exchange is an inherent part of the model, once the bidentate (and tridentate in Model VI) sites have been postulated. It will be recalled that the abundances of the multidentate sites were estimated on geometrical grounds, assuming a random disposition of sites on the surface of a sphere (Section 9.6). The finding that the model gives exchange ratios quite close to the observed ones supports the underlying assumptions about the nature of humic matter and its interactions with protons and metals. The ability to account for observed proton–metal exchange ratios was a major goal in formulating the NICCA–Donnan model, because the earlier NICA–Donnan model had given too low values.

10.8.2 The effect of metal ions on titration curves

As mentioned in Section 6.2.4, the determination of proton titration curves in the presence of metal ions is one of the classical methods for studying metal–ligand equilibria, but to apply it to humic substances requires a model for proton–metal competition reactions. The comprehensive models for humic matter ought to be able to account for titration data in the presence of metals. Examples for Models V and VI are shown in Fig. 10.26. The model can simulate the titration curves reasonably well, supporting the assumption that protons and metals compete for the major sites, which are principally carboxylic groups for the experimental conditions. The data do not provide any information about the rarer high-affinity sites, because only when high metal loadings are achieved (necessarily to the abundant weaker sites), are there measurable differences in the amounts of base required to attain a given pH.

10.9 Comparison of the NICCA–Donnan model and Model VI

The results and simulations covered in this chapter show that the two models provide useful quantitative descriptions of equilibrium data. They also go some way to explaining how the underlying reactions combine to govern cation binding by humic matter. The models both recognise the extreme heterogeneity of humic matter binding sites, but take it into account in different ways. In the NICCA–Donnan model, the heterogeneity is described with continuous functions, whereas in Model VI a collection of discrete, simple, conventional, chemical reactions is used. Several

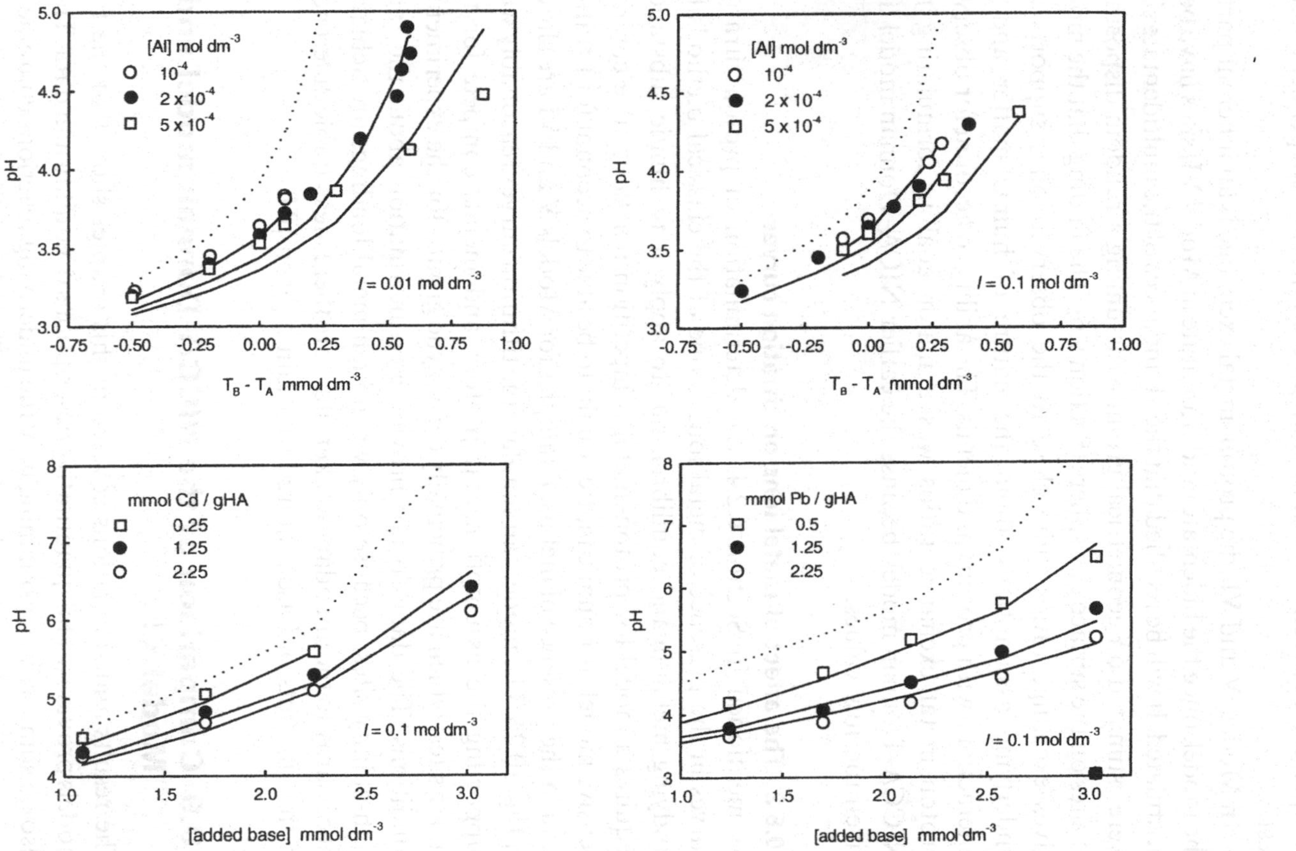

Figure 10.26. Proton displacement from humic acid by bound metals. The two top panels refer to Al (Tipping *et al.*, 1988b), the bottom panels to Cd and Pb (Stevenson, 1976). The solid lines are Model VI fits, using log K_{MA} values of 2.75 (Al), 1.27 (Cd) and 2.01 (Pb). Dotted lines are expected results in the absence of metals.

authors have stated that continuous functions are more realistic than discrete models, and better fits can indeed be achieved by the continuous approach. However, this is not the real difference in underlying principle between the two models, since it would be possible to increase the number of sites in Model VI to get closer to a continuous description, and without involving more parameters. In fact, the main difference between the two models lies in the use by the NICCA–Donnan model of equations that, although superficially like conventional equilibrium expressions, do not follow the Law of Mass Action, and so must be regarded as empirical. Model VI is based on conventional chemistry, and its intrinsic equilibrium constants for the abundant weak sites have values similar to those for simple compounds. However, the description of stronger sites in Model VI is more empirical.

In their full forms, the two models are very similar with regard to the number of adjustable parameters required to fit data. A feature of the development of Model VI has been to reduce the number of fitting parameters, by correlating one with another, and by fixing some at 'universal' values. This has the advantage of allowing parameter values to be extracted from small data sets. Thus it has been possible to apply Model VI to a large number of data sets, yielding 'best average' parameter values (Tables 10.1 and 10.5). To date, the NICCA–Donnan parameter values have been derived from a small number of extensive data sets, but work is in progress (C.J. Milne & D.G. Kinniburgh) to extend its applicability. For Model VI, parameter values for humic substances have been estimated from results for simple compounds, principally through the use of LFERs (Section 10.3.2). A similar strategy may also be possible for the NICCA–Donnan model.

Being a much simpler model in terms of the number of equations required for its description, the NICCA–Donnan model is computationally faster than Model VI. The continuous distribution function implies very many sites, but binding to them is computed as a distribution from the outset. In contrast, to compute metal binding in Model VI, each of its 80 binding sites has to be considered separately at each iteration.

10.10 Application of the models to field situations

The NICCA–Donnan model and Model VI can evidently encapsulate much of the information about cation–humic interactions that has been gained from laboratory studies on isolated humic substances. The models, together with their parameter values, therefore provide the means to predict interactions in the natural environment. A major goal of research into humic substances, carried out by numerous researchers over several decades, has therefore been attained. Of course, there are needs for more experimental work in the laboratory, and effort is required to establish averages and ranges of parameter values with more certainty. No doubt the models can be improved. However, much greater emphasis can now be placed on the gathering of field data to test model performance in the real world (see Chapters 13 and 14).

11

Predictive modelling

The comprehensive models described in Chapter 9, and evaluated in Chapter 10, depend upon fitting data to equations. In Model VI, assumptions are made about molecular size and shape in the electrostatic sub-model, and in the calculation of proximity factors (defining multidentate sites). These aspects of Model VI can therefore be regarded as predictive, i.e. not dependent upon data fitting. The purpose of this chapter is to discuss how predictive modelling can be taken further, the ultimate aim being to describe cation binding from knowledge about humic molecular properties, structures and sizes, without the need to extract parameters from binding data.

11.1 Electrostatic interactions

As has already been discussed in Section 9.5, electrostatic interactions can be simulated given information or assumptions about molecular size, shape, solvent permeability and net charge. Some work has been done to explore the influence of molecular size heterogeneity, which may influence the overall binding characteristics of a humic sample. Thus, a physically more realistic model is obtained by the strategy of Bartschat *et al.* (1992), i.e. postulating the humic matter to consist of molecules with a range of electrostatic characteristics, dependent upon molecular size. Implementation requires knowledge or assumptions about the size distribution of the sample.

As well as size, Coulombic effects depend upon the conformations of the humic molecules, the extent to which the structures are penetrated by water, and the degree to which conformations change as a function of molecular charge and solution conditions. Relevant experimental data are available (Sections 2.4.4 and 12.1), and can be used to guide simulation modelling of Coulombic effects. More ambitiously, the conformations of molecules might be calculated from their primary structure, using molecular dynamics methods. A prime example of the application of such methods is in the prediction of protein folding from peptide sequence, which is a major field of research in biophysical chemistry (e.g. Creighton, 1992; Dill *et al.*, 1995). Some preliminary work has been done with humic substances, based on hypothetical 'average' primary structures in systems with small numbers of water molecules (Schulten, 1996; Shevchenko *et al.* 1999). However, the reliable modelling of humic conformations in the fully hydrated state is yet to be achieved.

11.2 Binding sites

The chemical complexity of humic matter means that a very large number of arrangements of molecular groupings could provide binding sites for cations. Information on possible groupings is available, principally from spectroscopic methods (see Chapters 2 and 8). To use such information to predict cation binding by humic substances, some description is required of the way in which individual groups come together to form multidentate sites. This having been done, the binding properties of the sites can be predicted, either by comparison with the binding properties of individual groups or simple ligands, or by molecular modelling.

11.2.1 The RANDOM model

Murray & Linder (1983, 1984) devised a way to predict cation-binding sites in humic substances on the basis of chemical properties. Their RANDOM model uses a set of rules to generate random molecules of fulvic acid, by first defining a carbon skeleton, including aromatic and aliphatic structures, then by adding functional groups. The user of the model is required to provide the following input data:

(1) the fraction of aliphatic carbon atoms in methyl groups, other than methoxy;
(2) the ratio of ortho, meta and para rings, with respect to aliphatic side chains;
(3) the fraction of pairs of quinone groups that are ortho, as opposed to para;
(4) the fraction of the total carboxyls that occur on aromatic rings;
(5) the percentage C;

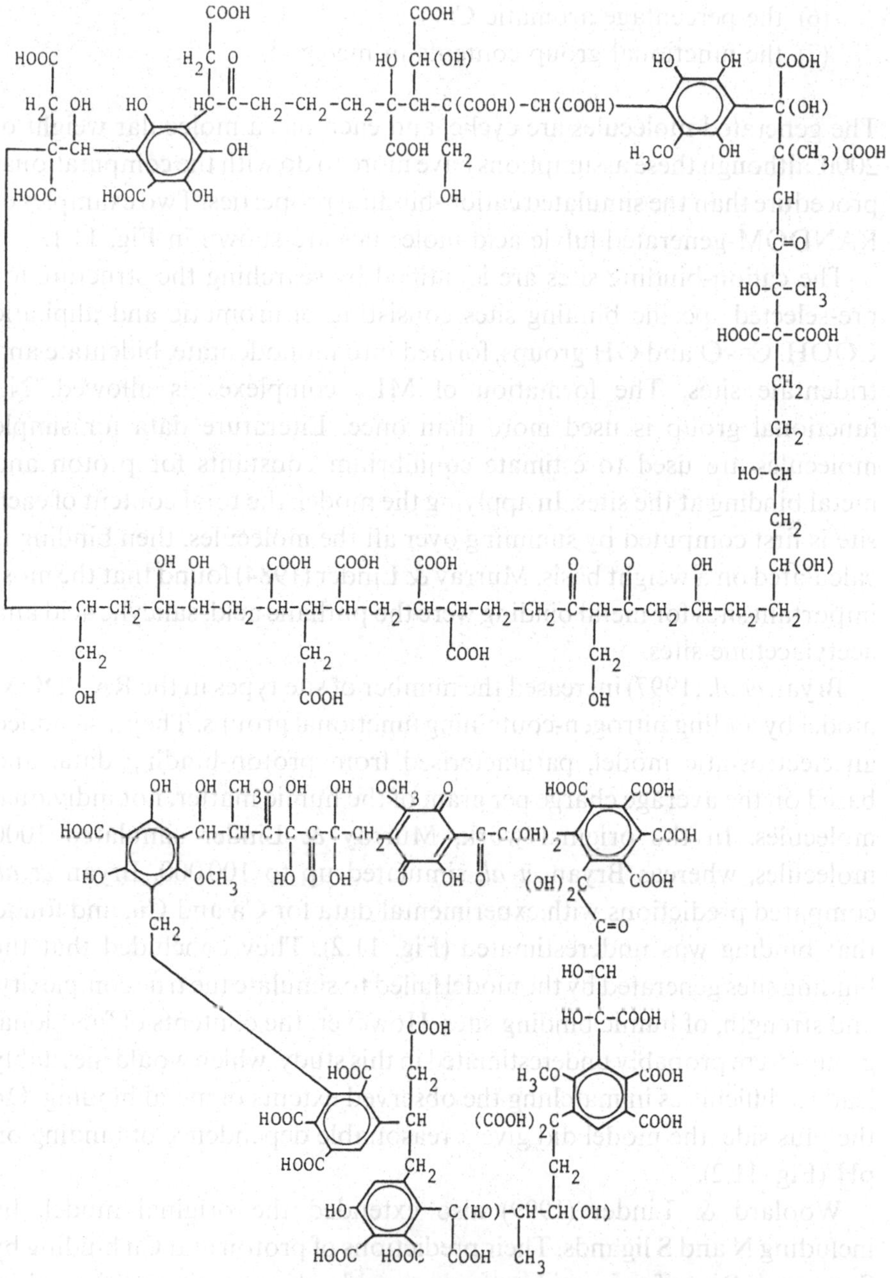

Figure 11.1. Typical computer-generated RANDOM structures. The upper structure has 15% aromaticity, the lower has 45%. [Reproduced from Murray, K. & Linder, P.W. (1983), Fulvic acids: Structure and metal binding. I. A random molecular model, *J. Soil Sci.* **34**, 511–523, with permission from Blackwell Science.]

(6) the percentage aromatic C;
(7) the functional group contents in meq g^{-1}.

The generated molecules are cyclic, and each has a molecular weight of 2000, although these assumptions have more to do with the computational procedure than the simulated cation-binding properties. Two examples of RANDOM-generated fulvic acid molecules are shown in Fig. 11.1.

The cation-binding sites are identified by searching the structure for pre-selected specific binding sites consisting of aromatic and aliphatic COOH, C═O and OH groups, formed into monodentate, bidentate and tridentate sites. The formation of ML_2 complexes is allowed. No functional group is used more than once. Literature data for simple molecules are used to estimate equilibrium constants for proton and metal binding at the sites. In applying the model, the total content of each site is first computed by summing over all the molecules, then binding is calculated on a weight basis. Murray & Linder (1984) found that the most important sites for metal binding were the phthalic acid, salicylic acid and acetylacetone sites.

Bryan *et al.* (1997) increased the number of site types in the RANDOM model by adding nitrogen-containing functional groups. They also added an electrostatic model, parameterised from proton-binding data, and based on the average charge per gram of the humic matter, not individual molecules. In the original work, Murray & Linder simulated 1000 molecules, whereas Bryan *et al.* simulated up to 100 000. Bryan *et al.* compared predictions with experimental data for Ca and Cu, and found that binding was underestimated (Fig. 11.2). They concluded that the binding sites generated by the model failed to simulate the true complexity, and strength, of humic binding sites. However, the contents of functional groups were probably underestimated in this study, which would inevitably lead to difficulties in matching the observed extents of metal binding. On the plus side, the model did give a reasonable dependence of binding on pH (Fig. 11.2).

Woolard & Linder (1999) also extended the original model, by including N and S ligands. Their predictions of proton and Cu binding by Suwannee River fulvic acid were reasonably close to observations (Fig. 11.3). However, the agreements were achieved by empirical adjustment of the carbonyl content of the fulvic acid, which influences the content of the acetylacetone site. Although such adjustment of the C═O content may be justified on the basis of analytical uncertainty, it is a form of parameterisation, and the 'purity' of prediction with the basic RANDOM approach is lost.

The RANDOM simulations have undoubtedly provided interesting results, and the model is useful as a way of drawing out the problems in predictive modelling. It deals with several important aspects of cation binding by humic matter, notably site heterogeneity, 'smeared' isotherms, pH dependence, multidentate sites, and competition effects, and it can be

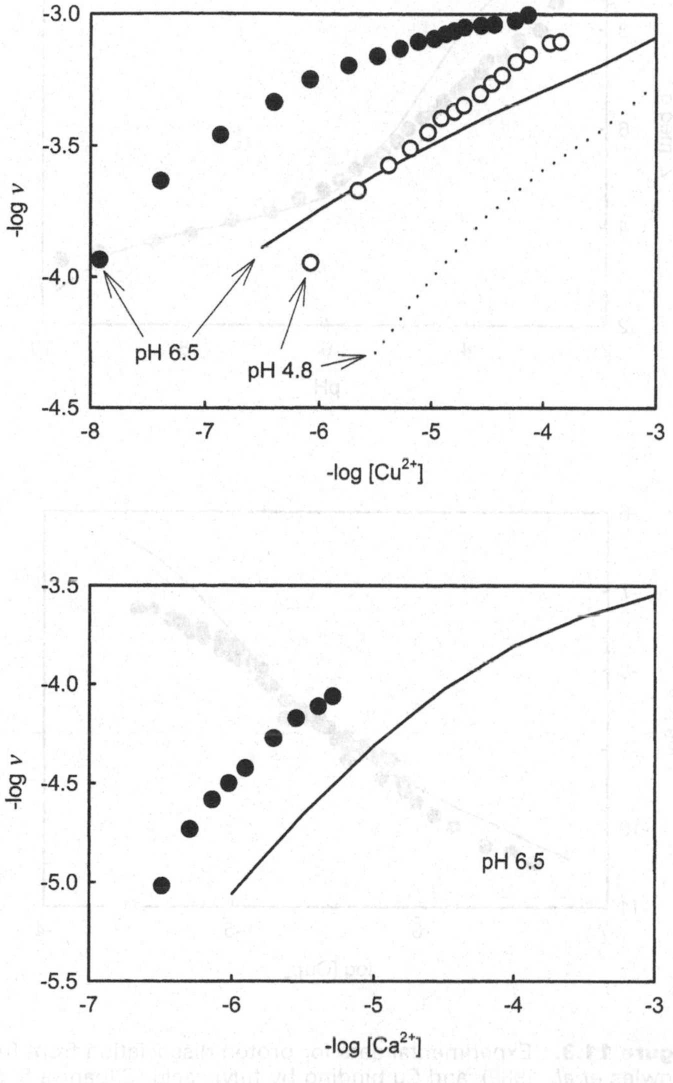

Figure 11.2. Binding of Ca and Cu by humic acids at $I = 0.06\,\mathrm{mol\,dm^{-3}}$. Points represent measured values, lines are simulations with the RANDOM model. [Redrawn from Bryan, N.D., Robinson, V.J., Livens, F.R., Hesketh, N., Jones, M.N. & Lead, J.R. (1997), Metal–humic interactions: a random structural modelling approach. *Geochim. Cosmochim. Acta* **61**, 805–820, with permission from Elsevier Science.]

adapted to include electrostatic effects. A limitation is that the model operates with pre-selected binding sites, for which there is, in general, no direct evidence. Although the sites are composed of groupings that are known to exist in humic matter, the device of equating them with simple

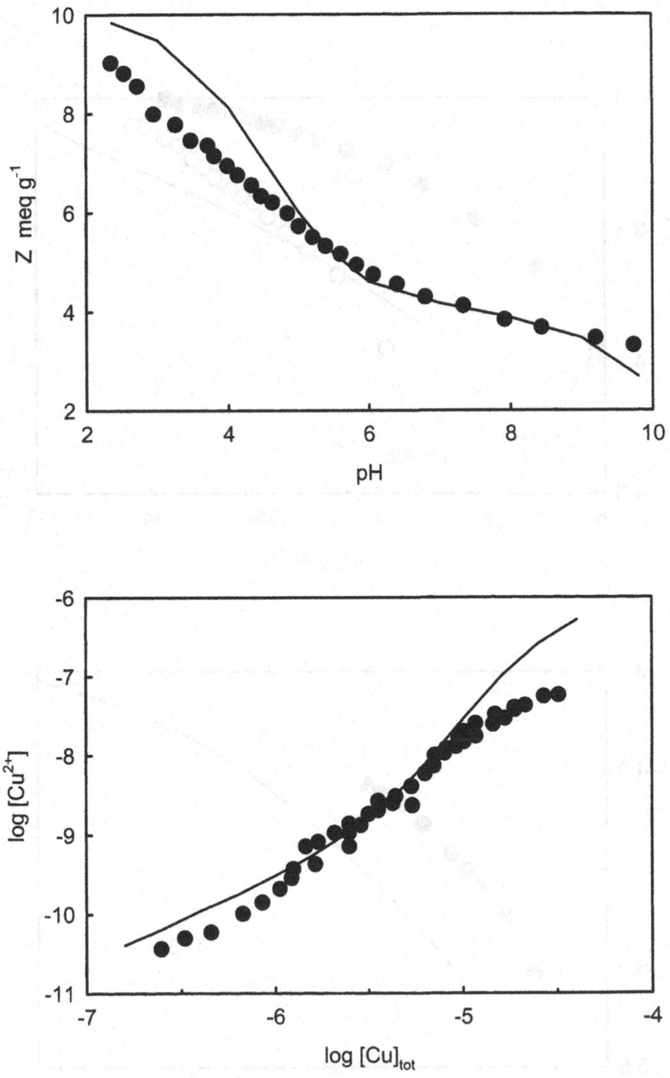

Figure 11.3. Experimental data for proton dissociation from fulvic acid (Bowles *et al.*, 1989), and Cu binding by fulvic acid (Cabaniss & Shuman, 1988a), shown by points, and RANDOM simulations, shown by lines. [Redrawn from Woolard, C.D. & Linder, P.W. (1999), Modelling of the cation binding properties of fulvic acids: An extension of the RANDOM algorithm to include nitrogen and sulphur donor sites, *Sci. Tot. Environ.* **226**, 35–46, with permission from Elsevier Science.]

mono-, bi- or tri-dentate ligands, for which thermodynamic data happen to be available, remains artificial, and inevitably limits the heterogeneity in binding sites that can be achieved.

11.2.2 Molecular modelling

Techniques are available to compute the properties and reactions of molecules and ions from first principles. In one example relevant to humic substances, Kubicki *et al.* (1996) used molecular orbital calculations to compute the structures, energetics and vibrational spectra of Al^{3+} and Al^{3+}–acetate complexes in the aqueous phase. Nantsis & Carper (1998a,b) performed molecular dynamics calculations to determine the molecular

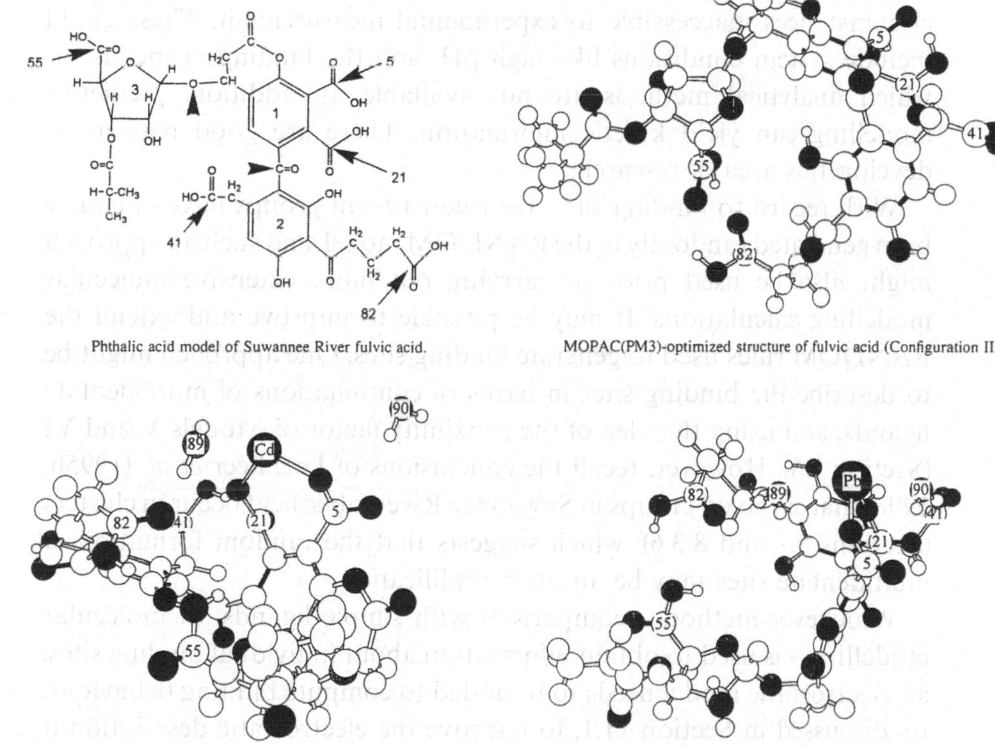

Phthalic acid model of Suwannee River fulvic acid.
MOPAC(PM3)-optimized structure of fulvic acid (Configuration III).
MOPAC(PM3)-optimized structure of hydrated Cd(site 1)-fulvic acid.
MOPAC(PM3)-optimized structure of hydrated Pb(site 1)-fulvic acid.

Figure 11.4. Molecular structures of metal–fulvic acid complexes, obtained by molecular modelling. [From Nantsis, E.A. & Carper, W.R. (1998b), Effects of hydration on the molecular structure of divalent metal ion–fulvic acid complexes: a MOPAC (PM3) study, *J. Mol. Struct. (Theochem)* **431**, 267–275, with permission from Elsevier Science.]

structures, enthalpies and entropies of cation–fulvic complexes, on the basis of one of the fulvic acid structures shown in Fig. 2.5. Examples of computed structures are shown in Fig. 11.4. The strengths of binding of the different cations could be estimated from computed enthalpies of interaction, since the entropies were similar. The order of computed affinity was $Mg^{2+} < Zn^{2+} < Pb^{2+} < Cd^{2+}$. The same sequence was computed for phthalic acid. It does not fully agree with measured data for either phthalic acid or fulvic acid, for both of which the sequence $Mg^{2+} < Zn^{2+} \sim Cd^{2+} < Pb^{2+}$ is found.

11.3 Prospects for predictive modelling

Predictive modelling of cation–humic interactions is potentially very valuable, since it allows binding properties to be determined for circumstances inaccessible to experimental measurement. These could include system conditions like high pH, and the binding of metals for which analytical methods are not available. In addition, predictive modelling can yield kinetic information. There are good reasons to develop this area of research.

With regard to binding sites, their constituent groups have of course been generated randomly in the RANDOM model, and such an approach might also be used prior to carrying out more extensive molecular modelling calculations. It may be possible to improve and extend the RANDOM rules used to generate binding sites. One approach might be to describe the binding sites in terms of combinations of monodentate ligands, and using the idea of the proximity factor of Models V and VI (Section 9.6). However, recall the conclusions of Leenheer *et al.* (1995b, 1998) that carboxyl groups in Suwannee River fulvic acid occur in clusters (Sections 8.1 and 8.3.6), which suggests that the random formation of multidentate sites may be an oversimplification.

Whichever method – comparison with simple ligands, or molecular modelling – is used to obtain information about individual binding sites, an electrostatic model needs to be added to compute binding behaviour. As discussed in Section 11.1, to improve the electrostatic description it might be necessary to treat the humic matter as a collection of molecules (with a given size distribution) rather than dealing in terms of a mass-average, the approach used in the comprehensive models of Chapters 9 and 10. If so, then rules would have to be found to apportion binding sites to individual molecules or fractions, while maintaining the overall, mass-based, behaviour of the system.

Predictive modelling is most usefully done at a level as close as possible to the true complexity of the system, and indeed it may help to reveal and describe that complexity, and to provide insight into the averaging that is taking place when macroscopic experimental measurements are made. Predictive modelling could therefore complement the more readily applicable parameterised models, even to the extent that binding data generated by predictive models might themselves be used to estimate model parameters.

12

Cation–humic binding and other physico-chemical processes

Humic substances undergo a number of physico-chemical interactions other than cation binding. They include aggregation, adsorption, the binding of hydrophobic organic compounds, photochemical reactions, and involvement in mineral precipitation and dissolution. In laboratory studies, conditions will usually be arranged to keep 'background' conditions simple, to focus on the interaction of interest. In the natural environment, however, some or all of the processes, including cation binding, proceed in parallel, and must influence one another. The purpose of the present chapter is to consider how cation binding influences these other interactions, and how they in turn may influence cation binding.

12.1 The conformation of humic matter

When cations are bound by humic substances, changes may occur in the conformation of the humic molecules. For example, it can be envisaged that the formation of multidentate complexes involving separate parts of the molecule may lead to a more compact structure. Another possibility is that, by reducing the net humic charge, cation binding leads to the reduction of repulsive interactions, again leading to compaction. The sensitivity of humic acid size and shape to cation binding has been emphasised by Swift (1989a), as discussed in Chapter 2. The more flexible and 'diffuse' is the conformation of the material before cation binding, the greater will be the changes in conformation on binding. Being smaller, fulvic acids have fewer opportunities to undergo conformational change.

Evidence for conformational change induced by proton binding comes from viscosity measurements (Ghosh & Schnitzer, 1980; Avena *et al.*,

1999). Figure 12.1 shows two examples of the variation of the intrinsic viscosity, $[\eta]$, of humic substances with pH and ionic strength. The results can be interpreted in terms of the following relationship, due to Einstein,

$$[\eta] = \frac{2.5 N_{Av} V_h}{M} \tag{12.1}$$

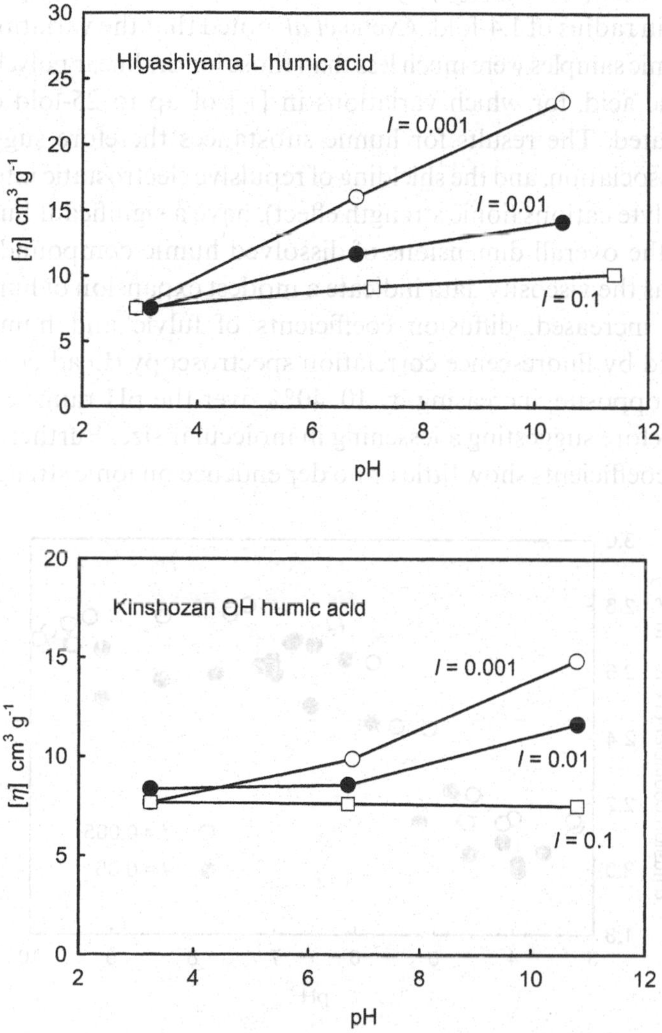

Figure 12.1. Dependence of the intrinsic viscosity of two humic acid samples on pH and ionic strength (in mol dm^{-3}). [Redrawn from Avena, M.J., Vermeer, A.W.P. & Koopal, L.K. (1999), Volume and structure of humic acids studied by viscometry pH and electrolyte concentration effects, *Coll. Surf. A* **151**, 213–224, with permission from Elsevier Science.]

Here, N_{Av} is Avogadro's number, V_h is the hydrodynamic volume of a humic molecule, and M is the molecular weight. The factor 2.5 applies to spherical molecules; larger values apply to ellipsoids. Thus the intrinsic viscosity is a direct measure of the volume of the molecule. The results in Fig. 12.1 show that the volume varies by up to a factor of 6, but the average ratio of maximum to minimum $[\eta]$ found by Avena *et al.* (1999) for nine samples of humic matter was smaller, 2.7. The hydrodynamic volume therefore varies, on average, by a factor of 2.7, which corresponds to a variation in radius of 1.4-fold. Avena *et al.* noted that the variations in $[\eta]$ for the humic samples were much less than those for the linear polyelectrolyte polyacrylic acid, for which variations in $[\eta]$ of up to 25-fold could be demonstrated. The results for humic substances therefore suggest that proton dissociation, and the shielding of repulsive electrostatic interactions by electrolyte cations (ionic strength effect), have a significant but modest effect on the overall dimensions of dissolved humic compounds.

Whereas the viscosity data indicate a modest expansion of humic acids as pH is increased, diffusion coefficients of fulvic and humic acids, determined by fluorescence correlation spectroscopy (Lead *et al.*, 2000), show the opposite, increasing by 10–40% over the pH range 3–10 (Fig. 12.2), therefore suggesting a lessening in molecular size. Furthermore, the diffusion coefficients show little or no dependence on ionic strength in the

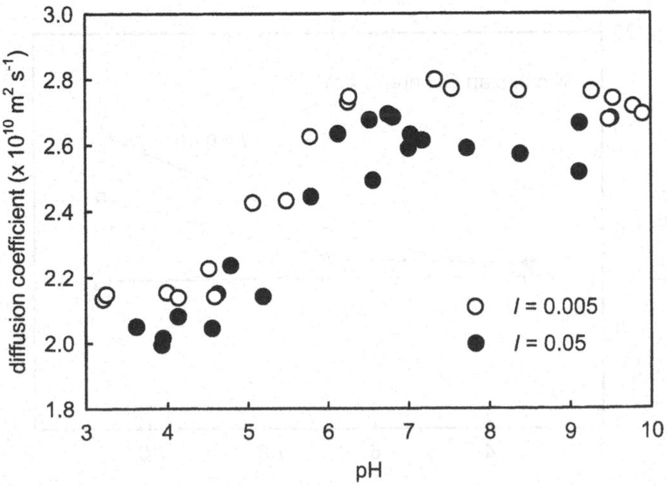

Figure 12.2. Dependence upon pH of the diffusion coefficient of fulvic acid, at two ionic strengths (mol dm^{-3}). [Redrawn from Lead, J.R., Wilkinson, K.J., Starchev, K., Canonica, S. & Buffle, J. (2000), Determination of diffusion coefficients of humic substances by fluorescence correlation spectroscopy: role of solution conditions, *Environ. Sci. Technol.* **34**, 1365–1369. Copyright 2000 American Chemical Society.]

range 0–0.1 mol dm^{-3}, which again is contrary to the conclusions from the viscosity results. According to Lead *et al.*, the smaller diffusion coefficients at low pH are due to aggregation (formation of dimers and trimers), rather than to changes in the sizes of individual molecules. The only firm conclusion that can be drawn from the viscosity and diffusion data, taken together, is that humic substances in solution do not undergo large changes in size as pH and ionic strength are varied.

There is evidence that conformational effects induced by cation binding affect the binding of non-polar micropollutants by humic substances. For example Schlautman & Morgan (1993) showed that the binding of polycyclic aromatic hydrocarbons by humic matter generally decreased with pH and with ionic strength, and increased with Ca^{2+} concentration, at neutral-to-high pH. They attributed the trends to variations in humic structure, in particular the dimensions and hydrophobicities of voids in the molecules. Ragle *et al.* (1997) concluded, from fluorescence studies of pyrene in solutions of humic acid, that cation binding brings about the formation of hydrophobic intramolecular domains in humic acid. Again from studies of pyrene fluorescence, Engebretson & von Wandruszka (1998) presented evidence that the interactions of cations with humic matter, and the associated conformational changes, take place on long time scales (tens of days). Figure 12.3 shows an example of the dependence

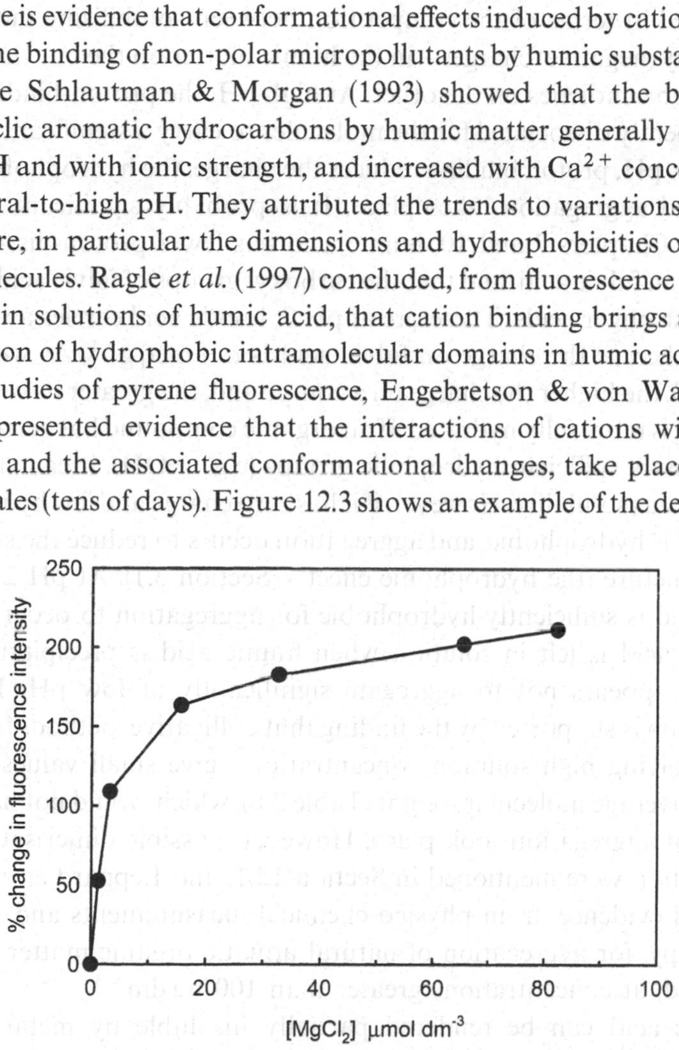

Figure 12.3. Variation with $MgCl_2$ concentration of the fluorescence intensity of pyrene, bound by humic acid (pH ~ 7). The measurements were made 5 h after the addition of the $MgCl_2$. [Redrawn from Engebretson, R.R. & Von Wandruszka, R. (1998), Kinetic aspects of cation-enhanced aggregation in aqueous humic acids, *Environ. Sci. Technol.* **32**, 488–493. Copyright 1998 American Chemical Society.]

of humic-bound pyrene fluorescence on metal cation concentration.

12.2 Aggregation of humic substances

The most familiar example of the aggregation of humic substances, due to interaction with a cation, is precipitation at low pH, a phenomenon which is exploited in the isolation of humic acid (Section 2.2). It can be interpreted in terms of the competition between solubilising moieties – principally (negative) charges – in the humic matter on the one hand, and hydrophobic moieties on the other. At high pH, the proton-binding sites are sufficiently dissociated for the molecules to carry a significant charge. But at low pH, proton binding reduces the charge, the hydrophobic effect 'wins', and aggregation takes place. It is probably significant that, as shown in Chapters 7 and 10, humic acid has fewer proton-dissociating groups than fulvic acid, and that the carboxyl groups of fulvic acid are, on average, stronger acids. Therefore at pH 2, humic acid carries a very low charge, whereas the charge on fulvic acid is *c.* 1 meq g^{-1}. As we saw in Chapter 3, the higher the charge on a compound, the greater its solubility, because it is more fully hydrated. If an organic compound has zero charge, but possesses sufficient hydrophilic groups (e.g. —OH), it can still have high aqueous solubility. However, if it has zero charge and few hydrophilic groups, it is hydrophobic, and aggregation occurs to reduce the strain on water structure (the hydrophobic effect – Section 3.1). At pH 2 or less, humic acid is sufficiently hydrophobic for aggregation to occur.

Fulvic acid is left in solution when humic acid is precipitated, and therefore appears not to aggregate significantly at low pH. Lack of aggregation is supported by the finding that colligative methods for fulvic acid, involving high solution concentrations, give small values for the number-average molecular weight (Table 2.6), which would not happen if significant aggregation took place. However, possible dimerisation and trimerisation were mentioned in Section 12.1, and Leppard *et al.* (1986) presented evidence, from physico-chemical measurements and electron microscopy, for aggregation of natural aquatic organic matter (mainly fulvic acid) at concentrations greater than 100 mg dm^{-3}.

Humic acid can be rendered partially insoluble by metal ions, a phenomenon used in some older fractionation schemes to separate 'grey' from 'brown' humic acid (Stevenson, 1994). An early systematic study of the phenomenon was made by Ong & Bisque (1968), some of whose results are shown in Table 12.1. According to Ong & Bisque, the binding of metal cations to the humic matter reduces intramolecular Coulombic repulsion, causing compaction of the humic structure, the expulsion of water and the creation of a hydrophobic colloid (cf. Section 3.2.3). The higher the charge on the cation, the more effectively is this achieved. Coagulation (aggregation) is then brought about by shielding of the remaining intermolecular electrostatic repulsion. The dependence upon electrolyte type follows the pattern established for conventional hydrophobic colloids, and again the higher the cation charge, the lower is the concentration required.

Tipping & Ohnstad (1984) studied the effects of pH and calcium concentration on the aggregation of aquatic humic substances. The results (Fig. 12.4) show that the extent of aggregation, as measured by the amount of humic matter removed by centrifugation, increases with the concentration – and therefore the extent of binding – of Ca. The greatest percentage removal is found at high pH, where Ca binding is sufficient to diminish substantially the net humic charge. An attempt to obtain a relationship between the extent of humic aggregation and net charge was made by Tipping *et al.* (1988b). Figure 12.5 (upper panel) shows how the removal by centrifugation depends upon pH in the presence of different concentrations of Al and Ca. The plot in the lower panel shows that percentage removal depends more-or-less directly upon the computed net charge, with no evidence of aggregation at (average) charges more negative than about $-2\,meq\,g^{-1}$. Tombacz & Meleg (1990) applied a simple surface ionisation model to humic acid in a 1:1 electrolyte, and estimated that the surface potential at which aggregation takes place is between -20 and -25 mV, over the pH range 2–9. Temminghoff *et al.* (1998) showed that the coagulation of humic acid by Ca^{2+} and Cu^{2+}

Table 12.1. Results of Ong & Bisque (1968) for the coagulation of peat humic substances.

Electrolyte	$mmol\,dm^{-3}$
LiCl	826
NaCl	598
KCl	335
$MgCl_2$	30
$CaCl_2$	7.2
$AlCl_3$	1.1
$FeCl_3$	1.52

The concentrations are the lowest at which coagulation could be detected visually (critical coagulation concentrations), at a humic concentration of $80\,mgC\,dm^{-3}$, pH 7.

could be explained in terms of the extent to which the humic acid is loaded with metal.

So far, the effects of cations have been discussed simply in terms of hydrophobicity, and insolubility. Bridging by cations is an alternative, or additional, mechanism. Underdown *et al.* (1985) interpreted the results of light-scattering studies with fulvic acid and Cu in terms of the formation of intermolecular bridging, or 'pseudochelation'. They viewed the aggregation as a cooperative phenomenon in which hydrogen bonding and hydrophobic interactions stabilise the bridged complex. If metal ions induce aggregation by bridging between humic molecules, there are implications for the strength of binding, since more metal–humic bonds may be formed. As discussed in Section 9.3.2, it would also follow that cation binding, expressed in terms of moles of metal bound per unit mass of humic matter, would depend not only on the free metal concentration but also on the humic concentration. A theoretical treatment has been given by Teasdale (1987), suggesting marked enhancement in the binding affinity of aggregated humic matter for copper at low metal–humic ratios, but experimental evidence is lacking. Another possibility is that the aggregation of humic matter might slow the rate at which metal ions can diffuse to humic binding sites, thereby giving rise to kinetic effects on metal binding.

There have been a number of studies of the physico-chemical properties

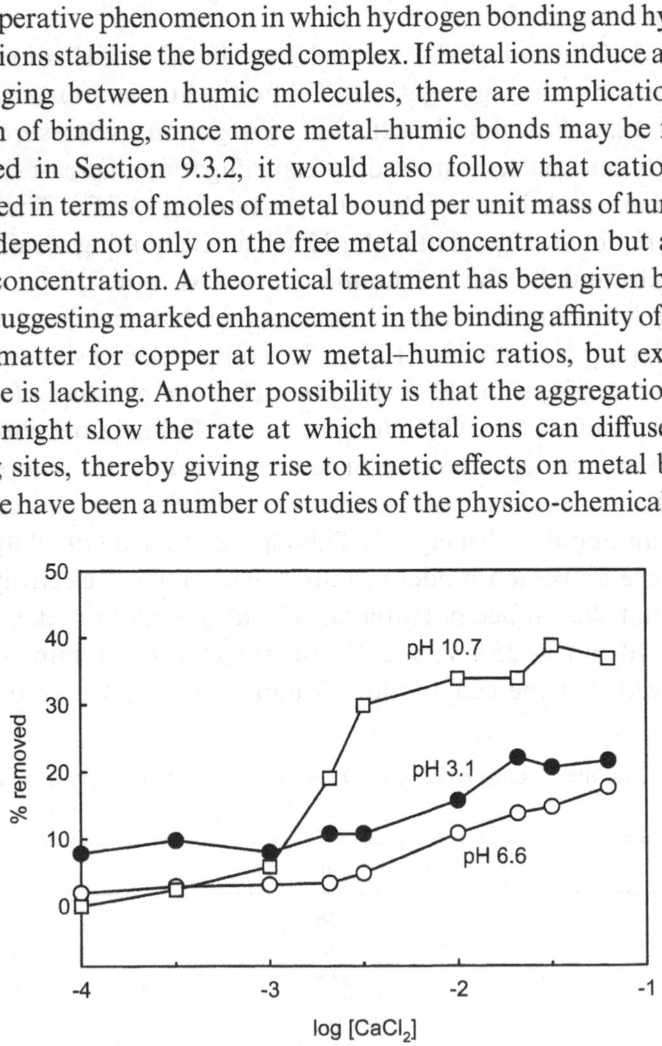

Figure 12.4. Effect of Ca on the aggregation of aquatic humic substances at three pH values. The aggregation was followed by determining the amount of humic matter removed by centrifugation, at a total concentration of humic substances of 61 mg dm^{-3}. [Redrawn from Tipping, E. & Ohnstad, M. (1984), Aggregation of aquatic humic substances, *Chem. Geol.* **44**, 349–357, with permission from Elsevier Science.]

of laboratory-formed humic aggregates, using viscometry (Hayano *et al.*, 1983), electron microscopy (Chen & Schnitzer, 1989), scattering methods (Österberg & Mortensen, 1992) and atomic force microscopy (Maurice & Namjesnik-Dejanovic, 1999). Aggregation has been likened to micellisation (Hayase & Tsubota, 1983; Wershaw, 1986; Guetzloff & Rice, 1994), and the fractal nature of humic aggregates has been emphasised by Österberg & Mortensen (1992), Österberg *et al.* (1995) and Senesi *et al.* (1997).

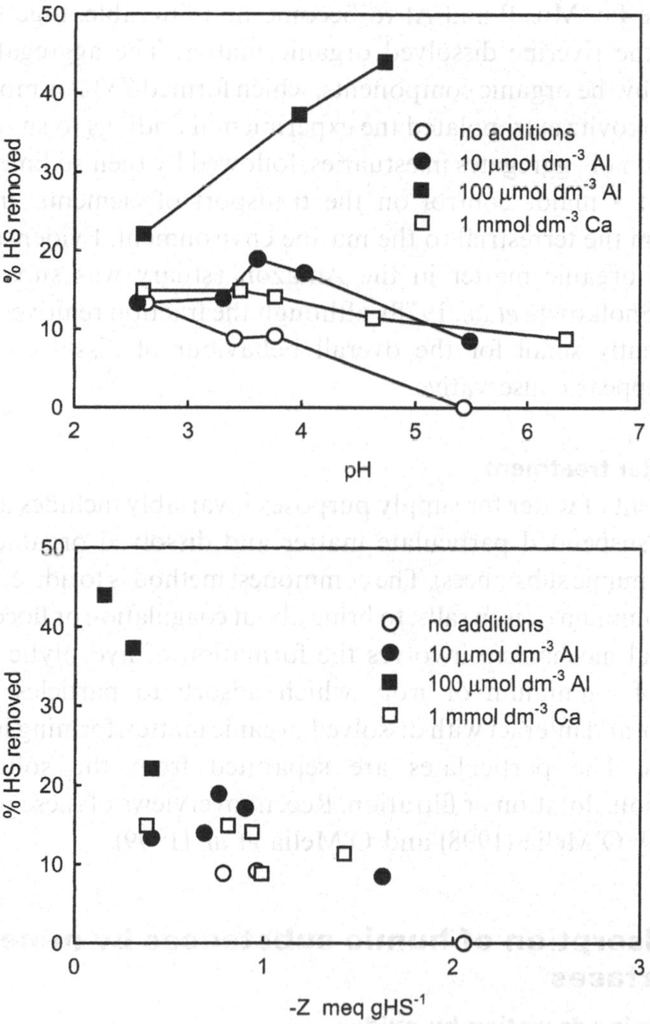

Figure 12.5. Percentage removal of the high-molecular-weight fraction of an aquatic humic sample as a function of pH at different Al and Ca concentrations (upper panel), and as a function of computed net electrical charge (lower panel). Plotted from data published by Tipping *et al.* (1988b).

12.2.1 Aggregation of humic substances in natural waters

Most of the studies of aggregation described above were conducted with humic acids, which are generally less soluble, and therefore more prone to aggregation, than fulvic acids, or hydrophilic acids. Nonetheless, Gjessing (1990) considered dissolved humic substances in freshwaters to be mainly in the form of aggregates. Sholkovitz (1976) carried out experiments in which filtered river waters from south-west Scotland, rich in humic matter, were mixed with filtered seawater. Rapid aggregation (termed flocculation by Sholkovitz) took place at the higher salinities, causing much of the Fe, Mn, P and Al to become non-filterable, together with 3–11% of the riverine dissolved organic matter. The aggregates were dominated by the organic components, which formed 75% or more of the mass. Sholokovitz extrapolated the experimental findings to suggest that the formation of aggregates in estuaries, followed by their sedimentation, would exert a major control on the transport of elements, including carbon, from the terrestrial to the marine environment. Evidence for the removal of organic matter in the Amazon estuary was subsequently presented (Sholkovitz *et al.*, 1978), although the fraction removed (3–6%) was sufficiently small for the overall behaviour of dissolved organic matter to appear conservative.

12.2.2 Water treatment

The treatment of water for supply purposes invariably includes a process to remove suspended particulate matter and dissolved organic matter (principally humic substances). The commonest method is to add controlled doses of aluminium or iron salts, to bring about coagulation or flocculation. The removal mechanism involves the formation of hydrolytic cationic polymers of aluminium or iron, which adsorb to particles, causing coagulation, and interact with dissolved organic matter, forming insoluble precipitates. The particulates are separated from the solution by sedimentation, flotation or filtration. Recent overviews of these processes are given by O'Melia (1998) and O'Melia *et al.* (1999).

12.3 Adsorption of humic substances by mineral surfaces

12.3.1 Humic adsorption by oxides

As discussed in Section 3.3.2, oxide surfaces interact with many solute species, including protons, metal cations, inorganic anions, and organic compounds. The most important oxides for humic adsorption in the natural environment are those of Al, Mn and Fe. The adsorption of macromolecular organic compounds at oxide surfaces involves a number of processes (Table 12.2) and cations play a significant rôle in three of them, ligand exchange, electrostatic interaction and coadsorption.

Humic substances and simple weak acids display a similar pH maximum in their adsorption to oxides (Fig. 12.6), and this can be explained in simple terms from the following reactions

$$\equiv SOH_2^+ = \equiv SOH + H^+ \qquad K_{SOH} = \frac{[\equiv SOH][H^+]}{[\equiv SOH_2^+]} \tag{12.2}$$

$$RH = R^- + H^+ \qquad K_{RH} = \frac{[R^-][H^+]}{[RH]} \tag{12.3}$$

$$\equiv SOH + RH = \equiv SR + H_2O \qquad K_{SR} = \frac{[\equiv SR]}{[\equiv SOH][RH]} \tag{12.4}$$

where $\equiv S$ represents the oxide surface and RH is a carboxylic acid. From the ligand-exchange reaction (12.4), it can be seen that most adsorption occurs, i.e. $[\equiv SR]$ is maximal, when the product $[\equiv SOH][RH]$ is

Table 12.2. Principal processes involved in the adsorption of organic macromolecules at mineral surfaces.

Process	Cause/significance
Hydrophobic effect	Removal of apolar moieties from contact with water
Ligand exchange	Exchange of surface hydroxyl groups for acid functional groups
Electrostatic interactions	Attraction and repulsion, and consequent changes in co-ion and counterion distributions
Van der Waals interactions	Important for large adsorbates, making many contacts
Hydrogen bonding	Exchange of hydrogen bonds with water for those with the surface, significant for polar groups in hydrophobic environments
Coadsorption	Adsorption of entities associated with the organic macromolecule, e.g. bound protons and metal cations
Competition	Diminishment of adsorption due to competition by another compound, e.g. SO_4^{2-}, HPO_4^{2-}
Conformational change	Change in size, shape, flexibility of the macromolecule
Fractionation	Adsorption of some components favoured over others
Element release	Dissolution of the solid engendered by the adsorbed organic matter

Based on the reviews of Norde (1980) and Tipping (1990).

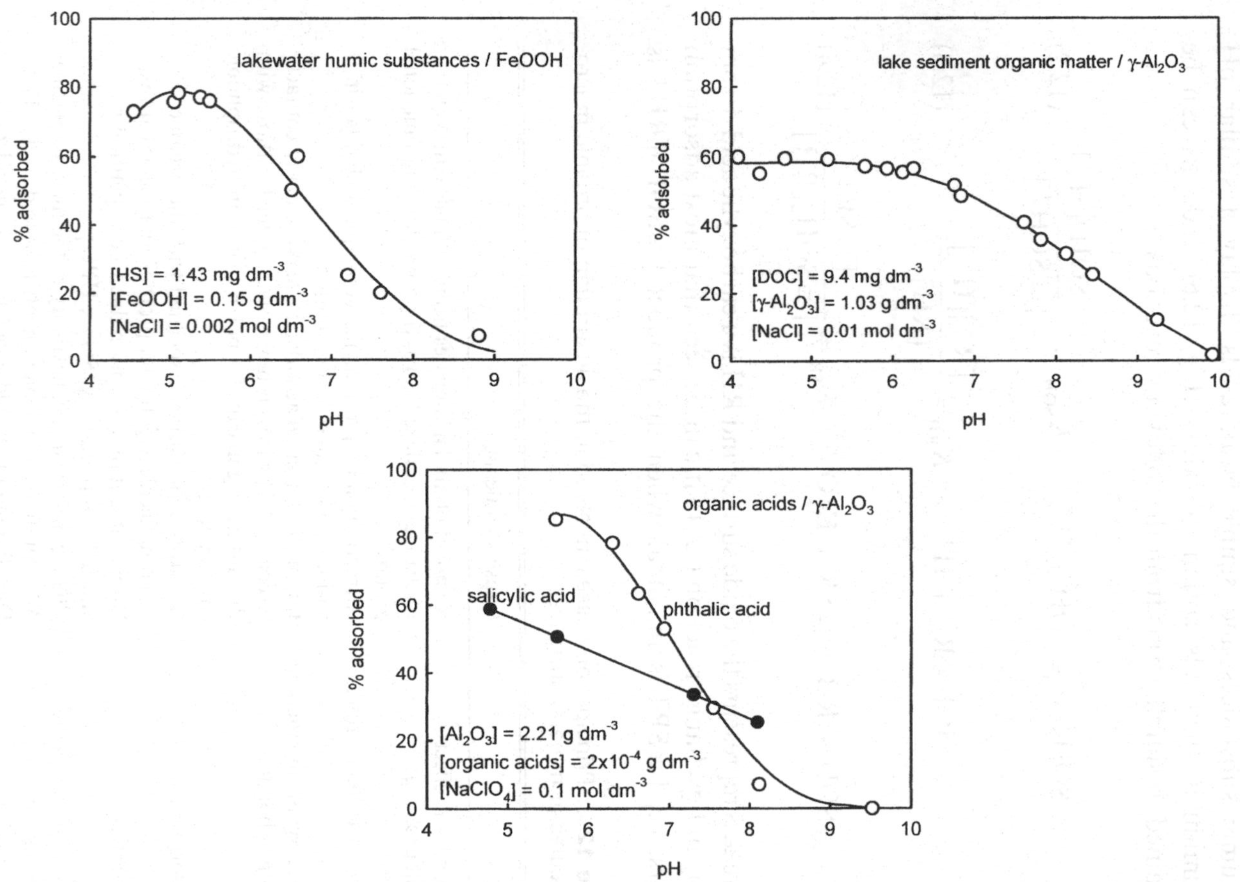

Figure 12.6. Adsorption of organic acids by oxides. The lines are for guidance only. [Redrawn from Tipping, E. (1981), The adsorption of aquatic humic substances by iron oxides, *Geochim. Cosmochim. Acta* **45**, 191–199, and Davis, J.A. (1982), Adsorption of natural dissolved organic matter at the oxide/water interface, *Geochim. Cosmochim. Acta* **46**, 2381–2393, both with permission of Elsevier Science, and from Kummert, R. & Stumm, W. (1980), The surface complexation of organic acids on hydrous γ-Al_2O_3, *J. Coll. Int. Sci.* **75**, 375–385, with permission of Academic Press.]

greatest. At low pH, the oxide surface is protonated according to reaction (12.2), and [$\equiv$SOH] is low. At high pH, the organic acid is deprotonated according to reaction (12.3), and [RH] is low. Thus maximum adsorption occurs at an intermediate pH, determined by the p*K* of the organic acid and the proton dissociation constant of the surface hydroxyl group.

Reactions (12.2)–(12.4) can also be used to show how adsorption of the organic acid by the oxide affects the binding of protons by both components. Thus, Fig. 12.7 shows calculated titration curves with and without reaction (12.4). The adsorption process means that the oxide and humic groups involved in ligand exchange are prevented from interacting with protons; they 'internally titrate' each other (Filius *et al.*, 2000). Thus adsorption is expected to reduce the buffering capacity of the system. This was observed in the haematite–humic acid system at pH 3–5 by Vermeer (1996).

Humic substances are of course much more complex than the simple organic acid of the preceding example, and additional proton interactions will occur as part of the adsorption process, involving acid groups not taking part in the ligand-exchange reaction *per se*. Thus, uptake of protons (and other cations – see below) will relieve electrostatic repulsion between adjacent adsorbed molecules (Tipping, 1981), and also between a negatively charged oxide surface and the humic anionic groups.

There are conflicting results with regard to the influence of monovalent cations on humic adsorption by oxides. Davis (1982) found that increasing the concentration of NaCl decreased natural organic matter adsorption to alumina, whereas Gu *et al.* (1994) reported no effect of NaCl concentration on adsorption to haematite, and Filius *et al.* (2000) observed a small decrease with NaCl concentration at low pH and a small increase at high pH. However, divalent cations have strong positive effects on adsorption (Fig. 12.8). One reason is the relief of repulsive electrostatic interactions between adjacent adsorbed humic molecules, as already discussed for protons. In addition, the formation of oxide–cation–humic bridges is possible (Tipping & Cooke, 1982).

12.3.2 Humic adsorption by other minerals

The most abundant mineral materials of high surface area in soils, sediments and aquatic suspended matter are aluminosilicates. The predominant surface is siloxane, composed of a nearly planar layer of oxygen atoms underlain by silicon atoms. In its pure form the siloxane surface carries no charge, but the isomorphous replacement of Si^{4+} by Al^{3+} creates permanent (pH invariant) negative charge. Adsorption of humic matter to the siloxane surface thus involves the coming together of two polyanionic entities, and is therefore strongly influenced by cations, which allow the electrostatic repulsion to be overcome. The higher the cationic charge, and the more strongly the cation binds to humic matter, the greater is its ability to promote the humic–siloxane interaction (Theng & Scharpenseel, 1975). Adsorption by the siloxane surface is much weaker than that by oxides of aluminium and iron. According to Hayes & Himes

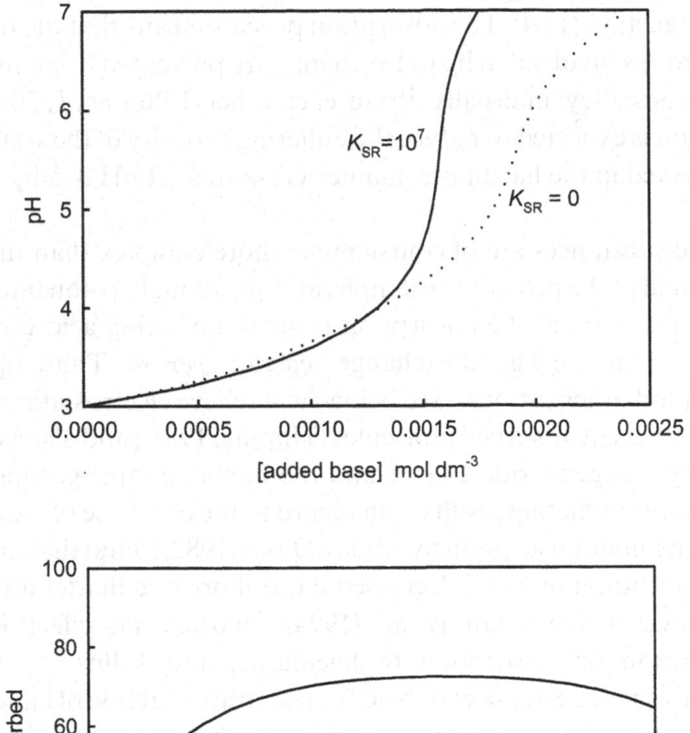

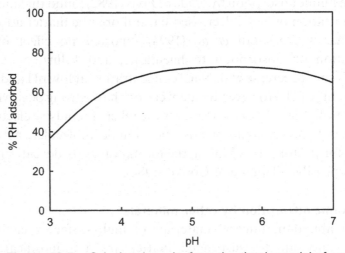

Figure 12.7. Calculated results from the simple model of equations (12.2)–(12.4) showing how adsorption affects proton buffering in humic–oxide systems. The results refer to equal total concentrations (10^{-3} mol dm^{-3}) of $\equiv$SOH and RH, and to $K_{SOH} = 10^{-7}$ and $K_{RH} = 10^{-4}$. The lower panel shows how adsorption depends upon pH. (When $K_{SR} = 0$ there is no adsorption.)

(1986), it involves, as well as cation bridging, van der Waals forces and hydrogen bonds. The hydrophobic effect must also play a rôle. Review articles about clay–humus complexes have been published by Tate & Theng (1980) and Hayes & Himes (1986).

12.3.3 Effects of humic adsorption on metal speciation

The speciation of trace metals in oxide–humic systems is complicated. The trace metal may be in several forms, as illustrated in Fig. 12.9. Metal distributions in such systems have been studied by a number of authors. An example from the work of Fairhurst *et al.* (1995), for the haematite–humic acid–Eu system, is shown in Fig. 12.10. The distribution of Eu depends

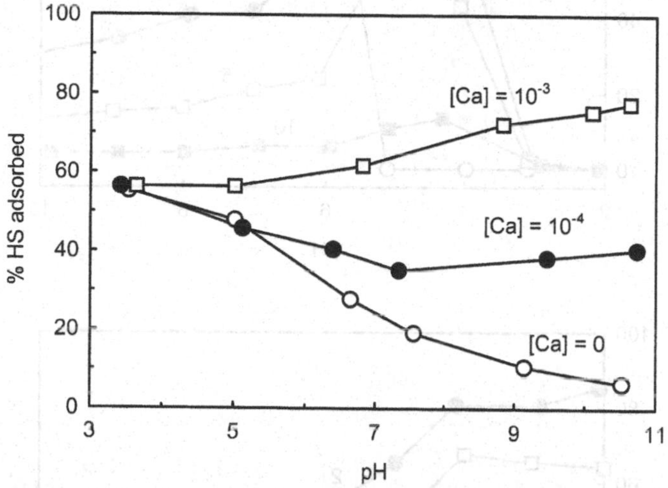

Figure 12.8. Effects of Ca on the adsorption of aquatic humic substances by an oxide of Mn ($MnO_{1.7}$). The concentrations of oxide and humic substances were 70 and 8 mg dm^{-3} respectively, and the background electrolyte was 0.01 mol dm^{-3} NaCl. Concentrations of Ca are given in mol dm^{-3}. [Redrawn from Tipping, E. & Heaton, M.J. (1983), The adsorption of aquatic humic substances by two oxides of manganese, *Geochim. Cosmochim. Acta* **47**, 1393–1397, with permission from Elsevier Science.]

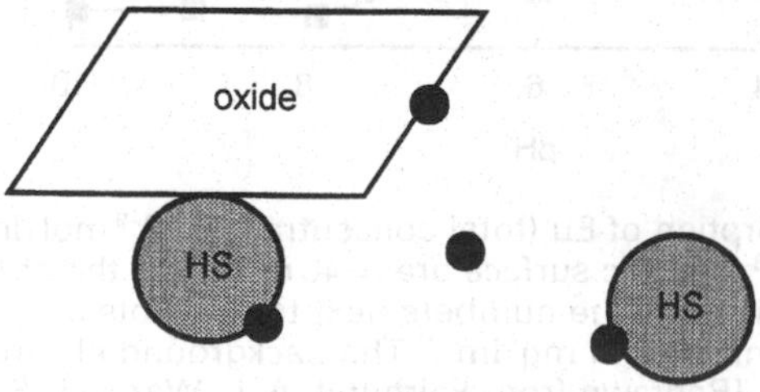

Figure 12.9. Schematic diagram of the distribution of a trace metal (black circle) in an oxide–humic system. Protons and other cations will also exert competitive effects.

upon the concentration of humic matter, and the extent to which the organic matter is adsorbed to the iron oxide. Adsorption of the humic acid effectively transfers metal-binding sites from the solution to the solid phase at low pH, while at higher pH dissolved humic acid maintains the metal in solution. Fairhurst *et al.* showed that the influence of humic acid

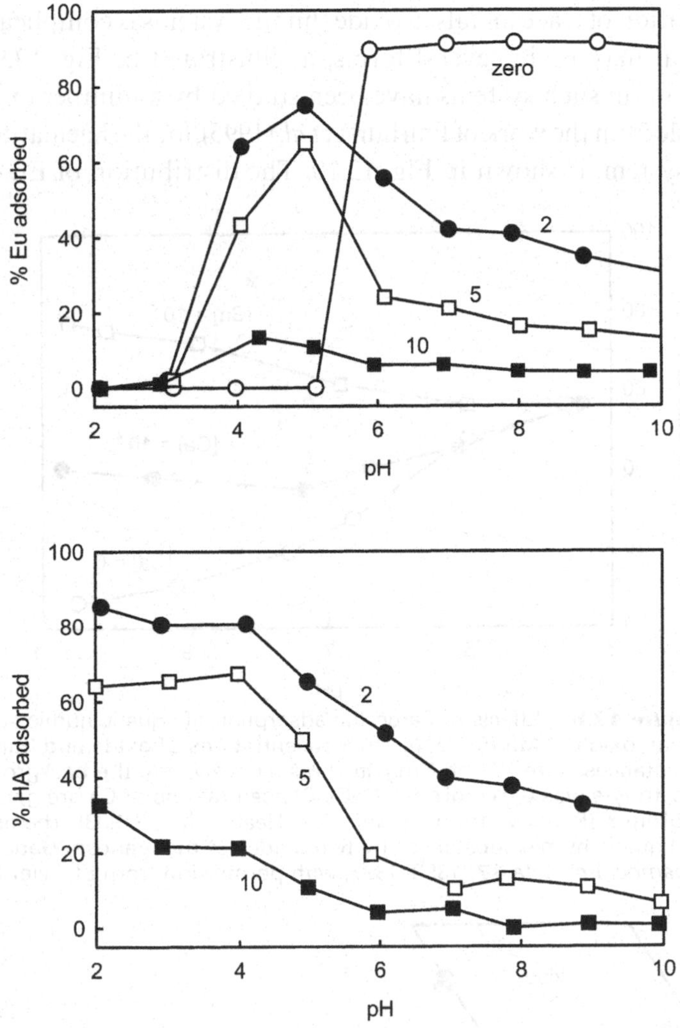

Figure 12.10. Adsorption of Eu (total concentration 10^{-9} mol dm^{-3}) by haematite (0.05 g dm^{-3}, specific surface area = 48 m^2 g^{-1}) in the absence and presence of humic acid. The numbers next to the plots are concentrations of humic acid in mg dm^{-3}. The background electrolyte was 0.05 mol dm^{-3} $NaClO_4$. [Redrawn from Fairhurst, A.J., Warwick, P. & Richardson, S. (1995), The influence of humic acid on the adsorption of europium onto inorganic colloids as a function of pH, *Coll. Surf. A* **99**, 187–199, with permission from Elsevier Science.]

on Eu binding varied among oxides (alumina, MnO_2, silica and TiO_2 as well as haematite) depending upon the adsorption of the humic acid. Ledin and coworkers (Ledin *et al.*, 1994; Düker *et al.*, 1995) reported qualitatively similar results for Eu and Zn binding in fulvic acid–oxide systems.

Is total metal binding in an oxide–humic system in which humic adsorption occurs significantly different from that expected from the contributions of the oxide and humic matter, acting independently? Decreased binding, compared to simple addition, has been observed in some studies, while others have shown greater binding of metal to take place. Table 12.3 summarises the results of several studies. Explanations for increased binding by the humic–oxide mixtures have been given in terms of electrostatic effects (Tipping *et al.*, 1983; Davis, 1984; Vermeer, 1996). Robertson (1996) suggested loss of sites (internal titration) to account for less-than-additive binding, while Vermeer (1996) proposed that the requirement of Cu to form bidentate complexes with humic substances makes it more sensitive than Cd to the influence of humic adsorption at the oxide surface.

Table 12.3. Metal binding in oxide–humic systems: comparison of observed metal binding with that expected assuming additive binding by the two components.

System	pH	log $[M^{2+}]$	Ratio	Reference
Goethite–aquatic HS–Cu	5.5	− 5.7	2	Tipping *et al.* (1983)
		− 4.3	1	
Alumina–aquatic NOM–Cu	4.5	~ − 6.5	4.5	Davis (1984)
	6	~ − 7.5	3.5	
Alumina–aquatic NOM–Cd	5.5–8.5	~ − 6.5	1	Davis (1984)
Goethite–HA–Cu	5.07	− 8	0.45	Robertson (1996)
		− 4	~ 1	
	6.07	− 9	0.8	
		− 4	~ 1	
Haematite–HA–Cd	4	− 5.8	1	Vermeer (1996)
		− 4.8	0.4	
	6	− 6	4	
		− 4	2	
	9	− 7	2.5	
		− 4	1.5	

The ratio is the *actual* bound metal concentration divided by the *expected* bound concentration for simple additivity (representative values). Thus a ratio > 1 indicates greater binding than expected, and a ratio < 1 less binding than expected.

Studies have also been made of the uptake of trace metals in mixtures of humic substances and non-oxide minerals. As with oxides, the adsorption of humic matter effectively transfers metal-binding sites to the solid phase, which thereby has metal-binding sites due both to the mineral surface and the adsorbed organic matter. Examples include kaolinite (Dalang *et al.*, 1984; Petrović *et al.*, 1999), saprolite (Zachara *et al.*, 1994), sand and calcite (Petrović *et al.*, 1999). Dalang *et al.* (1984) found that the uptake of Cu in kaolinite–fulvic acid mixtures followed simple additivity. Zachara *et al.* (1994) found the same for Co uptake in mixtures of humic acid and saprolite.

12.4 Binding of organic cations by humic substances

This book is concerned with the interactions of inorganic cations with humic matter, but organic cations (including pesticides) also occur in the natural environment. The binding of such compounds by humic substances involves electrostatic attraction, and also reactions characteristic of other organic compounds, especially the hydrophobic effect. That inorganic cations compete for binding is illustrated by the results of Burns *et al.* (1973) who reported that the binding of paraquat by humic acid was between two- and five-fold lower in the presence of $CaCl_2$ (0.335 mol dm^{-3}) than in its absence.

12.5 Colloid stability

According to the Deryagin–Landau–Verwey–Overbeek (DLVO) theory, the rate of aggregation of hydrophobic colloids (cf. Sections 3.2.3 and 3.3.7) is governed by two opposing processes. The first is attraction, due to van der Waals and other dispersion forces, the second is repulsion, due to electrostatic interactions between the diffuse double layers surrounding the particles. If the repulsive interaction is large, aggregation is slow and the suspension is said to be colloidally stable. Thus colloid stability depends upon the surface chemistry of the particles, and on the ionic strength. The rate of aggregation due to diffusive motion in a homogeneous suspension is given by the following equation due to von Smoluchowski

$$\frac{dN}{dt} = -kN^2 \tag{12.5}$$

where N is the number concentration of particles, and k is the rate constant. A maximum value for k (k_{max}) can be derived, corresponding to the situation where all collisions result in attachment, i.e. where the electrostatic repulsion is zero. Equation (12.5) can then be written

$$\frac{dN}{dt} = -\alpha k_{max} N^2 \tag{12.6}$$

where α, the collision efficiency, is the fraction of collisions that result in attachment. Thus colloid stability can be expressed in terms of α, which varies from 10^{-4} or less for a stabilised suspension to 1 for a fully destabilised one. Comprehensive accounts of colloid stability in theory and practice can be found in Hiemenz (1977) and Ives (1978).

Figure 12.11 shows a schematic plot of the collision efficiency vs. ionic strength for haematite particles. The particles are most colloidally stable at acid pH and low ionic strength. The stability at low pH can be accounted for by the higher (positive) charge. The increase of α with ionic strength is due to the compression of the diffuse part of the electrical double layer, and the consequent decrease in the repulsive force between particles.

12.5.1 Aggregation of particles in the presence of humic substances and cations

The adsorption of molecular species that alter the electrical charge of a particle surface can have substantial effects on colloid stability. In natural waters, humic substances, especially fulvic acids, are probably the most important adsorbates, along with cations (O'Melia, 1987; Wilkinson *et al.*,

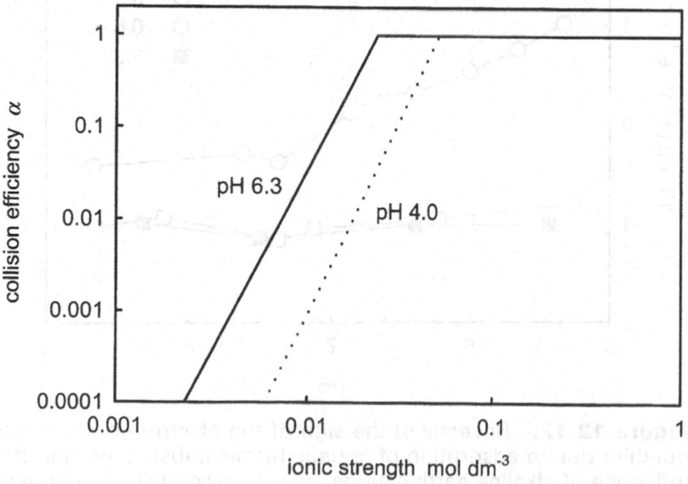

Figure 12.11. Dependence of the collision efficiency factor, α, on ionic strength, for pure haematite particles, shown schematically (based on results of Tiller & O'Melia, 1993).

1997). The surface charges of aluminium and iron oxides are positive at acid and neutral pH in simple electrolyte solutions, but they can be made negative by the adsorption of small amounts of humic substances, as illustrated by the electrophoresis data of Fig. 12.12. The electrophoretic mobility (u) provides a guide to the surface charge and colloid stability of the particles, large positive or negative values being associated with

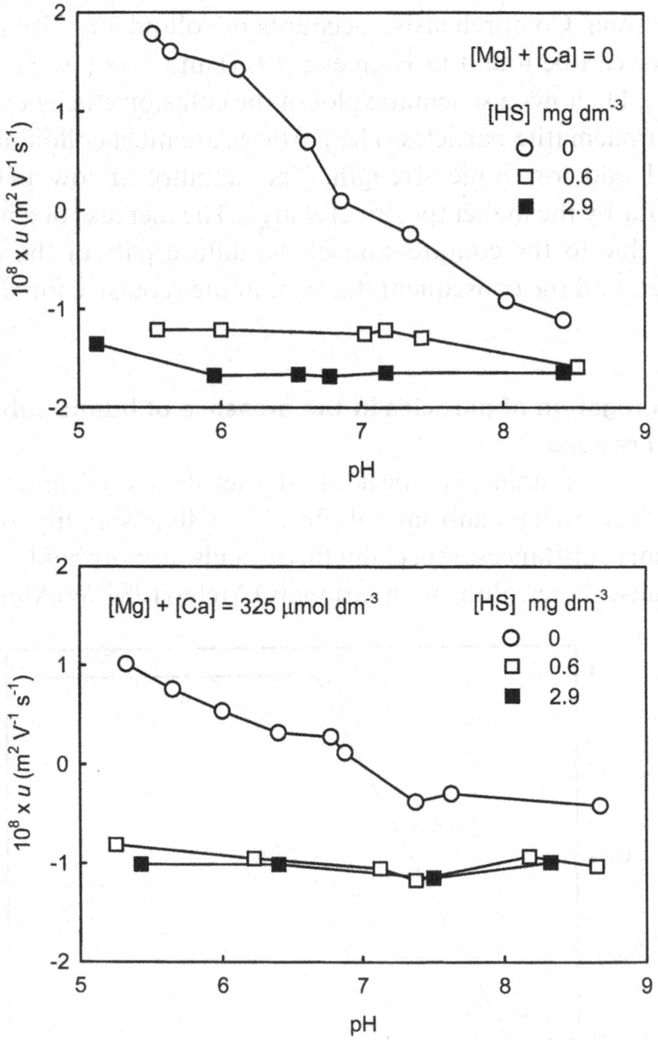

Figure 12.12. Reversal of the sign of the electrophoretic mobility of goethite due to adsorption of aquatic humic substances, and the influence of alkaline earth cations, at $I = 0.002$ mol dm^{-3}. [Redrawn from Tipping, E. (1981), The adsorption of aquatic humic substances by iron oxides, *Geochim. Cosmochim. Acta* **45**, 191–199, with permission of Elsevier Science.]

stability (strong electrostatic repulsion), and small values with rapid aggregation. Figure 12.13 shows how the colloid stability of humic-coated haematite particles is highly sensitive to Ca concentration, reflecting the binding of the cation to humic functional groups not involved in the adsorption reaction. The situation is actually somewhat more complicated, because there is a dependence of aggregation rate on humic adsorption density. Thus, Tipping & Higgins (1982) found that haematite particles were more colloidally stable at high adsorption densities, even though the electrophoretic mobilities were no higher than at lower adsorption densities. They attributed the phenomenon to steric stabilisation, i.e. particles coated with large amounts of humic matter were prevented from coming close enough together to aggregate. Jekel (1986) showed that humic substances could stabilise colloidal suspensions of silica and kaolinite.

12.5.2 Transport of colloids in porous media

The mechanisms controlling movements of mineral particles in porous media are essentially the same as those governing the stability of colloidal suspensions. The lower is α, the less likely is the attachment of a colloidal particle to the matrix of the porous medium, and the faster or further is migration.

The results of experiments investigating the influence of Ca^{2+} on the passage of haematite and silica particles through columns of washed sand

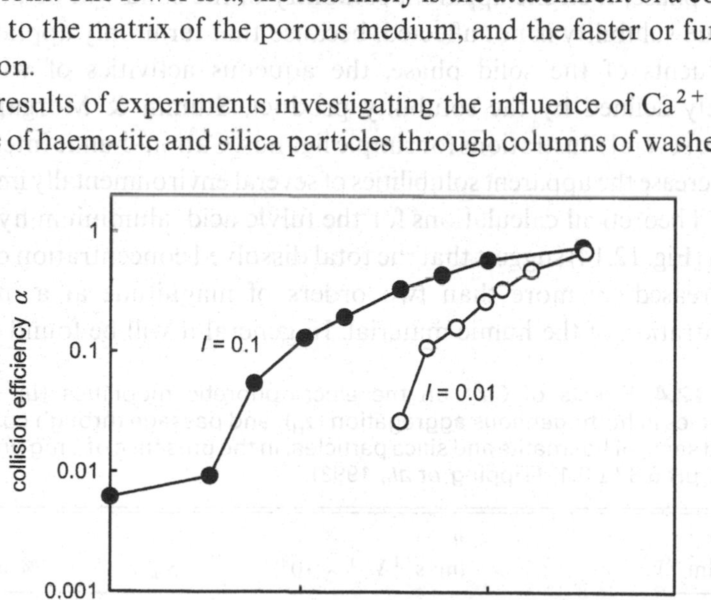

Figure 12.13. Collision efficiencies of haematite particles suspended in a solution of aquatic organic matter (10 mgC dm^{-3}). [Redrawn from Tiller, C.L. & O'Melia, C.R. (1993), Natural organic matter and colloidal stability: models and measurements, *Coll. Surf. A* **73**, 89–102, with permission from Elsevier Science.]

in the presence of fulvic acid are summarised in Table 12.4. The silica particles, which do not adsorb fulvic acid significantly, had high electrophoretic mobilities (u) in the absence and presence of Ca^{2+}, and were colloidally stable, as shown by the low values of the collision efficiency factor for homogeneous aggregation (α). Thus they were relatively unlikely to attach to the column material, and therefore high percentages passed through the columns. For the haematite particles, the presence of 10 mmol dm^{-3} Ca^{2+} decreased the magnitude of u and the value of α_H, due to binding of the cation to the adsorbed fulvic acid and the consequent lowering of net electrical charge. The interaction with the Ca^{2+} also increased the likelihood of attachment to the column material, as shown by the lack of detectable passage of the haematite particles.

12.6 Dissolution of minerals

12.6.1 Equilibrium

If a solution contains solutes that interact with the released constituents of a solid phase, then the apparent solubility of the solid will be increased. The true solubility is not affected, because that term only applies to the constituents of the solid phase, the aqueous activities of which are uniquely defined by the solubility product. Stumm & Morgan (1996) provide several illustrations for simple ligands. Cation–humic interactions may increase the apparent solubilities of several environmentally important solids. Theoretical calculations for the fulvic acid–aluminium hydroxide system (Fig. 12.14) suggest that the total dissolved concentration of Al can be increased by more than two orders of magnitude at a moderate concentration of the humic material. In general it will be found that the

Table 12.4. Effects of Ca^{2+} on the electrophoretic mobilities (u), collision efficiencies in homogeneous aggregation (α_H), and passage through columns of washed sand, of haematite and silica particles, in the presence of 1 mg dm^{-3} fulvic acid, at pH 6.8 (±0.1) (Tipping *et al.*, 1993).

$[Ca^{2+}]$* (mmol dm^{-3})		u ($m^2 s^{-1} V^{-1} \times 10^8$)	α_H	% elution
0	haematite	− 1.19	0.018	10.8
	silica	− 2.96	≤ 0.0002	35.0
10	haematite	− 0.78	0.031	0.0
	silica	− 1.73	≤ 0.0002	27.7

* Note that in the original publication the concentrations of Ca^{2+} were erroneously given as mol dm^{-3}.

stronger is the solution complexation of the metal cation, the greater will be the increase in apparent solubility. For example, considerable increases are expected for $Fe(OH)_3$ and HgS, but there will be little effect on $CaCO_3$.

12.6.2 Kinetics

The dissolution of minerals, otherwise known as chemical weathering, is a key geochemical process. There is much evidence that low molecular weight organic acids can increase the rate of dissolution of aluminosilicates (e.g. Huang & Keller, 1970; Blake & Walter, 1999) and alumina (Stumm & Furrer, 1987). The following mechanisms may operate:

(1) The addition of an organic acid to a mineral–water mixture may lower the pH, and in the acid pH range this usually leads to an increase in dissolution rate.
(2) The organic acid may form solution complexes with metal ions, notably Al, thereby both maintaining the disequilibrium between surface and solution, and preventing released metal from readsorbing to the mineral surface, where it may inhibit the dissolution process.
(3) The organic acid may facilitate the removal of metal ions from the mineral surface by forming surface complexes that can detach, thereby releasing the metal (Stumm & Furrer, 1987).

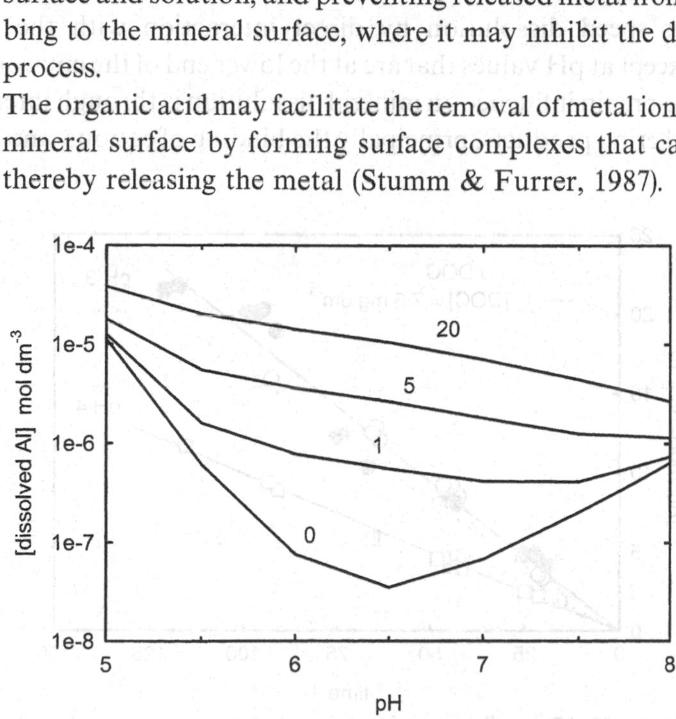

Figure 12.14. Theoretical enhancement of the total solubility of $Al(OH)_3$ due to solution complexation of Al by fulvic acid. The curves were calculated assuming a solubility product ($a_{Al^{3+}}/a^3_{H+}$) of $10^{9.6}$, and by using WHAM (Tipping, 1994) to compute solution speciation. The numbers next to the curves are fulvic acid concentrations in mg dm^{-3}.

The first two processes refer to solution conditions, and have nothing directly to do with the surface reaction. The third depends upon the interaction of the organic acid with the surface. In experiments aimed at exploring possibility (iii), Chin & Mills (1991) reported that the rate of dissolution of kaolinite was increased by low molecular weight organic ligands, while neither soil humic acid, nor dissolved organic matter from stream water, affected the rates of dissolution. Ochs (1996) presented evidence that humic acid can both speed and slow the dissolution of Al_2O_3, depending upon pH (Fig. 12.15). The results were explained by the formation of few humic–surface bonds at pH 3, allowing detachment of Al from the oxide surface to occur, while at higher pH, multiple bonding hinders detachment. In earlier work, Lundström & Öhman (1990) had shown that if soil dissolved organic matter was kept under non-sterile conditions, its rate-increasing effect on weathering was abolished, suggesting that the labile low molecular weight organic acids are the active components, rather than the humic substances, which are less susceptible to microbial degradation. Overall then, there is little evidence that humic substances speed dissolution by direct interaction with the mineral surface, except at pH values that are at the lower end of the environmental range. Their main influence on mineral dissolution is through interactions with weathering products, principally the binding of cations, and by their

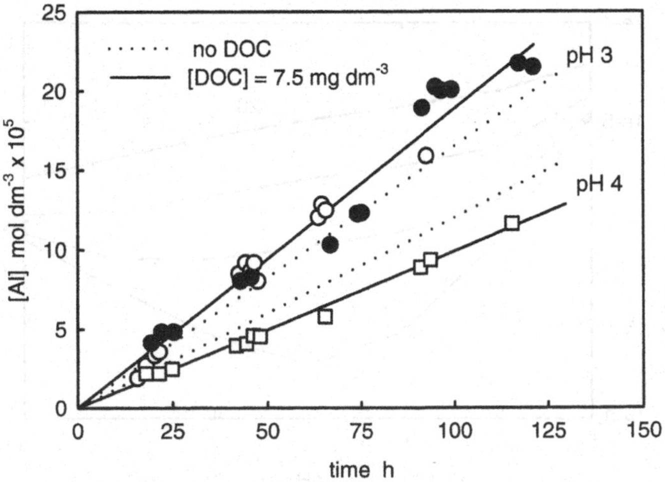

Figure 12.15. Influence of humic substances, measured as dissolved organic carbon (DOC), on the dissolution kinetics of γ-Al_2O_3. The different symbols for pH 3 indicate different sets of experiments. [From Ochs, M. (1996), Influence of humified and non-humified natural organic compounds on mineral dissolution, *Chem. Geol.* **132**, 119–124, with permission from Elsevier Science.]

influence on pH. Their reducing properties may also play a rôle in some circumstances.

12.7 Formation of mineral precipitates

The adsorption of humic substances to the surfaces of pre-formed minerals was discussed in Section 12.3. Under many circumstances in the natural environment, minerals form in the presence of humic substances, and consequently humic matter may influence particle formation. Thus humic substances may bind cations that are constituents of the mineral precipitate, or interact with the forming precipitate, or adsorb to seed crystals.

12.7.1 Iron oxides

By complexing Fe(III), humic substances may control the activity of Fe^{3+}, thereby determining which mineral solubility products are exceeded. This may favour the formation of goethite over haematite (Schwertmann *et al.*, 1986). Iron oxides are commonly formed by the oxygenation of ferrous iron. Davison & Seed (1983) found that humic substances had no effect on the rate of reaction at neutral pH. Tipping *et al.* (1989) found that humic substances were co-precipitated during formation of the oxide, and that they could be released by the addition of base without changing particle morphology. This suggests that the uptake process is simple adsorption, as opposed to more permanent incorporation into the particle structure. Neither Schwertmann *et al.* (1984) nor Tipping *et al.* (1989) found humic substances able to influence the crystal form of the iron oxide formed by oxygenation of ferrous iron (whereas dissolved silicate did, favouring ferrihydrite over lepidocrocite). Kodama & Schnitzer (1977) demonstrated that fulvic acid inhibited the transformation of ferrihydrite to haematite and goethite.

12.7.2 Aluminium minerals

As with Fe(III) and Fe^{3+}, the binding of Al by humic substances will influence the activity of Al^{3+}, thereby affecting the extent to which the solubility products of mineral phases are exceeded. Where precipitates are formed, fulvic acids (and low molecular weight acids), perturb the hydrolytic reactions of Al, and the crystallisation of their precipitation products (Huang & Violante, 1986). An important factor is the occupation of Al coordination sites by the organic ligands, which disrupts hydroxyl bridging between Al atoms. Further, the perturbing ligands distort the arrangement of unit sheets normally found in crystalline forms of $Al(OH)_3$. Aluminosilicates are also affected.

12.7.3 Metacinnabar (HgS)

Ravichandran *et al.* (1999) carried out experimental studies of the effects of humic substances on the formation of HgS. They found that at a low total concentration of Hg (50 nmol dm^{-3}), the presence of 10 mg dm^{-3} dissolved organic carbon (hydrophobic fraction) prevented the formation of HgS, due to complexation of the Hg^{2+}. At a higher total Hg concentration, HgS colloids were formed, but maintained in a highly dispersed state by adsorption of the organic matter. Experiments with natural water samples, and with model systems containing Ca^{2+}, demonstrated the importance of polyvalent cations in controlling the colloid stability of the HgS precipitates (cf. Section 12.5).

12.7.4 Calcite

Hoch *et al.* (2000) showed that quite low concentrations of aquatic humic substances could inhibit the growth of calcite crystals. The results of one set of experiments are shown in Fig. 12.16. The experimental conditions

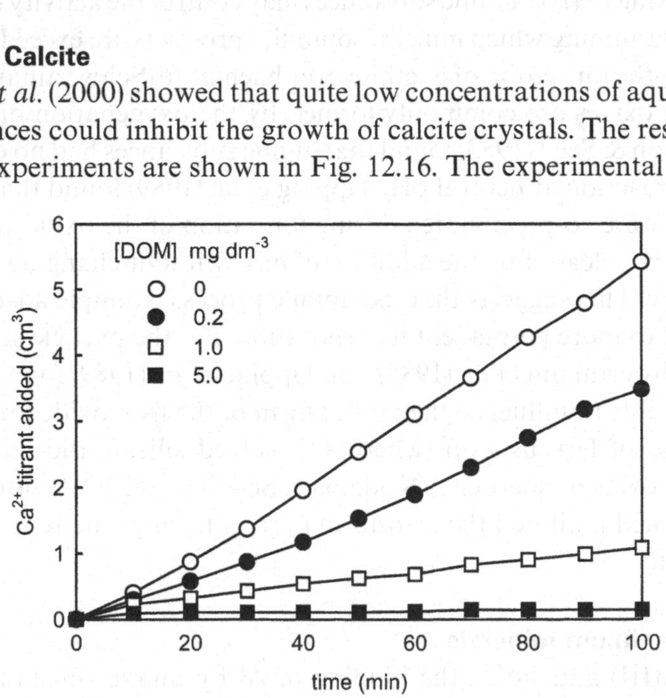

Figure 12.16. Inhibition of calcite precipitation by hydrophobic aquatic humic substances. The *y*-axis is the amount of $CaCl_2$ solution that had to be added (along with stoichiometric amounts of Na_2CO_3) in order to maintain constant solution conditions. The crystallisation took place on seeds of added calcite. For clarity, only some of the data are shown. [From Hoch, A.R., Reddy, M.M. & Aiken, G.R. (2000), Calcite crystal growth inhibition by humic substances with emphasis on hydrophobic acids from the Florida Everglades, *Geochim. Cosmochim. Acta* **64**, 61–72, with permission from Elsevier Science.]

were such that very little (less than 2%) of the Ca^{2+} in the solution phase was complexed by the humic matter, and so the inhibition could be attributed to adsorption of the humic compounds at the calcite surface, and the consequent prevention of nucleation and further crystal growth.

12.8 Other processes

There are other physico-chemical processes involving humic substances that are likely to be affected by the binding of protons and metal cations, but for which there is little available information about cation effects. One example is oxidation–reduction reactions. As mentioned in Section 3.3.4, hydroquinone and phenolic groups in humic matter have been implicated in changing the oxidation states of various metals. If such transformations involve binding interaction between the metals and the humic matter, it follows that the competitive binding of other (redox-insensitive) cations will affect the redox reactions. A less direct influence of competitive cation binding will be seen if humic matter can influence a redox reaction involving metals by stabilising one or more of the reactants or products.

Humic matter in natural waters is important in photochemistry (Section 3.3.8). Cation binding is known to influence the absorption and fluorescence spectra of humic substances, and therefore affects the absorption and emission of light energy. An important photochemical process in natural waters is the photo-oxidation of dissolved organic matter (principally humic substances) in the presence of iron, which involves adsorption of the organic matter at iron oxide surfaces (Voelker *et al.*, 1997). As we have already seen in Section 12.3, the adsorption reaction is strongly influenced by cations.

12.9 Concluding remarks

A number of examples of physico-chemical processes involving humic substances have been given, in which cations exert modifying effects, or which have an influence on cation binding. Therefore, experiments to explore the behaviour and influence of humic substances in the natural environment should include systematic variations of metal cation types and concentrations, as well as the more usual variations in pH and ionic strength. Furthermore, the development of quantitative models to describe humic behaviour and influence should take cation binding into account.

13

Cation binding by humic substances in natural waters

In this and the following chapter we shall see how knowledge about cation–humic reactions obtained in the laboratory can be used to interpret observations made on natural systems. Two approaches will be taken. Firstly, knowledge gained from studies and modelling of isolated humic substances will be used to predict, or speculate about, the circumstances in natural waters. Secondly, attempts will be made to interpret the results of measurements on natural water samples in which humic substances are suspected to exert a significant influence. For the most part, an equilibrium approach will be taken, for reasons discussed in Section 3.4.

13.1 Chemical speciation calculations

Cation–humic interactions in natural waters cannot be considered in isolation. As discussed in Chapter 3, natural waters contain many components other than humic substances, many of which combine with one another in parallel and competing reactions. Speciation calculations are therefore an essential part of understanding natural waters (and soils and sediments). Of course, a full speciation calculation, using a complicated model, is not always necessary, but the use of speciation codes such as those mentioned in Section 3.6.3 is often very helpful. WHAM (Windermere Humic Aqueous Model; Tipping, 1994) is a speciation code specifically designed to take into account interactions with humic matter, and will be used for most of the calculations in the present chapter. ECOSAT (Keizer & Van Riemsdijk, 1994), which includes the NICA model (Section 9.7), is an alternative.

WHAM is a combination of Humic Ion-Binding Model V (Section 9.6.1) with a model that allows the calculation of complexation reactions involving inorganic ligands. The model comes in two versions, one for waters and one for soils (see Chapter 14). In standard format, a data file is entered into the model, providing the total concentration of each reactant, including humic or fulvic acid. The pH may be specified or calculated. The model then uses a database of equilibrium constants and reaction stoichiometries to compute the concentrations of individual species such that the mass balance for each element is satisfied, and, if pH is to be calculated, closes the charge balance as well. A full account of the algorithm used in the calculations is given by Tipping (1994). WHAM is adequate for interactions involving cations present at relatively high concentrations, but underestimates the extent of binding to rare, strong sites in humic matter (Section 10.2). Therefore Humic Ion-Binding Model VI (Tipping, 1998b) will also be used, substituted for Model V in the WHAM framework, and referred to as WHAM/Model VI.

It should be recognised that equilibrium speciation calculations are subject to several kinds of error. Firstly, the model may be incorrect, by misrepresenting some of the reactions, omitting significant reactions, or falsely assuming the system to be at equilibrium. Secondly, the input data – total concentrations of chemical components etc. – will inevitably be subject to analytical errors. Thirdly, even if the model is structurally correct, its parameter values will be subject to uncertainty, arising from the experimental errors in their determination (see Section 10.3.5). Schecher & Driscoll (1988), Tipping *et al.* (1991) and Cabaniss (1997) discuss error propagation in detail.

The use of WHAM and WHAM/Model VI in this chapter and the next is sometimes done to fit data, and sometimes to show what might be expected. When used in the second way, the model is taken to provide a prediction, based on laboratory knowledge, as long as the reader can accept that the model does represent the laboratory knowledge correctly, as discussed in Chapter 10. If nothing else, the second approach may raise questions that might be tackled by further research.

13.1.1 The 'activity' of humic matter

A problem in the application of WHAM, or any other model in which natural organic matter is represented on the basis of data obtained with isolated humic matter, is knowing which values to choose for the concentrations of humic substances. There is no satisfactory way of determining the concentrations of humic substances *per se*; generally, only information on carbon contents is available. Therefore some assumption has to be made to estimate 'active' (in the sense of ion binding) humic concentrations from, for example, the concentration of dissolved organic carbon (DOC). It also has to be assumed that humic substances in natural samples interact with cations in the same way as their isolated counterparts. The maximum concentration of humic substances is approximately twice the DOC concentration, since humic matter is about 50% carbon by weight (Chapter 2). In general therefore, the concentration of active humic matter will be less than or equal to twice the DOC concentration.

13.2 Interactions with major ions and protons

For simple electrolytes, charge is not uniformly distributed at small distances. Because like charges repel and opposites attract, there tend to be more ions of opposite sign in the vicinity of a given ion. The effect is most commonly taken into account with activity coefficients, via ionic strength (Section 3.6.2). However, for larger ions, such as humic molecules, and particulates, the counterion accumulation is regarded as the diffuse part of the electrical double layer, as already discussed in Chapters 4, 5, 9 and 10. The way that WHAM approximates the situation for a solution of humic substances in a 1:1 electrolyte (MX) is represented schematically in Fig. 13.1 (see also Section 9.5). The bulk solution, free of humic substances, is denoted by S, the diffuse layer surrounding the humic molecules by D. The volumes V_S and V_D total 1 dm^3. They are each electrically neutral, therefore

$$[H^+]_S + [M^+]_S = [X^-]_S \tag{13.1}$$

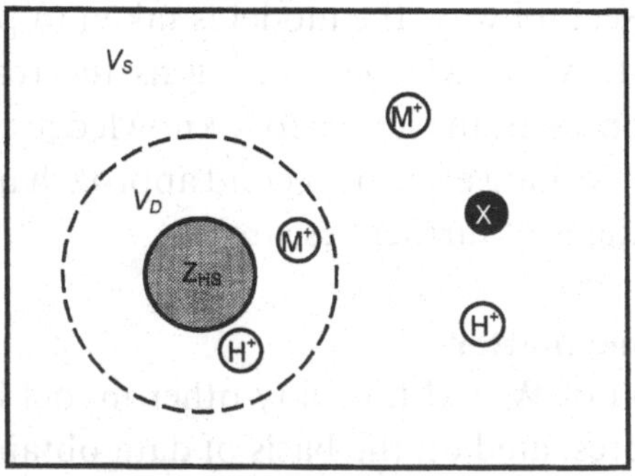

Figure 13.1. Schematic picture of the distribution of ions in a solution containing humic matter, according to Models V and VI. The zone enclosed by the dashed line is electrically neutral, as is the bulk solution outside it.

$$[H^+]_D + [M^+]_D + Z_{HS}[HS]_D = 0 \qquad (13.2)$$

where Z_{HS} is the net charge (negative) on the humic substances (HS) and square brackets indicate concentrations. The mass balances are given by

$$T_M = V_S[M^+]_S + V_D[M^+]_D \qquad (13.3)$$

$$T_X = V_S[X^-]_S \qquad (13.4)$$

$$T_{HS} = V_D[HS]_D \qquad (13.5)$$

where T_M, T_X and T_{HS} are the total concentrations of M, X and humic substances expressed in terms of dm^3 of total liquid. The equations become more complex when additional ions are introduced, and when metal cations bind to the humic material and modify its charge. The principles by which ions and water are distributed within the system remain the same however.

13.2.1 Predicted chemical state of fulvic acid in different surface waters

Here we consider the interactions of humic substances with major ions, and the net charge carried by humic substances. Examples for a range of waters, obtained from WHAM calculations, are shown in Table 13.1.

The net charge on the fulvic acid is calculated to be rather similar, in the range -2 to -3 meq g^{-1}, for all the examples except the acidic streamwater high in Al. The specific binding of Al reduces the negative charge to *c.* -0.5 meq g^{-1}; indeed, circumstances can be envisaged where the charge is reduced to zero, or even becomes positive, due to Al binding. In the absence of any specific binding, the charges would be much higher, especially at neutral pH values, being governed simply by pH, with modification due to ionic strength differences. Specific binding by Mg and Ca brings about near-constant charge among the neutral natural waters.

The counterions contributing most to charge neutralisation vary among the waters. In all except the hard freshwater, Na^+ is one of the two most important counterions. In the acid bogwater, H^+ is important, while Ca^{2+} plays a major rôle in the neutral freshwaters, and Mg^{2+} in seawater. The present calculations assume the counterions to have equal selectivities, i.e. accumulation depends only upon their solution concentration and ionic charge.

The volume of the fulvic acid diffuse layer is calculated to account for 13% of the total volume in the acid bog water, because of the high concentration of fulvic acid and the low ionic strength. In the other waters, the diffuse layer volume is less, being especially small in seawater. Note that the NICCA–Donnan model (Sections 9.5 and 9.7) would assign a 'Donnan volume', which is equivalent to the diffuse volume in Models V and VI, occupying only 1–2% of the total in the bog water (see also Section 9.5.4).

Table 13.1. The state of fulvic acid in different natural waters, calculated with WHAM.

		Acid bogwater	Acid streamwater	Neutral softwater	Hardwater	Seawater
TotNa	μmol dm^{-3}	100	200	500	600	468 000
TotMg	μmol dm^{-3}	20	20	100	500	53 000
TotAl	μmol dm^{-3}	1	30	0	0	0
TotK	μmol dm^{-3}	0	0	0	0	10 000
TotCa	μmol dm^{-3}	20	20	200	2000	10 000
TotCl	μmol dm^{-3}	115	250	600	700	545 000
TotNO$_3$	μmol dm^{-3}	10	30	30	30	0
TotSO$_4$	μmol dm^{-3}	40	40	40	40	28 000
[HCO_3^-]	μmol dm^{-3}	0	1	337	4190	2791
[CO_3^{2-}]	μmol dm^{-3}	0	0	1	104	24
FA	mg dm^{-3}	40	3	10	10	1
pH		4.00	4.88	7.69	8.76	8.40
I	mmol dm^{-3}	0.3	0.5	1.4	7.6	658
Z_{FA}	meq g^{-1}	−2.33	−0.56	−3.00	−2.74	−2.83
Z_{FA}*	meq g^{-1}	−2.36	−3.10	−4.85	−5.31	−5.47
ν_{Mg}	mmol g^{-1}	0.010	0.004	0.330	0.273	1.13
ν_{Al}	mmol g^{-1}	0.025	1.409	0	0	0
ν_{Ca}	mmol g^{-1}	0.010	0.004	0.659	1.086	0.21
DL% H^+		34	4	0	0	0
DL% Na^+		28	58	30	5	61
DL% Mg^{2+}		19	11	23	19	31
DL% Al^{3+}		0	14	0	0	0
DL% $AlOH^{2+}$		0	2	0	0	0
DL% K^+		0	0	0	0	1
DL% Ca^{2+}		19	11	46	75	6
% volume is DL		13	0.5	0.9	0.1	10^{-4}

Key: TotNa etc., total concentrations of inorganic components; Z_{FA}, net fixed charge on fulvic acid; Z_{FA}*, value for the same I but for a solution containing only NaCl (therefore no specific binding reactions); ν_X, binding at specific sites; DL%, % of fixed charge balanced by the ion in the diffuse layer; % volume is DL, percentage of the total volume occupied by the diffuse layer of the fulvic acid. All solutions are in equilibrium with atmospheric CO_2.

13.2.2 Contribution of humic substances to ionic charge balance
Following from the previous section, let us consider ionic charge concentrations in natural waters. All waters must be charge balanced, and humic substances will contribute. Charge balance is important in checking analytical results, and it forms the basis for simple models that assess the effects of acid deposition on waters (e.g. Posch *et al.*, 1993). Humic substances confound simple calculation, because their 'formal' charge (corresponding to Z^*_{FA} in Table 13.1) depends upon solution conditions. The greatest influence of humic substances is in acid solutions, which very often contain significant concentrations of Al. Table 13.2 shows calculations of ionic charge for such a solution, the composition of which satisfies charge balance according to the WHAM model. Ionic concentrations might be computed in several ways. If the contribution of humic matter (fulvic acid) is ignored, and if a charge of +3 is assigned to the aluminium, a very poor match between cations and anions is obtained (Balance-1). This is partly because the positive charge is overestimated, but mostly because the negative charge associated with the fulvic acid is ignored. A better approach is to take into account both the hydrolysis reactions of Al,

Table 13.2. Ionic charge balances for an acid water rich in humic substances (fulvic acid) and aluminium.

	Total	Balance-1 μeq dm^{-3}	Balance-2 μeq dm^{-3}	Balance-3 μeq dm^{-3}
H^+	(11)	11	11	11
Na^+	120	120	120	110
Mg^{2+}	30	60	60	46
Al^{3+}	25	75	56	0
$AlOH^{2+}$	—	—	9	0
$Al(OH)_2^+$	—	—	1	0
Ca^{2+}	30	60	60	46
Cl^-	100	100	100	113
NO_3^-	10	10	10	11
SO_4^{2-}	40	80	80	89
FA	30	0	94	0
Σ cations		326	305	213
Σ anions		190	284	213

Balance-1 is calculated ignoring both the contribution of fulvic acid, and the hydrolysis of Al. Balance-2 estimates the charge on fulvic acid from titration data in the absence of specifically binding cations, and assumes that Al undergoes hydrolysis but does not bind to fulvic acid. Balance-3 is from the application of WHAM.
The total concentration of H^+ is the value that would be obtained from the measured pH.
Total concentrations are in μmol dm^{-3}, except for fulvic acid (FA) which is in mg dm^{-3}.

which mean that the net charge on the metal is less than + 3, and also the degree of proton dissociation of the fulvic acid. The latter is calculated in the present case using WHAM, but other procedures could be used, for example the empirical equation of Oliver *et al.* (1983), which is discussed below. The computed ionic concentrations (Balance-2) come reasonably close to matching cations and anions. The error is less than 10%, which might be acceptable in analytical terms. In reality, the situation is more complicated than assumed in Balance-2, because the Al does not exist simply in inorganic forms (Al^{3+}, $AlOH^{2+}$, $Al(OH)_2^+$). According to WHAM, most of the Al is bound to the fulvic acid. In the WHAM picture of the solution, the fulvic acid is in an electrically neutral 'phase', because any residual negative charge after specific binding is neutralised by counterions (Fig. 13.1). The ions in the remaining solution then are perfectly balanced (Balance-3). Cation concentrations in the remaining solution are lower than the total concentrations for the whole system, because of binding to the fulvic acid, while anion concentrations are higher because they are excluded from the diffuse layer volume of the fulvic acid (cf. Fig. 13.1). Thus the WHAM picture is somewhat different from the 'standard' one in which all ionic species (including fulvic acid) are regarded as separate entities. Note that (a) in the example, the pH was calculated using WHAM, thereby forcing a charge balance, and (b) the WHAM-calculated pH is used in Balances 1 and 2, as if the concentration of H^+ (obtained from the pH) applies to the whole solution; in the WHAM picture, this concentration only applies to the volume of bulk solution, i.e. that not associated with the fulvic acid diffuse layer.

Tipping *et al.* (1991) assembled observations from a range of freshwaters containing significant concentrations of DOC. They used a forerunner of the WHAM model to investigate the contribution of humic substances to ionic concentrations. The same data, analysed with WHAM, are presented in Fig. 13.2. If humic substances are ignored, ratios of cations to anions tend to be appreciably greater than one. If 66% of the DOC is assumed to be due to fulvic acid, with average properties of isolated fulvic acid (and the remaining 34% to be inert with respect to ion binding), the charge balances are much better. The value of 66% was obtained by trial and error, aiming to make the average charge balance unity. It seems a reasonable value, being not greatly different from the maximum value of 100%, and provides support for the presumption that understanding the cation-binding properties of isolated humic matter provides useful insight into the operation of natural waters.

Oliver *et al.* (1983) tackled the problem of charge balance in DOC-rich waters differently, aiming simply to account for DOC-associated charge, without taking into account the detailed speciation of other cations. The 'Oliver model' describes the variation with pH of the apparent pK of dissolved organic matter by the empirical equation

$$\mathrm{p}K = a + b\mathrm{pH} + c(\mathrm{pH})^2 \qquad (13.6)$$

Equation (13.6) enables the degree of dissociation to be computed at any pH, and when this is coupled with the number of proton-dissociating groups per unit mass of organic matter, the charge on dissolved organic matter can be estimated. The values of the constants a, b and c were estimated to be 0.96, 0.90 and − 0.039 respectively by Oliver *et al.* (1983) from titration data for isolated humic substances. Driscoll *et al.* (1994) estimated values of 0.15, 1.41 and − 0.078 respectively by analysis of field data. The approach assumes that metal cations do not interact with dissolved organic matter; for example Al^{3+}, $AlOH^{2+}$ etc. are counted separately as cations, and the charge on dissolved organic matter (DOM) is taken to be the full value corresponding to the dissociation at that pH. The actual charge on the DOM is therefore overestimated (cf. Table 13.1), but a reasonable check of the charge balance is obtained. A related approach was developed by Kortelainen (1993), taking into account contributions by hydrophilic and hydrophobic acids.

Driscoll *et al.* (1994) compared the performance of the calibrated Oliver

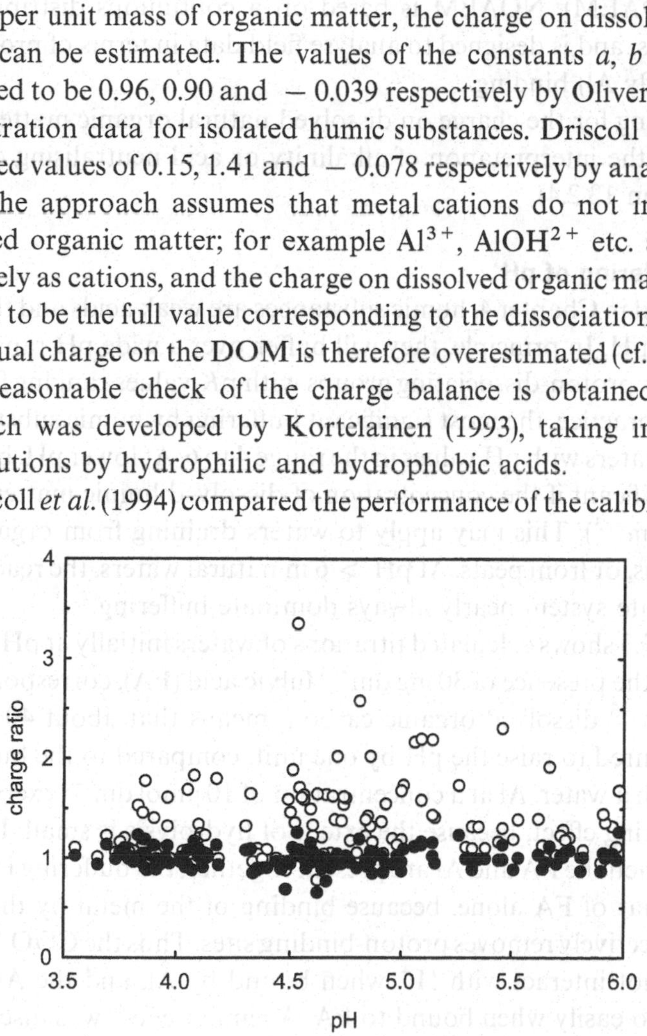

Figure 13.2. Charge ratio (cations/anions) in acidic surface water samples, rich in dissolved organic matter (DOM), from Canada, Europe and the USA. The open circles show ratios computed without taking DOM into account, the closed circles ratios computed with WHAM, assuming that 66% of the DOM behaved as isolated fulvic acid, the remainder being inert.

model with three simple representations (one-pK, two-pK and three-pK) of proton dissociation from organic matter, in the evaluation of a large field data set (Adirondack Lake Survey). None of the models took account of metal binding. The one-pK model was unsatisfactory, but the other two models performed acceptably. Köhler *et al.* (2000) applied a triprotic model to data from Swedish surface waters. Driscoll *et al.* (1994) noted the desirability of taking metal interactions into account, a problem addressed by Grzyb (1995) in developing the Natural Organic Anion Equilibrium Model (NOAEM). NOAEM is based on a continuous distribution of binding sites, and is designed to analyse field data in terms of proton and metal (chiefly Al) binding.

Accounting for the charge on dissolved natural organic matter is also relevant to the interpretation of alkalinity or acid neutralising capacity data (Section 13.2.4).

13.2.3 Buffering of pH

As explained in Chapter 4, humic substances are weak acids and therefore they buffer pH. In principle, they will buffer over a wide pH range, since they possess proton-dissociating groups with pK values ranging from c. 2 to c. 12. In practice, the most significant buffering by humic substances is in natural waters with pH values in the range 4 to 6. At lower pH, buffering will be significant if the concentration of dissolved humic matter is high ($> 25\,\mathrm{mg\,dm^{-3}}$). This may apply to waters draining from organic-rich soil horizons, or from peats. At pH > 6 in natural waters, the reactions of the carbonate system nearly always dominate buffering.

Figure 13.3 shows calculated titrations of waters initially at pH 4, 5 and 6. At pH 4, the presence of $30\,\mathrm{mg\,dm^{-3}}$ fulvic acid (FA), corresponding to *c.* $15\,\mathrm{mg\,dm^{-3}}$ dissolved organic carbon, means that about 40% more base is required to raise the pH by one unit, compared to the buffer-free case. In such a water, Al at a concentration of $10\,\mu\mathrm{mol\,dm^{-3}}$ exerts only a small buffering effect, because the extent of hydrolysis is small. It can be seen that when the FA and Al are present together, the buffering is slightly less than that of FA alone, because binding of the metal by the humic material effectively removes proton-binding sites. Thus the COO^- groups in FA cannot interact with H^+ when bound by Al, and the Al cannot hydrolyse so easily when bound to FA. A similar effect was discussed in relation to the adsorption of humic matter by oxides in Section 12.3. When the starting pH is 5, both FA and Al are more effective buffers, and their interaction causes a correspondingly greater loss of buffering. In waters of pH 6 or greater, Al concentrations are low, and the buffering by FA is of minor importance.

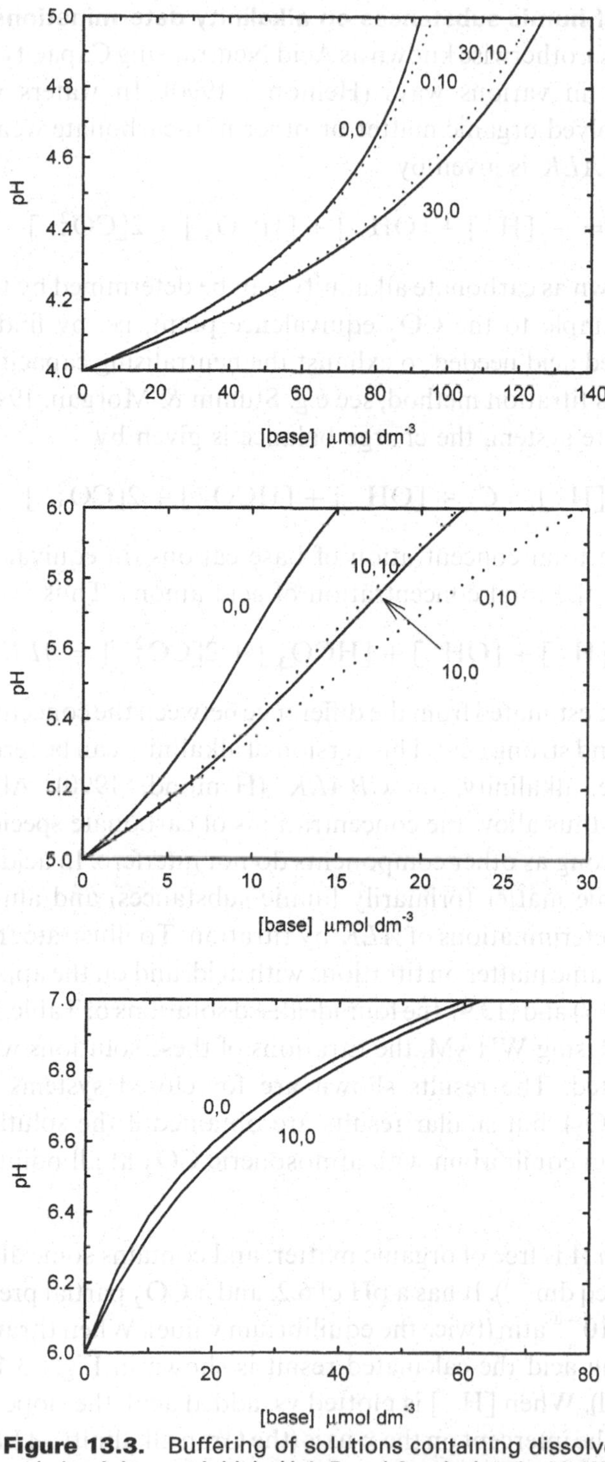

Figure 13.3. Buffering of solutions containing dissolved organic matter and aluminium, at initial pH 4, 5 and 6, calculated with WHAM. The pairs of values identifying the plots are the concentrations of fulvic acid and Al, in units of mg dm^{-3} and µmol dm^{-3} respectively. The solutions are assumed to be in equilibrium with atmospheric CO_2.

13.2.4 Effect of humic substances on alkalinity determinations

Alkalinity (*ALK*), otherwise known as Acid Neutralising Capacity (ANC), can be defined in various ways (Hemond, 1990). In waters without significant dissolved organic matter or other non-carbonate weak acids (e.g. $Al(OH)_2^+$), *ALK* is given by

$$ALK = -[H^+] + [OH^-] + [HCO_3^-] + 2[CO_3^{2-}] \quad (13.7)$$

This *ALK*, known as carbonate alkalinity, can be determined by titration of the water sample to the CO_2 equivalence point, i.e. by finding the amount of added acid needed to exhaust the neutralising capacity of the solution (Gran's titration method; see e.g. Stumm & Morgan, 1996). In a simple carbonate system, the charge balance is given by

$$C_B + [H^+] = C_A + [OH^-] + [HCO_3^-] + 2[CO_3^{2-}] \quad (13.8)$$

where C_B is the total concentration of base cations (in equivalents per dm^3) and C_A is the total concentration of acid anions. Thus

$$C_B - C_A = -[H^+] + [OH^-] + [HCO_3^-] + 2[CO_3^{2-}] = ALK \quad (13.9)$$

and *ALK* can be estimated from the difference between the concentrations of strong base and strong acid. This version of alkalinity can be termed the charge balance alkalinity, or *CBALK* (Hemond, 1990). Alkalinity determinations thus allow the concentrations of carbonate species to be determined, as long as other components do not interfere. In acid waters, dissolved organic matter (primarily humic substances) and aluminium interfere with determinations of *ALK* by titration. To illustrate the effect of dissolved organic matter on titrations with acid, and on the application of equations (13.7) and (13.9), the four idealised solutions of Table 13.3 can be considered. Using WHAM, the titrations of these solutions with acid will be simulated. The results shown are for closed systems (i.e. no degassing of CO_2), but similar results are obtained if the solutions are assumed to be in equilibrium with atmospheric CO_2 at all points in the titrations.

Solution A is free of organic matter, and contains some alkalinity (20 μeq dm^{-3}). It has a pH of 6.2, and a CO_2 partial pressure of 7×10^{-4} atm (twice the equilibrium value). When titrated with strong acid the calculated result is shown in Fig. 13.4 (upper panel). When $[H^+]$ is plotted vs. added acid, the slope is unity and the intercept on the *x*-axis (the Gran alkalinity, ALK_{Gran}) is 20 μeq dm^{-3} (Fig. 13.4, lower panel) which corresponds to

ALK (equation 13.7) and also to *CBALK* (equation 13.9).

Solution B is the same as A but with the addition of 20 mg dm^{-3} fulvic acid (corresponding to *c.* 10 mg dm^{-3} dissolved organic carbon), which lowers the pH to 4.5. The solution has the same *CBALK* (20 μeq dm^{-3}) as A, but *ALK* is now negative, at − 32 μeq dm^{-3}. The titration curve and Gran plot are shown in Fig. 13.4. The latter is slightly curved at low additions of acid, but when the added acid concentration exceeds 100 μeq dm^{-3} it is a straight line with a slope close to, although slightly less than, unity. The estimated ALK_{Gran} depends somewhat on the region of the plot chosen for regression, but a reasonable value obtained from the highest concentrations of added acid is − 23 μeq dm^{-3}.

Solution C is the same as A but with 20 mg dm^{-3} fulvic acid added, and sufficient Na to keep the pH and the HCO_3^- concentration the same as in A. *CBALK* is now 102 μeq dm^{-3}, while *ALK* remains at 20 μeq dm^{-3}. The Gran slope is 0.95, and a reasonable value for ALK_{Gran} is 54 μeq dm^{-3}.

Solution D is the same as B but with 10 μmol dm^{-3} Al added. *CBALK* remains at 20 μeq dm^{-3}, while *ALK* is now − 24 μeq dm^{-3}. The Gran slope is 0.95, and a reasonable value for ALK_{Gran} is − 17 μeq dm^{-3}.

Table 13.3. Hypothetical waters to illustrate the influences of fulvic acid (FA) and Al on the determination of alkalinity by titration with mineral acid.

	A	B	C	D
Na	200	200	282	200
Al	0	0	0	10
Ca	10	10	10	10
Cl	200	200	200	200
FA	0	20	20	20
pH	6.2	4.5	6.2	4.6
T_{CO_3}	58	38	58	38
[HCO_3^-]	21	0	21	1
ALK from eq. (13.7)	20	−32	20	−24
CBALK from eq. (13.9)	20	20	102	20
ALK_{Gran}	20	−23*	54*	−17*
Gran slope	1.00	0.97*	0.95*	0.95*

Values marked * depend somewhat on the range chosen for regression (cf. Fig. 13.4). Concentrations are in μmol dm^{-3}, except FA (mg dm^{-3}).

The above examples give some idea of how the presence of dissolved organic matter and dissolved aluminium in dilute waters can influence the measured alkalinity, and also how titration behaviour is affected, which relates back to Section 13.2.3. The slopes for B, C and D are sufficiently close to unity for them to be accepted as 'normal' in routine determinations of alkalinity. More detailed explanations of the interpretation of alkalinity in waters rich in dissolved organic carbon, and of practical aspects, have

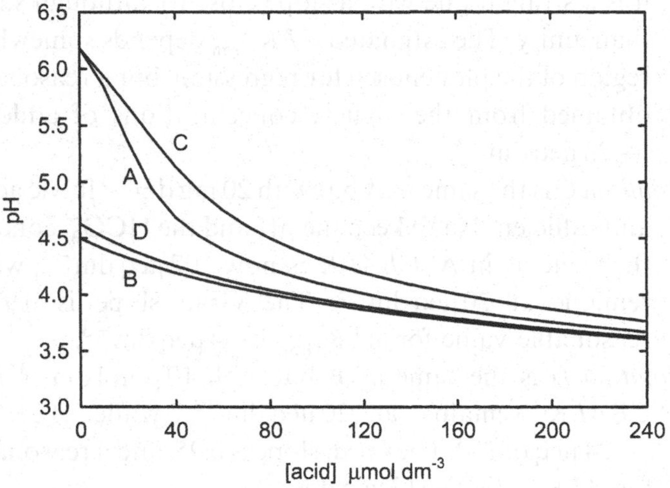

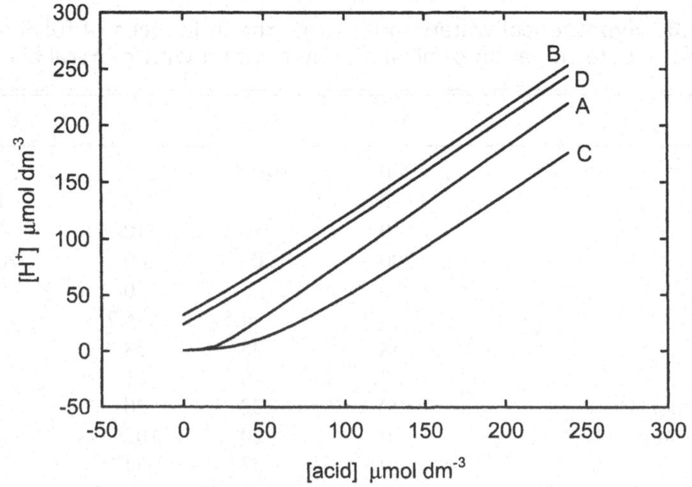

Figure 13.4. WHAM-simulated titrations with acid of the idealised surface waters of Table 13.3. The upper panel shows the 'raw' data from the titration, the lower panel shows Gran plots.

been given by Cantrell *et al.* (1990), Hemond (1990), Hongve (1990), Driscoll *et al.* (1994) and Cai *et al.* (1998).

Hemond (1990) pointed out that ALK_{Gran} values, which are routinely determined in acidification monitoring programmes, provide ambiguous information about the status of a natural water. He recommended that *CBALK* values should be used as the diagnostic variable. However, values of *CBALK* determined from equation (13.9), i.e. from the difference between the concentrations of strong acids and strong bases, may be imprecise. Therefore, he proposed the estimation of *CBALK* from ALK_{Gran} by application of the following equation

$$CBALK = ALK_{Gran} + \beta[\mathrm{DOC}] \tag{13.10}$$

where β is a constant. If equation (13.6) is used to estimate the charge on DOC, the value of β is 4.6 meq g^{-1}. Hemond obtained a value of 5.4 meq g^{-1} from field data for waters in the Adirondacks region (USA). These values of β provide reasonable explanations of *CBALK* and ALK_{Gran} in Table 13.3, assuming that fulvic acid is 50% carbon. A somewhat lower value of β applies to waters with high concentrations of Al. Neal *et al.* (1999) have extended this approach.

13.3 Interactions of humic substances with major cations

Here we focus on cations that can be present at high concentrations in natural waters. They include the alkali and alkaline earth cations, together with Al and Fe.

13.3.1 Alkali and alkaline earth cations

Generally, we can regard Na, Mg, Ca, and to some extent K, as major metals in environmental solutions. They are usually present at high concentrations relative to concentrations of humic binding sites, and their binding is weak. Therefore, in many natural waters, their reactivities are not greatly influenced by binding to humic matter. The influence is rather the other way around; the interactions influence the chemistry of the humic substances, through modification of net charge and by competition effects. For the same reasons, there are few data demonstrating the binding of these cations by humic matter in natural waters. The examples in Table 13.1 give some idea of the extent of such binding (Table 13.4). As already discussed, the binding of the divalent cations, Mg^{2+} and Ca^{2+}, affects the net charge on humic matter. It can be seen that, except for the acid bogwater, small or negligible percentages of the total metal are calculated to be associated with the organic matter. In the bogwater, the cations are calculated to be bound as counterions, neutralising the fulvic acid charge. Such binding is also anticipated for acid soil solutions, where concentrations of dissolved organic matter can be very high (several hundreds of mg dm^{-3}), and where most of the dissolved cations are held as counterions.

13.3.2 Aluminium

Aluminium is mobile in acid systems, where soil waters and surface waters may therefore contain quite high concentrations of the element. As shown by the computed speciation examples in Table 13.1 (acid bogwater and acid streamwater), Al is the metal exhibiting the greatest specific binding under such circumstances, and the binding has a strong effect on the net humic charge.

Probably the greatest interest in the binding of Al by humic matter in natural aqueous solutions came about because of the need to understand the overall chemical speciation of Al, in connection with its toxic effects, towards fish, other aquatic biota and plants, exacerbated by the effects of acid deposition. For example, Driscoll *et al.* (1980) showed that the complexation of Al by natural dissolved organic matter could eliminate the toxicity of the metal towards fish. In an influential paper, Johnson *et al.* (1981) reported the use of an analytical speciation scheme that allowed concentrations of different dissolved forms of Al to be estimated. The methodology had been developed by Driscoll, who subsequently published a detailed description (Driscoll, 1984). The key step in estimating organically bound metal is the passage of the sample through a column of cation-exchange resin (Section 6.5.4). The method has been widely used in acid rain research, following acceptance of the idea that the concentration of labile metal (i.e. the fraction that can be removed by the cation-exchange

Table 13.4. Amounts of alkali and alkaline earth metals bound to fulvic acid in the examples of Table 13.1 (WHAM calculations), expressed as percentages of the total metal.

	Acid bogwater	Acid streamwater	Neutral softwater	Hardwater	Seawater
Na	26.2	0.5	1.8	0.2	0.0004
Mg	46.3	0.5	6.9	1.1	0.004
K	—	—	—	—	0.0004
Ca	46.3	0.5	6.9	1.1	0.004

resin and is therefore not organically bound) provides the best guide to toxic effects. The speciation of the labile fraction is determined by equilibrium modelling, to obtain the concentrations of Al^{3+} and its complexes with OH^-, F^- and SO_4^{2-}. Other speciation methods for Al in field samples were subsequently developed including the use of equilibrium dialysis (LaZerte, 1984), fast separation in a flow-injection system (Clarke *et al.*, 1992) and high performance cation-exchange chromatography (Sutheimer & Cabaniss, 1995). Results from the work of Johnson *et al.* (1981) in Table 13.5 show that the distribution of Al among the different chemical forms is highly sensitive to the water composition. Notably, the concentration of Al^{3+} varied by two orders of magnitude over a fairly narrow pH range.

Although the aluminium system in surface waters is evidently quite complicated, once the analytical methodology had been developed it proved to be an excellent example of metal speciation to which chemical modelling could be applied. This arose firstly because concentrations of the metal are sufficiently high that measurements can be made in most laboratories without the need for very stringent anti-contamination procedures, secondly because a single Al species is rarely completely dominant, and thirdly because the widespread interest in the effects of acid deposition led to the performance of many analyses in a wide range of waters. The key aspect of any modelling effort was to account for the binding of the metal to natural organic matter, since the inorganic reactions were already adequately described. Models were developed by various authors to describe the organic complexation of Al (Table 13.6).

A data set for testing speciation models is summarised in Table 13.7. In the following exercise, it was assumed that all DOM was the same, irrespective of location. WHAM/Model VI was used to calculate concentrations of organically bound Al – [Al–org] – and the fraction of DOM active as isolated FA was optimised. Observed and calculated results are shown in Fig. 13.5. The active fraction of DOM is 0.67, which agrees very well with the value of 0.66 obtained from charge balance (Section 13.2.2). The agreement between observed and calculated [Al–org] is reasonable, given that only a single calibration parameter was used. In an earlier, similar, modelling exercise, Tipping *et al.* (1991) found that much of the discrepancy between observed and calculated values could be explained by the error in measured values, i.e. model inputs and [Al–org]. The modelling results suggest that isolated fulvic acid provides a reasonable representation of field DOM, if it is allowed that approximately one-third of the DOM is inert with respect to cation binding.

Table 13.5. Data of Johnson *et al.* (1981) for Al speciation at Falls Brook (Hubbard Brook Experimental Forest, USA), 4 May 1979.

Site	pH	$TotSO_4$	$[F^-]$	Al_r	Al_{mon}	Al_{org}	[DOC]	Al_{coll}	$Al_{mon,i}$	$\Sigma[Al\text{-}OH]$	$\Sigma[Al\text{-}F]$	$\Sigma[Al\text{-}SO_4]$	$[Al^{3+}]$
I	4.76	55	0.003	21.0	18.4	1.9	2.56	2.6	16.5	7.5	1.8	0.6	6.6
II	4.82	53	0.06	15.4	13.6	4.3	2.05	1.9	9.4	4.3	1.8	0.3	3.0
III	4.92	49	0.16	13.1	9.9	4.3	2.45	3.2	5.6	2.3	2.1	0.1	1.1
IV	5.00	48	0.42	9.2	7.0	3.9	2.37	2.2	3.1	0.9	1.8	0.0	0.3
V	5.30	50	2.2	8.0	6.5	4.6	2.50	1.6	1.9	0.3	1.6	0.0	0.0
VI	5.11	51	1.1	10.7	8.2	5.6	3.45	2.5	2.6	0.5	2.1	0.0	0.1
VII	5.10	52	1.2	11.9	9.5	6.8	3.88	2.4	2.7	0.4	2.2	0.0	0.1
VIII	5.15	51	1.3	11.3	8.8	7.0	3.61	2.5	1.8	0.3	1.4	0.0	0.1
IX	4.99	58	0.6	14.1	11.1	7.5	4.26	3.0	3.6	0.8	2.4	0.0	0.1

Concentrations are in $\mu mol\,dm^{-3}$ except for DOC ($mg\,dm^{-3}$).
Key: $[F^-]$ concentration of F^- determined with an ion-specific electrode, Al_r acid reactive Al, Al_{mon} monomeric Al, Al_{org} organically complexed (non-labile) Al, Al_{coll} colloidal Al, $Al_{mon,i}$ inorganic monomeric Al, $\Sigma[Al\text{-}OH]$, $\Sigma[Al\text{-}F]$, $\Sigma[Al\text{-}SO_4]$ total concentrations of Al complexes with OH^-, F^- and SO_4^{2-} respectively.

Table 13.6. Models for Al–organic complexation, applicable to field samples.

Model description	Reference
Formation of AlL_2, AlL_3 and $Al(OH)L_2$ complexes, in competition with H^+	Young & Bache (1985)
Empirical multivariate equation linking $[Al^{3+}]$, pH and [DOC]	Backes & Tipping (1987b)
Two-site model of humic matter, binding of protons, bidentate binding of Al, electrostatic term	Backes & Tipping (1987b)
Scatchard quasiparticle with three acidic groups, binding of Al^{3+} and $Al(OH)_2^+$	Blaser & Sposito (1987)
Diprotic acid (LH_2), formation of AlL^+, AlL_2^- and $AlLH_{-1}$	Lövgren *et al.* (1987)
Extension of Backes & Tipping (1987b) model to include COOH heterogeneity, empirical electrostatic sub-model	Tipping *et al.* (1988b)
Extension of Tipping *et al.* (1988b) model to account explicitly for diffuse layer accumulation	Tipping & Hurley (1988)
Humic Ion-Binding Model V (Section 9.6)	Tipping & Hurley (1992)
Scatchard model (Section 9.2), with empirical (polynomial) description of the variation of K with v, for pH 4, 4.5, 5 and 6	Gerke (1994)
Triprotic acid (AH_3), formation of AlA, $Al(H)A^+$ (Section 9.3.5)	Schecher & Driscoll (1995)
Natural Organic Anion Equilibrium Model (NOAEM); Gaussian distributions of functional groups	Grzyb (1995)
Single site model, pH dependent equilibrium constant	Sutheimer & Cabaniss (1997)
Humic Ion-Binding Model VI (Section 9.6)	Tipping (1998b)
NICCA–Donnan model (Section 9.7)	Kinniburgh *et al.* (1999)

13.3.3 Iron

The issue of the nature of iron in natural waters has been on the research agenda for many years, and is yet fully to be resolved. Concentrations of the metal in the 'dissolved' (as judged by filtration) fraction are generally much higher than expected from the solubilities of ferric oxyhydroxides. The likeliest explanations are that the extra metal is present as (a) ferrous iron, which is much more soluble than the ferric form, (b) as colloidal ferric oxyhydroxide particles, and/or (c) as organic (humic) complexes.

High concentrations of ferric iron were explained in terms of colloid stabilisation (peptisation) by Shapiro (1966), on the basis of laboratory experiments with natural aquatic organic matter and iron salts. Cameron & Liss (1984) drew similar conclusions from the results of experiments with a variety of stabilising agents, including tannic acid, detergents and phosphate, as well as humic acid. Perdue *et al.* (1976) showed that concentrations of Fe (and Al) in waters of the Satilla River system (USA) were correlated with concentrations of dissolved organic matter (Fig. 13.6), and proposed that organic complexation was the explanation. They also showed that the correlation between the sum of the concentrations of the two metals and DOM concentration was stronger than the individual correlations, which led to the conclusion that they are competing for the

Table 13.7. Field data for aluminium speciation. See Fig. 13.5.

Field site		pH	DOC	Al–tot	Al–org
1 R. Duddon	UK	4.9–5.7	0.3–1.1	1.4–20.3	0.1–1.7
2 Falls Brook	USA	4.8–5.4	2.1–4.3	3.1–13.6	0.7–9.0
3 Paint Lake inflow	Canada	5.3–5.8	1.9–2.8	2.7–7.9	2.0–4.0
4 Plastic Lake inflow	Canada	4.3–4.8	6.0–11.5	5.9–10.2	2.8–7.6
5 Plastic Lake outflow	Canada	4.9–5.9	2.0–3.6	0.4–4.3	0.4–2.7
6 R. McDonald	Canada	5.3–6.6	5.2–11.5	2.2–4.7	1.8–3.6
7 R. aux Rochers	Canada	5.4–6.4	5.1–9.2	2.7–5.0	2.1–3.1
8 R. de la Trinité	Canada	5.2–6.5	4.5–8.1	1.7–4.7	1.3–3.1
9 Birkenes Brook	Norway	4.2–4.7	3.7–4.8	16.1–25.7	4.2–6.9
10 R. Swale	UK	4.2–5.6	2.7–13.8	1.3–7.5	0.9–5.8
11 Whitray Beck	UK	3.9–5.8	2.5–27.5	3.0–10.1	1.6–5.1
12 Llyn Brianne stream	UK	4.5–4.7	1.1–3.4	13.7–25.6	3.9–7.1
13 Adirondack lakes	USA	5.3–7.3	2.7–7.4	0.1–10.7	0.0–7.5
All sites		3.9–7.3	0.3–27.5	0.1–25.7	0.0–9.0

There are nine samples from each site, except for the Adirondack lakes (eight samples). Concentrations of Al are in $\mu mol\,dm^{-3}$, those of DOC in $mg\,dm^{-3}$.

Source: References and further details are given by Tipping *et al.* (1991), except for the Adirondack lake samples, which are from Sutheimer & Cabaniss (1997).

same sites, and that their total solubility is determined by the total complexing capacity. The implied saturation of sites by Al and Fe(III) is unreasonable, in terms of the ideas about humic binding sites and binding strengths developed in this book (see also Section 13.4.5), but the influence of DOM on the two metals is beyond doubt. The peptisation and complexation mechanisms could operate simultaneously, since they are not mutually exclusive.

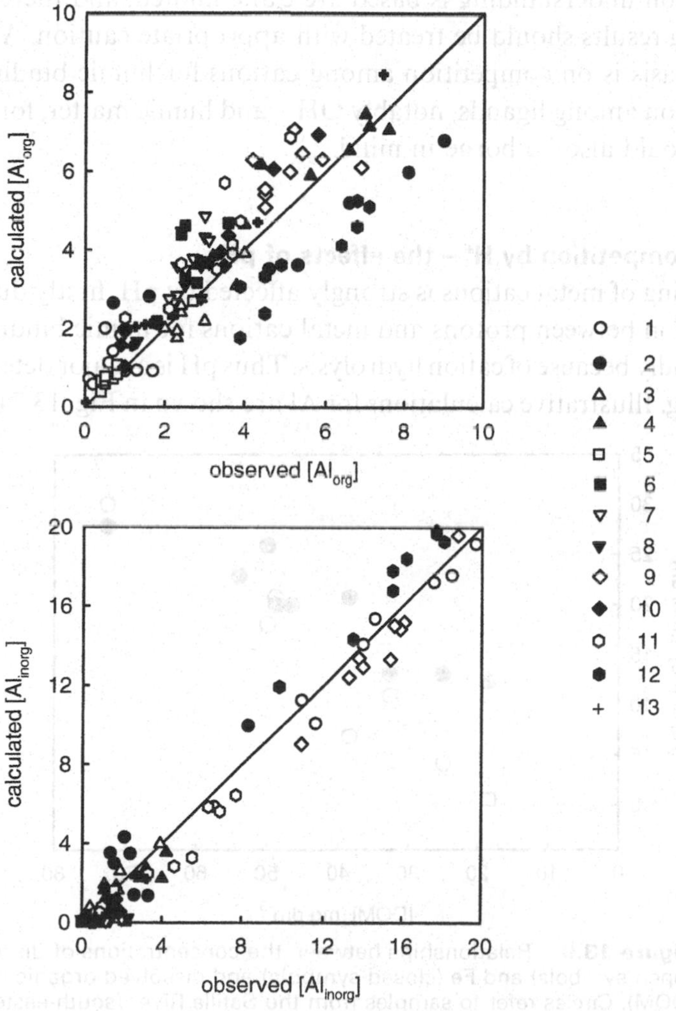

Figure 13.5. Comparison of observed and calculated concentrations of organically complexed and inorganic Al in surface waters. The key refers to the source data (116 samples in all), summarised in Table 13.7. 1: 1 lines are shown for guidance.

13.4 Competition

In principle, competition can involve all cations. However, the main competition effects in natural systems involve protons (pH dependence) and the major cations (Na, Mg, Al, K, Ca, Fe). There are few, if any, field data available to demonstrate competition effects systematically, although of course competition effects must be operating in all natural waters. Competition was discussed in relation to laboratory systems in Section 10.7. Here, the intention is to explore the implications for natural waters, by modelling. It must be borne in mind that the data on which competition understanding is based are quite limited, and therefore the modelling results should be treated with appropriate caution. Although the emphasis is on competition among cations for humic binding sites, competition among ligands, notably OH^- and humic matter, for a given cation should also be borne in mind.

13.4.1 Competition by H^+ – the effects of pH

The binding of metal cations is strongly affected by pH, firstly due to the competition between protons and metal cations for humic binding sites, and secondly because of cation hydrolysis. Thus pH is a major determinant of binding. Illustrative calculations for Al are shown in Fig. 13.7 (see also

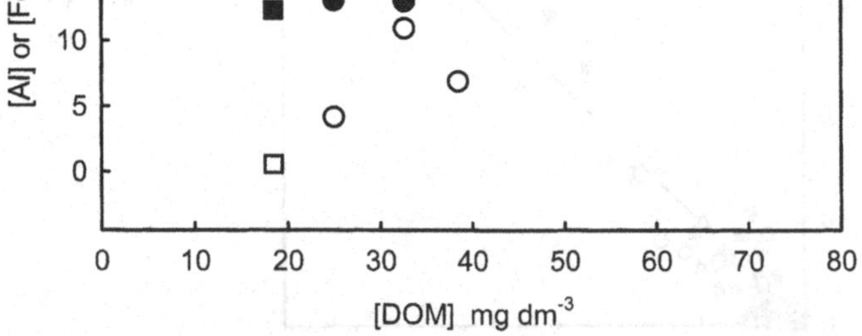

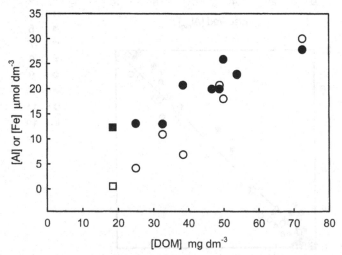

Figure 13.6. Relationships between the concentrations of dissolved Al (open symbols) and Fe (closed symbols) and dissolved organic matter (DOM). Circles refer to samples from the Satilla River (south-eastern USA), squares are from the composition of the 'World Average River'. [Redrawn from Perdue E.M., Beck K.C., Reuter J.H. (1976), Organic complexes of iron and aluminium in natural waters, *Nature* **260**, 418–420, with permission from *Nature* and the authors.]

Fig. 7.10). Here we see that in the low pH range, metal binding increases with pH, as the concentration of the competing protons decreases. However, at higher pH, humic binding is calculated to decrease, because the species $Al(OH)_2^+$ and $Al(OH)_4^-$ are assumed not to bind. Thus $Al(OH)_4^-$ becomes dominant, an example of the formation of complexes with competing ligands.

Figure 13.8 shows calculated results for low concentrations of Cu and Zn. Again, as $[H^+]$ decreases, binding increases, and the free metal ion concentrations decrease. Because Cu is bound more strongly than Zn, its free metal ion concentration is more sensitive to pH than that of Zn. On the other hand, the percentage of Zn bound is more sensitive to pH than that of Cu.

13.4.2 Ionic strength effects

Figure 13.9 shows the calculated effects of ionic strength on the binding of Cu and Zn by fulvic acid. The plots for $NaNO_3$ show how increasing ionic strength in a 1:1 electrolyte at constant pH gives rise to increases in free metal ion concentrations. This appears to be a competition effect, but is interpreted as a less direct process, in which the electrolyte diminishes the electrostatic attraction of the humic surface for the cation (cf. Section 9.5). In the model, in principle, there is also direct competition between the monovalent cation and Cu and Zn for binding as counterions, but such

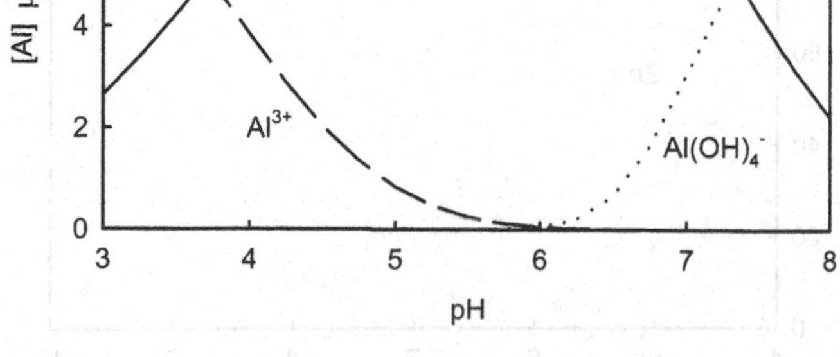

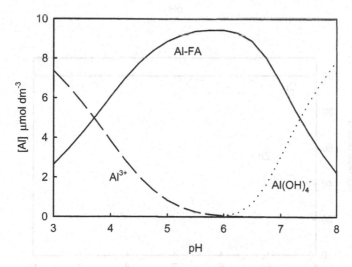

Figure 13.7. Aluminium binding by fulvic acid (FA) as a function of pH, at $I = 0.001$ mol dm^{-3}, calculated with WHAM/Model VI. The total concentrations of FA and Al are 10 mg dm^{-3} and 10^{-5} mol dm^{-3} respectively. Note that $AlOH^{2+}$ and $Al(OH)_2^+$ are omitted for clarity, since they make only minor contributions in this system.

binding does not contribute significantly to the binding of the trace metals for the conditions considered. The ionic strengths of Fig. 13.9 cover the range from dilute freshwater to seawater, and so cover most natural waters. In terms of the relative change in the concentration of the free metal ion, the effect of ionic strength is similar for Cu and Zn. However in terms of the fraction of total metal that is bound, Zn is much more sensitive. Thus the fraction of Zn bound varies from *c.* 99% at low ionic strength to less than 10% at high ionic strength, whereas more than 99.9% of the Cu is bound at all ionic strengths.

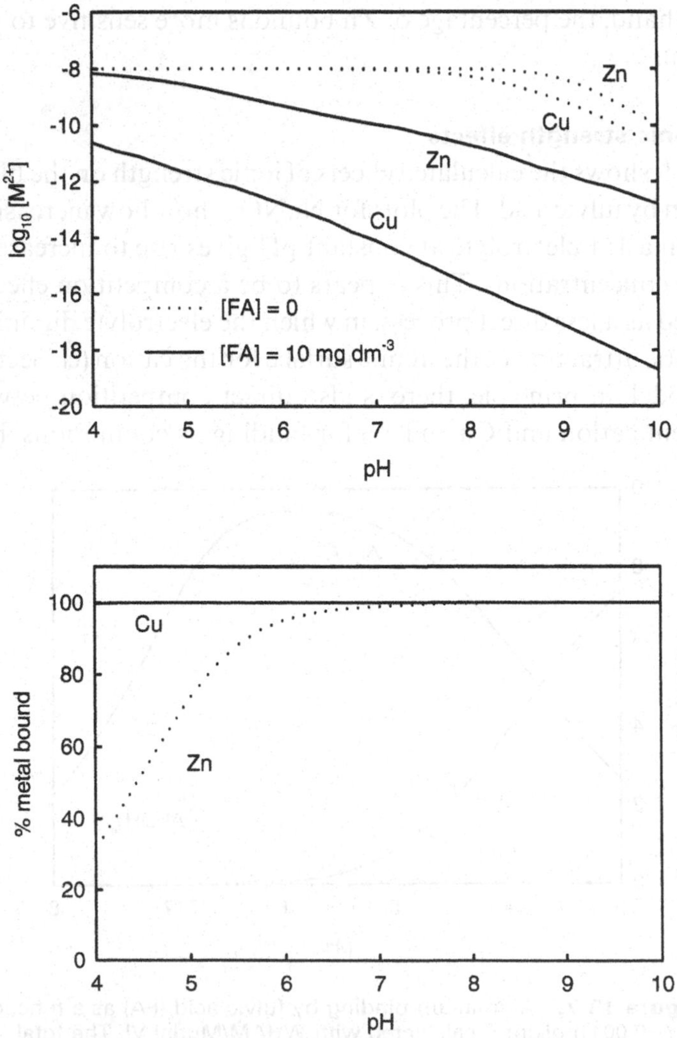

Figure 13.8. Copper and zinc binding by fulvic acid (FA) as a function of pH, at $I = 0.001$ mol dm^{-3}, calculated with WHAM/Model VI. The total concentrations of Cu and Zn are 10^{-8} mol dm^{-3}, and that of FA is 10 mg dm^{-3}.

13.4.3 Competition by alkali and alkaline earth cations

In Models V and VI (also in the NICCA–Donnan model), the alkali metal cations (Na^+, K^+ etc.) are assumed not to bind at specific sites. Therefore, as explained above, they exert competitive effects only through the electrostatic effect and as non-specifically bound counterions. The alkaline earth cations, on the other hand, also bind at the specific sites on the humic molecules and enter into direct competition. Figure 13.9 shows that Ca^{2+} is calculated to decrease Cu and Zn binding appreciably more than monovalent cations, at the same ionic strength. The relative binding strengths of different cations vary with the extent of loading, as discussed in Section 10.3, so Cu^{2+} (and other metal cations with affinity for N atoms) are highly preferred over Ca^{2+} by the strongest sites, but less so by the weaker sites.

13.4.4 Competition by aluminium

As we have seen in Section 13.3.2, Al can be present at high concentrations in acid systems, and its strong binding to humic matter diminishes the binding of other metals. Illustrations are provided by the model predictions of Fig. 13.10, for a simplified low ionic strength surface water. The relative effect is greater for Cu than for Zn, because Cu is bound to a far greater extent under the conditions considered.

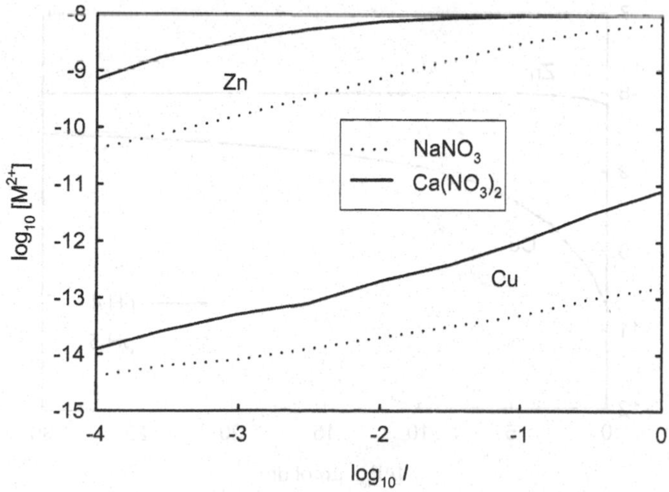

Figure 13.9. Variations with ionic strength of the unbound concentrations of Cu and Zn in fulvic acid solutions, calculated with WHAM/Model VI. The results refer to pH 7, with metal and fulvic acid concentrations of 10^{-8} mol dm^{-3} and 10 mg dm^{-3} respectively.

13.4.5 Competition by iron

The nature and extent of binding of Fe(III) by humic substances in natural waters remains unclear, as discussed in Section 13.3.3. We know, mainly from spectroscopic evidence (Chapter 8), that Fe(III) forms complexes with humic substances, and within the Model VI picture, strong binding of Fe(III) at the rare sites influenced by N is expected (Table 10.5). Most or all of these sites may be occupied by Fe(III) in natural waters. Figure 13.11 shows the results of calculations with Model VI, firstly assuming Fe to be absent, and secondly assuming the solutions to be in equilibrium with iron oxide phases of differing solubilities. The predictions are that Fe(III) could exert a strong competitive effect under natural water conditions. The effect is greater for Cu, because it is bound to a greater extent than Zn, so that its displacement from the strong sites causes a greater relative change in the solution concentration of the metal. Such competition is a feature of a system in which there are small numbers of strong sites.

The amounts of Fe calculated to be bound to the fulvic acid under the conditions of the waters in Fig. 13.11, for the more soluble Fe oxide ($K_{so} = 10^4$ at 25°C), are 2×10^{-4} mol g^{-1} at pH 5 and 3×10^{-5} mol g^{-1} at pH 8. The first value is similar to that of 5×10^{-4} mol g^{-1} that can be derived from the results of Perdue *et al.* (1976) plotted in Fig. 13.6. These values represent quite high proportions (up to 10%) of the total binding sites in humic matter.

A strong competition effect by Fe(III) seems plausible, but counter

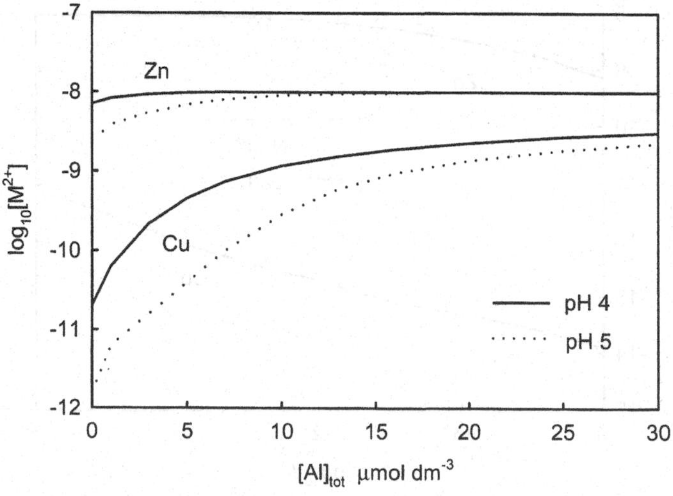

Figure 13.10. Influence of aluminium on the unbound concentrations of Cu and Zn in fulvic acid solutions, calculated with WHAM/Model VI. The results are for 10^{-8} mol dm^{-3} total concentrations of Cu and Zn, 10 mg dm^{-3} fulvic acid, and 2×10^{-4} mol dm^{-3} $NaNO_3$.

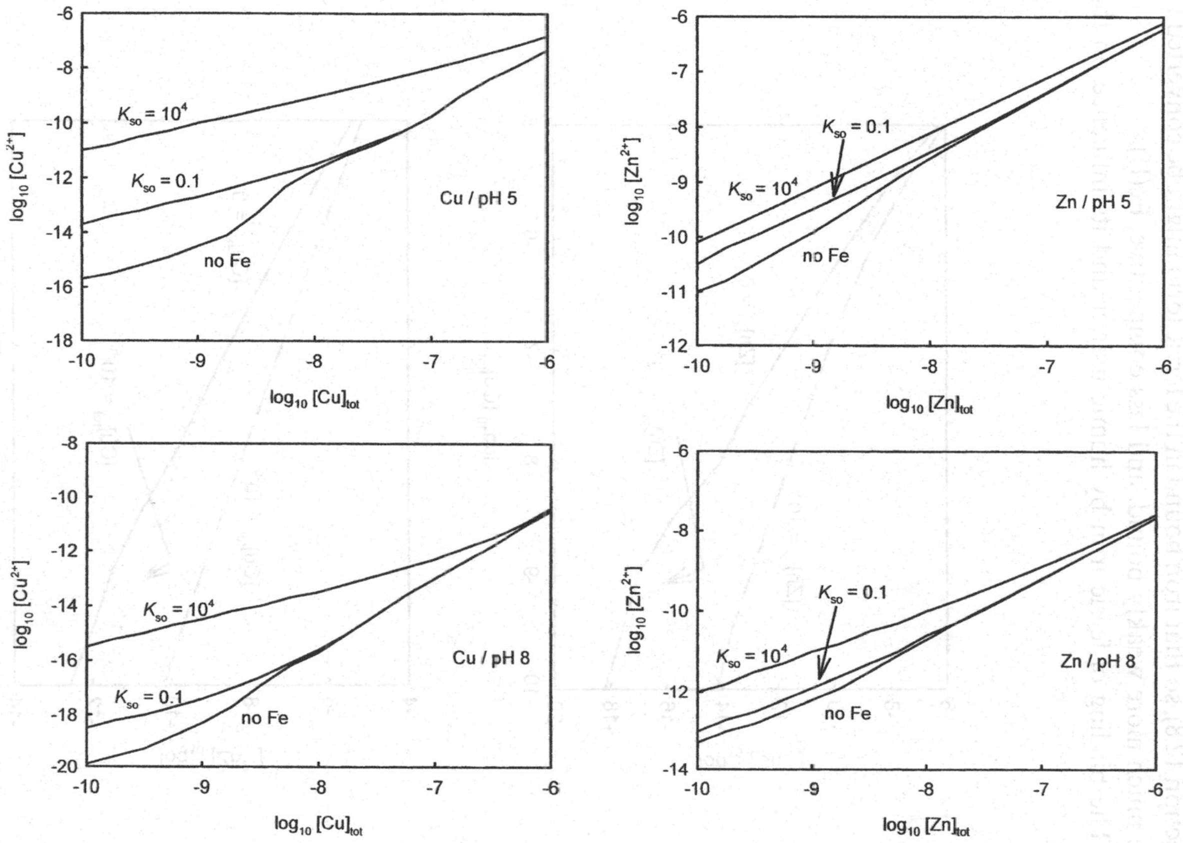

Figure 13.11. Plots of Cu and Zn binding by fulvic acid (10 mg dm^{-3}), as affected by Fe(III), calculated with WHAM/Model VI. The solubility products (K_{so}) are those at 25 °C, for goethite (0.1) and ferrihydrite (10^4), although the calculations refer to 10 °C and so the effective values are ~ 1 and ~ 10^5 respectively. The background electrolyte is 0.001 mol dm^{-3} $NaNO_3$.

arguments can be advanced. Firstly, there are as yet no data to show directly that Fe(III) does in fact bind strongly at the rare sites, nor that rare sites for Cu are the same as those for Fe(III). Secondly, humic substances reduce Fe(III) to Fe(II), especially under the influence of light (Section 12.8), so that iron bound in the ferric form might be converted to the much more weakly bound, and less competitive, Fe(II).

The binding of ferric iron by humic matter and its influence on the

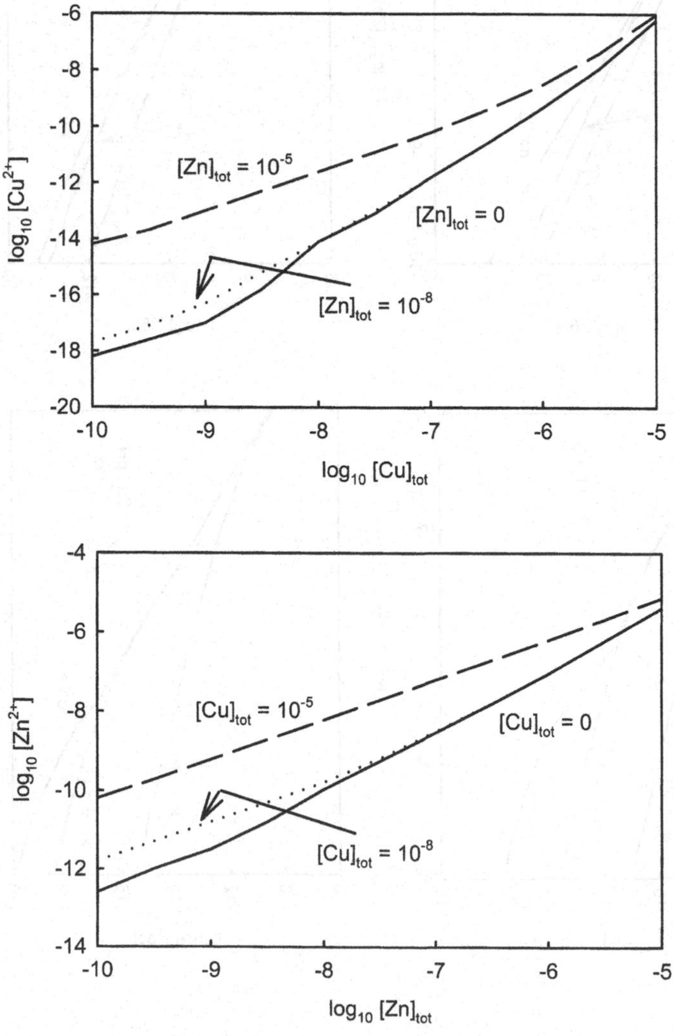

Figure 13.12 Competition between Cu and Zn for binding by fulvic acid at pH 7, $I = 0.001\,mol\,dm^{-3}$, calculated with WHAM/Model VI.

binding of trace metals is an important issue, and one that requires research. The system is complicated, and presents severe experimental difficulties. However, understanding of trace metal binding by humic matter in the natural environment will not be complete as long as the problem is unresolved.

13.4.6 Competition among trace elements

Figure 13.12 shows predicted competition effects between Cu and Zn for binding to fulvic acid. If the competing metal is present at high concentration, there is a significant competitive effect, with the free concentration of the other metal being increased by two to three orders of magnitude. However, if the competing metal is present at low (i.e. trace) concentration, the competitive effect is quite small.

13.5 Interactions of heavy metals with humic substances in natural waters

The speciation of metals such as Ni, Cu, Zn, Cd, Hg and Pb in natural waters is a key determinant of their transport, retention and bioavailability. A good general introduction to the subject, emphasising measurement techniques, was given by Florence & Batley (1980). Humic substances are recognised to play a major rôle in metal speciation. Indeed, this topic is probably the most frequently cited reason for performing research on cation–humic interactions.

13.5.1 Field evidence for heavy metal binding by humic substances

Many measurements have been carried out to determine heavy metal speciation in natural waters. Table 13.8 lists some examples. In each case there is evidence for organic complexation, with Cu, Hg and Pb tending to be complexed to greater extents than Ni, Zn and Cd, as would be expected from their affinities for organic ligands. In many cases, but not all, the authors have concluded that humic matter is primarily responsible for the binding. For example, Bruland (1992) presented evidence for the presence of Cd-specific ligands in the surface waters of the Pacific Ocean. Wells *et al.* (1998) reported that Cu, Zn, Cd and Pb are bound by different organic ligands, distinguishable by size fractionation. Xue and coworkers (Xue *et al.*, 1995; Xue & Sigg, 1999) interpreted their results for lakes in terms of 'background' binding by humic matter, together with algal-derived, high-affinity ligands in surface waters.

Table 13.8. Examples of field studies providing evidence for or against the binding of heavy metals by natural organic matter.

Type of water	Method	Results/conclusions	Reference
Nickel			
Seawater	CLE–CSV, CRCP–GFAAS	33–50% complexed by strong organic ligands	Donat *et al.* (1994)
Landfill leachates	Size fractionation, ion-exchange	20–40% in dissolved or colloidal organic forms	Jensen *et al.* (1999)
Copper			
River–estuary–sea	ASV	Up to 80% in organic forms	Duinker & Kramer (1977)
Estuarine and coastal waters	ASV	0–23% in organic (colloidal) complexes	Batley & Gardner (1978)
River water	ISE and modelling	Added Cu largely bound by fulvic acid type material	Cabaniss & Shuman (1988b)
Soil water	Column cation-exchange	90–98% of Cu organically bound	Berggren (1992)
River and lake waters	ISE and modelling	Results can be explained by humic binding	Dwane & Tipping (1998)
Coastal seawater	Physical and chemical separations	Metal binding by specific ligands	Wells *et al.* (1998)
Rice field waters	DPASV and CLE–CSV	$\geq$ 99% of Cu bound by high-affinity organic ligands	Witter *et al.* (1998)
Landfill leachates	Size fractionation, ion-exchange	60–90% in dissolved or colloidal organic forms	Jensen *et al.* (1999)
Lake waters	CLE–CSV and ASV	Low concns. of strong ligands + humic substances	Xue & Sigg (1999)
Zinc			
River–estuary–sea	ASV	Up to 60% in organic forms	Duinker & Kramer (1977)
Contaminated soil solutions	Cation-exchange separations	> 96% of metal present as Zn^{2+} or labile complexes	Holm *et al.* (1995a)
Lake waters	CLE–CSV	Low concns. of strong ligands + humic substances	Xue *et al.* (1995)
Coastal seawater	Physical and chemical separations	Metal binding by specific ligands	Wells *et al.* (1998)
Landfill leachates	Size fractionation, ion-exchange	10–40% in dissolved or colloidal organic forms	Jensen *et al.* (1999)

Cadmium			
Estuarine and coastal waters	ASV	0–10% in organic (colloidal) complexes	Batley & Gardner (1978)
Sewage-polluted river	ASV and UV irradiation	None present in organically bound forms	Iyer & Sarin (1992)
Seawater	DPASV	70% bound by Cd-specific organic ligands	Bruland (1992)
Soil solution	Dialysis and ion-exchange	20% present as organic complexes	Holm *et al.* (1995b)
Polluted groundwater	Dialysis and ion-exchange	98% present as organic complexes	Holm *et al.* (1995b)
Industrial leachates	Dialysis and ion-exchange	17–63% present as organic complexes	Holm *et al.* (1995b)
Coastal seawater	Physical and chemical separations	Metal binding by specific ligands	Wells *et al.* (1998)
Landfill leachates	Size fractionation, ion-exchange	90–100% in dissolved or colloidal organic forms	Jensen *et al.* (1999)
Lake waters	CLE–CSV and ASV	Low concns. of strong ligands + humic substances	Xue & Sigg (1999)
Mercury			
Lakes	Field monitoring and modelling	[Hg] and [CH_3Hg] increase with [DOC]	Driscoll *et al.* (1995)
Fresh and saline waters	OSWASV	Speciation dominated by organic complexes	Wu *et al.* (1997)
Everglades (USA)	Ultrafiltration	Strong binding of CH_3Hg by low mol. wt. DOC, total Hg bound by higher mol. wt. organics	Cai *et al.* (1999)
Upland streams	Field monitoring	Hg associated mainly with POC, also DOC	Kolka *et al.* (1999)
Lead			
River–estuary–sea	ASV	Up to 87% in organic forms	Duinker & Kramer (1977)
Estuarine and coastal waters	ASV	0–18% in organic (colloidal) complexes	Batley & Gardner (1978)
Sewage-polluted river	ASV and UV irradiation	Up to 40% present in organically bound forms	Iyer & Sarin (1992)
River water	DPASV	Strong binding by DOC, affinity falls with Pb loading	Botelho *et al.* (1994)
Soil solutions	DPASV	30–90% bound by DOC, depending upon pH	Sauvé *et al.* (1998)
Coastal seawater	Physical and chemical separations	Metal binding by specific ligands	Wells *et al.* (1998)
Landfill leachates	Size fractionation, ion-exchange	70–90% in dissolved or colloidal organic forms	Jensen *et al.* (1999)

ASV anodic stripping voltammetry; CLE–CSV competitive ligand equilibration – cathodic stripping voltammetry; CRCP–GFAAS chelating resin column partitioning – graphite furnace atomic absorption spectrometry; DOC dissolved organic carbon; DPASV differential pulse anodic stripping voltammetry; ISE ion-specific electrode; OSWASV Osteryoung square wave anodic stripping voltammetry; POC particulate organic carbon.

13.5.2 Influence of speciation on bioavailablity

Some heavy metals are essential to life, some are toxic, and some exhibit both characteristics depending upon concentration. A good general review is given by Fraústo da Silva & Williams (1997), and a review focusing on aquatic toxicity by Allen & Hansen (1996). As already noted

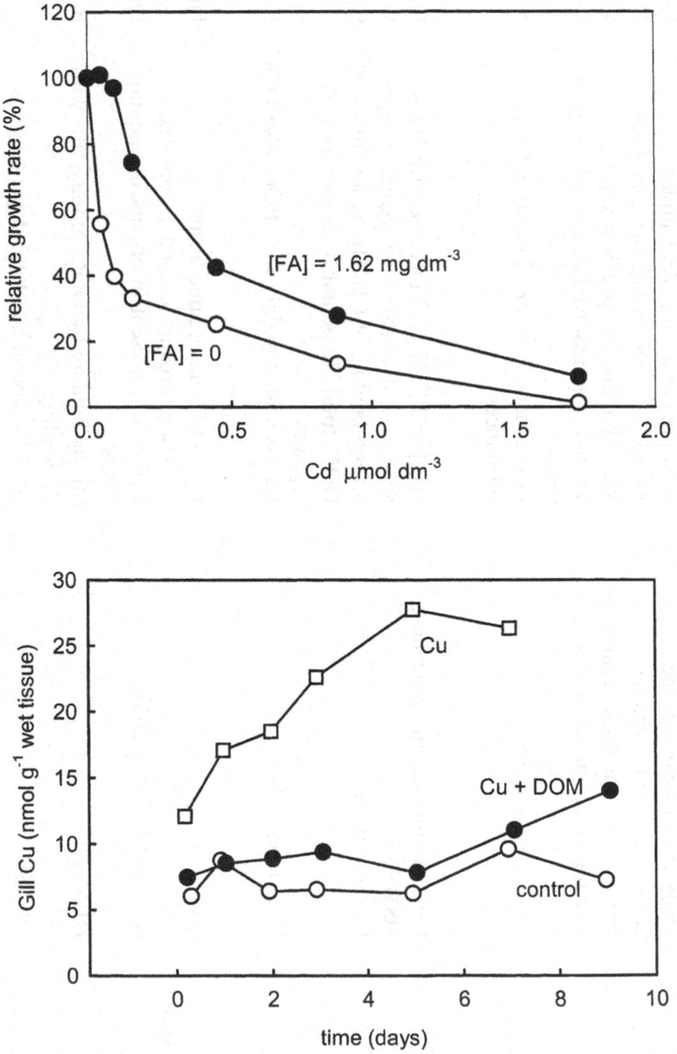

Figure 13.13. Examples of metal interactions with aquatic biota, and the influence of humic substances. The upper panel shows the toxic effect of Cd on a freshwater alga (Sedlacek *et al.*, 1983). The lower panel shows the extent of binding of Cu by the gills of rainbow trout [from Hollis, L., Muench, L. & Playle, R.C. (1997), Influence of dissolved organic matter on copper binding, and calcium on cadmium binding, by gills of rainbow trout, *J. Fish Biol.* **50**, 703–720, with permission from Academic Press Ltd].

with respect to aluminium (Section 13.3.2), bioavailability depends upon the chemical speciation of the metal, and in natural waters humic substances play a major rôle in determining speciation. Figure 13.13 shows two laboratory illustrations of the effects of humic matter on heavy metal toxicity and uptake by aquatic organisms.

Equilibrium speciation has been related directly to biological uptake by the 'Free Ion Activity Model' (FIAM). The FIAM was described by Morel (1983), and restated in an influential review by Campbell (1995), who listed the following key assumptions:

(1) The plasma membrane is the primary site for metal interaction.
(2) The interaction can be described by a 1:1 complexation reaction.
(3) Metal transport towards the membrane, and the subsequent complexation step, occur sufficiently rapidly that a pseudo-equilibrium is established between the metal species in solution and those at the membrane surface, i.e. faster than metal uptake (internalisation) and biological response.
(4) The biological response is strictly dependent on the concentration of the metal–biological ligand complex.
(5) In the range of metal concentration of interest, the concentration of free biological ligand sites remains virtually constant (i.e. is in excess), so that variations in the concentration of the metal–biological ligand complex follow directly those of the metal in solution.
(6) During exposure to the metal, the nature of the biological surface does not change, i.e. the bound metal does not alter the nature of the plasma membrane.

The FIAM predicts that biological response is directly proportional to the

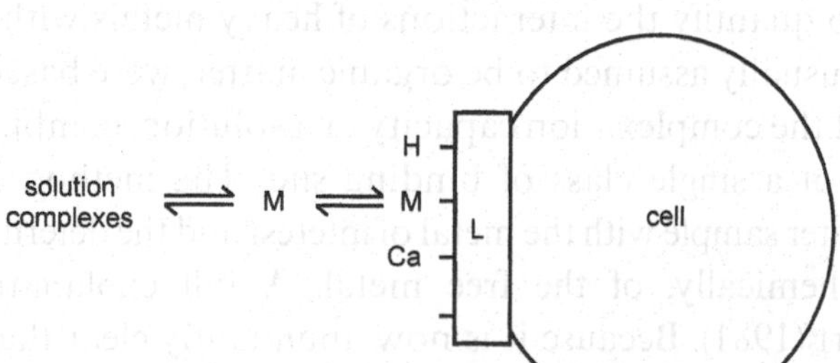

Figure 13.14. Schematic illustration of the principles of the Free Ion Activity Model for metal uptake by organisms. The free metal ion competes with protons and other cations for binding to a receptor (biological ligand), denoted by L, at the cell surface. See Section 13.5.2 for further explanation.

activity of the free metal ion, providing that pH, ionic strength and concentrations of cations that can compete for the biological ligand are constant. Thus, it can explain how complexation by solution-phase ligands (including humic substances) can diminish the nutrient or toxic effect of a metal. If the biological ligand is assumed to bind protons and other metals, then variations in bioavailability with pH and concentrations of competing metals (notably alkaline earth cations) are predicted. These points are illustrated schematically in Fig. 13.14. The Biotic Ligand Model (BLM) (Paquin *et al.*, 2000), is equivalent to the FIAM as described by both Morel (1983) and Campbell (1995). However, according to Meyer *et al.* (1999), there is a widespread belief that the FIAM simply relates toxicity to the free ion activity, with no attention paid to competing effects of other cations at the biological (or biotic) ligand. They therefore distinguish the 'current general interpretation of the FIAM' from the BLM.

In his review, Campbell (1995) presents numerous examples that are consistent with the FIAM, and also some that are not, because kinetic factors are important. These include cases where reduction of the metal has to take place at the cell surface, and where neutral complexes are taken up directly, without binding at the surface. Hudson (1998) explains in detail how the interplay between reaction kinetics and diffusion can lead to violations of the FIAM. When the FIAM/BLM applies, equilibrium speciation provides a direct link to bioavailability. When it does not, knowledge about equilibrium speciation still provides valuable information for the interpretation of metal uptake and biological effects.

13.6 Modelling heavy metal speciation in natural waters

First attempts to quantify the interactions of heavy metals with natural aquatic ligands, usually assumed to be organic matter, were based on the determination of the complexation capacity of a solution, combined with the assumption of a single class of binding site. The method involves titration of the water sample with the metal of interest and the determination, usually electrochemically, of the free metal. A full explanation was provided by Hart (1981). Because it is now abundantly clear that humic substances, together with any other ligands present in a natural water, provide a wide range of binding sites of different strengths, the complexation capacity approach is of little use in predicting metal speciation, and can be considered obsolete.

Modelling has also been done by fitting field data, or experimental data obtained from titrations of field samples. Because the data that can be obtained are often limited in range, and pH has often been kept constant, simple models can successfully be fitted. Thus, Town & Filella (2000) have concluded that the success of the two-site ('L1 and L2') model in describing metal complexation in natural samples arises from the presence of many types of ligand group, only some of which are responsible for binding over the rather narrow range accessible to analytical determinations. They question the interpretation of L1 and L2 in terms of specific compounds or ligand classes.

Cabaniss & Shuman (1988b) were among the first authors to attempt to model metal binding in natural water samples using a model parameterised on the basis of laboratory studies on isolated humic matter. They carried out titrations with copper, using an ion-selective electrode, and compared the observed concentrations of Cu^{2+} with those predicted using the five-site model described in Section 9.3.3. They achieved good agreements between observed and predicted $[Cu^{2+}]$, after adjusting the 'active concentration' of the DOM, to about 50% of the actual value.

Generally, the application of laboratory-based models to natural waters requires information about the water composition, including the DOC content and its nature. The discussion of competition effects in Section 13.4 highlights the need to take competition into account, and the potential difficulties in dealing with Fe binding, especially at low concentrations of the metal of interest. In the remainder of this section, examples of the application of WHAM/Model VI to heavy metal chemistry in a range of natural waters will be described, to illustrate the judgements that have to be made in defining the speciation problem, to show how modelling can help in the interpretation of field observations, and to provide some insight into the predictive capability of the model. Equilibrium is necessarily assumed, and we shall be trying to interpret the observations in terms only of 'simple' inorganic ligands such as OH^-, HCO_3^-, CO_3^{2-}, SO_4^{2-}, Cl^- etc., together with humic substances. In the absence of reliable equilibrium constants and stoichiometries for solution complexes with sulphide and thiols, we shall not be considering interactions with such species, but their potential importance, at least under some circumstances, is increasingly recognised (Al-Farawati & van den Berg, 1999; Rozan *et al.*, 2000) and should be borne in mind. The general modelling approach is to fix all the model's parameters at their default values (Chapter 10), and attempt to fit the data by adjusting the concentrations of 'active' humic substances.

13.6.1 Copper speciation in UK surface waters

Dwane & Tipping (1998) carried out studies in which natural water samples were acidified, spiked with Cu, and then titrated with base. The pH and Cu^{2+} concentration were monitored with electrodes during the

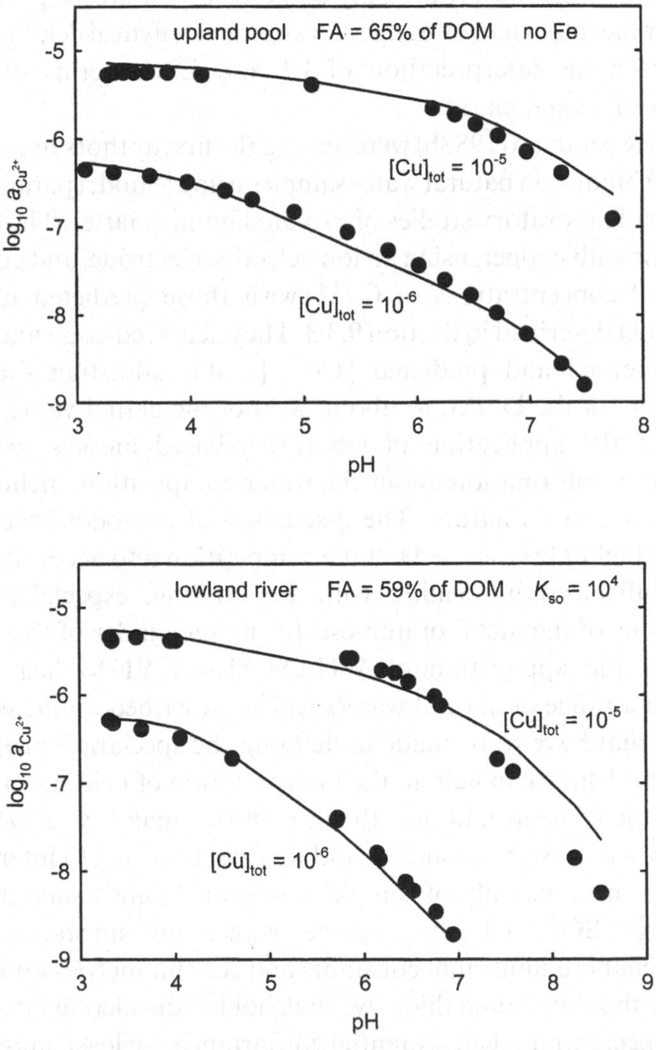

Figure 13.15. Variation of Cu^{2+} activity with pH in natural water samples spiked with Cu: measurements (points) and WHAM/Model VI fits. The upper panel refers to a softwater pool with [DOC] = 8.4 mg dm^{-3}, the lower panel to a hard river water with [DOC] = 13.9 mg dm^{-3}. For each water, a single adjustment of the 'active' concentration of fulvic acid was made. The fit for the pool assumed Fe(III) to be absent, while for the river the dissolved concentration of Fe(III) was assumed to be controlled by an iron (oxy)hydroxide with a solubility product of 10^4 at 25°C. The experimental data are from Dwane & Tipping (1998).

titrations. The chemical compositions of the waters were determined, so that competition effects could be taken into account. In the orginal work, Model V was applied; here we use Model VI. Optimisation is done by adjusting the fraction of DOM that consists of average fulvic acid, while assuming the remainder to be inert with respect to cation binding. Various solubility products are assumed for Fe(III) oxide.

Figure 13.15 shows two illustrative examples of the observed and simulated activities of Cu^{2+}. In both, the model provides a reasonably good simulation of the experimental results over a wide pH range, and for two different total Cu concentrations. For the pool water, the best fit in the absence of Fe(III) was obtained with 65% of the DOM as isolated fulvic acid (FA). Assuming a value of K_{so}(25 °C) of 10^2 for Fe(III) solubility made no difference to the fit, while a value of 10^4 required the FA content of the DOM to be 68%, and slightly worsened the goodness of fit. The optimised fulvic acid content of the lowland river DOM was 53% in the absence of Fe(III) and 59% for a K_{so} of 10^4. The fit was slightly better with Fe(III) present. The effects of Fe(III) are calculated to be small in these examples, because the concentrations of Cu are too high for Fe(III) competition to be significant.

Table 13.9. Summary of the data of Xue and coworkers for metal chemistry in Swiss and Italian lakes and rivers.

	Rivers	High-ALK lakes	Low-ALK lakes
pH	8.2–8.4	7.8–8.5	5.3–7.5
[DOC]	0.9–5.8	1.0–5.7	0.3–1.1
ALK	2.1–4.3	2.0–4.3	0.0–0.14
$[Cu]_{tot}$	10–68	4–17	3–71
$[Cu^{2+}]$	10^{-15}–10^{-13}	10^{-16}–10^{-13}	10^{-12}–10^{-9}
$[Cu^{2+}]/[Cu]_{tot}$	2×10^{-8}–5×10^{-6}	2×10^{-8}–3×10^{-6}	10^{-4}–0.09
$[Zn]_{tot}$	—	11–75	—
$[Zn^{2+}]$	—	10^{-9}–10^{-8}	—
$[Zn^{2+}]/[Zn]_{tot}$	—	0.07–0.10	—
$[Cd]_{tot}$	0.2–0.4	0.03–0.09	0.7
$[Cd^{2+}]$	10^{-11}–10^{-10}	10^{-12}–10^{-11}	10^{-9}
$[Cd^{2+}]/[Cd]_{tot}$	0.10	0.01–0.04	0.8–0.9

Alkalinity is denoted by ALK, and is in meq dm^{-3}, DOC concentrations are in mg dm^{-3}, metal concentrations in mol dm^{-3}.
In the high-ALK lakes and the rivers, the ALK is balanced mainly by Mg^{2+} and Ca^{2+}.

13.6.2 Cu, Zn and Cd in Swiss and Italian surface waters

Xue, Sigg and coworkers used the competitive ligand/DPASV technique (Section 6.7.4) to determine the concentrations of Cu^{2+}, Zn^{2+} and Cd^{2+} in surface water samples (Xue & Sigg, 1994; Xue *et al.*, 1996; Xue & Sigg, 1998). Table 13.9 summarises the chemistries of the different water

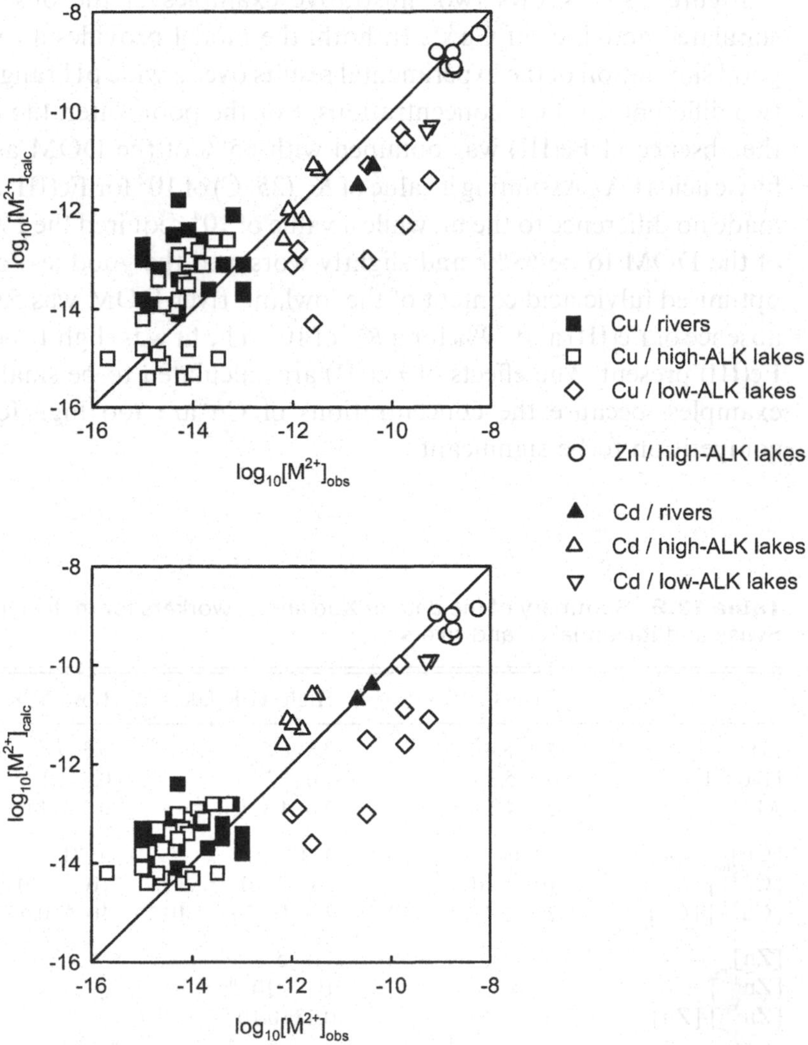

Figure 13.16. Results of Xue and coworkers for metals in Swiss and Italian lakes and rivers (cf. Table 13.9 and Section 13.6.2). The calculated results in the upper panel (Fe assumed to be absent) were obtained by assuming the dissolved organic matter to be 85% fulvic acid and 15% humic acid. In the lower panel (equilibrium with Fe-oxide, $K_{so} = 10^2$ assumed), the dissolved organic matter was assumed to be 60% fulvic acid and 40% humic acid.

samples studied. Note that Cu is strongly bound, Zn and Cd less so, as shown by the ratios of free to total metal.

WHAM/Model VI was applied 'globally', i.e. adjustments were made of the concentration of active humic matter to optimise for the entire data set. The binding of Cu was too strong to be accounted for simply by fulvic acid (FA). However, if some of the DOM was allowed to be present as humic acid (HA), then acceptable concentrations of humic substances could be obtained. For simplicity, it was assumed that all the measured DOM consisted of humic substances, and the fulvic/humic ratio was adjusted to optimise the model.

The upper panel of Fig. 13.16 compares the observed concentrations of free metal ions (Cu^{2+}, Zn^{2+} and Cd^{2+}) with the calculated values obtained assuming Fe(III) to be absent, with an optimised ratio of fulvic acid to humic acid of 85:15. The lower panel assumes the waters to be equilibrated with Fe(III) oxide (solubility product 10^2), with an optimised fulvic: humic ratio of 60:40. Higher K_{so} values required even more HA, which seems unrealistic. The model agrees with the observations by ranking the three metals in the same order, and there is a reasonable correlation between the observations and the predictions. On average, the observed and calculated values differ by *c.* 1 log unit, much of the discrepancy being due to the results for Cu in the low-alkalinity lakes (diamonds in Fig. 13.16). A possible explanation is that the 'active' humic proportion of the DOM is relatively low for these lakes. Optimisation of the model for the waters studied by Xue, Sigg and colleagues requires the dissolved organic matter to be more 'active' than found for the UK waters of Section 13.6.1, especially when Fe(III) is assumed to be present. This may reflect the presence of strong non-humic ligands (see Section 13.5.1).

13.6.3 Cu speciation in solutions from copper-contaminated soils

Vulkan *et al.* (2000) obtained 69 solutions from copper-contaminated soils in Chile, China and the UK, and analysed them for major solutes, together with concentrations of total dissolved copper and Cu^{2+} (using an ion-selective electrode). The pH values ranged from 4.5 to 8.0, the DOC concentrations from 6 to 102 mg dm^{-3}, and total Cu concentrations from 6.6×10^{-8} to 3.8×10^{-4} mol dm^{-3}. WHAM/Model VI was optimised by adjusting the fraction of DOC due to 'active' fulvic acid, the best value being 69%, similar to the values for the UK waters of Section 13.6.1. There was good agreement between observed and predicted values of log [Cu^{2+}] (Fig. 13.17). Assuming that Fe(III) was present with K_{so} values up to $10^{2.5}$ had negligible influence on the calculated results, because the copper

Table 13.10. Contaminated subsurface waters studied by Holm *et al.* (1995b).

	Soil solution	Compost leachate	Polluted groundwater	Industrial leachate 1	Industrial leachate 2	Industrial leachate 3
pH	6.6	6.6	6.8	7.2	7.0	6.5
Na $mmol\,dm^{-3}$	1.0	1.0	1.4	96	104	1.7
Mg $mmol\,dm^{-3}$	1.0	0.5	1.5	3.8	3.3	0.4
Ca $mmol\,dm^{-3}$	10	3.0	1.4	7.5	6.8	2.3
Cl $mmol\,dm^{-3}$	1.3	1.6	18	82	90	1.7
SO_4 $mmol\,dm^{-3}$	0.9	0.3	0	0.2	0.2	0.1
DOC $mg\,dm^{-3}$	64	13	264	79	34	3200
$[Cd]_{tot}$ $nmol\,dm^{-3}$	56	78	82	634	79	5890
obs% as Cd^{2+}	70	45	1	19	23	32
calc% as Cd^{2+} (no Fe)	25	52	1	13	14	0
calc% as Cd^{2+} ($K_{so} = 10^2$)	40	59	4	14	16	0
calc% as Cd^{2+} ($K_{so} = 10^4$)	52	67	7	16	18	0

Note that the compost leachate and the polluted groundwater were spiked with Cd. The K_{so} values refer to the solubility of Fe(III) oxide.

concentrations were such that the metal was sufficiently highly loaded onto the organic matter to overcome any competition by Fe(III).

13.6.4 Cd speciation in contaminated subsurface waters

Holm *et al.* (1995b) used a combination of size fractionation (dialysis) and adsorption to an ion exchange resin to determine the speciation of Cd in a soil solution, a compost leachate, a groundwater and three industrial leachates, six samples in all. Their results, and the results of WHAM/Model VI calculations, are shown in Table 13.10. The best overall fit was achieved if 60% of the DOC was assumed to be due to 'active' fulvic acid, a value similar to those found for the UK surface waters (Section 13.6.1) and copper-contaminated soils (Section 13.6.3). For the first five of the samples, the model predictions are satisfactory, whether or not equilibrium with an iron oxide phase is assumed. For the sixth (industrial leachate 3), the model predicts that essentially all the Cd is complexed by organic matter, whereas a substantial amount of Cd^{2+} was found experimentally. The disagreement suggests that the organic matter in this leachate is much less able to bind Cd than typical fulvic acid.

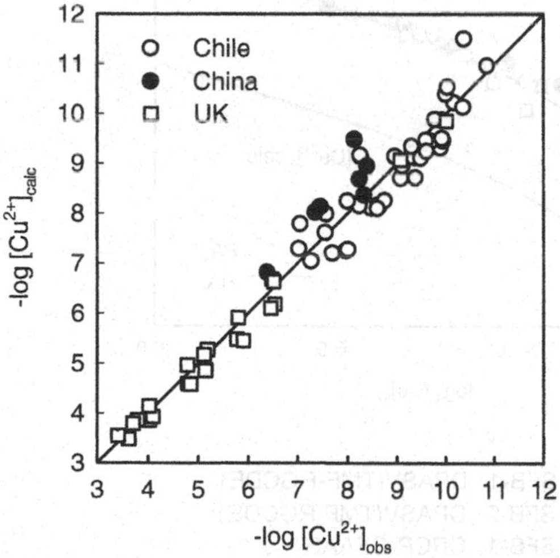

Figure 13.17. Observed and WHAM/Model VI-fitted log [Cu^{2+}] for copper-contaminated soil solutions; the 1:1 line is shown. [Redrawn from Vulkan, R., Zhao, F.-J., Barbosa-Jefferson, V., Preston, S., Paton, G.I., Tipping, E. & McGrath, S.P. (2000), Copper speciation and impacts on bacterial biosensors in the pore water of copper-contaminated soils, *Environ. Sci. Technol.* **34**, 5115–5121. Copyright 2000 American Chemical Society.]

13.6.5 Ni, Cu and Cd speciation in seawater

Donat *et al.* (1994) used electrochemical and chromatographic methods to determine the concentrations of labile Ni and Cu (approximately equivalent to inorganic forms) in the waters of San Francisco Bay. Figure 13.18 shows plots of labile metal concentration vs. total metal for Ni and Cu, and simulations with WHAM/Model VI. The copper data

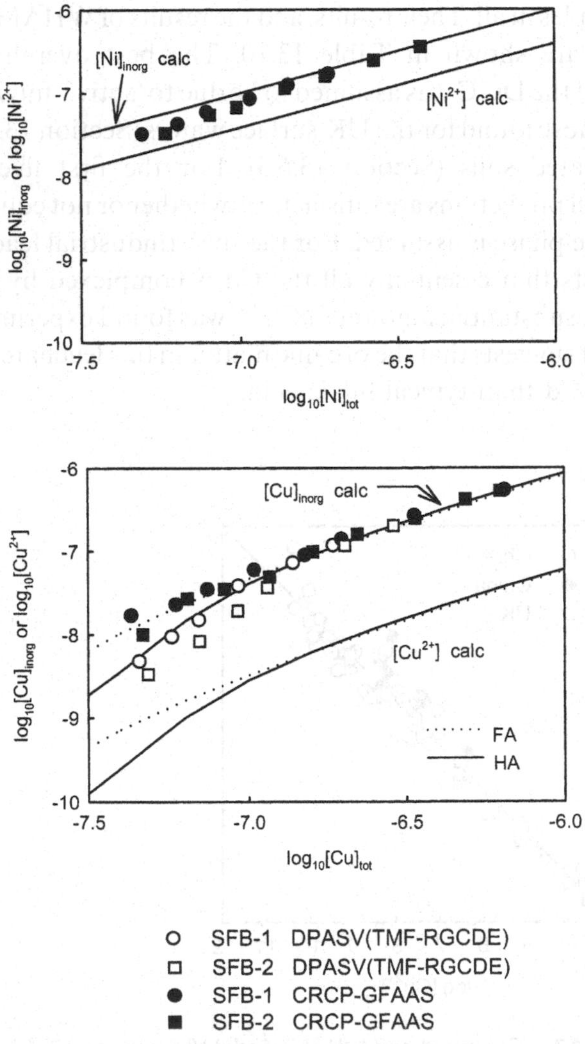

Figure 13.18. Dissolved metal speciation in samples of San Franciso Bay water, measured by Donat *et al.* (1994) (points) and simulated with WHAM/Model VI (lines). Determinations were made of electrochemically labile metal concentration following additions of known amounts of metal to filtered water samples. Only for values of K_{so} greater than 10^3 are the results affected by Fe(III) competition.

can be explained reasonably well by assuming humic substances to be present at concentrations of 1 mg dm^{-3} (fulvic acid) or 0.35 mg dm^{-3} (humic acid), which accord with typical levels of DOC in coastal waters (Section 2.5.3). Using the same concentrations of humic matter, essentially no binding of Ni by the organic matter is predicted, which agrees with the observations.

Bruland (1992) reported Cd speciation measurements in the Pacific Ocean, obtained by differential pulse anodic stripping voltammetry. As shown in Fig. 13.19, at depths of 200 m or greater, the concentration of Cd^{2+} is explained quite well simply by complexation with Cl^- (dotted line). A near-perfect match can be obtained by assuming the presence of 5 mg dm^{-3} humic acid, and the absence of Fe(III), as shown by the full line in Fig. 13.19, but such conditions are unrealistic (the DOC content of open ocean water is only *c.* 1 mg dm^{-3}) and cannot be taken to explain the small discrepancy between the observations and the chloride-only complexation picture. Furthermore, the presence of 5 mg dm^{-3} humic acid does not explain the observations for depths less than 200 m, where Cd^{2+} concentrations are substantially lower than predicted with WHAM/Model VI. The failure of the model to account for the observations is consistent with Bruland's assertion that specific, high-affinity ligands of biological origin are responsible for Cd binding in the surface waters.

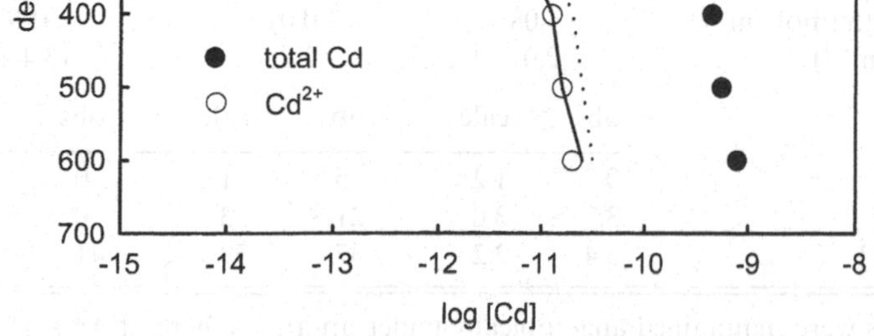

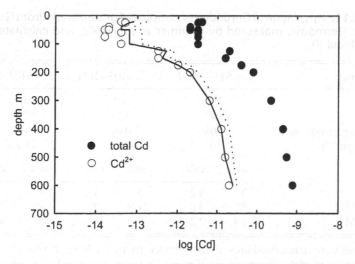

Figure 13.19. Cadmium speciation in the Pacific Ocean, measured by Bruland (1992) (points) and simulated (full and dotted lines). The dotted line is the predicted Cd^{2+} concentration assuming humic matter to be absent, the full line assumes 5 mg dm^{-3} humic acid, and no competition by Fe(III).

13.6.6 Conclusions from modelling heavy metals

The examples discussed in the previous sections show that a comprehensive binding model, parameterised with data from laboratory experiments, can be used to interpret field observations with some success. The examples cover a wide range of natural conditions, thereby providing a test of the model's ability to deal with the effects of pH, competing cations and competing ligands over a range of total heavy metal concentrations. Overall, it can be concluded that the dissolved organic matter in most of the samples considered behaves approximately as expected from knowledge about the properties of isolated humic substances. The best results are obtained with soil waters and freshwaters containing high concentrations of DOM, for which the 'active' humic matter concentration (assuming default model parameters to apply) corresponds to 60–70% of the total DOM, in agreement with the estimates from charge balance (Section 13.2.2) and aluminium speciation (Section 13.3.2). The predictive ability for low metal concentrations in lake waters with low concentrations of DOM (Section 13.6.2) is less convincing, while Cd speciation in seawater cannot be explained in terms of complexation with humic matter (Section 13.6.5). The issue of competition by Fe(III) presents a problem in the application of the model, and urgently requires research. Future analytical studies of metal speciation in natural waters would benefit from parallel modelling work, which would help to identify key data requirements, as well as providing a means to interpret the analytical results.

Table 13.11. Speciation of Cm added to samples of groundwater from Gorleben (Northern Germany), measured by Wimmer *et al.* (1992), and calculated with WHAM/Model VI.

Groundwater:	S113–1092		S102–2131		S107–1281	
$[Cm]_{tot}$	7.8		63		41	
pH	7.41		7.17		7.53	
Ionic strength ($mol\,dm^{-3}$)	1.08		0.01		0.09	
DOC ($mg\,dm^{-3}$)	2.0		2.8		13.4	
	obs	calc	obs	calc	obs	calc
$[Cm^{3+}]$	0.7	1.2	5	1	0	0
$[CmCO_3^{+}]$	3.7	3.0	21	3	0	2
[Cm–DOM]	3.4	3.2	37	59	41	40

The samples were maintained anaerobically under an atmosphere of Ar + 1% CO_2. In applying the model, concentrations of major ions were estimated from information published by Dearlove *et al.* (1991), and 50% of the dissolved organic matter was assumed to be 'active' as fulvic acid.
The concentrations of Cm are in $nmol\,dm^{-3}$.

13.7 Interactions with metallic radionuclides

In principle, these are no different from interactions with heavy metals, except that the metal concentrations will be very low, unless spiking experiments are performed. There have been few measurements of speciation in field samples that give data suitable for modelling. One example is provided by the study of curium speciation in groundwaters conducted by Wimmer *et al.* (1992), who used time-resolved laser-induced fluorescence spectroscopy to measure the concentrations of the species Cm^{3+}, $CmOH^{+}$ and $CmCO_3{}^{+}$, together with the concentration of Cm bound to dissolved organic matter. The analytical results, and the speciation computed with WHAM/Model VI, are shown in Table 13.11. The agreements are reasonably satisfactory.

13.8 Binding by dissolved humic matter compared to adsorption by suspended particulates

Natural waters contain suspended particulate matter (SPM) that is capable of adsorbing cations, and such adsorption must be taken into account in judging the importance of humic matter in the environmental behaviour of the cations. A full treatment of adsorption is beyond the scope of this book, but some simple comparisons can be made on the basis of published data for the constituents of SPM. The main binding components of the SPM are taken to be the oxides of manganese and iron, and organic matter, represented by humic acid.

Table 13.12 shows values of ν for the binding of Cu and Zn by humic substances and by oxides of manganese and iron(III) oxides, at pH 5 and a

Table 13.12. Comparison of Cu and Zn binding by humic substances, and oxides of manganese and iron.

	ν_{Cu}	ν_{Zn}
Fulvic acid	4×10^{-4}	2×10^{-5}
Humic acid	7×10^{-4}	7×10^{-5}
Manganese dioxide	3×10^{-4}	7×10^{-5}
Iron(III) oxide	2×10^{-6}	2×10^{-7}

Values of ν (mol g^{-1}) are shown in each case, for pH 5 and a free ion concentration of 10^{-6} mol dm^{-3}.
For humic substances, the values are calculated with WHAM/Model VI, those for manganese dioxide are estimated from Catts & Langmuir (1986), those for iron oxide from Dzombak & Morel (1990).
The oxide samples to which the results refer were synthetic materials with specific surface areas of several hundred m^2 g^{-1}.

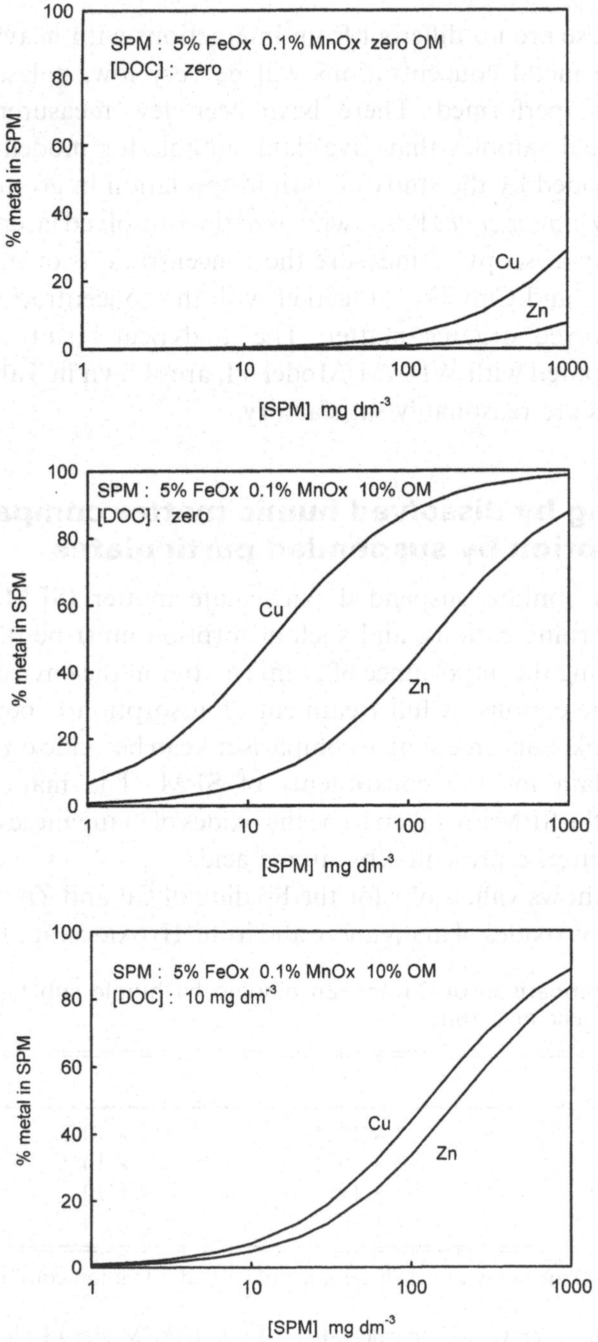

Figure 13.20. Predicted sorption at pH 5 of Cu and Zn by suspended particulate matter (SPM), as influenced by SPM composition and DOC concentration. Binding to SPM is estimated from values of ν given in Table 13.12, assuming the solid-phase organic matter to consist of humic acid. All the dissolved organic carbon (DOC) is assumed to be due to fulvic acid. The free ion concentrations are fixed at 10^{-6} mol dm^{-3}, which means that the total metal concentrations are increasing as [SPM] increases.

free metal concentration of 10^{-6} mol dm^{-3}. The pH of 5 is chosen because experimental data for metal adsorption by the oxides at higher pH values are scarce. On a weight basis, humic substances and manganese dioxide bind the two metals similarly, and to a greater extent than does iron oxide. To assess sorption by SPM, it was assumed that the material comprised 5% iron oxide and 0.1% manganese dioxide, which are reasonably typical values (Eisma, 1993). The effect of organic matter was explored by setting its content to zero and 10%. The rest of the SPM is assumed to consist of aluminosilicate materials that bind heavy metals only weakly, and/or 'inert' organic matter. The effect of complexation by dissolved organic matter was tested by including fulvic acid in the solution phase.

The plots in Fig. 13.20 show how the percentage of total metal varies with SPM concentration. In the absence of organic matter in either the SPM or in solution, sorption becomes significant at [SPM] > 100 mg dm^{-3}, although even a concentration of 1000 mg dm^{-3} is insufficient to sorb 50% of either metal. When organic matter (humic acid) is included in the SPM, but [DOC] is set to zero, sorption is much more significant, 50% binding of Cu being achieved at [SPM] values a little greater than 10 mg dm^{-3}. Copper is bound appreciably more strongly than zinc by the SPM under these circumstances. When 10 mg dm^{-3} DOC is included in the system, the sorption of Cu is noticeably reduced, as the dissolved fulvic acid competes effectively with the SPM. Also, the sorption of Cu and Zn are now quite similar, although the overall binding (solution complexation plus sorption) of Cu is much greater than that of Zn.

These simple examples demonstrate how different factors contribute to determining distributions of metals between the solid and solution phases. Particulate organic matter makes a large contribution to metal sorption by SPM, being more important than the oxides for the conditions considered. However, general conclusions cannot be drawn from the limited analysis presented here, and there may be circumstances under which suspended oxides are dominant. Papers by Radovanovic & Koelmans (1998), Lofts & Tipping (1998, 2000) and Davis *et al.* (1998) provide more detailed discussion of this topic, from the modelling viewpoint. Although there have been many studies of metal sorption by oxides and other minerals, the data available to describe environmental situations are less extensive than those for humic substances, and more information is required, especially at low adsorption densities and with regard to competition effects.

14

Cation binding by humic substances in soils and sediments

The non-living part of the soil comprises mineral solids derived from physical and chemical weathering, the decay products of living material, water and its solutes, and air. Rimmer (1998) gives an excellent short introductory account of soils and their properties, emphasising the importance of the solid–water interface. The textbooks of White (1987) and Rowell (1994) provide more detailed modern accounts. Bolt (1982), Sposito (1989) and McBride (1994) provide advanced accounts of soil chemistry, while the book edited by Alloway (1995) focuses on metals in soils. Soils are highly heterogeneous natural systems, and their cation-binding properties are determined by the combined effects of a variety of phases, comprising mineral solids (aluminosilicates, oxides, carbonates etc.) and natural organic matter, notably humic substances.

Cations in soils are important for several reasons. Several major plant nutrients are cations, notably Mg^{2+}, K^{+}, Ca^{2+} and NH_4^{+}. Micronutrient metals such as Mn, Fe, Cu and Zn also exist as cations in soils, as do phytotoxic elements such as Cd and Pb. Soil acidity is of concern with regard to crop growth, due to the effects of H^{+} itself, and the toxicity of cationic aluminium species. A number of pesticides are cationic. Soil exerts a major control on the supply of cations to surface and groundwaters, and consequently there is much interest in soil–solution interactions, in relation to surface water acidification, and the transport of heavy metals and radionuclides. Furthermore, cation binding by organic matter in soils has consequences for the behaviour of the organic compounds themselves, influencing aggregation, adsorption and solubility, and thereby soil structure, translocation of organic matter within soils, and transport to waters. Other processes involving soil organic matter, for example metabolic transformations, weathering, water holding and interactions with enzymes, are also influenced by cation binding.

The aim of the present chapter is to examine the importance or otherwise of cation–humic interactions in these mixtures, in particular the way in which the interactions influence the composition of the soil solution.

14.1 Components of the soil system

14.1.1 Soil water

The space in soil not occupied by the soil solids is occupied by water and air, and is known as the pore space. The porosity is the fraction of total soil volume occupied by pores, and it varies from *c.* 0.25 in sandy and compacted clay soils to 0.8 or more in peats. Soil is said to be saturated when all the pore space is occupied by water, and unsaturated when some air is present. Water is present in soil because of (a) incomplete drainage, (b) adsorption, due to hydrogen bonding at polar surfaces, and (c) capillarity, which depends upon the pore size. Thus the water experiences a range of tensions, or matric potentials, and is physically heterogeneous. Water is able to move downward through the soil when the gravitational force exceeds the upward suction due to capillary forces. Plant root uptake also influences water movements in soil.

Solute transport within soil is governed by water movement, diffusion and root uptake. Since the soil is a dynamic system, with continuous inputs and outputs of water, and since the water is subject to a range of matric potentials, the solutes are not uniformly distributed. Non-uniformity also arises from spatial heterogeneity of the soil solid surfaces, from localised microbial activity and from slow chemical reactions (Sposito, 1989). For these reasons, the 'soil solution', by which is meant the aqueous part of the soil and its solutes, is a somewhat elusive entity. There are various methods to obtain soil solution. In the field, the simplest method is to collect freely draining water, for example by inserting a flat plate horizontally into an exposed soil face. Devices are also available to collect water by suction at a known negative pressure. In the laboratory, soil water can be collected by centrifugation or by displacement with an immiscible solvent, aided by centrifugation. To carry out systematic chemical experiments, soil suspensions can be prepared, in which the solid:solution ratio is low, compared to the soil *in situ*, and so physical uniformity is more readily achieved. These various methods tend to provide solutions of differing compositions. For dissolved organic matter, the situation in soil is made more complicated by the possibility of transformations, occurring to different extents in pores of different size (Zsolnay, 1996; see Section 2.5.2).

14.1.2 Mineral components of soils

Minerals in soils influence cation concentrations in two ways, by dissolution (and precipitation) reactions, and by surface chemistry. In many soils, cations are derived principally from weathering reactions, i.e. dissolution of unstable minerals. Carbonates, oxides, phosphates, sulphates and sulphides may be thermodynamically stable in soils, and may control solution cation concentrations by reversible quasi-equilibrium reactions. Thus, dissolved Al concentrations may be controlled by equilibrium with gibbsite, $Al(OH)_3$, or jurbanite, $AlOHSO_4$, while dissolved Ca concentrations may depend upon equilibrium with calcite or gypsum. Redox reactions may also contribute, for example the formation and dissolution of iron and manganese oxides.

In predominantly mineral soils, the solids are classified in terms of sand (size range 50–2000 μm), silt (2–50 μm) and clay (< 2 μm). The clay fraction contains aluminosilicates, either originating from the parent soil-forming rock (e.g. feldspars, mica), or formed as the result of weathering (e.g. illite, kaolinite, vermiculite), together with oxides, principally of silicon, aluminium, iron and manganese, and humic substances, adsorbed to or aggregated with the mineral solids. The clay fraction is usually the most important for cation adsorption reactions, due to its high surface area and consequently large number of available binding sites.

Aluminosilicates are the most abundant mineral particles in the clay fractions of many soils. They owe their (negative) surface charge to a deficit of cationic charge within the crystal structure, brought about for example by the substitution of Si^{4+} by Al^{3+}, or Al^{3+} by Mg^{2+}. Soils in which these minerals are dominant are termed constant-charge soils, because the charge associated with the solids tends not to vary with pH. The content of negative charge is the cation exchange capacity of the soil, abbreviated to CEC, which has units of equivalents per gram dry soil. The CEC of a constant-charge soil thus depends upon the surface charge densities ($eq\,m^{-2}$), the surface areas, and the relative amounts of the aluminosilicates. Values of CEC for mineral soils dominated by aluminosilicates are generally of the order of $10^{-4}\,eq\,g^{-1}$. Much research effort has gone into the measurement and interpretation of cation exchange on aluminosilicates. To maintain electroneutrality, the negative surface charges must be balanced by adsorbed cations. This leads to the following type of formulation for the interactions of Na^+ and Ca^{2+} at the surface of a cation exchanger

$$2NaX + Ca^{2+} = CaX_2 + 2Na^+ \qquad (14.1)$$

(see equation 3.8 for an alternative expression). Various exchange selectivity coefficients can be postulated. For example, the mass action law due to Vanselow (1932) is written in terms of mole fractions of adsorbed species (x_{Na}, x_{Ca}) and solution activities (a_{Na}, a_{Ca}) as follows

$$K = \frac{a_{Na}^2 x_{Ca}}{a_{Ca} x_{Na}^2} \tag{14.2}$$

The equation allows the adsorption of the two cations to be calculated, depending upon their solution concentrations. Other formulations have been discussed in the textbooks mentioned at the start of the chapter. One approach is the Donnan model, as described for non-specific accumulation of counterions in Section 4.6.7 (see also Section 9.5). The Donnan model readily lends itself to the description of competitive adsorption of many different ions. In its simplest form, all cations of a given charge are assumed to be equally attracted to the surface, but selectivity coefficients can be included to account for the relative preferences of the exchanger for different cations. Simple expressions like equation (14.2) provide approximate descriptions of cation binding at constant pH by soils containing aluminosilicates and organic matter, but exchange site heterogeneity has to be invoked to obtain precise simulations (e.g. Vulava *et al.*, 2000).

Oxides of Al, Si, Mn and Fe are present in most soils. They all interact with cations, binding being strongest for Mn and Fe. Al and Fe oxides are amphoteric, and are important for the binding of anions, notably PO_4, and humic substances (Section 12.3). Reactions at oxide surfaces are usually described in terms of surface complexation (Section 3.3), using site densities and appropriate equilibrium constants. The extent of binding of cations generally increases with pH, as illustrated by Fig. 3.3. Oxides of Al and Fe dominate cation (and anion) binding in highly weathered tropical soils, and in the spodic horizons of podzols. Such soils are prominent examples of variable charge soils.

14.1.3 Organic matter

As discussed in Chapter 2, soil organic matter consists of a variety of entities, among which humic and fulvic acids contribute by far the most to the cation-binding properties. In the most extreme case of peats, the soil consists almost entirely of organic matter, of which 20% is typically present as humic substances. This gives a charge of *c.* 10^{-3} eq (g dry soil)$^{-1}$ at neutral pH, appreciably greater than is typical for aluminosilicate-dominated soils (Section 14.1.2). In non-peat soils, the cation exchange capacity increases with organic matter content (Sposito, 1989). The importance of organic matter in determining cation exchange capacity has been known for well over a century (Hargrove & Thomas, 1981).

14.1.4 Estimating cation binding by soil solids *in situ*

Much is known about the cation-binding properties of individual soil components, under well-defined laboratory conditions. Ideally, we would like to use such knowledge to predict the cation-binding properties of soils *in situ*. However, this is by no means straightforward, for the following reasons.

(1) The masses of individual soil components are difficult to determine, chemical extraction procedures providing only crude guides.
(2) Surface areas of individual components are not readily determined. A full quantitative description would require an elaborate separation procedure that did not alter the components, followed by gas adsorption measurements. This would not give meaningful results for organic matter (Section 2.4.5).
(3) The different components may interact with one another, thereby modifying their individual cation-binding properties (as discussed in Section 12.3).
(4) Cation-binding properties at the high solids concentrations found *in situ* are likely to differ from those at the lower concentrations of laboratory experiments. In particular, the diffuse electrical layers around the particles, or aggregates of particles, cannot develop to the same extent, and diffuse layers associated with individual components will overlap and interact with one another.
(5) For humic substances, much information has been derived from experiments with solutions or suspensions in which the materials are fully hydrated, which may not always apply to soils *in situ*.
(6) The heterogeneity of soil water (see above) means that even if a good description could be made for one soil sample, the field situation would be more complex.

Thus, while useful calculations can be performed to understand and predict the properties of real soils, several assumptions usually need to be made about the states of the solid components. Furthermore, soils are difficult to sample representatively and in such a way that structural integrity is maintained, making it problematic to obtain data directly relevant to the field situation. Generally therefore, the fundamentally based chemical predictions can only be tested against data obtained (a) from measurements on homogenised slurries of soil samples, or (b) from analyses of soil water collected by lysimetry.

14.1.5 Biological processes

Cation behaviour in soil is influenced by plant root uptake of nutrient cations, and by the release of those cations during decomposition. Microorganisms – bacteria and fungi – also take up and release cations.

14.2 Sorption of major cations by organic-rich acid soils

There is a long-standing interest in the aluminium chemistry of acid soils, due to concern about phytotoxicity, the transfer of the toxic metal to surface waters, and the need to understand translocation processes within the soil column in relation to podzolisation (see Section 14.5.3). In organic-rich acid soils, there is overwhelming evidence that Al concentrations in the soil solution are controlled by solid-phase soil organic matter. One demonstration, by Hargrove & Thomas (1981), is summarised by the plots in Fig. 14.1. Walker *et al.* (1990) determined concentrations of dissolved Al in equilibrium with solids from organic soils that had been

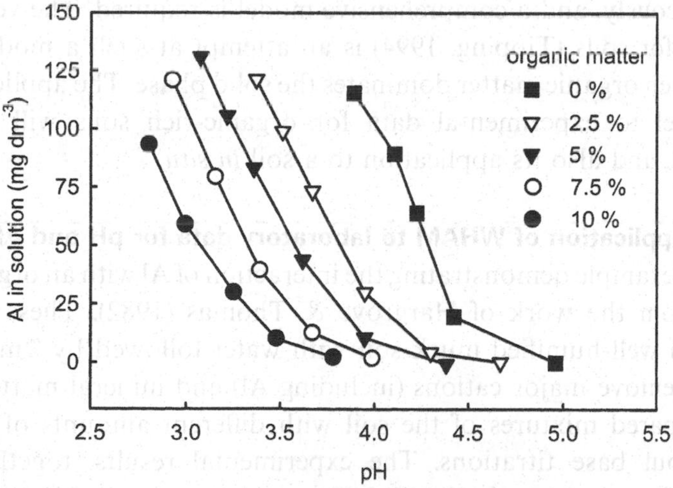

Figure 14.1. Solution concentrations of Al after equilibrating 10 g of a mineral subsoil with different amounts of H-saturated peat to achieve different percentage organic matter values. The background electrolyte was 0.1 mol dm^{-3} KCl. [Redrawn from Hargrove, W.L. & Thomas, G.W. (1981), Effect of organic matter on exchangeable aluminium and plant growth in acid soils. In *Chemistry in the Soil Environment*, ed. R.H. Dowdy, J.A. Ryan, V.V. Volk & D.E. Baker, pp. 151–166, with permission from the American Society of Agronomy and the Soil Science Society of America.]

first saturated with H^+ and then reacted with Al. They were found to depend upon pH and Al loading, and to be lower than would be expected for equilibrium with either crystalline or amorphous $Al(OH)_3$, implying that solution Al concentrations were controlled by Al complexed to particulate organic matter.

Concentrations of H^+ in acid organic-rich soils are controlled by the total concentrations of strong acid and strong base, together with buffering by the functional groups of organic matter and the hydrolysis reactions of Al, most of which is bound to the organic matter (Hargrove & Thomas, 1982; Skyllberg, 1999). In the absence of significant amounts of mineral cation-exchange surfaces, the binding of other major cations (Na^+, Mg^{2+}, K^+, Ca^{2+}) in such soils is due either to specific complexation by functional groups in the organic matter, or to non-specific counterion accumulation, the extents of which depend upon the proton and aluminium binding. Another reactant which may be significant is Fe(III). Thus, the chemistry of the soil–water system involves a suite of competitive interactions. In experimental studies, it may be possible to control the concentrations of some of the reactants to examine interrelationships among others, for example by working with a constant background electrolyte solution or by operating a pH-stat. In general however, and certainly in the natural environment, all the reactions must be considered simultaneously, and a comprehensive model is required. The version of WHAM for soils (Tipping, 1994) is an attempt at such a model, most useful when organic matter dominates the solid phase. The application of the model to experimental data for organic-rich soils will now be discussed, and also its application to a soil *in situ*.

14.2.1 Application of WHAM to laboratory data for pH and Al

A simple example demonstrating the interaction of Al with an organic soil comes from the work of Hargrove & Thomas (1982). These authors washed a well-humified muck soil with water followed by 2 mol dm^{-3} HCl to remove major cations (including Al) and mineral matter. They then prepared mixtures of the soil with different amounts of Al, and carried out base titrations. The experimental results, together with WHAM fits, are shown in Fig. 14.2. The model was applied by holding its parameters at their default values and optimising the soil content of humic substances. The model accounts for the main trends in the data, although it does not provide an exact fit. The important point is that the behaviour of soil organic matter can approximately be described on the basis of information derived from studies with isolated humic substances. The model requires 47.5% of the total solid material to consist of humic substances, the remaining 52.5% being assumed inert with respect to cation binding.

WHAM was used by de Wit *et al.* (1999) to interpret their data from experiments with the Oe horizon of a forest soil. They added aluminium to samples of the soil to obtain different loadings of the metal, then performed batch titration experiments in which acid or base was added to soil suspensions, followed by the measurement of pH and 'quickly reacting Al' (Al_{QR}) in the supernatant after centrifugation. Quickly reacting Al is a measure of the combined concentrations of inorganic species of Al, except for fluoride complexes (Clarke *et al.*, 1992). The plots in Fig. 14.3 show that the model is able to account for the variations of both pH and $[Al_{QR}]$ with added acid or base. The model was fitted by optimising the soil contents of humic and fulvic acids, the best values being 0.23 and 0.02 $g\,g^{-1}$ respectively, i.e. 25% of the soil solids (*c.* 28% of the soil organic matter) were required to be due to humic matter.

Other applications of WHAM to describe the pH buffering and Al chemistry of organic rich soils have been made (Tipping *et al.*, 1995a; Lofts *et al.*, 2001a). The model has been calibrated by optimising the soil content of humic matter, and usually also the soil content of 'active' Al, i.e. the soil Al that participates in solid–solution interactions. Although both soil attributes can be estimated analytically, the measured values cannot be used directly for modelling because the model response is too sensitive to

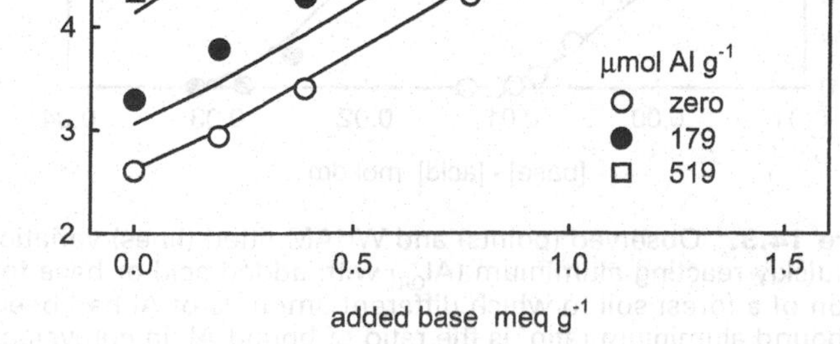

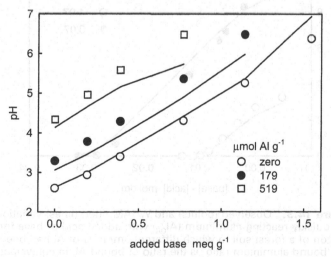

Figure 14.2. Base titrations of a muck soil loaded with different amounts of aluminium. The points are the experimental data of Hargrove & Thomas (1982). The lines are WHAM fits, obtained assuming the humic acid to fulvic acid ratio to be 9:1.

variations in their values. From applications of the model to 25 different organic-rich soils, including peats and forest O horizons, the soil content of humic substances is calibrated to be between 6% and 29% of the soil organic matter (average 14%). The range is similar to that of 9% to 30% (average 19%) estimated by extraction with base, but otherwise there is not a close correspondence between the two sets of values. Tipping *et al.* (1995a) found that the optimised soil contents of 'active' Al in 16 organic moorland soils were between 10% and 82% (average 39%) of the soil Al extracted by 0.1 mol dm^{-3} HNO_3. Lofts *et al.* (2001a) found that the

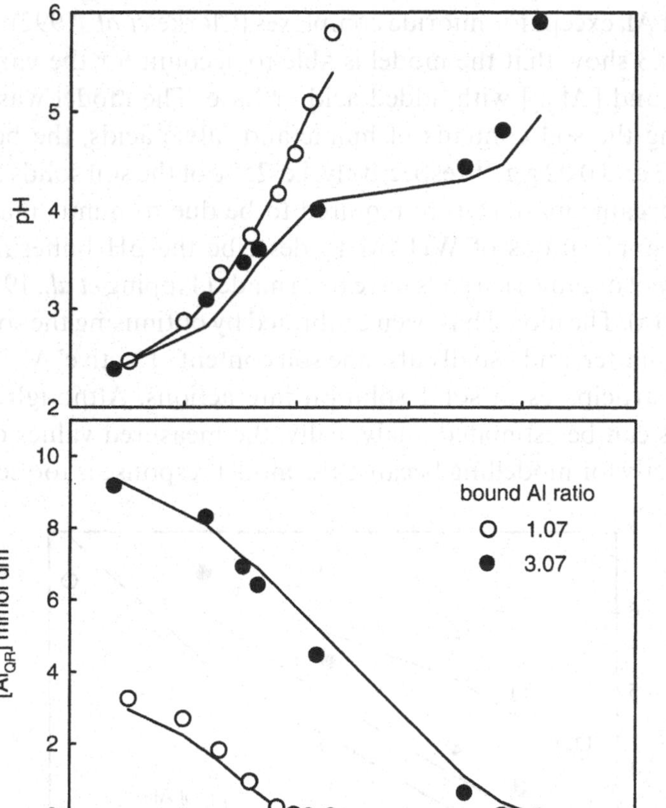

Figure 14.3. Observed (points) and WHAM-fitted (lines) variations of pH and quickly reacting aluminium (Al_{QR}) with added acid or base for the Oe horizon of a forest soil to which different amounts of Al had been added. The 'bound aluminium ratio' is the ratio of bound Al, in equivalents, to the carboxyl content of the soil. [Redrawn from de Wit, H.A., Kotowski, M. & Mulder, J. (1999), Modeling aluminium and organic matter solubility in the forest floor using WHAM, Soil Sci. Soc. Am. J. **63**, 1141–1148, with permission from the Soil Science Society of America.]

optimised soil contents of 'active' Al in three organic forest soil horizons were between 90% and 181% (average 124%) of the values estimated by extraction with 0.1 mol dm^{-3} $CuCl_2$.

The modelling results show that variations in pH and the solution concentration of Al can be described by WHAM with adjustment of the total soil contents of the two major reactants, i.e. humic substances and aluminium. The optimised values of these contents are not too far from analytical estimates, which are themselves only approximations of the true 'active' contents. The model simultaneously predicts pH and solution Al concentration, a necessary feature if the effects of changes in acid or base inputs are to be predicted for soils *in situ*. Thus it can justifiably be used to estimate soil responses to, for example, changes in the acidity of atmospheric deposition, or in rates of input of new humic matter, or to soil treatments such as the addition of lime.

14.2.2 Base cations

In comparison to aluminium, there have been rather few studies of the interactions of base cations with organic soil solids. Tipping *et al.* (1995a) found that, with some adjustment of the selectivity coefficients for counterion accumulation (Section 9.6), WHAM could approximately describe the behaviours of Mg^{2+}, K^+ and Ca^{2+} when soils were titrated with NaOH. They obtained the best results with selectivity coefficients as follows: H^+ 1, Na^+ 0.25, Mg^{2+} 0.75, Al^{3+}, $AlOH^{2+}$ and $Al(OH)_2{}^+$ 0.5, K^+ 1.0, Ca^{2+} 1.0 (values adjusted to the nearest 0.25). However, these values are very approximate, and there is a pressing need for more extensive studies to obtain more secure ones. According to the model, counterion accumulation is the mechanism of base cation binding for Na^+, K^+, and, at low pH, Mg^{2+} and Ca^{2+}. At higher pH, specific complexation reactions become more significant for the alkaline earth cations.

14.2.3 Application of WHAM to an organic soil *in situ*

Table 14.1 summarises a calculation of the distribution of ions in a typical moorland soil in the UK uplands. The active components of the system are taken to be humic and fulvic acids, together with metals and anions. The sources of the inorganic components are taken to be either atmospheric deposition or the dissolution of mineral matter, but the latter does not present significant surface sites to affect cation binding by the soil solids.

Panel (a) defines the speciation problem by specifying the total

Table 14.1. Application of WHAM (soils version) to a typical moorland organic-rich soil from the UK uplands.

(a) Reactant concentrations (total water)

	$g\,dm^{-3}$		$mol\,dm^{-3}$
FA	17	Na	5×10^{-4}
HA	70	Mg	3×10^{-3}
		Al	0.1
		Ca	3×10^{-3}
$FA_{aq} = 50\,mg\,dm^{-3}$		Cl	2×10^{-4}
$pCO_2 = 0.01$ atm		SO_4	3×10^{-5}

(b) Fraction distribution of water

FA diffuse layer	0.2110
HA diffuse layer	0.0385
Bulk water	0.7505
Bulk water + dissolved FA diffuse layer	0.7510

(c) Composition of bulk water

	$mol\,dm^{-3}$		$mol\,dm^{-3}$
H^+	3.12×10^{-5}	Cl^-	2.66×10^{-4}
Na^+	2.21×10^{-4}	SO_4^{2-}	3.84×10^{-5}
Mg^{2+}	2.41×10^{-5}	HCO_3^-	6.44×10^{-6}
Al^{3+}	3.84×10^{-6}	H_2CO_3	5.26×10^{-4}
Ca^{2+}	1.83×10^{-5}		

pH = 4.52

Ionic strength = $4.42 \times 10^{-4}\,mol\,dm^{-3}$

(d) Composition of the aqueous phase

	$mol\,dm^{-3}$		$mol\,dm^{-3}$
H^+	3.12×10^{-5}	Cl^-	2.66×10^{-4}
Na-tot	2.22×10^{-4}	SO_4^{2-}	3.84×10^{-5}
Mg-tot	2.97×10^{-5}	HCO_3^-	6.44×10^{-6}
Al-tot	7.47×10^{-5}	H_2CO_3	5.26×10^{-4}
Ca-tot	2.38×10^{-5}		
			$mg\,dm^{-3}$
		FA	50

(e) Binding by humic matter

Specific sites	$mol\,gFA^{-1}$	$mol\,gHA^{-1}$
H^+	2.44×10^{-3}	1.68×10^{-3}
Na^+	0	0
Mg^{2+}	7.59×10^{-6}	5.57×10^{-7}
Al^{3+}	1.12×10^{-3}	9.63×10^{-4}
$AlOH^{2+}$	5.78×10^{-5}	5.91×10^{-5}
Ca^{2+}	5.82×10^{-6}	5.57×10^{-7}

Diffuse layer	$mol\,gFA^{-1}$	$mol\,gHA^{-1}$
H^+	8.34×10^{-6}	6.67×10^{-7}
Na^+	1.48×10^{-5}	1.18×10^{-6}
Mg^{2+}	1.04×10^{-4}	1.51×10^{-5}
Al^{3+}	2.38×10^{-4}	6.23×10^{-5}
$AlOH^{2+}$	1.14×10^{-6}	1.64×10^{-7}
Ca^{2+}	1.05×10^{-4}	1.52×10^{-5}

(f) Net charge and diffuse layer composition

	FA	HA
$Z\,eq\,g^{-1}$	-1.15×10^{-3}	-2.49×10^{-4}
% H^+	0.7	0.3
% Na^+	1.3	0.5
% Mg^{2+}	17.9	12.1
% Al^{3+}	61.7	74.9
% $AlOH^{2+}$	0.2	0.1
% Ca^{2+}	18.2	12.2

(g) Distributions of master species

	bulk solution	FA specific	FA diffuse	HA specific	HA diffuse
Na	0.332	0.000	0.502	0.000	0.166
Mg	0.006	0.043	0.587	0.013	0.351
Al	0.000	0.200	0.041	0.716	0.044
Ca	0.005	0.033	0.595	0.013	0.355
Cl	1.000	0.000	0.000	0.000	0.000
SO_4	0.969	0.000	0.023	0.000	0.007

concentrations of all the reactants, expressed as moles or grams per dm^3 of total water. In the example, the total concentration of humic matter (fulvic acid, FA, plus humic acid, HA) is 87 g per dm^3 of total water. This corresponds to a soil bulk density of 0.28 g cm^{-3} (a typical value) if the humic matter is assumed to make up 25% of the soil dry weight, and if the density of organic matter is taken to be 1.5 g cm^{-3}. In addition, the concentration of dissolved FA is specified, together with the partial pressure of CO_2. The values are entered into the WHAM input file, and the model then calculates the distributions of water and chemical species within the soil.

Panel (b) shows how water is distributed. This raises the first major problem for the model, because of the overlap of diffuse layers, referred to in Section 9.5.4. In the example, the concentrations of humic substances are very high, and the ionic strength is low, so that if the diffuse layers were unconstrained, their total volume would be 152 dm^3! WHAM gets over the difficulty by forcing overlap of diffuse layers, restricting their total volume to be 0.25 dm^3 per dm^3 of the total water. The model further restricts the size of the diffuse layer when the net charge on the humic matter is low, as is the case here for HA. The main result of these restrictions is to favour the accumulation in the diffuse layers of those (cat)ions with higher charges.

Panel (c) shows the composition of the 'bulk water', by which is meant the water other than that in the diffuse zones. The solution is, necessarily, charge balanced. Note that the concentrations of Cl^- and SO_4^{2-} are greater than the total concentrations of panel (a) because anions are unable to enter the diffuse layer. The composition of the 'aqueous phase', shown in panel (d), differs from that of the bulk solution because of the presence of dissolved FA. Because the dissolved FA has cations associated with it, the total concentrations of the metals, especially Al, are greater than in the bulk solution.

Panels (e) and (f) summarise how different cations are bound by humic matter. The specific sites of both FA and HA are occupied principally by H^+ and Al^{3+}, with minor contributions from the alkaline earth cations and $AlOH^{2+}$. The amounts of H^+, Al^{3+} and $AlOH^{2+}$ bound per gram of FA and HA are similar, whereas Mg^{2+} and Ca^{2+} are more strongly bound by FA, because the electrostatic attraction is calculated to be greater, due to the higher net charge of FA. The main constituent of the diffuse layers is Al^{3+}, accounting for 62% of the counterion charge in FA and 75% in HA.

Panel (g) shows how the master species are distributed. The simplest is

Cl, which is confined to the bulk solution because it exists only as an anion. Sulphate is also largely anionic; the *c.* 3% present in the diffuse layers is due to the $AlSO_4^+$ complex. The metals are present mainly as counterions, except for Al, most of which is bound at the specific sites on FA and HA.

This picture of the organic soil *in situ* is reasonably consistent with observations, insofar as the aqueous phase composition is similar to that found in drainage water sampled by lysimeters, and the bulk soil composition is similar to that obtained by analysis. The application of the model highlights the difficulties of dealing with ionic distributions in concentrated systems at low ionic strength.

14.3 Sorption of major cations by mineral soils

In the preceding sections we were concerned with soils containing so much organic matter that there is little doubt that it exerts essentially complete control over the soil solution chemistry. When it comes to 'mineral' soil horizons, aluminosilicates and oxides might contribute to solid–solution interactions, and the dissolution and precipitation of oxide phases, notably those of Al and Fe(III), may also exert solubility control.

14.3.1 $[Al^{3+}]$–$[H^+]$ relationships in acid mineral soils

The widely used model of acid soils, formulated by Reuss (1983) places equilibrium with $Al(OH)_3$ at the centre of the soil reactions, controlling the concentrations not only of Al species but also H^+. Base cation concentrations are controlled by exchange reactions at a fixed charge cation exchanger. The advantage of the model is its simplicity, and this has led to its use in dynamic hydrochemical models for assessing the impacts of acid rain on terrestrial and freshwater ecosystems, most notably the Model of Acidification of Groundwaters In Catchments (MAGIC; Cosby *et al.*, 1985). The Reuss model predicts that $[Al^{3+}]/[H^+]^3$ is a constant, i.e. a plot of $\log[Al^{3+}]$ vs. $\log[H^+]$ should be a straight line with a slope of 3. Table 14.2 shows that in some instances the cubic relationship is approximately followed, and the solubility product of gibbsite (a crystalline form of $Al(OH)_3$) is exceeded. In other cases, however, $\log[Al^{3+}]$ vs. $\log[H^+]$ plots have slopes of less than 3 and the solutions are appreciably undersaturated with respect to gibbsite.

If solution Al concentrations are not controlled by solid phase $Al(OH)_3$, the alternatives are (a) control by another mineral phase, e.g. $AlOHSO_4$ (Ludwig *et al.*, 1998), (b) kinetic control (Matzner & Prenzel,

Table 14.2. Summary of Al solubility data for acid mineral soils.

Soil type	% OM of solids	Temp. (°C)	pH range	Slope	Apparent log K_{so}	log K_{so}* gibbsite	Reference
Topsoil	5.3	20	3.5–5.0	0.9	9.4	8.1	Bache (1974)
Topsoil	5.6	20	4.0–5.0	1.7	7.5	8.1	Bache (1974)
Subsoil	0.3	20	4.5–5.0	2.2	8.6	8.1	Bache (1974)
Subsoil	1.0	20	4.0–5.0	1.5	8.3	8.1	Bache (1974)
Podzol Bh	2.4	8	3.3–4.0	1.6	6.6	8.9	Berggren & Mulder (1995)
Podzol Bs	2.8	8	4.1–4.6	2.3	9.1	8.9	Berggren & Mulder (1995)
Brown earth A	1.9	8	3.8–4.7	1.8	8.2	8.9	Berggren & Mulder (1995)
Brown earth B	6.8	8	4.2–4.9	2.9	9.4	8.9	Berggren & Mulder (1995)
Podzol E	0.6–1.4	4	3.4–4.0	0.25	6.0	9.2	van Hees *et al.* (2000b)
Podzol B	0.3–2.3	4	4.5–6.0	0.46	10.1	9.2	van Hees *et al.* (2000b)

* The log K_{so} values for gibbsite at the different temperatures are based on a value of 7.74 at 25°C (Palmer & Wesolowski, 1992). The value of apparent log K_{so} is for mid-range of observations, calculated as (3pH − pAl).

1992), or (c) some sort of adsorption reaction. The latter could be cation exchange at mineral surfaces, sorption by oxides, or binding by organic matter. A simple cation-exchanger binding both Al^{3+} and H^+ would give a cubic relationship. Binding by oxides and organic matter would not. Lövgren *et al.* (1990) determined Al binding by FeOOH as a function of pH, and their data give slopes of less than 2 in the pH range 3–4.4. Figure 14.4 shows WHAM-predicted relationships for different ratios of Al to humic substances. At the lowest ratio (10^{-4} mol g^{-1}), the slope is 1.76, reflecting binding of the Al mainly at bidentate sites. At higher ratios the slope decreases, so that at a ratio of 2×10^{-3} mol g^{-1} the slope is only about 0.3. Thus, in principle, binding by organic matter offers a way to explain slopes of less than 3 in mineral soils containing sufficient organic matter.

The possibility that organic matter does control aluminium solution concentrations in mineral soil horizons was examined by Berggren & Mulder (1995) and Tipping *et al.* (1995a). The same optimisation approach was adopted as used for the organic soils (Section 14.2.1), i.e. analytically estimated soil contents of organic matter and Al were used as guides, but the contents of 'active' reactants were adjusted, to match the predicted values to the observations. An example of a successful model application is shown in Fig. 14.5. In this case, the optimised soil content of

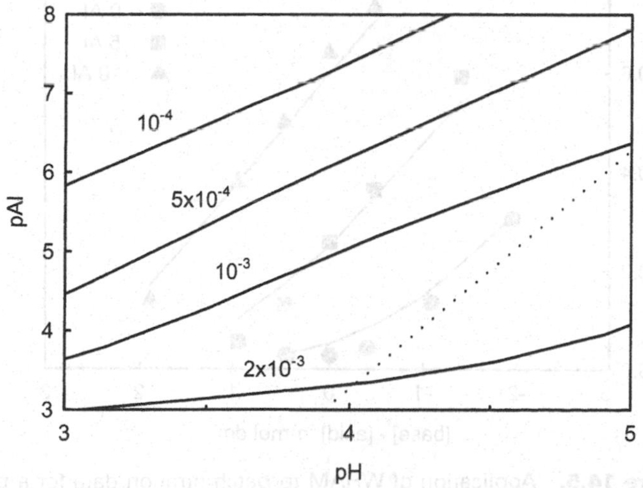

Figure 14.4. WHAM-calculated plots of $-\log [Al^{3+}]$ vs. pH for Al in a 50:50 mixture of fulvic and humic acid (total concentration 4 g dm^{-3}). The results refer to an ionic strength of 0.01 mol dm^{-3} and a temperature of 10°C. The numbers next to the lines are the ratios of Al to humic substances in mol g^{-1}. The dotted line represents the equilibrium relation for gibbsite (log K_{so} = 7.74 at 25°C).

humic substances corresponded to 49% of the organic matter extractable with sodium pyrophosphate, while the optimised soil content of Al corresponded to 71% of the $CuCl_2$ extractable Al. In other examples, better fits were obtained if solubility control of Al by an $Al(OH)_3$ phase

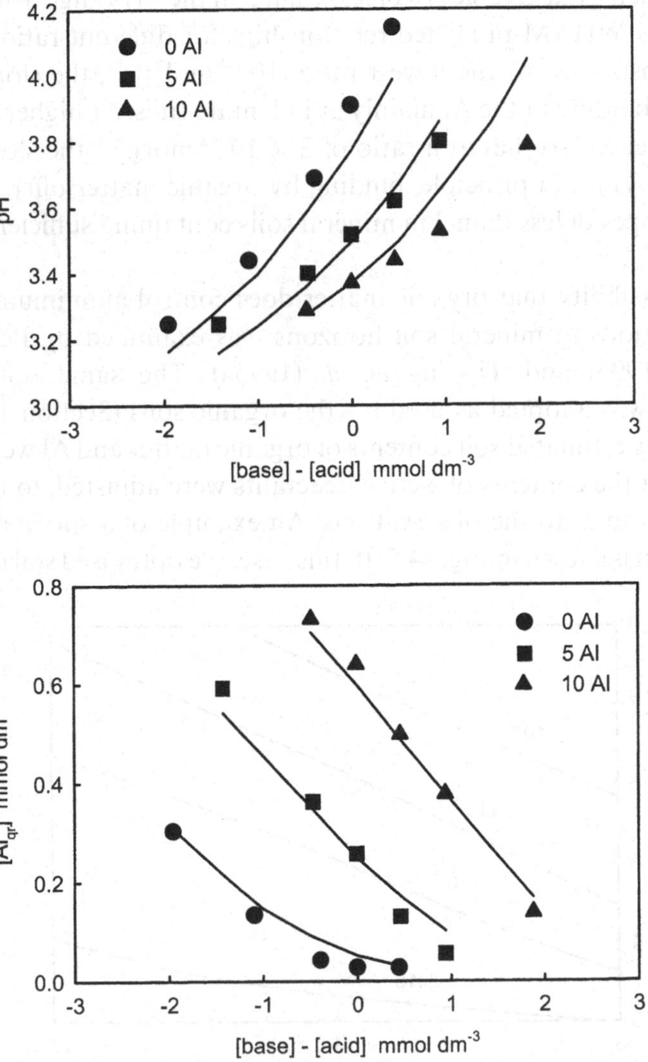

Figure 14.5. Application of WHAM to batch titration data for a podzol Bh horizon. The results refer to a soil concentration of 154 g dm^{-3}. The original soil Al content was 15 μmol g^{-1}, the figures next to the symbols indicate additions of extra Al (μmol g^{-1}). Quickly reacting Al (Al_{QR}) comprises inorganic species. [Redrawn from Berggren, D. & Mulder, J. (1995), The role of organic matter in controlling aluminium solubility in acidic mineral soil horizons, *Geochim. Cosmochim. Acta* **59**, 4167–4180, with permission from Elsevier Science.]

was invoked (Tipping *et al.*, 1995a). Thus, in mineral soils with low Al contents, solid-phase organic matter may control Al solution concentrations, but where Al/organic matter ratios are high, an oxide phase may play a rôle.

If organic matter is controlling solution Al concentrations, there must be a source of the metal if an approximately constant soil composition is to be maintained. Mineral dissolution is the obvious source, taking place too slowly for the dissolution reaction itself to control solution concentrations.

14.3.2 Cation exchange of base cations in acid mineral soils

Gustafsson *et al.* (2000) measured the sorption of Ca by the B horizon of a sandy forest podzol at different pH values. The fall in Ca solution concentration with increasing pH could be described fairly well with WHAM (Fig. 14.6), the optimised soil humic matter content (0.6 mg g^{-1}) being *c.* 25% of the total organic matter. Under such circumstances, the metal is bound almost entirely in the diffuse layer, and so the Donnan sub-model is responsible for describing the solution concentrations. Tipping *et al.* (1995a) found that WHAM could account approximately for base cation binding by the mineral horizon of a moorland soil, the 'active' humic content of which (18 mg g^{-1}) had been estimated by fitting Al and pH data. In these cases therefore, it appears that the

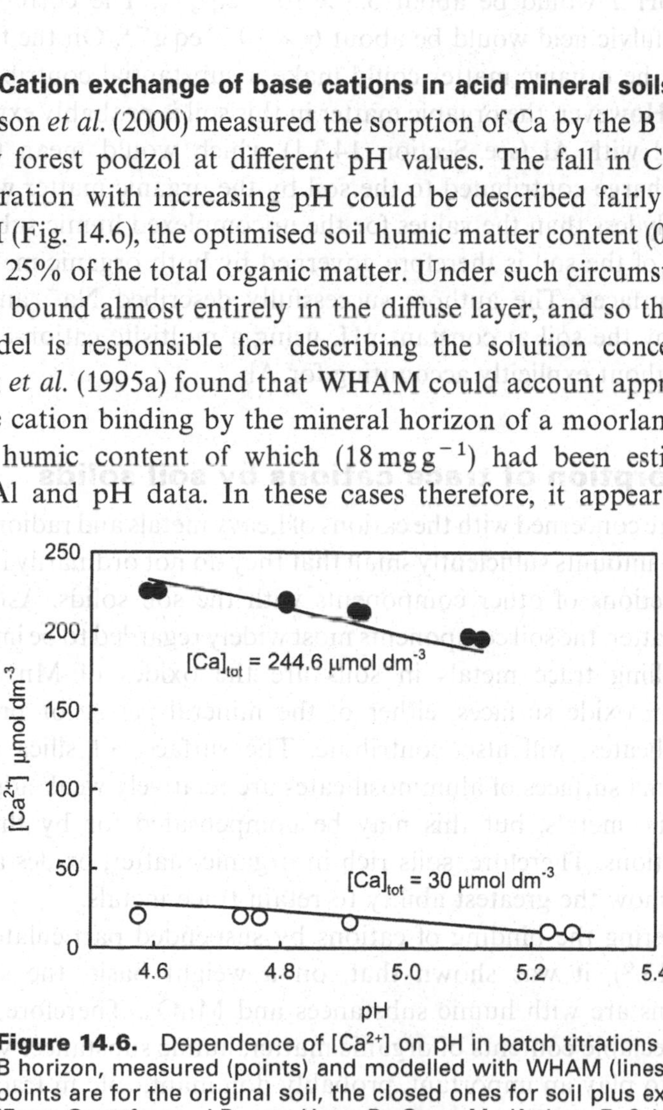

Figure 14.6. Dependence of $[Ca^{2+}]$ on pH in batch titrations of a podzol B horizon, measured (points) and modelled with WHAM (lines). The open points are for the original soil, the closed ones for soil plus extra Ca. [From Gustafsson, J.P., van Hees, P., Starr, M., Karltun, E. & Lundström, U. (2000), Partitioning of base cations and sulphate between solid and dissolved phases in three podzolised forest soils, *Geoderma* **94**, 311–333, with permission from Elsevier Science.]

humic matter is the principal cation-binding component of the soil.

Vulava *et al.* (2000) studied the exchange between Na^+ and Ca^{2+} on the fine sand fraction of an acid B horizon sample. The material had a cation-exchange capacity (CEC) of 6×10^{-5} eq g^{-1}. The main minerals in the clay fraction were vermiculite, illite and kaolinite, and the sample contained 6 mg g^{-1} organic carbon. If all the organic carbon were present as humic acid, and if all the COOH groups were able to ionise, then the CEC at pH 7 would be about 3.5×10^{-5} eq g^{-1}. The corresponding figure for fulvic acid would be about 6×10^{-5} eq g^{-1}. On the face of it therefore, the organic matter could make a substantial contribution to the CEC. However, the organic matter in this soil is probably extensively complexed with Al (see Section 14.3.1) which would mean that the negative charge contributed to the soil by the organic matter would be appreciably less than the values for the uncomplexed humic substances. The CEC of the soil is therefore governed by both organic matter and mineral surfaces. The authors successfully described Na^+ and Ca^{2+} sorption by the soil at constant pH, using a multisite cation-exchange model, without explicitly accounting for Al.

14.4 Sorption of trace cations by soil solids

Here we are concerned with the cations of heavy metals and radionuclides, present in amounts sufficiently small that they do not ordinarily influence the interactions of other components with the soil solids. Aside from organic matter, the soil components most widely regarded to be important in controlling trace metals in soils are the oxides of Mn and Fe. Aluminium oxide surfaces, either of the mineral *per se* or present in aluminosilicates, will also contribute. The surfaces of silica and the non-alumina surfaces of aluminosilicates are relatively weak adsorbents for cationic metals, but this may be compensated for by their high concentrations. Therefore, soils rich in organic matter, oxides and clay minerals show the greatest ability to retain trace metals.

Considering the binding of cations by suspended particulate matter (Section 13.8), it was shown that, on a weight basis, the strongest interactions are with humic substances and MnO_2. Therefore, in soils with appreciable contents of organic matter, humic substances would be expected to play an important, probably dominant, rôle in trace cation binding. However, there are complicating factors, notably interactions of humic substances with mineral surfaces (Section 12.3), competition for binding by major cationic solutes, notably Al^{3+}, Fe^{3+} and their hydrolysis products, and the trapping of cations within growing oxide phases (Brümmer *et al.*, 1988). These may alter the balance between the organic matter and mineral phases in their uptake of cations.

A point which should be appreciated is that the term 'total soil metal' usually does not mean the true total, but rather the metal that can be released by moderately concentrated mineral acids (typically HNO_3 in the concentration range 0.1–1 mol dm^{-3}). This pool of metal is assumed to be able to partition between the soil solids and the solution. The soils invariably also contain residual metal, which requires more drastic conditions (e.g. concentrated HF) for its release, and which does not participate in solid–solution partitioning. The residual metal may be present in mineral matrices, or it may be bound metal that has undergone some sort of ageing process, rendering it less available. The analytical uncertainty hinders the definition of the speciation problems that we would like to be able to solve (cf. Section 14.1.4).

In the following sections, evidence for the importance or otherwise of humic matter in the binding of trace cations in soils is considered.

14.4.1 Heavy metals

McBride *et al.* (1997) analysed the data of Sauvé *et al.* (1997) for 70 soils varying in pH (3.4 to 7.6), total copper content (Cu_T = 14 to 3083 μg g^{-1}), organic matter (OM = 0.4 to 55.5% C) and Cu^{2+} activity ($\log\{Cu^{2+}\} = -12.2$ to -6.2). They showed that $\{Cu^{2+}\}$ could be described well ($r^2 = 0.80$) by the following empirical equation

$$\log\{Cu^{2+}\} = a + b\mathrm{pH} + c\log Cu_T + d\log OM \qquad (14.3)$$

The values of the constants were as follows: $a = -1.28$, $b = -1.37$, $c = 1.95$, $d = -1.95$. Thus, the solution activity decreases with pH, increases with total soil metal, and decreases with OM. As concluded by McBride *et al.*, the variations in $\{Cu^{2+}\}$ are as expected if soil organic matter is the principal adsorbent of Cu. However in subsequent work, Sauvé *et al.* (2000) found that it was not necessary to include the term in OM to describe Cd^{2+} activities, for a set of soils varying in OM from 0.8 to 10.8% C.

Janssen *et al.* (1997) carried out a survey of 20 Dutch soils contaminated with Cd, Cr, Cu, Ni, Pb and Zn, determining soil properties, together with distribution (or partition) coefficients, K_D, defined by

$$K_D = \frac{\text{amount of metal bound per unit mass of dry solids}}{\text{amount of metal per unit volume of solution}} \qquad (14.4)$$

They found that K_D varied widely with metal and soil type and correlated most strongly ($r^2 = 0.54$ to 0.85) and significantly ($p \leq 0.001$) with pH. The soil content of iron, and the solution DOC concentration also contributed to explaining variations in K_D, but total soil organic matter content did not.

Lee *et al.* (1996) measured the binding of Cd added, at quite high concentrations, to suspensions of 15 mineral soils. They found that, at a given pH, variations among the soils could be accounted for very well by differences in organic matter content, while variations in oxide content were without significant influence. They concluded that 'surficial adsorption sites' in the soils were principally composed of organic matter. Applications of WHAM to data for two of the soils studied by Lee *et al.* are shown in Fig. 14.7. For the Boonton Union County soil, which is fairly rich in organic matter (8.6% by weight), the model provides a faithful simulation of the variation of binding with pH, on the assumption that 32% of the organic matter behaves like isolated humic substances (half fulvic acid, half humic acid). The result supports the supposition that organic matter is dominant in Cd binding under the experimental conditions. For the sandy loam (0.2% organic matter), the best fit (which is not very good) requires 250% of the organic matter to behave like isolated humic substances, which is clearly unrealistic; it is likely that in this case mineral surfaces dominate Cd sorption.

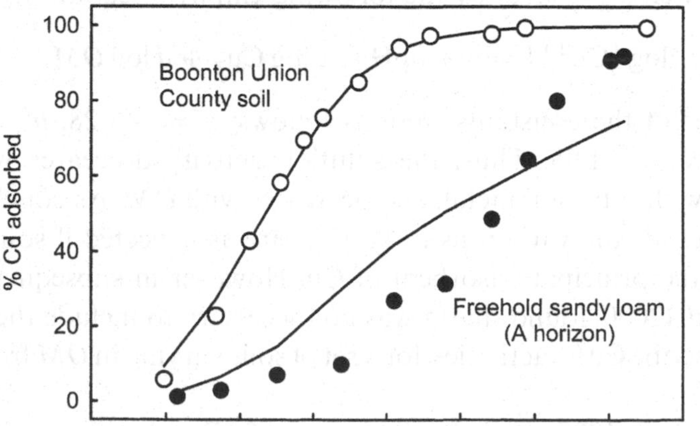

Figure 14.7. Observed (points) and WHAM-fitted (lines) adsorption of Cd by two soil samples. The observations are from experiments by Lee *et al.* (1996) in which 10^{-4} mol dm^{-3} $Cd(NO_3)_2$ was added to suspensions (10 g dm^{-3}) of soil in 0.01 mol dm^{-3} $NaNO_3$.

Another study in which the uptake of Cd by a soil was studied was reported by Benedetti *et al.* (1996), who measured binding by the organic surface horizon of a podzol as a function of [Cd^{2+}], and used the NICA–Donnan model to interpret the results (Fig. 14.8). The soil content of humic substances was estimated by calibrating the model to proton-binding data. In the calibration, all model parameters except the site densities were held constant at values that had been estimated from fitting the model to laboratory data for isolated humic acid. The optimised total site density corresponded to 55% of the value for isolated humic acid, and yielded a reasonably good prediction of Cd binding by the soil.

Benedetti *et al.* (1996) also used the NICA–Donnan model to simulate the solid–solution distributions (K_D values; equation 14.4) of Cu and Cd at different points in the profile of a podzol (Fig. 14.9). The modelling took account of binding to organic matter in both the solid and solution phases. Very good simulations of K_D were achieved for the upper, highly organic, part of the profile, but the model overestimated K_D for the lower depths, where the organic matter content of the soil was < 1%. The authors suggested the underestimation to be due to interaction of the organic matter with mineral matter, causing 'neutralization' of organic sites active in metal binding. An interesting feature of the results is that the

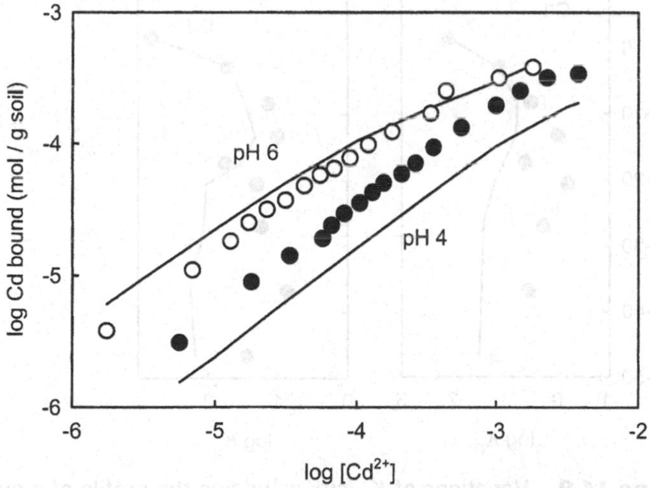

Figure 14.8. Binding of Cd by the organic horizon of a podzol, suspended in 0.01 mol dm^{-3} $CaCl_2$. The points are measured values, the lines are predictions of the NICA–Donnan model, calibrated by fitting proton-binding data for the same soil. The concentration of Cd^{2+} is in mol dm^{-3}. [Adapted from Benedetti, M.F., van Riemsdijk, W.H., Koopal, L.K., Kinniburgh, D.G., Gooddy, D.C. & Milne, C.J. (1996), Metal ion binding by natural organic matter: from the model to the field. *Geochim. Cosmochim. Acta* **60**, 2503–2513, with permission of Elsevier Science.]

K_D values for Cd are somewhat higher than those for Cu, despite the fact that Cu is more strongly bound by the soil solids. This arises because the dissolved organic matter also has a greater affinity for Cu. For example, at the top of the profile, approximately 50% of the Cd in solution is bound by dissolved organic matter, whereas essentially all of the Cu is bound (see Section 14.5.2).

In the examples presented above, soil organic matter was dominant in binding heavy metals in three cases, but in two it did not appear to play a significant rôle, suggesting the participation of other solid-phase sorbents. When soil organic matter is dominant, models based on results with isolated humic substances can reproduce observations quantitatively, for reasonable levels of 'active' humic matter.

14.4.2 Radionuclides

Sanchez *et al.* (1988) suspended a peat soil, low in mineral matter and aluminium, in 'artificial acid rain' (final pH 4.0), and added trace amounts of radioactive Co, Sr and Cs. The distributions of the radionuclides between the solid and solution phases were determined by centrifugation-depletion, and the results expressed in terms of K_D (equation 14.4). Values

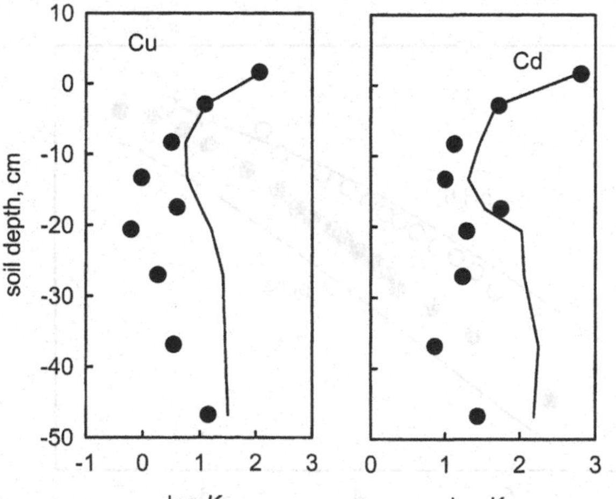

Figure 14.9. Variations of K_D ($cm^3\ g^{-1}$) down the profile of a podzol. The points show the measured values of Gooddy *et al.* (1995) for successive slices of soil, not just the depth plotted. The lines are simulations with the NICA–Donnan model. The pH increased from *c.* 3.5 to *c.* 4.1 from top to bottom. Soil depths are given relative to the base of the organic horizon. [Redrawn from Benedetti, M.F., van Riemsdijk, W.H., Koopal, L.K., Kinniburgh, D.G., Gooddy, D.C. & Milne, C.J. (1996), Metal ion binding by natural organic matter: from the model to the field, *Geochim. Cosmochim. Acta* **60**, 2503–2513, with permission from Elsevier Science.]

of K_D thus obtained are shown in Table 14.3, where they are compared with values calculated using WHAM. The model is able to reproduce the observations quite well on the assumption that *c.* 15% of the peat solids consist of humic substances, which is in general agreement with the values estimated from fitting Al and pH data (Section 14.2.1).

Another study in which trace amounts of radionuclides were added to soil suspensions was carried out by Tipping *et al.* (1995c), using organic and mineral horizons containing substantial amounts of adsorbed aluminium. The WHAM model was calibrated on the basis of variations in pH and dissolved Al in batch titrations, and then used to predict the solid–solution distributions of Co, Sr, Cs and Am. The observed and predicted K_D values are shown in Fig. 14.10. The model provides good predictions for Co, Sr and Am added to the organic horizons, but severely underestimates the K_D values for Cs. For the mineral horizons, the predicted values of K_D for Co, Sr and Am follow the correct trends, but are too low by a factor of about 3 (0.5 log units), and again the predictions are much too low for Cs. The discrepancies with regard to Cs are explained by the high affinity of Cs^+ for clay minerals, especially the 'frayed-edge' sites of illite (Sawhney, 1972), and provide a clear example of a situation in which humic substances contribute little to the sorption properties of a soil for a particular metal cation. The underestimation of sorption by the mineral horizons may indicate that solid-phase components other than humic substances are involved, or it may reflect the approximate nature of the calibration procedure.

The uptake of Co and Sr by the soil samples is, according to the model, mainly due to the accumulation of Co^{2+} and Sr^{2+} as counterions in the humic diffuse layers. Because the binding of these cations is relatively weak, the K_D values are sensitive to other major ions, especially Al^{3+}. For example, Tipping *et al.* (1995c) calculated that increasing the bound

Table 14.3. Sorption of trace amounts of radioelements by peat, measured by Sanchez *et al.* (1988) and calculated with WHAM (Tipping *et al.*, 1995c).

Radionuclide	Obs. range of $\log K_D$	Calc. $\log K_D$ for $CHS = 0.1$	Calc. $\log K_D$ for $CHS = 0.2$
Co	3.4–3.9	3.4	3.8
Sr	3.3–3.9	3.5	4.0
Cs	1.8–3.3	2.8	3.1

The results refer to pH 4.0, with a background electrolyte of 'artificial acid rain' (i.e. a dilute solution of electrolytes).
The soil contents of humic substances (*CHS*) assumed for modelling are in $g\,g^{-1}$.

aluminium ratio (cf. Fig. 14.3) from 0 to 1 at constant pH (4.0) caused a fall in K_D for Co from 10^4 to $20\,cm^3\,g^{-1}$. This explains why the K_D values for Sr in the high-aluminium soils of Fig. 14.10 are so much lower than those for the low-aluminium peat of Table 14.3.

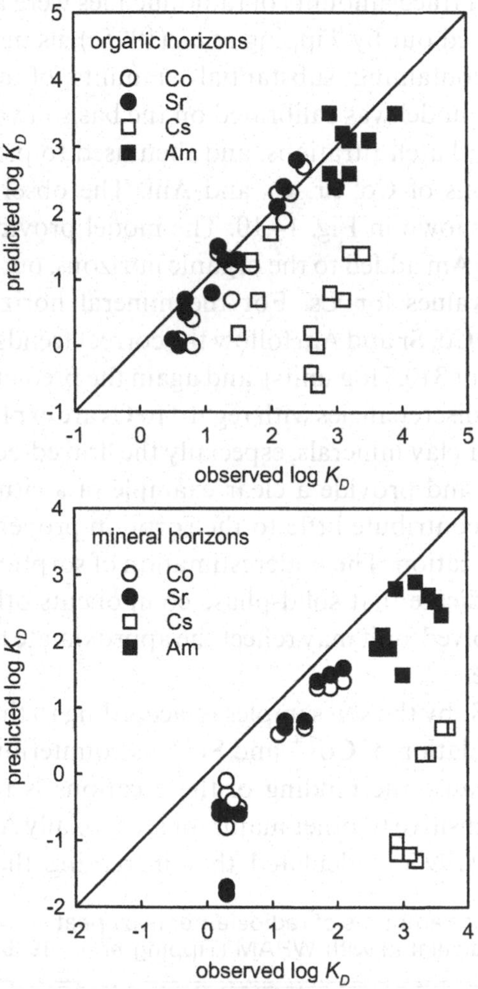

Figure 14.10. Observed and predicted values of K_D ($cm^3\,g^{-1}$) for the adsorption of trace radionuclides by the organic and mineral horizons of three UK moorland soils. The results refer to batch titration experiments in which the pH of the soil suspensions was varied in the range 3.2–4.5, in some cases with the addition of $CaCl_2$. The predicted values of K_D take account of the presence of dissolved organic matter in the aqueous phase; [DOC] ranged from 3.2 to 34.9 mg dm^{-3}. [Redrawn from Tipping, E., Woof, C., Kelly, M., Bradshaw, K. & Rowe, J.E. (1995c), Solid–solution distributions of radionuclides in acid soils: application of the WHAM chemical speciation model. *Environ. Sci. Technol.* **29**, 1365–1372. Copyright 1995. Am. Chem. Soc.]

In contrast to Co and Sr, WHAM calculates that Am is bound almost exclusively at specific sites on the humic matter, in both the solid and aqueous phases, with binding so strong that only a small fraction of the total metal is present in solution as Am^{3+}. Thus, the solid–solution distribution of Am is determined mainly by the concentration of dissolved humic substances. If there were no dissolved humic matter, the K_D values for Am would be in the range 10^4–10^5 $cm^3\,g^{-1}$. A similar situation was seen for Cu in the podzol profile described in Section 14.4.1 and Fig. 14.9 (see also Section 14.5.2).

In agreement with the conclusions drawn for heavy metals, the examples presented here provide strong evidence for the participation of humic substances in controlling solution concentrations of radionuclides in soils, while also showing that in some instances other soil components play significant rôles.

14.5 Dissolved organic matter (DOM) in soil solution

Soil solution DOM, most of which consists of humic substances, is a significant, often dominant, factor in the speciation of cationic metals, and is also important in the chemistry of other soil components such as pesticides and organic contaminants. In the soil environment *per se*, this has implications for bioavailability and the distribution of the metals between the solid and solution phases (see Section 14.4). The transport of DOM, and associated components, between soil horizons, and into drainage water is another important issue. Here, we are concerned with the ways in which the interactions of cations with humified DOM influence the behaviour of both reactants in the soil–water system.

14.5.1 Distribution of potential DOM between solid and solution

By 'potential DOM' is meant organic material that may pass into soil solution. Whether it does so depends upon its interactions with the solid phase of the soil. The fundamental processes governing the transfer were discussed with respect to humic substances in Chapter 12. Broadly speaking, we can divide them into adsorption onto mineral surfaces, and absorption into solid-phase soil organic matter. Both are influenced by cations.

Tipping & Woof (1991) assumed solution concentrations of DOM in suspensions of organic soil horizons to be governed by the incorporation of humic substances into soil organic matter aggregates. They described the reaction using a simple partition function, but incorporating a dependence on the net humic charge; the greater is the net charge, the less is the tendency of the humic matter to be absorbed by the soil. The assumed relationship was

$$K_P = \frac{CFA_i}{[FA_i]} = \exp\{\beta(Z_i| - |Z|)\} \tag{14.5}$$

where CFA_i is the soil content (e.g. in $g\,g^{-1}$) of absorbed humic matter (assumed to be fulvic acid), $[FA_i]$ is the concentration in solution, $|Z|$ is the magnitude of net charge on the FA (summed over the specific binding sites), $|Z_i|$ is the magnitude of a 'characteristic' charge, and β is a scaling constant. When $|Z|$ is less than $|Z_i|$, K_P is greater than unity, i.e. absorption by the soil organic matter is relatively strong. When K_P is less than unity, absorption is weak. Thus, the extent of partitioning is assumed to depend upon the competition between hydrophobicity, which favours absorption, and the solvation of charged groups, which promotes water solubility. The value of $|Z_i|$ can be regarded as a measure of the hydrophobicity of the fulvic acid; the greater is $|Z_i|$, the higher is the value of $|Z|$ required to overcome the hydrophobic interactions. The fulvic acid is postulated to consist of a series of fractions, differing in their values of $|Z_i|$, thus accounting for heterogeneity in the humic matter. To apply the model, values of $|Z|$ are obtained from the extents of cation binding, calculated with WHAM. Thus, the contributions to $|Z|$ of all specifically bound major cations (H^+, Mg^{2+}, Al^{3+}, $AlOH^{2+}$, Ca^{2+}) can be taken into account. In essence, the greater is the extent of complexation, and therefore the lower is $|Z|$, the greater is the extent of absorption. The model is optimised by adjusting the relative amounts of the different fulvic acid fractions. It provides reasonably good fits of experimental data describing, for example, the variation of [DOM] in suspensions of a soil adjusted to different pH values. Figure 14.11 shows the results of applying the model to the different horizons of a podzol (Lofts *et al.*, 2001b).

Although the partitioning model can describe the interactions of humic substances with both organic horizons and mineral ones, many authors believe that adsorption onto mineral surfaces, especially oxides, is the main sorption process in mineral soils (see e.g. Kaiser & Zech, 1999, and references quoted therein). The mechanisms of such adsorption are as described in Section 12.3. Vance & David (1992), studying spodosol mineral horizons, found a slight fall in sorption as pH was increased from 4 to 5.5, and Kaiser & Zech (1999) observed increasing desorption as the pH was raised from 4 to 10 in suspensions of a subsoil low in organic

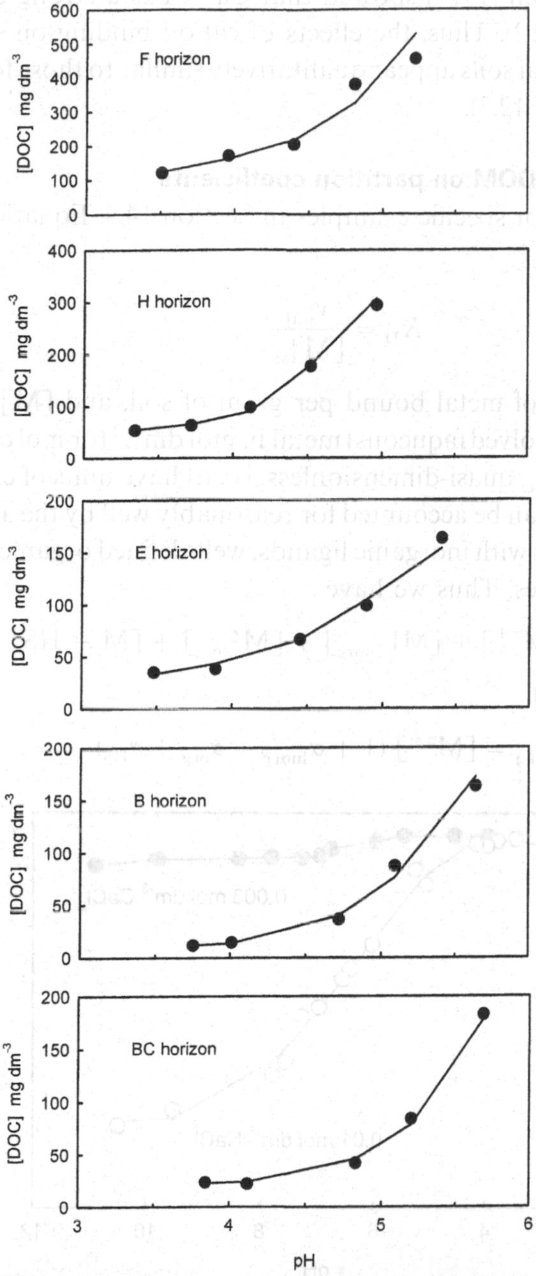

Figure 14.11. The release of DOM from different horizons of a Norwegian podzol. Soil suspensions (10 g moist soil + 25 cm^3 10^{-3} mol dm^{-3} NaCl) were adjusted to different pH values by adding acid or base, and the supernatant DOC concentrations were measured after equilibration. The points are observations, the lines are simulations using the model described in Section 14.5.1. Note the different *y*-axis scales. [Reproduced from Lofts, S., Simon, B.M., Tipping, E. & Woof, C. (2001b), Modelling the solid–solution partitioning of organic matter in European forest soils, *Eur. J. Soil Sci.* **52**, 215–226, with permission of Blackwell Science.]

matter (0.7% C). Shen (1999) showed that Ca^{2+} has a strong effect on adsorption (Fig. 14.12). Thus, the effects of cation binding on sorption reactions with mineral soils appear qualitatively similar to those found for pure oxides (Section 12.3).

14.5.2 Influence of DOM on partition coefficients

This was discussed for specific examples in Section 14.4. Equation (14.4) can be written

$$K_D = \frac{v_{soil}}{[M]_{aq}} \tag{14.6}$$

where v_{soil} is moles of metal bound per gram of soil, and $[M]_{aq}$ is the concentration of dissolved (aqueous) metal in mol dm^{-3} (or mol cm^{-3} if it is desired to make K_D quasi-dimensionless, i.e. to have units of cm^3 g^{-1}). The aqueous metal can be accounted for reasonably well by the aquo ion, M^{z+}, plus complexes with inorganic ligands, well-defined organic ligands, and humic substances. Thus we have

$$[M]_{aq} = [M^{z+}] + [ML_{inorg}] + [ML_{org}] + [M - HS] \tag{14.7}$$

which can be written

$$[M]_{aq} = [M^{z+}]\,(1 + \alpha_{inorg} + \alpha_{org} + \alpha_{HS}) \tag{14.8}$$

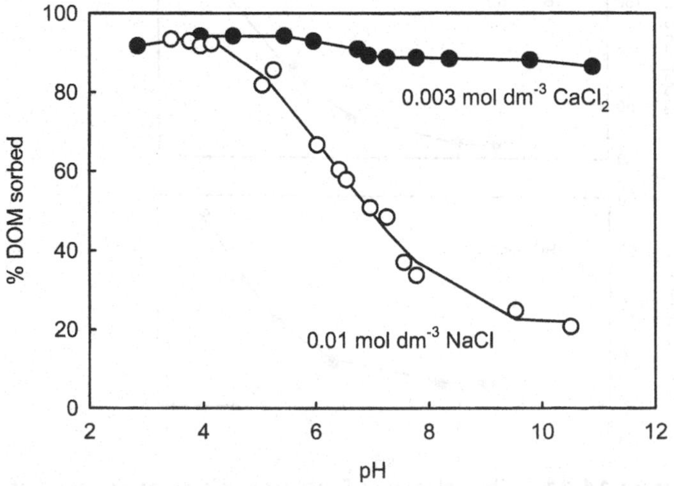

Figure 14.12. Adsorption of DOM as a function of pH by a mineral soil from Taiwan, in different electrolyte solutions with similar ionic strengths. The total concentration of DOC was 65 mg dm^{-3}, that of the soil was 26 g (dry weight) dm^{-3}. The lines are for guidance only. [Redrawn from Shen, Y.-H. (1999), Sorption of natural dissolved organic matter on soil, *Chemosphere* **38**, 1505–1515, with permission from Elsevier Science.]

where the α terms are appropriate collections of equilibrium constants and ligand concentrations. An alternative partition coefficent, K_D^*, can be defined in terms of $[M^{z+}]$

$$K_D^* = \frac{\nu_{soil}}{[M^{z+}]} \tag{14.9}$$

This is more chemically meaningful than K_D, being a direct measure of the strength of the interaction of the metal ion with the solid material. We can then relate K_D and K_D^* by the equation

$$K_D = \frac{K_D^*}{(1 + \alpha_{inorg} + \alpha_{org} + \alpha_{HS})} \tag{14.10}$$

Thus we see that the more complexed the metal is in solution, the lower is K_D. A metal with a high affinity for solution ligands (of which HS will generally be dominant) can therefore have highly variable values of K_D, while maintaining the same K_D^*, depending upon the dissolved ligand concentrations. One metal can have a lower K_D^*, but a higher K_D, than another, if the metal with higher affinity is appreciably complexed in solution. The simple K_D is best regarded as a convenient variable, expressing no more than the solid–solution distribution under a given set of circumstances. Expressions like K_D^* can be more useful, although they do not take into account the effects of pH and competing metals, for which more sophisticated models of binding by the solid phase (e.g. WHAM or NIC(C)A–Donnan) are needed.

14.5.3 Podzolisation

Podzols are a major class of soils, prevalent in cool, humid climates under forest or heath vegetation (Lundström *et al.*, 2000). Their chief characteristics are an organic-rich surface horizon, underlain by a weathered eluvial horizon, which, relative to the parent material, is rich in Si and poor in Al, Fe and base cations. Below the E horizon are B horizons enriched in Al, Fe and organic matter. The mechanism of formation of podzols remains a matter of debate (Browne, 1995; Lundström *et al.*, 2000), but an important contributory process is generally accepted to be the movement of DOM in percolating water downwards through the soil profile. During this passage, the organic matter binds increasing amounts of Al and Fe, becoming progressively less soluble and/or prone to adsorb onto mineral surfaces. The formation of the B horizons is due to the removal of the metal-complexed organic matter from solution, and its subsequent accumulation. According to Browne (1995), the increase in pH from the O and E horizons to the B is responsible for the formation of polynuclear hydroxy Al ions, which dissociate from the Al–organic complexes and precipitate, subsequently reacting with dissolved Si to form the aluminosilicate minerals allophane and imogolite. Thus, it can be appreciated that the formation and dissociation of cation–DOM complexes is central to the formation of podzols.

14.5.4 Soil-derived DOM and the acidity of surface waters

The rôle of soil organic matter, especially DOM, in the acid status of surface waters is a long-standing issue in acid rain research. Referring to waters in southern Norway, Rosenqvist (1978) claimed that 'the greater part of the acidity in the fresh waters is the result of ion-exchange reactions in raw humus in the soil profiles of the catchment areas'. An important part of Rosenqvist's argument was that soil organic matter has a high content of 'exchangeable protons', much greater than amounts deposited in precipitation. Krug & Frink (1983) argued that the effect of increased atmospheric deposition of sulphuric acid was to decrease soil pH, which led to a reduction in the solubility of soil DOM, and consequently lower DOM concentrations in drainage and surface waters. The effect of acid deposition was claimed to increase the sulphate flux through the soil, but to decrease that of DOM, leading to surface water little different in pH, but with organic acid anions substituted by inorganic ones (i.e. sulphate). The organic acids consequently accumulating in the soil were then supposed to be oxidised to CO_2 and H_2O. According to Krug & Frink: 'Humic materials act as weak acids; they dissolve in water and colour and acidify it'. These points have been widely debated, but misunderstandings remain about the rôles of soil organic matter and dissolved organic acids in surface water acidification. While much effort has gone into defining the acid–base properties of DOM (Section 13.2), less emphasis has been placed on the production and dissolution of organic matter. The time scale of change in response to increases (and decreases) in acid deposition should also be appreciated. Some insights can be obtained by application of the model of a soil–water system schematised in Fig. 14.13.

In the model, the soil consists only of organic matter, i.e. it is an approximation of the surface horizon of a podzol, or the acrotelm of a peat. The organic components are the mixture of plant litter, its partial decomposition products (humin), microbial biomass, humic acid (HA), potential DOM (PDOM), and DOM. The inorganic components of the system are H^+, Na^+ and a mobile anion, X^-, which enter in rainfall.

Within the soil, HA and PDOM are formed from the litter/humin/biomass pool, and are lost either by mineralisation to CO_2 or, in the case of PDOM, partly as DOM in drainage water. The interactions of the inorganic components with HA, PDOM and DOM are calculated using WHAM, both PDOM and DOM being represented by fulvic acid. The model is, of course, highly simplified, and ignores numerous biogeochemical processes that contribute to soil and water acidification (see e.g. Reuss *et al.*, 1987; de Vries & Breeuwsma, 1987). For simplicity, evaporation is also ignored. The point is to isolate the fundamental reactions involving organic matter.

The soil depth is set at 15 cm, the bulk density at 0.25 g cm^{-3}, the porosity at 0.833, and the soil content of HA at 0.20 g g^{-1}. In all the scenarios, the system runs at steady state for 100 years, with slightly acid rain ($[Na^+] = 200$ μmol dm^{-3}, $[H^+] = 20$ μmol dm^{-3}, $[X^-] = 220$ μmol dm^{-3}, pH 4.70) entering at a constant rate (1500 mm a^{-1}). Over the next 50 years, the rainfall acidity increases linearly ($[H^+]$ rises to

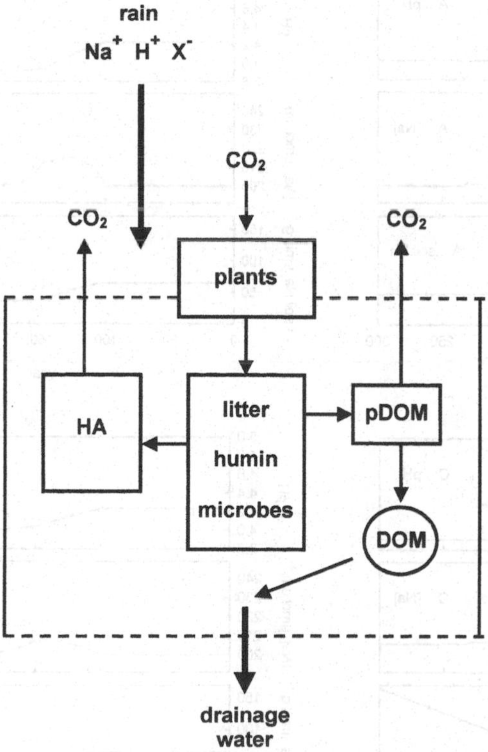

Figure 14.13. A simple model of organic soil, with rainfall input, and drainage water containing dissolved organic matter (DOM). Key: HA, humic acid; PDOM, potential DOM.

100 μmol dm^{-3}, [X$^-$] to 300 μmol dm^{-3}, [Na$^+$] stays constant, pH decreases to 4.01). The rainfall composition then remains constant for a further 150 years. The change in [X$^-$] with time is shown in the top panel of Fig. 14.14. Four scenarios are considered, with different assumptions about inputs and outputs of PDOM (Table 14.4). The concentration of

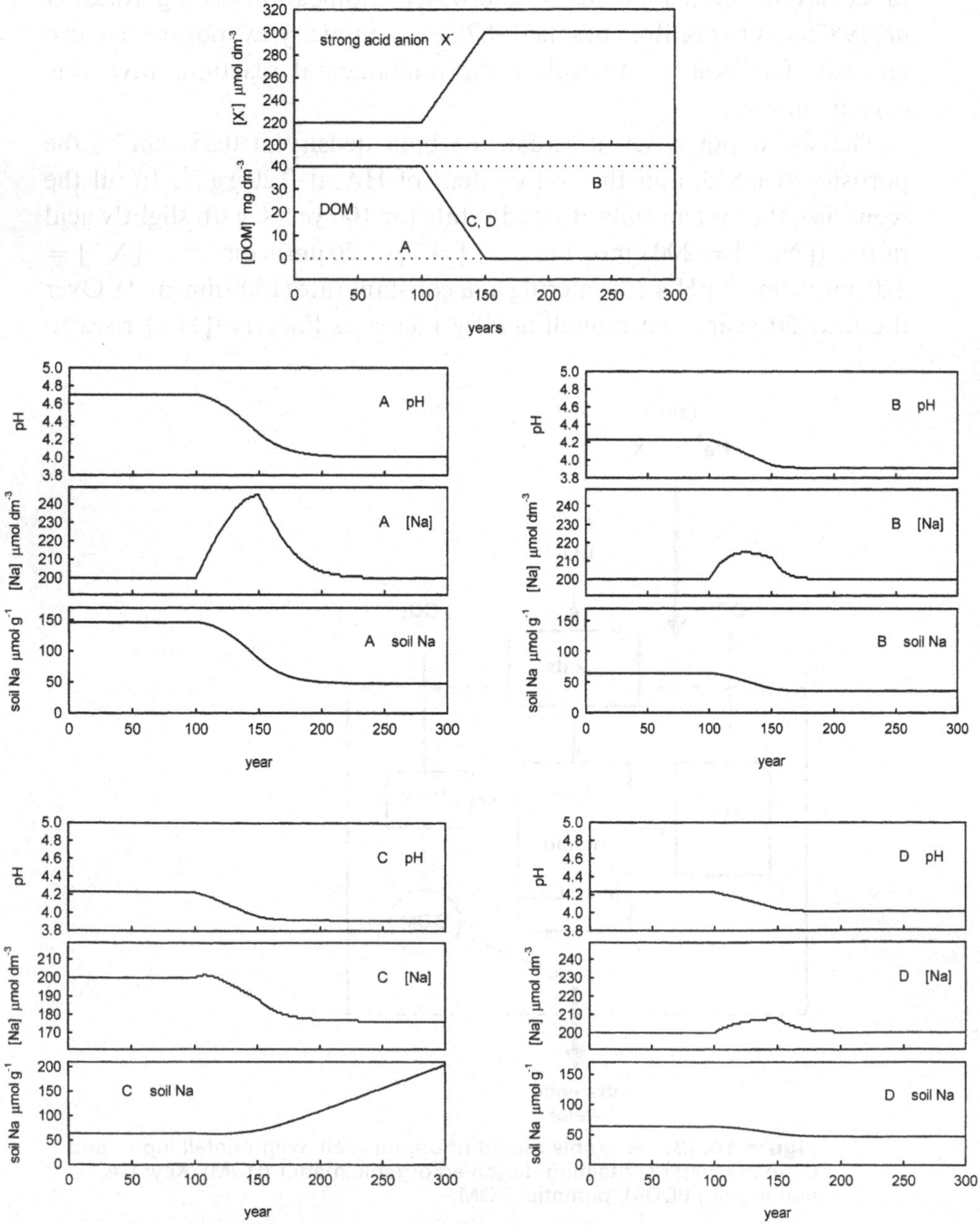

Figure 14.14. Outputs from the organic soil/acid rain model of Fig. 14.13. Note the different *y*-axes scales in C.

DOM in drainage water varies between zero and 40 mg dm^{-3} among the different scenarios, and with time; note that the range corresponds to [DOC] values between zero and 20 mg dm^{-3}. Results of the simulations are shown in Fig. 14.14.

In Scenario A, there is neither PDOM nor DOM, and the soil composition with respect to organic components is constant throughout the 300-year period. For the first 100 years, the drainage water has the same composition as the rainfall. When the rainfall becomes more acid, the drainage water does so as well, but the pH lags behind that of the rain because of buffering by the soil HA. The new steady-state pH of 4.01 is not approached until about 50 years after the change in rainfall composition is complete. The concentration of Na in the drainage water increases on acidification, because of release from the soil as the additional protons in deposition bind to HA. According to WHAM, this arises because the binding of protons at specific sites on the HA reduces the net charge, so that fewer Na^+ ions are needed as counterions (see Section 14.2). The soil Na content undergoes a substantial depletion, from *c.* 150 to 50 µmol g^{-1}, as a result of the acidification. This example shows that the pH of drainage water is not simply dependent upon ion-exchange processes, as suggested by Rosenqvist (see above). In the absence of any net inputs from the soil, the drainage water at steady state has the same composition as the incoming rain. The significance of the large store of protons in the soil

Table 14.4. Scenarios of organic soil acidification used to run the model described in Section 14.5.4. See Figs. 14.13 and 14.14.

Scenario	Period (years)	Soil HA content (g g^{-1})	PDOM input (g m^{-2} a^{-1})	PDOM loss (g m^{-2} a^{-1}) as CO_2	as DOM
A	0–100	0.20	0	0	0
	101–150	0.20	0	0	0
	151–300	0.20	0	0	0
B	0–100	0.20	60	0	60
	101–150	0.20	60	0	60
	151–300	0.20	60	0	60
C	0–100	0.20	60	0	60
	101–150	0.20	60	0	→0
	151–300	0.20	60	0	0
D	0–100	0.20	60	0	60
	101–150	0.20	60	→60	→0
	151–300	0.20	60	60	0

(which can be estimated by measuring the amount of base required to raise the pH to a given value, typically 7 or 8) to which Rosenqvist referred, is not that it governs drainage water pH at steady state, but rather that it buffers the soil against short-term increases in pH. Similarly, the large soil content of unprotonated sites buffers against short-term decreases in pH, as seen in the present example. The significant point is that soil organic matter is a very powerful buffer, but ultimately it cannot, by ion-exchange reactions, control steady-state drainage water pH.

Scenario B is the same as A except that PDOM is formed in the soil, and immediately becomes DOM. Thus the soil acts as a source of DOM to the drainage water. The pre-acidification steady-state pH of 4.23 is lower than that in A because of the formation and partial dissociation of the (P)DOM. Some of the Na in the drainage water is present bound (as counterions) to the DOM, but the total concentration of Na is necessarily (because of steady state) the same as the input. Because the pH is lower than in Scenario A, the soil requires less Na^+ as a counterion, and so the pre-acidification soil Na content is lower. In B, acidification is assumed to have no effect on the production of PDOM, nor on its conversion to DOM. When acidification occurs, the system responds qualitatively in the same way as in A, but the relative changes are smaller. The final steady-state pH is 3.91, which corresponds to an increase in a_{H^+} following acidification of 64 μmol dm^{-3}. This is less than the 78 μmol dm^{-3} change in the rainfall a_{H^+}, and is due to the binding of protons by the (P)DOM.

Scenario C is the same as B except that the PDOM is assumed to become unable to leave the soil following acidification; a linear change from complete conversion to DOM to zero conversion is assumed to take place during the transitional acidification period. After acidification, the PDOM accumulates continuously in the soil (i.e. there is no conversion to CO_2), and so the system as a whole cannot reach a new steady state. However the drainage water does reach one, having the same final pH as in B, but a lower [Na]. Thus we find that, in terms of pH, the acidifying effect on drainage water associated with (P)DOM is due to its production, but not its dissolution. The difference between B and C is in the location of Na. In the acidified quasi-steady state in C, the Na is accumulating, along with the PDOM, in the soil, whereas in B it is leaving in the drainage water. Note that the drainage water in C is less well buffered than in B, because of the absence of DOM (cf. Section 13.2.3).

Scenario D is like B, except that the net production of PDOM following acidification is assumed to be zero (i.e. all PDOM is converted to CO_2). The net production is assumed to fall linearly from the pre-acidification value to zero during the 50-year acidification period.

This example shows the smallest changes in steady-state pH (from 4.23 to 4.01) and soil Na content (a fall of only 26%). It accords with the theory of Krug & Frink (see above) that acidification due to the production of humic matter can be replaced by acidification due to strong acids. For such a situation to arise, it is essential that (i) appreciable 'organic acidification' was going on before the change in the acidity of deposition, (ii) a fall in soil pH leads to a substantial decrease in DOM solubility, and (iii) the immobilised (P)DOM is efficiently mineralised. Condition (ii) is difficult to envisage, because if 'organic acidification' is proceeding, then the soil pH will be comparatively low, any further acidification from atmospheric deposition will not change it greatly, and therefore there is little scope for DOM solubility to change very much (see Section 14.5.1). Condition (iii) also seems unlikely because the fact that the soil is rich in organic matter tells us that decomposition, including mineralisation, is proceeding slowly, so that for PDOM not to accumulate it would have to be much more susceptible to the processes of decomposition than the rest of the soil organic matter.

14.5.5 Transport of trace metals

Metals bound to mobile DOM may be transported through soils and aquifers. The phenomenon is of importance in heavy metal pollution, in contaminated soils, and in upland soils receiving atmospheric deposition of metals. It is also significant with regard to radioactive waste disposal, where the concern is that mobile radionuclides, emanating from buried waste, may be transported to the biosphere by DOM and/or dispersed colloids.

However, the transport process is not simply a matter of the metal hitching a ride on a mobile carrier. As we have seen already in this chapter, and also in Chapter 12, humic matter undergoes numerous physico-chemical interactions, which can affect both its transport and metal-binding properties. Insight into the complexities of metal transport is provided by the work of Temminghoff *et al.* (1998), who studied the release of Cu from columns of copper-contaminated sandy soil, through which different solutions were passed. Figure 14.15 shows the results of one of their experiments. The observations were interpreted as follows:

Period I Elution with $NaNO_3$ removes organic matter from the soil in the form of DOM, and Cu is co-transported. As the concentration of DOC in the eluate decreases, so does that of the Cu.

Period II The addition of DOM to the eluant results in a rapid increase in the eluate DOC concentration, indicating little retention of DOM by the soil. Copper concentration in the eluate increases, due to co-transport with the DOM.

Period III The addition of $Ca(NO_3)_2$ to the eluant, while maintaining the input of DOM, causes the DOC concentration in the eluate to decrease markedly, because the binding of Ca^{2+} renders the DOM insoluble (due to aggregation and/or adsorption). Because the DOM is immobilised, so too is bound Cu.

Period IV When the eluant concentration of $Ca(NO_3)_2$ is

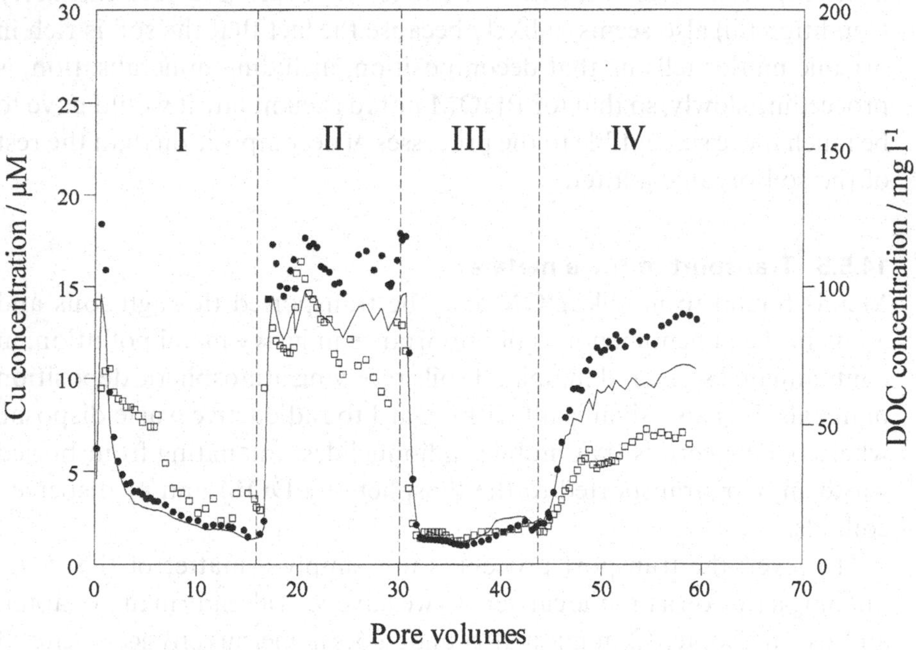

Figure 14.15. Influence of humic acid on the transport of Cu in a copper-contaminated sandy soil. The eluants were as follows:

I	3 mmol dm^{-3} $NaNO_3$
II	3 mmol dm^{-3} $NaNO_3$ + 100 mg dm^{-3} DOC
III	0.5 mmol dm^{-3} $Ca(NO_3)_2$ + 1.5 mmol dm^{-3} $NaNO_3$ + 100 mg dm^{-3} DOC
IV	0.1 mmol dm^{-3} $Ca(NO_3)_2$ + 2.7 mmol dm^{-3} $NaNO_3$ + 100 mg dm^{-3} DOC

The DOC was added as purified soil humic acid. The ionic strength is constant throughout. The pH varied between *c.* 4.7 and *c.* 5.5. Concentrations of DOC are shown by filled circles, concentrations of Cu by open squares. The line is the Cu concentration predicted using the NICA model. [Reproduced from Temminghoff, E.J.M., van der Zee, S.E.A.T.M. & De Haan, F.A.M. (1998), Effects of dissolved organic matter on the mobility of copper in contaminated sandy soil, *Eur. J. Soil Sci.* **49**, 617–628, with permission from Blackwell Science.]

decreased, the DOM becomes less prone to removal from solution, the eluate DOC concentration rises, and consequently Cu is mobilised.

The authors simulated Cu concentrations in the eluate using ECOSAT (Keizer & van Riemsdijk, 1994) incorporating the NICA model, which allowed the binding of both Ca and Cu by the DOM to be simulated, and also competition between the two metals. The simulations of Cu concentration are in accord with the observations. This example, which is quite simple compared to many field situations, shows how cation–humic interactions influence the transport of both the cations and the organic matter.

14.6 Colloids in soil and aquifer porewaters

Colloids exist in suspension in soil and aquifer porewaters, and, if mobile, will contribute to the transport of adsorbed trace metals, together with any other adsorbates (see e.g. McDowell-Boyer *et al.*, 1986; Kretzschmar *et al.*, 1999). It is desirable, if only to aid discussion, to distinguish colloids from DOM, but this is not always possible. For example, colloids present in groundwater from a glacial sand/silt aquifer system at the Gorleben salt dome in Germany consist principally of humic matter, and have sizes ranging from less than 1.5 nm to 1000 nm (Dearlove *et al.*, 1991). The colloids are responsible for binding, and presumably transporting, the majority of the actinides and lanthanides present. In other systems, the larger colloids are primarily mineral in nature, but adsorbed humic matter may exert a strong influence on their colloidal properties, and may also furnish metal-binding sites. The binding of major cations by organic colloids may affect mobility, by promoting aggregation and adsorption (Sections 12.2 and 12.3). The mobilities of mineral colloids may also be affected by interactions with major cations, which control colloid stability either by direct binding to the mineral surface, or by binding to adsorbed humic matter (Section 12.5).

14.7 Cation–humic interactions in sediments

The rôle of suspended sediments in cation binding in natural waters was discussed in Section 13.8. Here we are concerned primarily with the binding of cations to solids in deposited sediments, in lakes, rivers and oceans. Some of what was said about soils will apply to sediments, insofar as their solid fractions are mixtures and humic substances may or may not exert a significant control on pore water concentrations of cations. Sediments differ from soils in that they exchange materials with overlying water, and often have little internal water movement, diffusive transport being dominant.

Sediment composition obviously depends upon the nature of suspended particulate matter (SPM) in the water column, together with the relative sedimentation behaviour of the different SPM types, and any reworking or other post-depositional changes. Sediments vary in particle size, from muds in low-energy environments to gravels in high-energy ones. Organic matter content depends on particle size, as well as on source terms. It is therefore difficult to generalise about sediment organic content in freshwater and estuarine systems. In lakes, the organic content can vary from 1% or less to 60% or more. Coastal marine sediments can have organic matter contents of 4% or more, while sediments underlying the open oceans contain less than 1% organic matter (Huc, 1988).

The organic matter enters the sediments as dead phytoplankton, zooplankton and macrophytes, and their associated bacteria, in adsorption complexes with mineral particles, as soil organic matter and as sewage solids. Humic substances are formed *in situ*, but may also be terrestrially derived. As noted in Section 2.5.4, the majority of the humic matter in lake and marine sediments is usually in the humin fraction, although Wafica (1994) found that as much as 85% of the organic matter in a sediment sample from Lake Edku (Egypt) was extractable as humic and fulvic acids. According to Ishiwatari (1985), both the humin and humic acid of lake sediments may consist of complex aggregates of different humified components. The two fractions differ in their functional group contents, but although the humin has fewer oxygen-containing functional groups than the humic acid, it can be expected to possess significant cation-binding capability.

In systems where the overlying water varies in pH, humic matter in sediments will contribute to buffering properties, by binding and releasing protons. However, the main interest in cation–humic interactions in sediments is to do with heavy metals. The accumulation of heavy metals in sediments, the degree to which they can be remobilised, and their bioavailability and toxicity, are important issues in pollution research.

14.7.1 Organic matter – metal associations in sediments demonstrated by chemical analysis

Humic substances extracted from sediments carry with them significant amounts of metals. The greatest enrichment factors, defined as the ratio of the metal content of extracted humic substances to the metal content of the sediment as a whole, were found for Cu (21–40) and Pb (1.3–2.9) in Lake Ontario sediments (Nriagu & Coker, 1980). In the marine sediments studied by Nissenbaum & Swaine (1976), the greatest enrichment factors were for Ni (1.5–10), Cu (4–50), Zn (3–80), Mo (10–75) and Pb (1–60). These values are qualitatively consistent with metal-binding strengths to humic matter (see e.g. Chapter 10). However, as noted by Nissenbaum & Swaine, the extraction methods for humic substances involve rather harsh pH conditions, and so any humic metal contents reported are minimum values. More sophisticated analytical approaches are required to investigate the sediment speciation of metals *in situ*.

14.7.2 Selective and sequential extraction of metals from sediments

Many studies have been conducted in which the chemical forms of metals within sediments (also soils and other solid media) have been investigated by extraction with reagents that are specific towards certain phases, or metal fractions. The approach is referred to as selective and/or sequential extraction. Kersten & Förstner (1995) provide a recent comprehensive review, in which they discuss both analytical methodology and interpretation. The results of two studies on lake sediments (Tables 14.5 and 14.6) show that extraction techniques identify organic matter as an important location of heavy metals. Extraction results are interpreted in terms of the lability of the metals, i.e. the ease with which they can be released from the sediment, especially with regard to bioavailability.

The extraction methodology and interpretation are not fully compatible with the theory of cation–humic interactions developed in this book. A key assumption in the use of extraction techniques is that the reagents are

Table 14.5. Sequential extraction of surficial sediments from the central part of Lake Vegoritis, Greece (Fytianos *et al.*, 1995). The results are expressed as percentages of total metal.

Fraction	Cr	Ni	Cu	Zn	Cd	Pb
1	1	5	8	3	54	10
2	2	6	4	6	18	12
3	61	65	83	62	14	40
4	36	24	6	30	14	39

Fraction 1 (acetic acid), metal on weak exchange sites and in carbonates
Fraction 2 (acidic NH_2OH), metal bound to easily and moderately reducible Fe and Mn oxides
Fraction 3 (acidic H_2O_2), metal bound to organic matter and present as sulphides
Fraction 4, detrital metal

indifferent to phases that are not removed by the extractant. But simple acidification, caused by the acetic acid and acid hydroxylamine steps in the sequential extraction scheme of Table 14.5, can, according to WHAM, bring about the desorption of metals from binding sites on humic substances, without extracting the organic matter. Table 14.7 shows simulated extraction data for two hypothetical sediments in which the metals are entirely bound to humic matter. In Case A, the sediment is assumed to consist only of organic matter with the cation-binding properties of average humic acid. In Case B, humic acid forms 10% of the sediment, the rest being inert with respect to metal binding. In both cases the 'field' pH is near to neutral, which means that the humic matter must possess bound base cations. The sediment in Case A has a high buffering capacity and so when treated with acetic acid (0.11 mol dm^{-3}) at a sediment concentration of 25 g dm^{-3} (a typical routine value) the final pH comes to 4.3, which causes little desorption of the heavy metals. Buffering also occurs when the residue from the acetic acid step is treated with 0.01 mol dm^{-3} hydroxylamine hydrochloride plus 0.01 mol dm^{-3} HNO_3, and the final pH here is 4.0, so again there is little desorption. The result is therefore that nearly all ($\geq 97\%$) of the heavy metals appear in the organic fraction. A quite different result is obtained if the sediment is a weaker pH buffer, as in Case B. Here the pH values in the extraction steps are 3.5 (acetic acid) and 2.1 (acid hydroxylamine), so that the metals are subject to much greater competition by protons, and substantial losses occur during the two extractions. In Case B, only Cu survives the

Table 14.6. Non-sequential chemical extraction of the surficial sediments of two lakes in Ontario, Canada (Young & Harvey, 1992). The results (means of three samples) are presented as percentages of the total non-detrital metal.

	Ni	Cu	Zn	Cd
Harp Lake				
ER	38	25	32	21
R	54	10	61	71
Org	8	65	7	8
Crosson Lake				
ER	14	9	25	7
R	43	5	58	70
Org	43	86	17	24

ER, easily reducible fraction, i.e. exchangeable metal + metal bound to Mn oxides
R, reducible fraction, i.e. metal bound to Fe oxides
Org, alkali-extractable metal, i.e. metal bound to organic matter

extractions, with 98% appearing in the organic fraction. Less than 15% of the other metals appear to be associated with the organic matter. Thus, even though in both cases the heavy metals were originally present bound to organic matter (they could be nowhere else), the simulated extractions provide different interpretations. The modelling suggests that, in general, the sequential extraction scheme will place metal desorbed from humic

Table 14.7. Metal extractions of sediments dominated by humic acid, simulated with WHAM/Model VI. See Section 14.7.2.

Case A: sediment consisting of 100% humic acid

	'Field' pH = 6.6	Acetic acid final pH = 4.3		Acid NH_2OH final pH = 4.0		Organic fraction
	Solid before extraction $\mu mol\,g^{-1}$	Solid after extraction $\mu mol\,g^{-1}$	% of total metal extracted	Solid after extraction $\mu mol\,g^{-1}$	% of total metal extracted	% of total metal
Na	338	55	84	13	12	4
Ca	1330	830	38	600	17	45
Ni	1	0.98	2	0.97	1	97
Cu	1	1	0	1	0	100
Zn	1	0.99	1	0.99	0	99
Cd	1	0.99	1	0.98	1	98
Pb	1	1	0	1	0	100

Case B: sediment consisting of 10% humic acid, 90% inert material

	'Field' pH = 6.6	Acetic acid final pH = 3.5		Acid NH_2OH final pH = 2.1		Organic fraction
	Solid before extraction $\mu mol\,g^{-1}$	Solid after extraction $\mu mol\,g^{-1}$	% of total metal extracted	Solid after extraction $\mu mol\,g^{-1}$	% of total metal extracted	% of total metal
Na	34	2	94	0	6	0
Ca	133	34	74	1	25	1
Ni	1	0.73	27	0.03	70	3
Cu	1	1	0	0.98	2	98
Zn	1	0.88	12	0.05	82	6
Cd	1	0.83	17	0.04	79	4
Pb	1	0.99	1	0.13	86	13

The dry solids concentration was $25\,g\,dm^{-3}$.
The two extractants employed are $0.11\,mol\,dm^{-3}$ acetic acid, and $0.01\,mol\,dm^{-3}$ $NH_2OH.HCl/0.01\,mol\,dm^{-3}$ HNO_3.
The pH values differ between the two cases because the buffering capacity in A is 10 times that in B.

matter into the easily exchangeable category, which is reasonable, and the easily reducible category, i.e. bound by Mn and Fe oxides, which is unreasonable. It is possible that the importance of Mn and Fe oxides in metal binding in sediments has been overestimated, and that of humic matter correspondingly underestimated.

14.7.3 Solid-solution partitioning of heavy metals in sediments

Knowledge about the pore water concentrations of heavy metals is needed to understand metal bioavailability, potential for remobilisation, and diagenesis within the sediment. The pool of sediment-bound metal can potentially exert the primary control on pore water concentrations, and the simplest approach to describing such control is by the use of K_D values (equation 14.4). A number of studies have been performed in which sediment samples have been equilibrated with added heavy metals and the solid-solution distributions measured.

Mahony *et al.* (1996) studied the binding of Cu, Cd and Pb added to lake sediments with organic matter contents in the range 1.8 to 23.8% (as organic C). They fitted the results to Langmuir isotherms at pH 6, 7 and 8. The results for the different sediments at a given pH could be condensed onto the same isotherm if binding was expressed in terms of sediment organic carbon content. Lin and Chen (1998) carried out experiments with bed sediments ranging in organic matter content from 1.4 to 63%, taken from rivers in Taiwan. The sediment samples were suspended in solutions of the heavy metals (Cr, Cu, Zn and Pb) under standardised conditions, and the amounts of metal taken up were measured. Metal uptake was quite strongly correlated with organic matter content, with r^2 ranging from 0.36 to 0.90 (mean 0.61).

Fu *et al.* (1992) studied proton and cadmium binding by three riverine sediments, and by humic acids extracted from those sediments. They used a multiple-site monoprotic model to describe proton binding, and in each case found very similar pK values for the sediment and the isolated humic acid. On average, 60% of the sediment content of proton-binding groups could be accounted for by the humic matter. The humic acid accounted almost exactly for the Cd binding properties of the sediment (Table 14.8). The results confirmed an earlier conclusion of Gardiner (1974), who deduced, from comparative experiments with radioactive Cd, the likely importance of humic acid in the sorption of the metal by river muds.

In each of the studies mentioned so far, heavy metal was added to the sediment samples. The results therefore refer mainly to conditions of high loading, and may not be fully representative of field conditions. Direct measurements of the solid–solution partitioning of Cd under field conditions were made by Tessier *et al.* (1993), in a detailed study of 49 sediments from 38 lakes in Ontario and Québec (Canada), including examples of lakes polluted by the emissions of metal smelters. The data set covered a wide range of conditions (Table 14.9). Tessier and his colleagues analysed their data using surface complexation concepts, and concluded that Cd binding sites on organic matter, and to a much lesser extent iron oxyhydroxides, controlled the partitioning of the metal. The data can also be analysed with WHAM, to see how well the partitioning can be explained in terms of the cation-binding properties of isolated humic matter. Figure 14.16 compares observed values of K_D (as defined by equation 14.4) with predictions of WHAM/Model VI. The model can be made to account quite well for the data, with the assumptions listed in the figure legend. A noteworthy point is the sensitivity of the calculated K_D

Table 14.8. Cadmium binding by riverine sediments, compared with binding by humic acid (HA) isolated from the sediments (Fu *et al.*, 1992).

Sample	Sediment HA ($g\,g^{-1}$)		R_{mean} ($g\,g^{-1}$)	$HA_{required}/HA_{measured}$
RS3	0.0116	pH 4.5	0.0107	0.922
		pH 6.5	0.0102	0.879
RS4	0.0215	pH 4.5	0.0201	0.935
		pH 6.5	0.0200	0.930
RS6	0.0207	pH 4.5	0.0197	0.952
		pH 6.5	0.0201	0.971

The ratio R is the amount of Cd bound per g sediment at a given free Cd concentration, divided by the amount of Cd bound per g of HA at the same free concentration. The final column compares the sediment content of HA required to explain the Cd binding properties with the measured value.

Table 14.9. Ranges of conditions in 49 lake sediments sampled by Tessier *et al.* (1993).

Variable	Unit	Range
pH	—	4.07–8.10
Dissolved [Na]	$\mu mol\,dm^{-3}$	20–1780
Dissolved [Ca]	$\mu mol\,dm^{-3}$	47–700
Alkalinity	$\mu mol\,dm^{-3}$	0–1430
[DOC]	$mg\,dm^{-3}$	4.1–14.6
Dissolved [Cd]	$nmol\,dm^{-3}$	0.018–17.5
Solid-phase Cd	$nmol\,g^{-1}$	1.34–108
Solid-phase organic matter	$mgC\,g^{-1}$	3.1–194

values to the choice of log K_{MA}. The default values (Table 10.5) do not give very good agreement with the observations. However, if binding by humic acid (in the solid phase of the sediment) is slightly strengthened, and that by fulvic acid (solution phase) is slightly weakened, then good agreement is obtained. Such an adjustment is justified by the sizes of the standard deviations in log K_{MA} shown in Table 10.5. The analysis overestimates K_D for the three acid lakes, but the values can be made to agree with the observations by assuming the presence of modest amounts of Al in the system. Aluminium competes quite strongly with Cd for both solution complexation and sorption, but more so for the solid phase, thereby lowering K_D relative to the Al-free case. Overall, it appears that binding by natural organic matter can account for the partitioning of Cd in these lake sediments, on the assumption that essentially all the organic matter is active in binding metals.

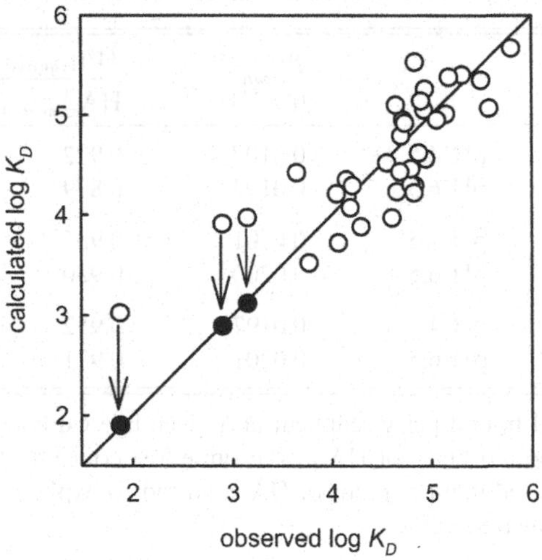

Figure 14.16. Observed and calculated K_D values ($cm^3 g^{-1}$), according to equation (14.4), for Cd sorption by surficial lake sediments. The observations are from Tessier *et al.* (1993), and the calculations were performed with WHAM/Model VI. The following assumptions and adjustments were made in applying the model:

(a) sedimentary organic matter is 100% humic acid (HA);
(b) the systems are in equilibrium with $Fe(OH)_3$, with $K_{so} = 10^{2.5}$;
(c) 65% of the dissolved organic matter is fulvic acid (FA), the rest is inert;
(d) log K_{MA} values for Cd were changed from the defaults of 1.3 (HA) and 1.6 (FA) to 1.4 and 1.5.

The calculated values shown as filled circles were obtained by assuming the sediment to contain Al, at contents of 169, 114 and 138 μmol g^{-1} in increasing order of log K_D.

14.7.4 Mineral sediment phases

The preceding sections provide strong evidence for the importance of organic matter in general, and humic substances in particular, in the binding of cationic metals by sediment solids. However, mineral solid-phase components may also contribute to, and sometimes dominate, the binding. In oxic sediments, as in soils, the most important are the oxides of Mn and Fe (see e.g. Tessier *et al.*, 1996). Aluminosilicates may also play a rôle. In anoxic sediments, sulphide is invariably present in significant concentrations and metals may then be controlled by the formation of solid phase metal sulphides (see e.g. Morse *et al.*, 1987; Ankley *et al.*, 1996).

15

Research needs

In this closing chapter, ways are suggested to advance knowledge about cation–humic interactions, and to apply that knowledge more widely. Laboratory measurements on isolated materials are discussed, then field measurements, and finally modelling at the ecosystem level.

15.1 Research needs for isolated humic materials

The material covered in Chapters 7–11 demonstrates that much work has been done on humic substances and their interactions with cations. Many cations have been studied, and a substantial body of knowledge has been built up. Furthermore, there have been many studies of the chemical structures and physico-chemical properties of humic substances. Of course, any new information on humic substances is potentially valuable in improving our understanding and ability to predict cation–humic interactions, but several key areas can be identified. Work with isolated materials under well-defined conditions is essential to understand the details of the reactions.

15.1.1 Isolation of humic matter

The understanding and prediction of cation binding by isolated humic substances is a step on the way to describing how cations interact with organic matter in the natural environment. We would be foolish to ignore the available information on isolated humic and fulvic acids, but the information is valuable only if it can be used to describe the real world. Therefore we have to take account of whether the isolated materials are (a) significantly unaltered from their natural states, apart from being freed of 'contaminating' entities, and (b) are sufficiently representative of natural organic matter as a whole.

With regard to point (a), it is uncomfortable to be dealing with materials obtained by such crude and violent isolation methods, involving pH values more extreme than the materials will normally experience in the natural environment (cf. Section 2.2). Although modelling work (Chapters 13 and 14) indicates that natural organic matter has similar properties to those of the isolated materials, calibration is always necessary in attempting to describe interactions in the field. From the point of view of determining cation binding, it would be helpful to have materials to work on that have been freed from interfering entities by milder means. Thus, concentrating dissolved organic matter in a natural water sample, combined with the removal or replacement of associated cations, would provide a sample whose cation-binding properties could be studied, probably providing a closer approximation to the natural situation than isolated fulvic acid from the same source. This approach would also address point (b), by dealing with all the organic matter in a sample, not just the fulvic and humic acids. Improved isolation strategies for the natural organic matter of soils and sediments are less easy to devise.

15.1.2 Measurement techniques

A characteristic feature of the available information on cation binding by humic matter is that there are large numbers of data for some metals, (Ca, Cu, Cd, Eu, Pb) but very few, or none, for others. The situation is largely explained by the fact that some metals are easier to measure than others – for Ca, Cu, Cd and Pb, ion-selective electrodes are available. Techniques applicable to many metals, and covering wide ranges of metal loadings and concentrations of unbound metal, are very much needed, and their development is a major challenge to experimentalists. Separation methodology probably offers the most hope. For humic isolates, or fractions, of sufficiently large molecular size, equilibrium dialysis is probably the most general method available. However, its use at low levels of metal depends upon the availability of a sufficiently sensitive analytical technique, except when radiotracers can be used. The need to work with materials of relatively large molecular size also applies to Diffusion Gradients in Thin films (DGT; Section 6.5.5), but DGT has the advantage that it can be used to determine low concentrations of unbound metal by accumulating the metal in the 'sink' resin. Separation methods such as dialysis and DGT have the advantage that a number of metals can be studied simultaneously, thereby making the acquisition of data more rapid, and providing greater insight into competition effects than techniques which simply follow the effect of an unmeasured cation on the binding of the cation that can be detected. In Donnan dialysis (Section 6.5.2), the separation is based on charge not size, and the technique holds promise, although again it requires highly sensitive analytical methods if binding data at low free metal concentrations are to be obtained.

15.1.3 Equilibrium binding

Based on the data summarised and discussed in the earlier chapters, a number of gaps in knowledge about the equilibrium interactions of cations with humic substances can be identified.

Missing metals Binding data are available for many cations of environmental concern, but there are few for Fe^{3+}, Ag^{+} and Hg^{2+}. Data for Fe^{3+} are urgently required to establish the importance or otherwise of the competitive effects of ferric iron (Section 13.4.5). Data for Ag and Hg binding at low concentrations are needed, given the high toxicities of the two metals, and their preference for sulphur ligands. All three metals present severe difficulties to the experimentalist.

Ranges of loadings and unbound concentrations As mentioned in the previous section, technique limitations mean that only for a few cations are there data covering wide ranges of loadings and unbound concentrations. Therefore much of the insight into site heterogeneity in humic matter has come from measurements on a small number of metals. There is a pressing need for information about the binding of a wider range of metals at the low-abundance, high-affinity sites.

Ranges of pH Most binding data refer to pH values of less than 7, principally because researchers have wanted to avoid the complications of metal hydrolysis reactions and carbonate complexation. However, many natural environments have pH values greater than 7, and extrapolation is currently necessary to estimate the extents of cation–humic interactions under such circumstances. Fortunately, the outstanding experiments of Benedetti and co-workers (Section 7.2) did cover wide ranges of both pH and free cation concentrations, for Ca, Cu and Cd, and the results can be simulated satisfactorily with comprehensive models (Chapter 10). Nonetheless, data over wider ranges of pH for other metals are desirable.

Ionic strength effects Most of the information about ionic strength effects on cation binding comes from proton data. Models calibrated with proton data can account quite well for the effects of ionic strength on metal cation binding (Section 10.5), but systematic studies for a range of metals are lacking.

Competition effects There are a number of published data sets describing experiments in which the binding of one metal has been measured, for different added amounts of another metal. The comprehensive models of Chapter 10 are able to account reasonably satisfactorily for the observations, on the assumption that the different metals share the same binding sites. However, the true situation is likely to be more complicated. We cannot even be certain whether metals with the same preferred coordination number share the same sites, and it seems unlikely that metals with different preferred coordination numbers do so. Therefore, more competition data are desirable, especially if experiments can be devised in which simultaneous measurements are made of the binding of more than one metal. Data on proton–metal competition could usefully be extended to metals other than those discussed in Section 10.8. Competition experiments involving pairs of metals with high affinities for the low-abundance sites could be especially revealing.

Mixed complexes There is very little information about the formation of mixed complexes involving a humic molecule, a metal cation, and another ligand. The simplest case is the binding of metal hydrolysis products. Models V and VI (Section 9.6) assume that the first hydrolysis product of a metal binds with the same intrinsic association constant as the parent cation, whereas the NICA and NICCA models permit only the parent cation to bind, but both modelling approaches can fit available data. The few data sets describing the formation of mixed complexes with low molecular weight ligands were summarised in Section 7.2.6.

Comparison of different humic samples Although cation-binding results obtained with different humic samples are broadly consistent with one another, derived parameter values exhibit appreciable variations, as shown, for example, by standard deviations in the values of log K_{MA} (Table 10.5). We do not know how representative are the samples of humic and fulvic acid used to obtain the currently available binding data. There is thus a need for the systematic study of the binding properties of samples obtained from different environments, and at different times. Studies of the hydrophilic acid fraction (cf. Section 2.2.1) could usefully be included.

Temperature effects There is an almost complete absence of information on the temperature dependence of metal binding by humic substances. Data would be useful to permit the effects of temperature to be taken into account in predictions of field behaviour. In addition, enthalpy and entropy values might provide insight into the natures of ligand groups.

15.1.4 Kinetics

There are relatively few data on kinetics, and so this is an area where there is much scope for gathering new information. Most of the presently available data refer to quite high metal loadings, whereas information about the rates of association and dissociation at the less-abundant, strong sites would be more relevant to field situations. However, the required measurements would be technically very demanding. Developments in kinetics would be helped by an improved theoretical framework (see below).

15.1.5 Modelling

The comprehensive models of Chapter 10 provide the currently most successful descriptions of equilibrium cation binding by humic matter. As more experimental data accrue, the models will be tested more fully, and improvements may be required. As far as parameterised modelling is concerned, Model VI and the NIC(C)A models represent differing approaches, the former being based on conventional binding sites and equilibrium constants, the latter taking a more distributional, empirical approach. Each has its advantages and disadvantages. Better versions of the two models are likely to involve modifications, rather than radical changes. As discussed in Chapter 11, the development of predictive models, with no or few parameters, may be possible, and ultimately would permit the utilisation of information about humic chemical structure, physical properties and heterogeneity to describe cation binding.

It should be recognised that the models are mathematical expressions of testable hypotheses, and therefore they can – and should – be used in experimental design. Exploratory calculations, conducted before experiments are done, can identify experimental conditions likely to provide clear-cut answers. Such calculations can also suggest topics for investigation, especially when model predictions are not intuitively obvious. An example is the prediction by both Model VI and the NIC(C)A model that competitive effects among metal cations are greatest at low ionic strength (Section 10.7).

With regard to kinetics, it would be helpful to develop a more fundamental approach, since most of the modelling done to date has been little more than curve fitting (Section 7.3). Use could be made of data already available to obtain initial descriptions and explore field consequences, thereby focusing effort to areas where kinetics are significant. A possible approach is to use knowledge about simple metal–ligand reactions, combined with an equilibrium model for humic matter, to describe kinetic effects. Models V and VI may be good candidates for the equilibrium model, since they deal in conventional chemical reactions, which could be described by conventional kinetic theory.

15.2 Research needs in field studies

The examples of Chapters 13 and 14 show that, in a number of cases, the chemical speciation of an environmental system can be explained largely by the cation-binding properties of humic substances, as described by the comprehensive models of Chapter 10, combined with the 'side reactions' of inorganic complexation. Examples were also given where speciation is dominated by other reactants such as specific ligands, metal oxides and aluminosilicates. In general, field speciation investigations should provide data that can be used not just to test the rôle of humic substances, but rather to determine the contributions of all the possible reactants. However, more elaborate models are then needed, dealing with dissolved, colloidal and particulate complexants, and probably also the interactions among them that lead to non-additive effects. Simple attempts to formulate such 'assemblage models' have been made (Lofts & Tipping, 1998; Radovanovic & Koelmans, 1998), but they are less advanced than the models for humic substances alone. Therefore, in the immediate future most progress with cation–humic interactions will be made by choosing field situations where humic matter is likely to be dominant in governing cation behaviour.

Comparison of field speciation data with the predictions of the comprehensive models can be regarded either as a model testing procedure or as an aid to interpreting the analytical results, for example to identify ligands other than humic materials in the natural system. Thus, it can be advantageous to perform model calculations in parallel with the analytical work.

15.2.1 Measurements on field samples

In principle, any field measurement, or measurement on a minimally altered field sample, that reveals something about the chemical speciation of the system, is useful for model testing. Perhaps the simplest determination is that of pH, which can be used when humic matter plays a major rôle in buffering, notably in acid waters draining bogs and peats, and in acid soils. In some cases, concentrations of free metal ions may be measurable, in others the combined concentrations of a defined collection of species of the metal. Comparison of the measured concentration(s) with those predicted by the model provides either a test of the model or a means to calibrate it.

A number of methods for determining speciation in water samples transferred to the laboratory were mentioned in Chapter 13, and to these can be added the Donnan dialysis method (Sections 6.5.2 and 15.1.2). A recent book edited by Buffle & Horvai (2000) provides a comprehensive review of techniques available for *in situ* monitoring of aquatic systems, including methods that provide speciation information. They include continuous-flow analysis (CFA) methods, *in situ* permeable liquid membranes, and diffusional techniques, notably DGT. Given this array of methods, the next decade will undoubtedly see a substantial growth in information about speciation in natural waters, including complexation by humic matter. Sediments can also be studied by *in situ* techniques, but soils are less readily investigated, because of their greater spatial heterogeneity and varying water contents.

Apart from direct measurements on nearly unaltered samples, experimental manipulation of natural samples is possible. This can include titrations with acid and base, or metals, and the addition of competing metals. The manipulation approach is a good general way to test model predictions, being a compromise between relatively inefficient measurements on single samples, and the gathering of large numbers of data with isolated materials.

15.2.2 Auxiliary data

In the interpretation of analytical data on speciation, and the application of the comprehensive models, information about the chemical composition of the water, soil or sediment under investigation is essential. Thus there is a need to analyse for major solutes, pH, aluminium, iron, trace metals, organic matter, and possibly mineral components (oxides, aluminosilicates etc). Of these, organic matter and iron deserve particular attention. Where possible, specific ligands should also be determined.

As discussed in Chapters 13 and 14, organic matter presents problems for modelling because there are no generally applicable ways to convert simple measurements of concentration (DOC, soil or sediment C content etc.) to the 'active' concentrations of humic matter required as inputs to the comprehensive models. At present, *ad hoc* calibration is used, yielding results that are acceptable in that (a) the 'active' concentrations are not too far from measured values, and (b) once calibrated, the models give predictions that show the same trends as the measured data. However, it would be more satisfactory to have analytical methods, or modelling approaches, that yield the input concentrations more directly. As noted in Chapter 14, for the solid phases of soils and sediments, the problem is exacerbated by the high concentrations of the components and their mutual interactions, as well as by the imprecision of extraction methodology. But the problem should be more tractable for dissolved organic matter. The challenge is to account for variations in the cation-binding 'activity' of DOM among field locations, and temporally at a given location. Systematic studies are required, coupled with standardised, preferably simple, characterisation techniques for the organic matter.

A second issue to be resolved is the chemistry of ferric iron, in relation to the likely occupancy of strong binding sites by Fe^{3+} and its hydrolysis products. In some of the model applications of Chapter 13, the simplifying assumption was made that the activity of Fe^{3+} was controlled by equilibrium with iron (hydr)oxide, and this demonstrated the probable importance of the competition effects of Fe(III). To make progress, measurements of the concentrations of free iron species in natural samples are much needed. This is a formidable task because the free aqueous concentrations are low; the total solubility of $Fe(OH)_3$ with $K_{so} = 10^{2.5}$ is only *c.* 10^{-11} mol dm^{-3} at pH 7.

15.3 Cation–humic interactions in catchments

In Chapter 12, the ways in which cation binding might affect, and be affected by, other functional physico-chemical properties of humic substances were reviewed. In Chapters 13 and 14, mention was made of the importance of cation–humic interactions in moderating the bioavailability and toxicity of metals, with illustrations based largely on interactions occurring in small, defined, samples of the natural environment. At this molecular level, knowledge about cation–humic interactions can be used to interpret and predict the natural environment, as long as the concentrations of the reactants in the sample are available as inputs to model calculations. For example, in applying the Biotic Ligand Model (Section 13.5.2), which uses WHAM to take account of the toxicity-lowering effect of metal binding by dissolved organic matter, the pH and solute concentrations would be provided by measurement or by prior calculations or assumptions. The application then simply requires the speciation calculation to be performed, the key output being the degree to which the biotic ligand is loaded with the toxic metal of interest.

However, we can also use knowledge about cation–humic interactions, as encapsulated in a speciation model, to understand and predict processes at 'higher' levels, in more dynamic circumstances. For example, we might attempt to explain how humic matter in soils governs the composition of drainage waters, as was done in simple terms with the acidification model of Section 14.5.4. The most useful level at which to operate is probably that of the catchment, i.e. the smallest natural unit of the landscape that combines linked terrestrial and aquatic ecosystems (Hornung & Reynolds, 1995). To apply knowledge about humic substances at the catchment scale, it must be combined with knowledge about other materials and processes, also represented by appropriate models.

A number of catchment models have been developed over the last two decades, to interpret and predict the effects of acid deposition. Well-known examples include the Birkenes Model (Christophersen *et al.*, 1982), MAGIC (Cosby *et al.*, 1985), and the SAFE model (Sverdrup *et al.*, 1995). The models account for chemical and biological processes postulated to determine the acid–base status of soil and the chemical composition of drainage water. Each model recognises the sorption and release of cations by soil solids to be a key process, the reactions being described in terms of 'cation exchange', i.e. by postulating a fixed-charge solid phase exchanger. Aluminium chemistry is described in terms of the dissolution of $Al(OH)_3$, and this reaction also exerts a significant control on H^+ concentrations. As discussed in Chapter 14, the chemistry of soils rich in organic matter (which occur in the vast majority of locations impacted by acid deposition) is more properly represented by reactions involving solid-phase humic substances. Therefore, the models based on the cation exchange/$Al(OH)_3$ picture are incorrect, although they can be calibrated to reproduce field observations. In principle, better results, and more reliable predictions, should be obtained from models that include a description of the cation-binding properties of humic substances. Such a model has been formulated, by embedding the WHAM model, the humic sorption model of Section 14.5.1, and several other sub-models describing processes not involving humic matter, into a simple hydrological framework (Tipping,

1996a). The resulting 'CHemistry of the Uplands Model' (CHUM) has achieved reasonable simulation of major-solute stream water chemistry on a daily time-step (Fig. 15.1). Radionuclide behaviour has also been investigated (Tipping, 1996b). The Soil Acidification/Soil Development (SASD) model, formulated by Santore *et al.* (1995), operates at much longer time scales, and simulates the formation of a podzol, starting with a bare landscape following glaciation. It uses the triprotic model of Section 9.3.5, coupled with a simple adsorption model.

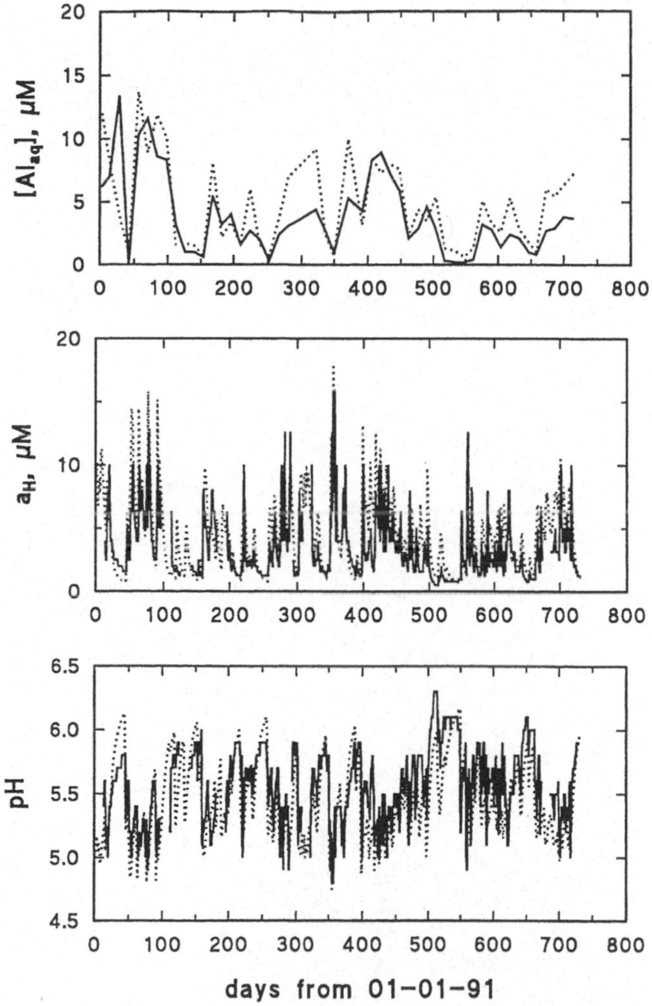

Figure 15.1. Simulation of stream water chemistry using the CHemistry of the Uplands Model (CHUM). The full lines are observations, the dotted lines model simulations. [From Tipping, E. (1996), CHUM: a hydrochemical model for upland catchments, *J. Hydrol.* **174**, 304–330, with permission from Elsevier Science.]

CHUM and the SASD model are first attempts to transfer molecular-level information about cation–humic interactions to the catchment scale. They could be enhanced by including more detailed descriptions of processes in streams, lakes and sediments. Links to carbon-cycling models, including descriptions of humic matter formation, would complete the picture. By such an integrative approach, a large body of detailed information about humic substances, accumulated by many researchers over many decades, could be brought more fully to bear in explaining and predicting the natural environment.

References

Aiken, G.R. (1985). Isolation and concentration techniques for aquatic humic substances. In *Humic Substances in Soil, Sediment, and Water*, ed. G.R. Aiken, D.M. McKnight, R.L. Wershaw & P. MacCarthy, pp. 363–385. New York: Wiley.

Aiken, G.R. & Malcolm, R.L. (1987). Molecular weight of aquatic fulvic acids by vapor pressure osmometry. *Geochim. Cosmochim. Acta* **51**, 2177–2184.

Aiken, G.R., McKnight, D.M., Wershaw, R.L. & MacCarthy, P. (1985a). *Humic Substances in Soil, Sediment, and Water*. New York: Wiley.

Aiken, G.R., McKnight, D.M., Wershaw, R.L. & MacCarthy, P. (1985b). An introduction to humic substances in soil, sediment, and water. In *Humic Substances in Soil, Sediment, and Water*, ed. G.R. Aiken, D.M. McKnight, R.L. Wershaw & P. MacCarthy, pp. 1–9. New York: Wiley.

Alberts, J.J. & Filip, Z. (1998). Metal binding in estuarine humic and fulvic acids: FTIR analysis of humic–metal complexes. *Environ. Technol.* **19**, 923–931.

Al-Farawati, R. & van den Berg, C.M.G. (1999). Metal-sulfide complexation in seawater. *Marine Chem.* **63**, 331–352.

Allen, H.E. & Hansen, D.J. (1996). The importance of trace metal speciation to water quality criteria. *Water Environ. Res.* **68**, 42–54.

Allison, J.D., Brown, D.S. & Novo-Gradac, K.J. (1991). *MINTEQA2/PRODEFA2, A Geochemical Assessment Model for Environmental Systems* (EPA/600/3-91/021). Athens, GA: Environmental Protection Agency.

Alloway, B.J. (1995). *Heavy Metals in Soils*. London: Blackie Academic Press.

Altmann, R.S. & Buffle, J. (1988). The use of differential equilibrium functions for interpretation of metal binding in complex ligand systems: its relation to site occupation and site affinity distributions. *Geochim. Cosmochim. Acta* **52**, 1505–1519.

Ankley, G.T., Di Toro, D.M., Hansen, D.J. & Berry, W.J. (1996). Technical basis and proposal for deriving sediment quality criteria for metals. *Environ. Toxicol. Chem.* **15**, 2056–2066.

Avdeef, A., Zabronsky, J. & Stutling, H.H. (1983). Calibration of copper ion selective electrode response to pCu 19. *Anal. Chem.* **55**, 289–304.

Avena, M.J., Vermeer, A.W.P. & Koopal, L.K. (1999). Volume and structure of humic acids studied by viscometry pH and electrolyte concentration effects. *Coll. Surf. A* **151**, 213–224.

Averett, R.C., Leenheer, J.A., McKnight, D.M. & Thorn, K.A. (1989). *Humic Substances in the Suwannee River, Georgia: Interactions, Properties and Proposed Structures*. USGS Open File Report 87–557. Denver, CO: US Geological Survey.

Bache, B.W. (1974). Soluble aluminium and calcium–aluminium exchange in relation to the pH of dilute calcium chloride suspensions of acid soils. *J. Soil Sci.* **25**, 320–332.

Backes, C.A. & Tipping, E. (1987a). An evaluation of the use of cation-exchange resin for the determination of organically-complexed Al in natural acid waters. *Int. J. Environ. Anal. Chem.* **30**, 135–143.

Backes, C.A. & Tipping, E. (1987b). Aluminium complexation by an aquatic humic fraction under acidic conditions. *Wat. Res.* **21**, 211–216.

Barak, P. & Chen, Y. (1992). Equivalent radii of humic macromolecules from acid–base titration. *Soil Sci.* **154**, 184–195.

Baron, J. & Bricker, O.P. (1987). Hydrologic and chemical flux in Loch Vale watershed, Rocky Mountain National Park. In *Chemical Quality of Water and the Hydrologic Cycle*, ed. R.C. Averett & D.M. McKnight, pp. 141–155. Chelsea, MI: Lewis Publishers.

Bartschat, B.M., Cabaniss, S.E. & Morel, F.M.M. (1992). An oligoelectrolyte model for cation binding by humic substances. *Environ. Sci. Technol.* **26**, 284–294.

Bates, R.G. (1973). *Determination of pH, Theory and Practice*, 2nd edn. New York: Wiley.

Batley, G.E. & Gardner, D. (1978). A study of copper, lead and cadmium speciation in some estuarine and coastal marine waters. *Est. Coast. Mar. Sci.* **7**, 59–70.

Beck, A.J. & Jones, K.C. (1993). Natural organic substances and contaminant behaviour: progress, conflicts and uncertainty. In *Organic Substances in Soil and Water: Natural Constituents and Their Influence on Contaminant Behaviour*, ed. A.J. Beck, K.C. Jones, M.H.B. Hayes & U. Mingelgrin, pp. 184–194. Cambridge: Royal Society of Chemistry.

Beck, A.J., Jones, K.C., Hayes, M.H.B. & Mingelgrin, U. (1993). *Organic Substances in Soil and Water: Natural Constituents and Their Influence on Contaminant Behaviour*. Cambridge: Royal Society of Chemistry.

Beckett, R., Jue, Z. & Giddings, C.J. (1987). Determination of molecular weight distributions of fulvic and humic acids using flow field-flow fractionation. *Environ. Sci. Technol.* **21**, 289–295.

Benedetti, M.F., Milne, C.J., Kinniburgh, D.G., van Riemsdijk, W.H. & Koopal, L.K. (1995). Metal-ion binding to humic substances – application of the nonideal competitive adsorption model. *Environ. Sci. Technol.* **29**, 446–457.

Benedetti, M.F., van Riemsdijk, W.H., Koopal, L.K., Kinniburgh, D.G., Gooddy, D.C. & Milne, C.J. (1996). Metal ion binding by natural organic matter: from the model to the field. *Geochim. Cosmochim. Acta* **60**, 2503–2513.

Berdén, M. & Berggren, D. (1990). Gel filtration chromatography of humic substances in soil solutions using HPLC-determination of the molecular weight distribution. *J. Soil Sci.* **41**, 61–72.

van den Berg, C.G.M. (1984). Determination of the complexing capacity and conditional stability constants of complexes of copper(II) with natural organic ligands in seawater by cathodic stripping voltammetry of copper-catechol complex ions. *Mar. Chem.* **15**, 1–18.

Berggren, D. (1992). Speciation of copper in soil solutions from podzols and cambisols of S. Sweden. *Wat. Air Soil Pollut.* **62**, 111–123.

Berggren, D. & Mulder, J. (1995). The role of organic matter in controlling aluminium solubility in acidic mineral soil horizons. *Geochim. Cosmochim. Acta* **59**, 4167–4180.

Bertha, E.L. & Choppin, G.R. (1978). Interaction of humic and fulvic acids with Eu(III) and Am(III). *J. Inorg. Nucl. Chem.* **40**, 655–658.

Bidoglio, G., Omenetto, N. & Roubach, P. (1991). Kinetic studies of lanthanide interactions with humic substances by time resolved laser induced fluorescence.

Radiochim. Acta **52/53**, 57–63.

Birkett, J.W., Jones, M.N., Bryan, N.D. & Livens, F.R. (1997). Computer modelling of partial specific volumes of humic substances. *Eur. J. Soil Sci.* **48**, 131–137.

Blake, R.E. & Walter, L.M. (1999). Kinetics of feldspar and quartz dissolution at 70–80 degrees C and near-neutral pH: Effects of organic acids and NaCl. *Geochim. Cosmochim. Acta.* **63**, 2043–2059.

Blaser, P. & Sposito, G. (1987). Spectrofluorometric investigation of trace metal complexation by an aqueous chestnut leaf litter extract. *Soil Sci. Soc. Am. J.* **51**, 612–619.

Bloom, P.R. & Leenheer, J.A. (1989). Vibrational, electronic, and high-energy spectroscopic methods for characterizing humic substances. In *Humic Substances II. In Search of Structure*, ed. M.H.B. Hayes, P. MacCarthy, R.L. Malcolm & R.S. Swift, pp. 409–446. Chichester: Wiley.

Bloom, P.R. & McBride, M.B. (1979). Metal ion binding and exchange with hydrogen ions in acid-washed peat. *Soil Sci. Soc. Am. J.* **43**, 687–692.

Bolt, G.H. (1982). *Soil Chemistry B. Physico-Chemical Models*, 2nd edn. Amsterdam: Elsevier.

Bonifazi, M., Pant, B.C. & Langford, C.H. (1996). Kinetic study of the speciation of copper(II) bound to humic acid. *Environ. Technol.* **17**, 885–890.

Bonn, B.A. & Fish, W. (1993). Measurement of electrostatic and site-specific associations of alkali metal cations with humic acid. *J. Soil Sci.* **44**, 335–345.

Borggaard, O.K. (1974). Experimental conditions concerning potentiometric titration of humic acid. *J. Soil Sci.* **25**, 189–195.

Botelho, C.M.S., Boaventura, R.A.R. & Gonçalves, M.L.S.S. (1994). Interactions of lead(II) with natural river water: part I. Soluble organics. *Sci. Tot. Environ.* **149**, 69–81.

Bowles, E.C., Antweiler, R.C. & MacCarthy, P. (1989). Acid–base titration and hydrolysis of fulvic acid from the Suwannee River. In *Humic Substances in the Suwannee River, Georgia: Interactions, Properties and Proposed Structures*, USGS Open File Report 87–557, ed. R.C. Averett, J.A. Leenheer, D.M. McKnight & K.A. Thorn, pp. 205–229. Denver, CO: US Geological Survey.

Boyd, S.A., Sommers, L.E., Nelson, D.W. & West, D.X. (1981). The mechanism of copper(II) binding by humic acid: an electron spin resonance study of a copper(II)–humic acid complex and some adducts with nitrogen donors. *Soil Sci. Soc. Am. J.* **45**, 745–749.

Boyd, S.A., Sommers, L.E., Nelson, D.W. & West, D.X. (1983). Copper(II) binding by humic acid extracted from sewage sludge: an electron spin resonance study. *Soil Sci. Soc. Am. J.* **47**, 43–46.

Breslow, E. (1973). Metal–protein complexes. In *Inorganic Biochemistry Vol. 1*, ed. G.L. Eichhorn, pp. 227–249. Amsterdam: Elsevier.

Bresnahan, W.T., Grant, C.L. & Weber, J.H. (1978). Stability constants for the complexation of copper(II) ions with water and soil fulvic acids measured by an ion selective electrode. *Anal. Chem.* **50**, 1675–1679.

Brezonik, P.L. (1994). *Chemical Kinetics and Process Dynamics in Aquatic Systems.* Boca Raton, FL: Lewis Publishers.

Brown, G.K., Cabaniss, S.E., MacCarthy, P. & Leenheer, J.A. (1999). Cu(II) binding by a pH-fractionated fulvic acid. *Anal. Chim. Acta.* **402**, 183–193.

Browne, B.A. (1995). Toward a new theory of podzolisation. In *Carbon Forms and Functions in Forest Soils*, ed. W.W. McFee & J.M. Kelly, pp. 253–273. Madison, WI: Soil Science Society of America.

Browne, B.A. & Driscoll, C.T. (1993). pH-dependent binding of aluminium by a fulvic acid. *Environ. Sci. Technol.* **27**, 915–922.

Browne, B.A., Driscoll, C.T. & McColl, J.G. (1990). Aluminium speciation using morin: II.

Principles and procedures. *J. Environ. Qual.* **19**, 73–82.

Bruland, K.W. (1992). Complexation of cadmium by natural organic ligands in the central North Pacific. *Limnol. Oceanogr.* **37**, 1008–1017.

Brümmer, G.W., Gerth, J. & Tiller, K.G. (1988). Reaction kinetics of the adsorption and desorption of nickel, zinc and cadmium by goethite. I. Adsorption and diffusion of metals. *J. Soil Sci.* **39**, 37–52.

Bryan, N.D., Robinson, V.J., Livens, F.R., Hesketh, N., Jones, M.N. & Lead, J.R. (1997). Metal–humic interactions: a random structural modelling approach. *Geochim. Cosmochim. Acta* **61**, 805–820.

Bryan, N.D., Hesketh, N., Livens, F.R., Tipping, E. & Jones, M.N. (1998). Metal ion – humic substance interaction. A thermodynamic study. *J. Chem. Soc. Farad. Trans.* **94**, 95–100.

Bryan, N.D., Jones, D.M., Appleton, M., Livens, F.R., Jones, M.N., Warwick, P., King, S. & Hall, A. (2000). A physicochemical model of metal–humate interactions. *Phys. Chem. Chem. Phys.* **2**, 1291–1300.

Buckau, G., Kim, J.I., Klenze, R., Rhee, D.S. & Wimmer, H. (1992). A comparative spectroscopic study of the fulvate complexation of trivalent transuranium ions. *Radiochim. Acta* **57**, 105–111.

Buffle, J. (1984). Natural organic matter and metal–organic interactions in aquatic systems. In *Metal Ions in Biological Systems. Vol* 18. *Circulation of Metals in the Environment*, ed. H. Sigel, pp. 165–221. New York: Marcel Dekker.

Buffle, J. (1988). *Complexation Reactions in Aquatic Systems: An Analytical Approach.* Chichester: Ellis-Horwood.

Buffle, J. & Altmann, R.S. (1987). Interpretation of metal complexation by heterogeneous complexants. In *Aquatic Surface Chemistry: Chemical Processes at the Particle–Water Interface*, ed. W. Stumm, pp. 351–383. New York: Wiley.

Buffle, J. & Horvai, G. (2000). *In Situ Monitoring of Aquatic Systems; Chemical Analysis and Speciation.* IUPAC Series in Analytical and Physical Chemistry of Environmental Systems, Vol. 6. Chichester: Wiley.

Buffle, J., Greter, F.-L. & Haerdi, W. (1977). Measurement of complexation properties of humic and fulvic acids in natural waters with lead and copper ion-selective electrodes. *Anal. Chim. Acta* **49**, 216–222.

Buffle, J., Deladoey, P., Greter, F.L. & Haerdi, W. (1980). Study of the complex formation of copper(II) by humic and fulvic substances. *Anal. Chim. Acta* **116**, 255–274.

Buffle, J., Altmann, R.S. & Filella, M. (1990). Effect of physico-chemical heterogeneity of natural complexants. Part II. Buffering action and role of their background sites. *Anal. Chim. Acta* **232**, 225–237.

Buffle, J., Wilkinson, K.J., Stoll, S., Filella, M. & Zhang, J. (1998). A generalised description of aquatic colloidal interactions: the three-colloid component approach. *Environ. Sci. Technol.* **32**, 2887–2899.

Burba, P. (1994). Labile/inert metal species in aquatic humic substances – an ion exchange study. *Fres. J. Analyt. Chem.* **348**, 301–311.

Burns, I.G., Hayes, M.H.B. & Stacey, M. (1973). Some physico-chemical interactions of paraquat with soil organic materials and organic compounds. II. Adsorption and desorption equilibria in aqueous suspensions. *Weed Res.* **13**, 79–90.

Byler, D.M., Gerasimowicz, W.V., Susi, H. & Schnitzer, M. (1987). FT-IR spectra of soil constituents: fulvic acid and fulvic acid complex with ferric ions. *Appl. Spectrosc.* **41**, 1428–1430.

Cabaniss, S.E. (1990). pH and ionic strength effects on nickel–fulvic acid dissociation kinetics. *Environ. Sci. Technol.* **24**, 583–588.

Cabaniss, S.E. (1991). Carboxylic acid content of a fulvic acid determined by potentiometry and aqueous Fourier transform infrared spectrometry. *Anal. Chim. Acta* **255**, 23–30.
Cabaniss, S.E. (1992). Synchronous fluorescence spectra of metal–fulvic acid complexes. *Environ. Sci. Technol.* **26**, 1133–1139.
Cabaniss, S.E. (1997). Propagation of uncertainty in aqueous equilibrium calculations: non-Gaussian output distributions. *Anal. Chem.* **69**, 3658–3664.
Cabaniss, S.E. & Shuman, M.S. (1988a). Copper binding by dissolved organic matter; I. Suwannee River fulvic acid equilibria. *Geochim. Cosmochim. Acta* **52**, 185–193.
Cabaniss, S.E. & Shuman, M.S. (1988b). Copper binding by dissolved organic matter; II. Variation in type and source of organic matter. *Geochim. Cosmochim. Acta* **52**, 195–200.
Cacheris, W.P. & Choppin, G.R. (1987). Dissociation kinetics of thorium–humate complex. *Radiochim. Acta* **42**, 185–190.
Cai, W-J., Wang, Y. & Hodson, R.E. (1998). Acid–base properties of dissolved organic matter in the estuarine waters of Georgia, USA. *Geochim. Cosmochim. Acta* **62**, 473–483.
Cai, Y., Jaffé, R. & Jones, R.D. (1999). Interactions between dissolved organic carbon and mercury species in surface waters of the Florida Everglades. *Appl. Geochem.* **14**, 395–407.
Cameron, A.J. & Liss, P.S. (1984). The stabilization of 'dissolved' iron in freshwaters. *Wat. Res.* **18**, 179–185.
Cameron, R.S. & Posner, A.M. (1974). Molecular weight distribution of humic acid from density gradient, ultracentrifugation profiles corrected for diffusion. *Trans. 10th Int. Congr. Soil Sci., Moscow* **2**, 325–331.
Cameron, R.S., Thornton, B.K., Swift, R.S. & Posner, A.M. (1972). Molecular weight and shape of humic acid from sedimentation and diffusion measurements on fractionated extracts. *J. Soil Sci.* **23**, 394–408.
Campbell, P.G.C. (1995). Interactions between trace metals and aquatic organisms: a critique of the free-ion activity model. In *Metal Speciation and Bioavailability in Aquatic Systems*, ed. A. Tessier & D.R. Turner, pp. 45–102. Chichester: Wiley.
Cantrell, K.J., Serkiz, S.M. & Perdue, E.M. (1990). Evaluation of acid neutralizing capacity data for solutions containing natural organic acids. *Geochim. Cosmochim. Acta* **54**, 1247–1254.
Carlsen, L., Bo, P. & Larsen, G. (1984). Radionuclide–humic acid interactions studied by dialysis. In *Geochemical Behaviour of Radioactive Waste*, ed. G.S. Barney, J.D. Navratil & W.W. Schulz, pp. 167–178. Washington, DC: American Chemical Society.
Catts, J.G. & Langmuir, D. (1986). Adsorption of Cu, Pb and Zn by δMnO_2: applicability of the site binding–surface complexation model. *Appl. Geochem.* **1**, 255–264.
Chen, Y. & Schnitzer, M. (1989). Sizes and shapes of humic substances by electron microscopy. In *Humic Substances II. In Search of Structure*, ed. M.H.B. Hayes, P. MacCarthy, R.L. Malcolm & R.S. Swift, pp. 621–638. Chichester: Wiley.
Cheshire, M.V., Berrow, B.A., Goodman, B.A. & Mundie, C.M. (1977). Metal distribution and nature of some Cu, Mn and V complexes in humic and fulvic acid fractions of soil organic matter. *Geochim. Cosmochim. Acta* **41**, 1131–1138.
Chin, P-K.F. & Mills, G.L. (1991). Kinetics and mechanisms of kaolinite dissolution: effects of organic ligands. *Chem. Geol.* **90**, 307–317.
Chin, Y.P., Aiken, G. & O'Loughlin, E. (1994). Molecular weight, polydispersity, and spectroscopic properties of aquatic humic substances. *Environ. Sci. Technol.* **28**, 1853–1858.

Chiou, C.T., Lee, J.-F. & Boyd, S.A. (1990). The surface area of soil organic matter. *Environ. Sci. Technol.* **24**, 1164–1166.

Chiou, C.T., Lee, J.-F. & Boyd, S.A. (1992). Reply to the Comment on "The Surface Area of Soil Organic Matter" (Pennell & Rao, 1992). *Environ. Sci. Technol.* **26**, 404–406.

Choppin, G.R. & Clark, S.B. (1991). The kinetic interactions of metal ions with humic acids. *Marine Chem.* **36**, 27–38.

Choppin, G.R. & Labonne-Wall, N. (1997). Comparison of two models for metal–humic interactions. *J. Radioanal. Nucl. Chem.* **221**, 67–71.

Christensen, J.B., Tipping, E., Kinniburgh, D.G., Grøn, C. & Christensen, T.H. (1998). Proton binding by groundwater fulvic acids of different age, origins, and structure modeled with the Model V and NICA–Donnan Model. *Environ. Sci. Technol.* **32**, 3346–3355.

Christophersen, N., Seip, H.M., & Wright, R.F. (1982). A model for streamwater chemistry at Birkenes, Norway. *Wat. Resour. Res.* **18**, 977–996.

Chung, K.H., Rhee, S.W., Shin, H.S. & Moon, C.H. (1996). Probe of cadmium(II) binding on soil fulvic acid investigated by ^{113}Cd NMR spectroscopy. *Can. J. Chem.* **74**, 1360–1365.

Clapp, C.E., Emerson, W.W. & Olness, A.E. (1989). Sizes and shapes of humic substances by vicosity measurements. In *Humic Substances II. In Search of Structure*, ed. M.H.B. Hayes, P. MacCarthy, R.L. Malcolm & R.S. Swift, pp. 497–514. Chichester: Wiley.

Clapp, C.E., Hayes, M.H.B. & Swift, R.S. (1993). Isolation, fractionation, functionalities, and concepts of structures of soil organic macromolecules. In *Organic Substances in Soil and Water: Natural Constituents and Their Influence on Contaminant Behaviour*, ed. A.J. Beck, K.C. Jones, M.H.B. Hayes & U. Mingelgrin, pp. 31–69. Cambridge: Royal Society of Chemistry.

Clark, S.B. & Choppin, G.R. (1990). Kinetics of rare earth metal binding to aquatic humic acids. In *Chemical Modeling of Aqueous Systems II*, ACS Symp. Ser. No. 416, ed. D.C. Melchior & R.L. Bassett, pp. 519–525. Washington, DC: American Chemical Society.

Clarke, N., Danielsson, L.-G. & Sparén, A. (1992). The determination of quickly reacting aluminium in natural waters by kinetic discrimination in a flow system. *Int. J. Environ. Anal. Chem.* **48**, 77–100.

Clarke, N., Danielsson, L.-G. & Sparén, A. (1995). Studies of aluminium complexation to humic and fulvic acids using a method for the determination of quickly reacting aluminium. *Wat. Air Soil Pollut.* **84**, 103–116.

Cleven, R.F.M.J., de Jong, H.G. & van Leeuven, H.P. (1986). Pulse polarography of metal/polyelectrolyte complexes and operation of the mean diffusion coefficient. *J. Electroanal. Chem.* **202**, 57–68.

Colston, B.J., van Elteren, J.T., Kolar, Z.I. & de Goeij, J.J.M. (1997). Kinetics in a Eu(III)–humic acid system by isotopic exchange with $^{152m}Eu^{3+}$ and size exclusion chromatography – a feasibility study. *Radiochim. Acta* **78**, 111–115.

Conte, P. & Piccolo, A. (1999). Conformational arrangement of dissolved humic substances. Influence of solution composition on association of humic molecules. *Environ. Sci. Technol.* **33**, 1682–1690.

Cooper, W.J., Zika, R.G., Petasne, R.G. & Fischer, A.M. (1989). Sunlight-induced photochemistry of humic substances in natural waters: major reactive species. In *Aquatic Humic Substances: Influence on Fate and Treatment of Pollutants*, Adv. Chem. Ser. No. 219, ed. I.H. Suffet & P. MacCarthy, pp. 333–362. Washington, DC: American Chemical Society.

Cosby, B.J., Hornberger, G.M., Galloway, J.N. & Wright, R.F. (1985). Modeling the effects of acid deposition: assessment of a lumped parameter model of soil water and streamwater chemistry. *Wat. Resour. Res.* **21**, 51–63.

Creighton, T.E. (1992). *Protein Folding.* New York: Freeman.

Czerwinski, K.R., Buckau, G., Scherbaum, F. & Kim, J.I. (1994). Complexation of the uranyl ion with aquatic humic acid. *Radiochim. Acta* **65**, 111–119.

Dai, K.H., David, M.B., Vance, G.F., McLaughlin, J.W. & Fernandez, I.J. (1996). Acidity characteristics of soluble organic substances in spruce-fir forest floor leachates. *Soil Sci.* **161**, 694–704.

Dalang, F., Buffle, J. & Haerdi, W. (1984). Study of the influence of fulvic substances on the adsorption of copper (II) ions at the kaolinite surface. *Environ. Sci. Technol.* **18**, 135–141.

David, M.B. & Vance, G.F. (1991). Chemical character and origin of organic acids in streams and seepage lakes of central Maine. *Biogeochem.* **12**, 17–41.

Davis, H. & Mott, C.J.B. (1981). Titrations of fulvic acid fractions I: Interactions influencing the dissociation/reprotonation equilibria. *J. Soil Sci.* **32**, 379–391.

Davis, J.A. (1982). Adsorption of natural dissolved organic matter at the oxide/water interface. *Geochim. Cosmochim. Acta* **46**, 2381–2393.

Davis, J.A. (1984). Complexation of trace metals by adsorbed natural organic matter. *Geochim. Cosmochim. Acta* **48**, 679–691.

Davis, J.A. & Leckie, J.O. (1980). Surface ionization and complexation at the oxide/water interface. III Adsorption of anions. *J. Coll. Int. Sci.* **74**, 32–43.

Davis, J.A., Coston, J.A., Kent, D.B. & Fuller, C.C. (1998). Application of the surface complexation concept to complex mineral assemblages. *Environ. Sci. Technol.* **32**, 2820–2828.

Davison, W. & Seed, G. (1983). The kinetics of the oxidation of ferrous iron in synthetic and natural waters. *Geochim. Cosmochim. Acta* **47**, 67–79.

Davison, W. & Zhang, H. (1994). *In situ* speciation measurements of trace components in natural waters using thin-film gels. *Nature* **367**, 546–548.

Dearlove, J.P.L., Longworth, G., Ivanovich, M., Kim, J.I., Delakowitz, B. & Zeh, P. (1991). A study of groundwater-colloids and their geochemical interactions with natural radionuclides in Gorleben aquifer systems. *Radiochim. Acta* **52/53**, 83–89.

DeBorger, R. & DeBacker, H. (1968). Détermination des poids moleculaire moyen des acides fulviques per cryoscopie en milieu aqueux. *C.R. Acad. Sci. Ser. D.* **266**, 2052–2055.

Deczky, K. & Langford, C.H. (1978). Application of water nuclear magnetic resonance relaxation times to study of metal complexes of the soluble soil organic fraction fulvic acid. *Can. J. Chem.* **56**, 1947–1951.

De Haan, H. (1992). Impacts of environmental changes on the biogeochemistry of aquatic humic substances. *Hydrobiologia* **229**, 59–71.

Del Rio, J.C. & Hatcher, P.G. (1996). Structural characterization of humic substances using thermochemolysis with tetramethylammonium hydroxide. In *Humic and Fulvic Acids: Isolation, Structure and Environmental Role*, ACS Symposium Series No. 651, ed. J.S. Gaffney, N.A. Marley & S.B. Clark, pp.78–95. Washington, DC: American Chemical Society.

Dempsey, B.A. (1981). *The protonation, calcium complexation and adsorption of a fractionated aquatic fulvic acid.* Ph.D. Thesis, Johns Hopkins University.

Dempsey, B.A. & O'Melia, C.R. (1983). Proton and calcium complexation of four fulvic acids fractions. In *Aquatic and Terrestrial Humic Materials*, ed. R.F. Christman & E.T. Gjessing, pp. 239–273. Ann Arbor, MI: Ann Arbor Science.

Dierckx, A., Maes, A. & Vancluysen, J. (1994). Mixed complex formation of Eu^{3+} with humic acid and a competing ligand. *Radiochim. Acta* **66/67**, 149–156.
van Dijk, H. (1959). Zur Kenntnis der Basenbindung von Huminsäuren. *Z. Pflanzenernähr. Düng. Bodenk.* **84**, 150–155.
Dill, K.A., Bromberg, S., Yue, K., Fiebig, K.M., Yee, D.P., Thomas, P.D. & Chan, H.S. (1995). Principles of protein folding – a perspective from simple exact models. *Protein Sci.* **4**, 561–602.
Dobbs, J.C., Susetyo, W., Knight, F.E., Castles, M.A. & Carreira, L.A. (1989). A novel approach to metal–humic complexation studies by lanthanide ion probe spectroscopy. *Int. J. Environ. Anal. Chem.* **37**, 1–17.
Donat, J.R., Lao, K.A. & Bruland, K.W. (1994). Speciation of dissolved copper and nickel in South San Francisco Bay: a multi-method approach. *Anal. Chim. Acta.* **284**, 547–571.
Driscoll, C.T. (1984). A procedure for the fractionation of aqueous aluminium in dilute acidic waters. *Int. J. Environ. Anal. Chem.* **16**, 267–283.
Driscoll, C.T., Baker, J.P., Bisogni, J.J. & Schofield, C.L. (1980). Effect of aluminium speciation on fish in dilute acidified waters. *Nature* **284**, 161–164.
Driscoll, C.T., Lehtinen, M.D. & Sullivan, T.J. (1994). Modeling the acid–base chemistry of organic solutes in Adirondack, New York, lakes. *Wat. Resour. Res.* **30**, 297–306.
Driscoll, C.T., Blette, V., Yan, C., Schofield, C.L., Munson, R. & Holsapple, J. (1995). The role of dissolved organic carbon in the chemistry and bioavailability of mercury in remote Adirondack lakes. *Wat. Air Soil Pollut.* **80**, 499–508.
Dubach, P. & Mehta, N.C. (1963). The chemistry of soil humic substances. *Soils Fert.* **26**, 293–300.
Duinker, J.C. & Kramer, C.J.M. (1977). An experimental study on the speciation of dissolved zinc, cadmium, lead and copper in River Rhine and North Sea water, by differential pulsed anodic stripping voltammetry. *Marine Chem.* **5**, 207–228.
Düker, A., Ledin, A., Karlsson, S. & Allard, B. (1995). Adsorption of zinc on colloidal (hydr)oxides of Si, Al and Fe in the presence of fulvic acid. *Appl. Geochem.* **10**, 197–205.
Dwane, G.C. & Tipping, E. (1998). Testing a humic speciation model by titration of copper-amended natural waters. *Environ. Int.* **24**, 609–616.
Dzombak, D.A. & Morel, F.M.M. (1990). *Surface Complexation Modeling; Hydrous Ferric Oxide.* New York: Wiley.
Dzombak, D.A., Fish, W. & Morel, F.M.M. (1986). Metal–humate interactions. 1. Discrete ligand and continuous distribution models. *Environ. Sci. Technol.* **20**, 669–675.
Eisma, D. (1993). *Suspended Particulate Matter in the Aquatic Environment.* Berlin: Springer.
Engebretson, R.R. & von Wandruszka, R. (1998). Kinetic aspects of cation-enhanced aggregation in aqueous humic acids. *Environ. Sci. Technol.* **32**, 488–493.
Ephraim, J. (1986) *Studies of the protonation and metal ion complexation equilibria of natural organic acids: fulvic acids.* Ph.D. Thesis, State University of New York at Buffalo.
Ephraim J. & Marinsky J.A. (1986). A unified physicochemical description of the protonation and metal ion complexation equilibria of natural organic acids (humic and fulvic acids). 3. Influence of polyelectrolyte properties and functional heterogeneity on the copper ion binding equilibria in an Armadale horizons Bh fulvic acid sample. *Environ. Sci. Technol.* **20**, 367–376.
Ephraim, J., Alegret, S., Mathuthu, A., Bicking, M., Malcolm, R.L. & Marinsky, J.A. (1986). A unified physicochemical description of the protonation and metal ion complexation equilibria of natural organic acids (humic and fulvic acids). 2.

Influence of polyelectrolyte properties and functional group heterogeneity on the protonation equilibria of fulvic acid. *Environ. Sci. Technol.* **20**, 354–366.
Ephraim, J.H., Borén, H., Pettersson, C., Arsenie, I. & Allard, B. (1989a). A novel description of the acid–base properties of an aquatic fulvic acid. *Environ. Sci. Technol.* **23**, 356–362.
Ephraim, J.H., Marinsky, J.A. & Cramer, S.J. (1989b). Complex-forming properties of natural organic acids. Fulvic acid complexes with cobalt, zinc and europium. *Talanta* **36**, 437–443.
Ephraim, J.H., Pettersson, C., Nordén, M. & Allard, B. (1995). Potentiometric titrations of humic substances: do ionic strength effects depend on molecular weight? *Environ. Sci. Technol.* **29**, 622–628.
Ephraim, J.H., Pettersson, C. & Allard, B. (1996). Correlations between acidity and molecular size distributions of an aquatic fulvic acid. *Environ. Int.* **22**, 475–483.
Esteves da Silva, J.C.G. & Machado, A.A.S.C. (1996). Interaction of fulvic acids with Al(III) studied by self-modeling curve resolution of second-derivative fluorescence spectra. *Appl. Spectr.* **50**, 436–443.
Esteves da Silva, J.C.G., Machado, A.A.S.C. & Garcia, T.M.O. (1995). Beryllium(II) as a probe for study of the interactions of metals and fulvic acids by synchronous fluorescence spectroscopy. *Appl. Spectr.* **49**, 1500–1506.
Eugster, H.P. & Hardie, L.A. (1978). Saline lakes. In *Lakes – Chemistry, Geology, Physics*, ed. A. Lerman, pp. 237–293. Berlin: Springer.
Fairhurst, A.J., Warwick, P. & Richardson, S. (1995). The influence of humic acid on the adsorption of europium onto inorganic colloids as a function of pH. *Coll. Surf. A* **99**, 187–199.
Filius, J.D., Lumsdon, D.G., Meeussen, J.C.L., Hiemstra, T. & van Riemsdijk, W.H. (2000). Adsorption of fulvic acid on goethite. *Geochim. Cosmochim. Acta* **64**, 51–60.
Fiol, S., López, A., Ramos, J.M. & Arce, F. (1999). Study of the acid–base properties of three fulvic acids extracted from different horizons of a soil. *Anal. Chim. Acta* **385**, 443–449.
Fish, W. & Morel, F.M.M. (1985). Propagation of error in fulvic acid titration data: a comparison of three analytical methods. *Can. J. Chem.* **63**, 1185–1193.
Fish, W., Dzombak, D.A. & Morel, F.M.M. (1986). Metal–humate interactions. 2. Application and comparison of models. *Environ. Sci. Technol.* **20**, 676–683.
Fitch, A. & Helmke, P.A. (1989). Donnan equilibrium/graphite furnace atomic absorption estimates of soil extract complexation capacities. *Anal. Chem.* **61**, 1295–1298.
Fitch, A., Stevenson, F.J. & Chen, Y. (1986). Complexation of Cu(II) with a soil humic acid: response characteristics of the Cu(II) ion-selective electrode and ligand concentration effects. *Org. Geochem.* **9**, 109–116.
Florence, T.M. & Batley, G.E. (1980). Chemical speciation in natural waters. *CRC Crit. Rev. Anal. Chem.* **9**, 219–296.
Fraústo da Silva, J.J.R. & Williams, R.J.P. (1997). *The Biological Chemistry of the Elements.* Oxford: Clarendon.
Freeze, R.A. & Cherry, J.A. (1979). *Groundwater*. Englewood Cliffs, NJ: Prentice-Hall.
Frenkel, A.I., Korshin, G.V. & Ankudinov, A.L. (2000). XANES study of Cu^{2+}-binding sites in aquatic humic substances. *Environ. Sci. Technol.* **34**, 2138–2142.
Fu, G., Allen, H.E. & Cao, Y. (1992). The importance of humic acids to proton and cadmium binding in sediments. *Environ. Toxicol. Chem.* **11**, 1363–1372.
Fukushima, M., Tanaka, S., Nakamura, H. & Ito, S. (1996). Acid–base characterization of molecular weight fractionated humic acid. *Talanta* **43**, 383–390.
Fuoss, R.M. (1958). Ionic association. III. The equilibrium between ion pairs and free ions.

J. Am. Chem. Soc. **80**, 5059–5061.

Furch, K. (1984). Water chemistry of the Amazon basin: the distribution of chemical elements among freshwaters. In *The Amazon*, ed. H. Sioli, pp. 167–199. Dordrecht: Junk.

Fytianos, K., Bovolenta, S. & Muntau, H. (1995). Assessment of metal mobility from sediment of Lake Vegoritis. *J. Environ. Sci. Health* **A30**, 1169–1190.

Gaffney, J.S., Marley, N.A. & Clark, S.B. (1996). *Humic and Fulvic Acids. Isolation, Structure, and Environmental Role.* ACS Symposium Series No. 651. Washington, DC: American Chemical Society.

Gamble, D.S. (1973). Na^+ and K^+ binding by fulvic acid. *Can. J. Chem.* **51**, 3217–3222.

Gamble, D.S., Schnitzer, M. & Skinner, D.S. (1977). Mn^{II}–fulvic acid complexing equilibrium measurements by electron spin resonance spectrometry. *Can. J. Soil Sci.* **57**, 47–53.

Gamble, D.S., Underdown, A.W. & Langford, C.H. (1980). Copper(II) titration of fulvic acid ligand sites with theoretical, potentiometric, and spectrophotometric analysis. *Anal. Chem.* **52**, 1901–1908.

Gardiner, J. (1974). The chemistry of cadmium in natural water – II. The adsorption of cadmium on river muds and naturally occurring solids. *Wat. Res.* **8**, 157–164.

Gerke, J. (1994). Aluminium complexation by humic substances and aluminium species in the soil solution. *Geoderma* **63**, 165–175.

Ghosh, K. & Schnitzer, M. (1980). Macromolecular structures of humic substances. *Soil Sci.* **129**, 266–276.

Gillam, A.H. & Riley, J.P. (1981). Correction of osmometric number-average molecular weights of humic substances for dissociation. *Chem. Geol.* **33**, 355–366.

Ginstrup, O. (1973). Experimental and computation methods for studying multicomponent equilibria – II. An automated system for precision emf-titrations. *Chem. Instrum.* **4**, 141–155.

Gjessing, E.T. (1976). *Physical and Chemical Characteristics of Aquatic Humus.* Ann Arbor, MI: Ann Arbor Science.

Gjessing, E.T. (1990). Mechanisms and effects of reactions of organic acids with anions. In *Organic Acids in Aquatic Ecosystems*, ed. E.M. Perdue & E.T. Gjessing, pp. 179–187. Chichester: Wiley.

Glaus, M.A., Hummel, W. & van Loon, L.R. (1995a). Equilibrium dialysis – ligand exchange: adaptation of the method for determination of conditional stability constants of radionuclide–fulvic acid complexes. *Anal. Chim. Acta* **303**, 321–331.

Glaus, M.A., Hummel, W. & van Loon, L.R. (1995b). Stability of mixed-ligand complexes of metal ions with humic substances and low molecular weight ligands. *Environ. Sci. Technol.* **29**, 2150–2153.

Goodman, B.A. & Cheshire, M.V. (1973). Electron paramagnetic resonance evidence that copper is complexed in humic acid by the nitrogen of porphyrin groups. *Nature* **244**, 158–159.

Goodman, B.A. & Cheshire, M.V. (1975). The bonding of vanadium in complexes with humic acid: an electron paramagnetic resonance study. *Geochim. Cosmochim. Acta* **39**, 1711–1713.

Gooddy, D.C., Shand, P., Kinniburgh, D.G. & van Riemsdijk, W.H. (1995). Field-based partition coefficients for trace elements in soil solutions. *Eur. J. Soil Sci.* **46**, 265–285.

Green, S.A., Morel, F.M.M. & Blough, N.V. (1992). Investigation of the electrostatic properties of humic substances by fluorescence quenching. *Environ. Sci. Technol.* **26**, 294–302.

Gregor, J.E. & Powell, H.K.J. (1987). Effects of extraction procedures on fulvic acid properties. *Sci. Tot. Environ.* **62**, 3–12.

Grzyb, K.R. (1995). NOAEM (natural organic anion equilibrium model): a data analysis algorithm for estimating functional properties of dissolved organic matter in aqueous environments: Part I. Ionic component speciation and metal association. *Org. Geochem.* **23**, 379–390.

Gu, B., Schmitt, J., Chen, Z., Liang, L. & McCarthy, J.F. (1994). Adsorption and desorption of natural organic matter on iron oxide: mechanisms and models. *Environ. Sci. Technol.* **28**, 38–46.

Guetzloff, T.F. & Rice, J.A. (1994). Does humic acid form a micelle? *Sci. Tot. Environ.* **152**, 31–35.

Gustafsson, J.P., van Hees, P., Starr, M., Karltun, E. & Lundström, U. (2000). Partitioning of base cations and sulphate between solid and dissolved phases in three podzolised forest soils. *Geoderma* **94**, 311–333.

Hancock, R.D. & Marsicano, F. (1980). Parametric correlation of formation constants in aqueous solution. 2. Ligands with large donor atoms. *Inorg. Chem.* **19**, 2709–2714.

Hansen, E.H. & Schnitzer, M. (1969). Molecular weight measurement of polycarboxylic acids in water by vapor pressure osmometry. *Anal. Chim. Acta* **46**, 247–254.

Hargrove, W.L. & Thomas, G.W. (1981). Effect of organic matter on exchangeable aluminium and plant growth in acid soils. In *Chemistry in the Soil Environment*, ed. R.H. Dowdy, J.A. Ryan, V.V. Volk & D.E. Baker, pp. 151–166. Madison, WI: American Society of Agronomy, Soil Science Society of America.

Hargrove, W.L. & Thomas, G.W. (1982). Titration properties of Al–organic matter. *Soil Sci.* **134**, 216–225.

Harrison, R.M. & de Mora, S.J. (1996). *Introductory Chemistry for the Environmental Sciences.* Cambridge: Cambridge University Press.

Hart, B. (1981). Trace metal complexing capacity of natural waters; a review. *Environ. Technol. Lett.* **2**, 95–110.

Harvey, G.R. & Boran, D.A. (1985). Geochemistry of humic substances in seawater. In *Humic Substances in Soil, Sediment, and Water*, ed. G.R. Aiken, D.M. McKnight, R.L. Wershaw & P. MacCarthy, pp. 233–247. New York: Wiley.

Hatcher, P.G., Breger, I.A., Maciel, G.E. & Szeverenyi, N.M. (1985). Geochemistry of humin. In *Humic Substances in Soil, Sediment, and Water*, ed. G.R. Aiken, D.M. McKnight, R.L. Wershaw & P. MacCarthy, pp. 275–302. New York: Wiley.

Hayano, S., Shinozuka, N. & Shinji, O. (1983). Solution properties of marine humic acid. I. Viscometric and pH-dependent behaviours. *J. Jap. Oil Chem. Soc.* (*Yukagaku*) **32**, 10–17.

Hayase, K. & Tsubota, H. (1983). Sedimentary humic acid and fulvic acid as surface active substances. *Geochim. Cosmochim. Acta* **47**, 947–952.

Hayes, M.H.B. (1985). Extraction of humic substances from soil. In *Humic Substances in Soil, Sediment, and Water*, ed. G.R. Aiken, D.M. McKnight, R.L. Wershaw & P. MacCarthy, pp. 329–362. New York: Wiley.

Hayes, M.H.B. & Himes, F.L. (1986). Nature and properties of humus–mineral complexes. In *Interactions of Soil Minerals with Natural Organics amd Microbes*, ed. P.M. Huang & M. Schnitzer, pp. 103–158. Madison, WI: Soil Science Society of America.

Hayes, M.H.B., MacCarthy, P., Malcolm, R.L. & Swift, R.S. (1989). *Humic Substances II. In Search of Structure.* Chichester: Wiley.

Hedges, J.I., Keil, R.G. & Benner, R. (1997). What happens to terrestrial organic matter in the ocean? *Org. Geochem.* **27**, 195–212.

van Hees, P.A.W., Lundström, U.S. & Giesler, R. (2000a). Low molecular weight organic acids and their Al-complexes in soil solution – composition, distribution and seasonal variation in three podzolized soils. *Geoderma* **94**, 173–200.

van Hees, P.A.W., Lundström, U.S., Starr, M. & Giesler, R. (2000b). Factors influencing aluminium concentrations in soil solution from podzols. *Geoderma* **94**, 289–310.

Hemond, H.F. (1990). Acid neutralizing capacity, alkalinity, and acid–base status of natural waters containing organic acids. *Environ. Sci. Technol.* **24**, 1486–1489.

Herbert, B.E. & Bertsch, P.M. (1995). Characterization of dissolved and colloidal organic matter in soil solution: a review. In *Carbon Forms and Functions in Forest Soils*, ed. W.W. McFee & J.M. Kelly, pp. 63–88. Madison, WI: Soil Science Society of America.

Hiemenz, P.C. (1977). *Principles of Surface and Colloid Chemistry*. New York: Marcel Dekker.

Higgo, J.J.W., Kinniburgh, D.G., Smith, B. & Tipping, E. (1993). Complexation of Co^{2+}, Ni^{2+}, $UO_2{}^{2+}$ and Ca^{2+} by humic substances in groundwaters. *Radiochim. Acta* **61**, 91–103.

Hintelmann, H., Welbourn, P.M. & Evans, R.D. (1995). Binding of methylmercury compounds by humic and fulvic acids. *Wat. Air Soil Pollut.* **80**, 1031–1034.

Hoch, A.R., Reddy, M.M. & Aiken, G.R. (2000). Calcite crystal growth inhibition by humic substances with emphasis on hydrophobic acids from the Florida Everglades. *Geochim. Cosmochim. Acta* **64**, 61–72.

Hoigné, J., Faust, B.C., Haag, W.R., Scully Jr., F.E. & Zepp, R.G. (1989). Aquatic humic substances as sources and sinks of photochemically produced transient reactants. In *Aquatic Humic Substances: Influence on Fate and Treatment of Pollutants*, Adv. Chem. Ser. No. 219, ed. I.H. Suffet & P. MacCarthy, pp. 363–381. Washington, DC: American Chemical Society.

Hollis, L., Muench, L. & Playle, R.C. (1997). Influence of dissolved organic matter on copper binding, and calcium on cadmium binding, by gills of rainbow trout. *J. Fish Biol.* **50**, 703–720.

Holm, P.E., Christensen, T.H., Tjell, J.C. & McGrath, S.P. (1995a). Speciation of cadmium and zinc with application to soil solutions. *J. Environ. Qual.* **24**, 183–190.

Holm, P.E., Andersen, S. & Christensen, T.H. (1995b). Speciation of dissolved cadmium: interpretation of dialysis, ion exchange and computer (GEOCHEM) methods. *Wat. Res.* **29**, 803–809.

Hongve, D. (1990). Shortcomings of Gran titration procedures for determination of alkalinity and weak acids in humic water. *Wat. Res.* **24**, 1305–1308.

van den Hoop, M.A.G.T., & van Leeuwen, H.P. (1997). Influence of molar mass distribution on the complexation of heavy metals by humic material. *Coll. Surf. A* **120**, 235–242.

van den Hoop, M.A.G.T., van Leeuwen, H.P. & Cleven, R.F.M.J. (1990). Study of the polyelectrolyte properties of humic acids by conductimetric titration. *Anal. Chim. Acta* **232**, 141–148.

Hornung, M. & Reynolds, B. (1995). The effects of natural and anthropogenic environmental changes on ecosystem processes at the catchment scale. *Trends Ecol. Evolut.* **10**, 443–449.

Huang, P.M. & Violante, A. (1986). Influence of organic acids on crystallization and surface properties of precipitation products of aluminium. In *Interactions of Soil Minerals with Natural Organics and Microbes*, ed. P.M. Huang & M. Schnitzer, pp. 159–221. Madison, WI: Soil Science Society of America.

Huang, W.L. & Keller, W.D. (1970). Dissolution of rock-forming silicate minerals in organic acids: simulated first-stage weathering of fresh mineral surfaces. *Am. Mineral.* **55**, 2076–2094.

Huc, A.Y. (1988). Sedimentology of organic matter. In *Humic Substances and Their Role in the Environment*, ed. F.H. Frimmel & R.F. Christman, pp. 215–243. Chichester: Wiley.

Hudson, R.J.M. (1998). Which aqueous species control the rates of trace metal uptake by biota? Observations and predictions of non-equilibrium effects. *Sci. Tot. Environ.* **219**, 95–115.

Hummel, J.P. & Dreyer, W.J. (1962). Measurement of protein-binding phenomena by gel filtration. *Biochim. Biophys. Acta* **63**, 530–532.

Ingri, N, Kakolowicz, W., Sillén, L.G. & Warnqvist, B. (1967). HALTAFALL, a general program for calculating the composition of equilibrium mixtures. *Talanta* **14**, 1261–1286.

Ishiwatari, R. (1985). Geochemistry of humic substances in lake sediments. In *Humic Substances in Soil, Sediment, and Water*, ed. G.R. Aiken, D.M. McKnight, R.L. Wershaw & P. MacCarthy, pp. 147–180. New York: Wiley.

Ives, K.J. (1978). *The Scientific Basis of Flocculation.* Alphen aan den Rijn: Sijthoff & Noordhoff.

Iyer, V.N. & Sarin, R. (1992). Chemical speciation and bioavailability of lead and cadmium in an aquatic system polluted by sewage discharges. *Chem. Spec. Bioavail.* **4**, 135–142.

James, R.O. & Healy, T.W. (1972). Adsorption of hydrolyzable metal ions at the oxide–water interface. III. A thermodynamic model of adsorption. *J. Coll. Int. Sci.* **40**, 65–81.

Janssen, R.P.T., Peijnenburg, W.J.G.M., Posthuma, L. & van den Hoop, M.A.G.T. (1997). Equilibrium partitioning of heavy metals in Dutch field soils. 1. Relationship between metal partition coefficients and soil characteristics. *Environ. Toxicol. Chem.* **16**, 2470–2478.

Jarvis, R.A., Bendelow, V.C., Bradley, R.I., Carroll, D.M., Furness, R.R., Kilgour, I.N.L. & King, S.J. (1984). *Soils and Their Use in Northern England.* Harpenden: Soil Survey of England and Wales.

Jekel, M.R. (1986). The stabilization of dispersed mineral particles by adsorption of humic substances. *Wat. Res.* **20**, 1543–1554.

Jenkinson, D.S. (1981). The fate of plant and animal residues in soil. In *The Chemistry of Soil Processes*, ed. D.J. Greenland & M.H.B. Hayes, pp. 505–561. Chichester: Wiley.

Jensen, D.L., Ledin, A. & Christensen, T.H. (1999). Speciation of heavy metals in landfill-leachate polluted groundwater. *Wat. Res.* **33**, 2642–2650.

Johnson, N.M., Driscoll, C.T., Eaton, J.S., Likens, G.E. & McDowell, W.H. (1981). 'Acid rain', dissolved aluminium and chemical weathering at the Hubbard Brook Experimental Forest, New Hampshire. *Geochim. Cosmochim. Acta* **45**, 1421–1437.

Jones, M.N. & Bryan, N.D. (1998). Colloidal properties of humic substances. *Adv. Coll. Int. Sci.* **78**, 1–48.

Jones, M.N., Birkett, J.W., Wilkinson, A.E., Hesketh, N., Livens, F.R., Bryan, N.D., Lead, J.R., Hamilton-Taylor, J. & Tipping, E. (1995). Experimental-determination of partial specific volumes of humic substances in aqueous-solutions. *Anal. Chim. Acta* **314**, 149–159.

Jones, R.I. (1992). The influence of humic substances on lacustrine planktonic food chains. *Hydrobiologia* **229**, 73–91.

Kaiser, K. & Zech, W. (1999). Release of natural organic matter sorbed to oxides and a subsoil. *Soil Sci. Soc. Am. J.* **63**, 1157–1166.

Keizer, M.G. & van Riemsdijk, W.H. (1994). *ECOSAT: Equilibrium Calculation of Speciation and Transport.* Wageningen: Wageningen Agricultural University.

Kemp, A.L.W. & Johnston, L.M. (1979). Diagenesis of organic matter in the sediments of

Lakes Ontario, Erie and Huron. *J. Great Lakes Res.* **5**, 1–10.

Kersten, M. & Förstner, U. (1995). Speciation of trace metals in sediments and combustion waste. In *Chemical Speciation in the Environment*, ed. A.M. Ure & C.M. Davidson, pp. 234–275. London: Blackie.

Kim, J.I. & Czerwinski, K.R. (1996). Complexation of metal ions with humic acid: metal ion charge neutralization model. *Radiochim. Acta* **73**, 5–10.

Kim, J.I., Rhee, D.S. & Buckau, G. (1991a). Complexation of Am(III) with humic acids of different origin. *Radiochim. Acta* **52/53**, 49–55.

Kim, J.I., Wimmer, H. & Klenze, R. (1991b). A study of curium(III) humate complexation by time resolved laser fluorescence spectroscopy (TRLFS). *Radiochim. Acta* **54**, 35–41.

Kinniburgh, D.G., Milne, C.J. & Venema, P. (1995). Design and construction of a PC-based automatic titrator. *Soil Sci. Soc. Am. J.* **59**, 417–422.

Kinniburgh, D.G., Milne, C.J., Benedetti, M.F., Pinheiro, J.P., Filius, J., Koopal, L. & van Riemsdijk, W.H. (1996). Metal ion binding by humic acid: application of the NICA–Donnan model. *Environ. Sci. Technol.* **30**, 1687–1698.

Kinniburgh, D.G., van Riemsdijk, W.H., Koopal, L.K., Borkovec, M., Benedetti, M.F. & Avena, M.J. (1999). Ion binding to natural organic matter: competition, heterogeneity, stoichiometry and thermodynamic consistency. *Coll. Surf. A* **151**, 147–166.

Kodama, H. & Schnitzer, M. (1977). Effect of fulvic acid on the crystallization of Fe(III)-oxides. *Geoderma* **19**, 279–291.

Köhler, S., Laudon, H., Wilander, A. & Bishop, K. (2000). Estimating organic acid dissociation in natural surface waters using total alkalinity and TOC. *Wat. Res.* **34**, 1425–1434.

Kolka, R.K., Grigal, D.F., Verry, E.S. & Nater, E.A. (1999). Mercury and organic carbon relationships in streams draining forested upland/peatland watersheds. *J. Environ. Qual.* **28**, 766–775.

Kononova, M.M. (1961). *Soil Organic Matter. Its Nature, Its Role in Soil Formation and in Soil Fertility*. Oxford: Pergamon.

Koopal, L.K., van Riemsdijk, W.H., de Wit, J.C.M. & Benedetti, M.F. (1994). Analytical isotherm equations for multicomponent adsorption to heterogeneous surfaces. *J. Coll. Int. Sci.* **166**, 51–60.

Kortelainen, P. (1993). *Contribution of Organic Acids to the Acidity of Finnish Lakes.* Publication 13 of the Water and Environment Research Institute, National Board of Waters and the Environment, Finland. Helsinki, Finland.

Kretzschmar, R., Borkovec, M., Grolimund, D. & Elimelech, M. (1999). Mobile subsurface colloids and their role in contaminant transport. *Adv. Agron.* **66**, 121–193.

Krug, E.C. & Frink, C.R. (1983). Acid rain on acid soil: a new perspective. *Science* **221**, 520–525.

Kubicki, J.D., Blake, G.A. & Apitz, S.E. (1996). Molecular orbital models of aqueous aluminium-acetate complexes. *Geochim. Cosmochim. Acta* **60**, 4897–4911.

Kummert, R. & Stumm, W. (1980). The surface complexation of organic acids on hydrous γ-Al_2O_3. *J. Coll. Int. Sci.* **75**, 375–385.

Kupsch, H., Franke, K., Degering, D., Tröger, W. & Butz, T. (1996). Speciation of aquatic heavy metals in humic acids by $^{111m}Cd/^{199m}Hg$ TDPAC. *Radiochim. Acta* **73**, 145–147.

Lakatos, B., Tibai, T. & Meisel, J. (1977). EPR spectra of humic acids and their metal complexes. *Geoderma* **19**, 319–338.

Lakshman, S., Mills, R., Fang, F., Patterson, H. & Cronan, C. (1996). Use of fluorescence polarization to probe the structure and aluminium complexation of three molecular weight fractions of a soil fulvic acid. *Anal. Chim. Acta* **321**, 113–119.

Langmuir, D. (1997). *Aqueous Environmental Geochemistry*. Englewood Cliffs, NJ: Prentice Hall.

Lavigne, J.E., Langford, C.H. & Mak, M.S. (1987). Kinetic study of speciation of nickel(II) bound to a fulvic acid. *Anal. Chem.* **59**, 2616–2620.

Law, I.A., Tuck, J.J., Graham, C.L. & Hayes, M.H.B. (1997). An automated procedure for studying sorption by humic substances using a combined continuous-flow and flow-injection analysis process. In *Humic Substances in Soils, Peats and Waters*, ed. M.H.B. Hayes & W.S. Wilson, pp. 189–198. Cambridge: Royal Society of Chemistry.

LaZerte, B.D. (1984). Forms of aqueous aluminium in acidified catchments of central Ontario: a methodological analysis. *Can. J. Fish. Aquat. Sci.* **41**, 766–776.

Lead, J.R., Hamilton-Taylor, J., Hesketh, N., Jones, M.N., Wilkinson, A.E. & Tipping, E. (1994). A comparative study of proton and alkaline earth binding by humic substances. *Anal. Chim. Acta* **294**, 319–327.

Lead, J.R., Hamilton-Taylor, J., Peters, A., Reiner, S. & Tipping, E. (1998). Europium binding by fulvic acids. *Anal. Chim. Acta.* **369**, 171–180.

Lead, J.R., Wilkinson, K.J., Starchev, K., Canonica, S. & Buffle, J. (2000). Determination of diffusion coefficients of humic substances by fluorescence correlation spectroscopy: role of solution conditions. *Environ. Sci. Technol.* **34**, 1365–1369.

Ledin, A., Karlsson, S., Düker, A. & Allard, B. (1994). The adsorption of europium to colloidal iron oxyhydroxides and quartz – the impact of pH and an aquatic fulvic acid. *Radiochim. Acta* **66/67** 213–220.

Lee, M.H., Choi, S.Y., Chung, K.H. & Moon, H. (1993). Complexation of cadmium(II) with humic acids: effects of pH and humic origin. *Bull. Korean Chem. Soc.* **14**, 726–732.

Lee, S.-Z., Allen, H.E., Huang, C.P., Sparks, D.L., Sanders, P.F. & Peijnenburg, W.J.G.M. (1996). Predicting soil-water partition coefficients for cadmium. *Environ. Sci. Technol.* **30**, 3418–3424.

Leenheer, J.A. (1981). Comprehensive approach to preparative isolation and fractionation of dissolved organic carbon from natural waters and wastewaters. *Environ. Sci. Technol.* **15**, 578–587.

Leenheer, J.A., McKnight, D.M., Thurman, E.M. & MacCarthy, P. (1989). Structural components and proposed structural models of fulvic acid from the Suwannee River. In *Humic Substances in the Suwannee River, Georgia: Interactions, Properties and Proposed Structures*, USGS Open File Report 87-557, ed. R.C. Averett, J.A. Leenheer, D.M. McKnight & K.A. Thorn, pp. 331–359. Denver, CO: US Geological Survey.

Leenheer, J.A., Wershaw, R.L. & Reddy, M.M. (1995a). Strong-acid, carboxyl-group structures in fulvic acid from the Suwannee River, Georgia. 1. Minor structures. *Environ. Sci. Technol.* **29**, 393–398.

Leenheer, J.A., Wershaw, R.L. & Reddy, M.M. (1995b). Strong-acid, carboxyl-group structures in fulvic acid from the Suwannee River, Georgia. 2. Major structures. *Environ. Sci. Technol.* **29**, 399–405.

Leenheer, J.A., Brown, G.K., MacCarthy, P. & Cabaniss, S.E. (1998). Models of metal binding structures in fulvic acid from the Suwannee River, Georgia. *Environ. Sci. Technol.* **32**, 2410–2416.

Leppard, G.G., Buffle, J. & Baudat, R. (1986). A description of the aggregation properties of aquatic pedogenic fulvic acids. Combining physico-chemical data and microscopical observations. *Wat. Res.* **20**, 185–196.

Li, J., Perdue, E.M. & Gelbaum, L.T. (1998). Using cadmium-113 spectrometry to study metal complexation by natural organic matter. *Environ. Sci. Technol.* **32**, 483–487.

Lin, C.F., Houng, L.M., Lo, K.S. & Lee, D.Y. (1994). Kinetics of copper complexation with dissolved organic matter using stopped-flow fluorescence technique. *Toxicol. Environ. Chem.* **43**, 1–12.

Lin, J.-G. & Chen, S.-Y. (1998). The relationship between adsorption of heavy metal and organic matter in river sediments. *Environ. Int.* **24**, 345–352.

Lofts, S. & Tipping, E. (1998). An assemblage model for cation-binding by natural particulate matter. *Geochim. Cosmochim. Acta.* **62**, 2609–2625.

Lofts, S. & Tipping, E. (2000). Solid-solution partitioning in the Humber rivers: application of WHAM and SCAMP. *Sci. Tot. Environ.* **251/252**, 381–399.

Lofts, S., Woof, C., Tipping, E., Clarke, N. & Mulder, J. (2001a). Modelling pH buffering and aluminium solubility in European forest soils. *Eur. J. Soil Sci.* **52**, 189–204.

Lofts, S., Simon, B.M., Tipping, E. & Woof, C. (2001b). Modelling the solid-solution partitioning of organic matter in European forest soils. *Eur. J. Soil Sci.* **52**, 215–226.

van Loon, L.R., Granacher, S. & Harduf, H. (1992). Equilibrium dialysis-ligand exchange: a novel method for determining conditional stability constants of radionuclide–humic acid complexes. *Anal. Chim. Acta* **268**, 235–246.

Lovelock, J.E. (1979). *Gaia: A New Look at Life on Earth.* Oxford: Oxford University Press.

Lövgren, L., Hedlund, T., Öhman, L.-O. & Sjöberg, S. (1987). Equilibrium approaches to natural water systems – 6. Acid–base properties of a concentrated bog-water and its complexation reactions with aluminium(III). *Wat. Res.* **21**, 1401–1407.

Lövgren, L., Sjöberg, S. & Schindler, P.W. (1990). Acid/base reactions and Al(III) complexation at the surface of goethite. *Geochim. Cosmochim. Acta* **54**, 1301–1306.

Lu, Y.J. & Chakrabarti, C.L. (1995). Kinetic studies of metal speciation using inductively-coupled plasma mass spectrometry. *Int. J. Environ. Anal. Chem.* **60**, 313–337.

Ludwig, B., Prenzel, J. & Khanna, P.K. (1998). Modelling cations in three acid soils with differing acid input in Germany. *Eur. J. Soil Sci.* **49**, 437–445.

Lundström, U.S. & Öhman, L.-O. (1990). Dissolution of feldspars in the presence of natural organic solutes. *J. Soil Sci.* **41**, 359–369.

Lundström, U.S., van Breemen, N. & Bain, D.C. (2000). The podzolisation process. A review. *Geoderma* **94**, 91–107.

Ma, H., Kim, S.D, Cha, D.K. & Allen, H.E. (1999). Effect of kinetics of complexation by humic acid on the toxicity of copper to *Ceriodaphnia Dubia. Environ. Toxicol. Chem.* **18**, 828–837.

MacCarthy, P. & Rice, J.A. (1985). Spectroscopic methods (other than NMR) for determining functionality in humic substances. In *Humic Substances in Soil, Sediment, and Water*, ed. G.R. Aiken, D.M. McKnight, R.L. Wershaw & P. MacCarthy, pp. 527–559. New York: Wiley.

MacCarthy, P. & Rice, J.A. (1991). An ecological rationale for the heterogeneity of humic substances: a holistic perspective on humus. In *Scientists on Gaia*, ed. S.H. Schneider & P.J. Boston, pp. 339–345. Cambridge, MA: MIT Press.

Machesky, M.L. (1993). Calorimetric acid–base titrations of aquatic and peat-derived fulvic and humic acids. *Environ. Sci. Technol.* **27**, 1182–1189.

Maes, A., Tits, J., Mermans, G. & Dierckx, A. (1992). Measurement of the potentially available charge and the dissociation behaviour of humic acid from cobaltihexammine adsorption. *J. Soil Sci.* **43**, 669–677.

Mahony, J.D., DiToro, D.M., Gonzalez, A.M., Curto, M., Dilg, M., DeRosa, L.D. & Sparrow, L.A. (1996). Partitioning of metals to sediment organic carbon. *Environ. Toxicol. Chem.* **15**, 2187–2197.

Malcolm, R.L. (1985). Geochemistry of stream fulvic and humic substances. In *Humic Substances in Soil, Sediment, and Water*, ed. G.R. Aiken, D.M. McKnight, R.L. Wershaw & P. MacCarthy, pp. 181–209. New York: Wiley.

Malcolm, R.L. (1989). Application of solid-state ^{13}C NMR spectroscopy to geochemical studies of humic substances. In *Humic Substances II. In Search of Structure*, ed. M.H.B. Hayes, P. MacCarthy, R.L. Malcolm & R.S. Swift, pp. 339–372. Chichester: Wiley.

Malcolm, R.L. (1990). The uniqueness of humic substances in each of soil, stream and marine environments. *Anal. Chim. Acta* **232**, 19–30.

Malcolm, R.L. (1993). Concentrations and composition of dissolved organic carbon in soils, streams and groundwaters. In *Organic Substances in Soil and Water: Natural Constituents and Their Influence on Contaminant Behaviour*, ed. A.J. Beck, K.C. Jones, M.H.B. Hayes & U. Mingelgrin, pp. 19–30. Cambridge: Royal Society of Chemistry.

Mandal, R., Sekaly, A.L.R., Murimboh, J., Hassan, N.M., Chakrabarti, C.L., Back, M.H., Grégoire, D.C. & Schroeder, W.H. (1999). Effect of the competition of copper and cobalt on the lability of Ni(II)–organic ligand complexes. Part I. In model solutions containing Ni(II) and a well-characterized fulvic acid. *Anal. Chim. Acta* **395**, 309–322.

Mandel, M. & Leyte, J.C. (1964a). Interaction of polymethacrylic acid and bivalent cations. I. *J. Polym. Sci.* **A2**, 2883–2899.

Mandel, M. & Leyte, J.C. (1964b). Interaction of polymethacrylic acid and bivalent cations. II. *J. Polym. Sci.* **A2**, 3771–3780.

Manning, G.S. (1978). The molecular theory of polyelectrolyte solutions with applications to the electrostatic properties of polynucleotides. *Quart. Rev. Biophys.* **11**, 179–246.

Mantoura, R.F.C. & Riley, J.P. (1975). The use of gel filtration in the study of metal binding by humic acids and related compounds. *Anal. Chim. Acta* **78**, 193–200.

Mantoura, R.F.C. & Woodward, E.M.S. (1983). Conservative behaviour of riverine dissolved organic carbon in the Severn Estuary: chemical and geochemical implications. *Geochim. Cosmochim. Acta* **47**, 1293–1309.

Marinsky, J.A. (1987). A two-phase model for the interpretation of proton and metal ion interaction with charged polyelectrolyte gels and their linear analogues. In *Aquatic Surface Chemistry: Chemical Processes at the Particle–Water Interface*, ed. W. Stumm, pp. 49–81. New York: Wiley.

Marinsky, J.A. & Ephraim, J. (1986). A unified physicochemical description of the protonation and metal ion complexation equilibria of natural organic acids (humic and fulvic acids). 1. Analysis of the influence of polyelectrolyte properties on protonation equilibria in ionic media: fundamental concepts. *Environ. Sci. Technol.* **20**, 349–354.

Marinsky, J.A., Gupta, S. & Schindler, P. (1982). The interaction of Cu(II) ion with humic acid. *J. Coll. Int. Sci.* **89**, 401–411.

Marinsky, J.A., Reddy, M.M., Ephraim, J.H. & Mathuthu, A.S. (1992). Unpublished manuscript.

Marinsky, J.A., Reddy, M.M., Ephraim, J.H. & Mathuthu, A.S. (1995) Computational scheme for the prediction of metal ion binding by a soil fulvic acid. *Anal. Chim. Acta* **302**, 309–322.

Marinsky, J.A., Mathuthu, A., Ephraim, J.H. & Reddy, M.M. (1999). Calcium ion binding to a soil fulvic acid using a Donnan potential model. *Radiochim. Acta* **84**, 205–211.

Marshall, S.J., Young, S.D. & Gregson, K. (1995). Humic acid–proton equilibria: a comparison of two models and assessment of titration error. *Eur. J. Soil Sci.* **46**,

471–480.

Martell, A.E. & Hancock, R.D. (1996). *Metal Complexes in Aqueous Solutions*. New York: Kluwer.

Martell, A.E. & Smith, R.M. (1977). *Critical Stability Constants. Vol 3: Other Organic Ligands*. New York: Plenum.

Marx, G. & Heumann, K.G. (1999). Mass spectrometric investigations of the kinetic stability of chromium and copper complexes with humic substances by isotope-labelling experiments. *Fres. J. Anal. Chem.* **364**, 489–494.

Mathur, S.P. & Farnham, R.S. (1985). Geochemistry of humic substances in natural and cultivated peatlands. In *Humic Substances in Soil, Sediment, and Water*, ed. G.R. Aiken, D.M. McKnight, R.L. Wershaw & P. MacCarthy, pp. 53–85. New York: Wiley.

Mathuthu, A.S. (1987). A unified physicochemical description of the protonation and metal ion complexation equilibria of natural organic acids (fulvic acid) in ionic media. Ph.D. Thesis, State University of New York at Buffalo.

Mathuthu, A.S. & Ephraim, J.H. (1995). Binding of cadmium to Laurentide fulvic acid. Justification of the functionalities assigned to the predominant acidic moieties in the fulvic acid molecule. *Talanta* **42**, 1803–1810.

Mathuthu, A.S., Marinsky, J.A. & Ephraim, J.H. (1995). Dissociation properties of Laurentide fulvic acid: identifying the predominant acidic sites. *Talanta* **42**, 441–447.

Mattigod, S.V. & Sposito, G. (1979). Chemical modelling of trace metal equilibria in contaminated soil solutions using the computer program GEOCHEM. In *Chemical Modeling in Aqueous Systems*, ed. E.A. Jenne, pp. 837–856. Washington, DC: American Chemical Society.

Matzner, E. & Prenzel, J. (1992). Acid deposition in the German Solling area: effects on soil solution chemistry and Al mobilization. *Wat. Air Soil Pollut.* **61**, 221–234.

Maurice, P.A. & Namjesnik-Dejanovic, K. (1999). Aggregate structures of sorbed humic substances observed in aqueous solution. *Environ. Sci. Technol.* **33**, 1538–1541.

McBride, M. (1978). Transition metal bonding in humic acid: an ESR study. *Soil Sci.* **126**, 200–209.

McBride, M. (1982). Electron spin resonance investigation of Mn^{2+} complexation in natural and synthetic organics. *Soil Sci. Soc. Am. J.* **46**, 1137–1143.

McBride, M. (1994). *Environmental Chemistry of Soils*. Oxford: Oxford University Press.

McBride, M., Sauvé, S. & Hendershot, W. (1997). Solubility control of Cu, Zn, Cd and Pb in contaminated soils. *Eur. J. Soil Sci.* **48**, 337–346.

McDowell-Boyer, L.M., Hunt, J.R. & Sitar, N. (1986). Particle transport through porous media. *Wat. Resour. Res.* **22**, 1901–1921.

McKeague, J.A., Cheshire, M.V., Andreux, F. & Berthelin, J. (1986). Organo-mineral complexes in relation to pedogenesis. In *Interactions of Soil Minerals with Natural Organics and Microbes*, ed. P.M. Huang & M. Schnitzer, pp. 549–592. Madison, WI: Soil Science Society of America.

McKnight, D.M. (1991). Feedback mechanisms involving humic substances in aquatic ecosystems. In *Scientists on Gaia*, ed. S.H. Schneider & P.J. Boston, pp. 330–338. Cambridge MA: MIT Press.

McKnight, D.M., Feder, G.L., Thurman, M., Wershaw, R.L. & Westall, J.C. (1983). Complexation of copper by aquatic humic substances from different environments. *Sci. Tot. Environ.* **28**, 65–76.

Meyer, J.S., Santore, R.C., Bobbitt, J.P., Debrey, L.D., Boese, C.J., Paquin, P.R., Allen, H.E., Bergman, H.L. & DiToro, D.M. (1999). Binding of nickel and copper to fish gills predicts toxicity when water hardness varies, but free-ion activity does not.

Environ. Sci. Technol. **33**, 913–916.
Milne, C.J., Kinniburgh, D.G., De Wit, J.C.M., van Riemsdijk, W.H. & Koopal, L.K. (1995). Analysis of proton binding by a peat humic acid using a simple electrostatic model. *Geochim. Cosmochim. Acta* **59**, 1101–1112.
Morel, F.M.M. (1983). *Principles of Aquatic Chemistry*. New York: Wiley.
Morel, F.M.M. & Hering, J.G. (1993). *Principles and Applications of Aquatic Chemistry*. New York: Wiley.
Morel, F.M.M., Westall, J.C., O'Melia, C.R. & Morgan, J.J. (1975). Fate of trace metals in Los Angeles County wastewater discharge. *Environ.Sci. Technol.* **9**, 756–761.
Morse, J.W., Millero, F.J., Cornwell, J.C. & Rickard, D. (1987). The chemistry of the hydrogen-sulfide and iron sulfide systems in natural-waters. *Earth Sci. Rev.* **24**, 1–42.
Mota, A.M. & Correia dos Santos, M.M. (1995). Trace metal speciation of labile chemical species in natural waters: electrochemical methods. In *Metal Speciation and Bioavailability in Aquatic Systems*, ed. A. Tessier & D.R. Turner, pp. 206–257. Chichester: Wiley.
Mota, A.M., Rato, A., Brazia, C. & Simões Gonçalves, M.L. (1996). Competition of Al^{3+} in complexation of humic matter by Pb^{2+}: a comparative study with other ions. *Environ. Sci. Technol.* **30**, 1970–1974.
Moulin, V., Robouch, P. & Vitorge, P. (1987). Spectrophotometric study of the interaction between americium(III) and humic materials. *Inorg. Chim. Acta* **140**, 303–306.
Moulin, V., Tits, J., Moulin, C., Decambox, P., Mauchien, P. & de Ruty, O. (1992). Complexation behaviour of humic substances towards actinides and lanthanides studied by time-resolved laser-induced spectrofluorometry. *Radiochim. Acta* **58/59**, 121–128.
Murray, K. & Linder, P.W. (1983). Fulvic acids: Structure and metal binding. I. A random molecular model. *J. Soil Sci.* **34**, 511–523.
Murray, K. & Linder, P.W. (1984). Fulvic acids: Structure and metal binding. II. Predominant metal binding sites. *J. Soil Sci.* **35**, 217–222.
Nagasawa, M., Murase, T. & Kondo, K. (1965). Potentiometric titration of stereoregular polyelectrolytes. *J. Phys. Chem.* **69**, 4005–4012.
Nantsis, E.A. & Carper, W.R. (1998a). Molecular structure of divalent metal ion – fulvic acid complexes. *J. Mol. Struct.* (*Theochem*) **423**, 203–212.
Nantsis, E.A. & Carper, W.R. (1998b). Effects of hydration on the molecular structure of divalent metal ion – fulvic acid complexes: a MOPAC (PM3) study. *J. Mol. Struct.* (*Theochem*) **431**, 267–275.
Nash, K.L. & Choppin, G.R. (1980). Interaction of humic and fulvic acids with Th(IV). *J. Inorg. Nucl. Chem.* **42**, 1045–1050.
Neal, C., Reynolds, B. & Robson, A.J. (1999). Acid neutralisation capacity measurements within natural waters: towards a standardised approach. *Sci. Tot. Environ.* **244**, 233–241.
Nissenbaum, A. & Swaine, D.J. (1976). Organic matter–metal interactions in recent sediments: the role of humic substances. *Geochim. Cosmochim. Acta* **40**, 809–816.
Norde, W. (1980). Adsorption of proteins at solid surfaces. In *Adhesion and Adsorption of Polymers, Part B*, ed. L.-H. Lee, pp. 801–825. New York: Plenum.
Nordén, M., Ephraim, J.H. & Allard, B. (1993). The binding of strontium and europium by an aquatic fulvic acid – ion exchange distribution and ultrafiltration studies. *Talanta* **40**, 1425–1432.
Nriagu, J.O. & Coker, R.D. (1980). Trace metals in humic and fulvic acids from Lake Ontario sediments. *Environ. Sci. Technol.* **14**, 443–446.
O'Brien, B.J. & Stout, J.D. (1978). Movement and turnover of soil organic matter as

indicated by carbon isotope measurements. *Soil Biol. Biochem.* **10**, 309–317.

Ochs, M. (1996). Influence of humified and non-humified natural organic compounds on mineral dissolution. *Chem. Geol.* **132**, 119–124.

Öhman, L.O. & Sjöberg, S. (1981). Equilibrium and structural studies of silicon (IV) and aluminium (III) in aqueous solution. 1. The formation of ternary mononuclear and polynuclear complexes in the system Al^{3+}–gallic acid–OH^-. A potentiometric study in 0.6 M Na(Cl). *Acta Chem. Scand. A* **35**, 201–212.

Olander, D.S. & Holtzer, A. (1968). The stability of the polyglutamic acid α helix. *J. Am. Chem. Soc.* **90**, 4549–4560.

Oliver, B.G., Thurman, E.M. & Malcolm, R.L. (1983). The contribution of humic substances to the acidity of colored natural waters. *Geochim. Cosmochim. Acta* **47**, 2031–2035.

Olson, D.L. & Shuman, M.S. (1983). Kinetic spectrum method for analysis of simultaneous, first-order reactions and application to copper(II) dissociation from aquatic macromolecules. *Anal. Chem.* **55**, 1103–1107.

O'Melia, C.R. (1987). Particle–particle interactions. In *Aquatic Surface Chemistry. Chemical Processes at the Particle–Water Interface*, ed. W. Stumm, pp. 385–403. New York: Wiley.

O'Melia, C.R. (1998). Coagulation and sedimentation in lakes, reservoirs and water treatment plants. *Wat. Sci. Tech.* **37**, 129–135.

O'Melia, C.R., Becker, W.C. & Au, K.-K. (1999). Removal of humic substances by coagulation. *Wat. Sci. Tech.* **40**, 47–54.

Ong, H.L. & Bisque, R.E. (1968). Coagulation of humic colloids by metal ions. *Soil Sci.* **106**, 220–224.

Österberg, R. & Mortensen, K. (1992). Fractal dimensions of humic acids. *Eur. Biophys. J.* **21**, 163–167.

Österberg, R., Mortensen, K. & Ikai, A. (1995). Direct observation of humic acid clusters, a nonequilibrium system with a fractal structure. *Naturwissenschaften* **82**, 137–139.

Österberg, R., Wei, S. & Shirshova, L. (1999). Inert copper complexes formed by humic acids. *Acta Chem. Scand.* **53**, 172–180.

Palmer, D. & Wesolowski, D.J. (1992). Aluminium speciation and equilibria in aqueous solution: II. The solubility of gibbsite in acidic sodium chloride solutions from 30 to 70°C. *Geochim. Cosmochim. Acta* **56**, 1093–1111.

Paquin, P.R., Santore, R.C., Wu, K.B., Kavvadas, C.D. & Di Toro, D.M. (2000). The biotic ligand model: a model of the acute toxicity of metals to aquatic life. *Environ. Sci. Pol.* **3**, S175–S182.

Parkhurst, D.L., Thorstenson, D.C. & Plummer, L.N. (1980). *PHREEQE – A Computer Program for Geochemical Calculations.* USGS Wat. Res. Invest. Rep. 80–96. Washington, DC: US Geological Survey.

Parton, W.J., Schimel, D.S., Cole, C.V. & Ojima, D.S. (1987). Analysis of factors controlling soil organic levels in Great Plains grasslands. *Soil Sci. Soc. Am. J.* **51**, 1173–1179.

Patterson, H.H., Cronan, C.S., Lakshman, S., Plankey, B.J. & Taylor, T.A. (1992). Comparison of soil fulvic acids using synchronous scan fluorescence spectroscopy, FTIR, titration and metal complexation kinetics. *Sci. Tot. Environ.* **113**, 179–196.

Paxeus, N. & Wedborg, M. (1985). Acid–base properties of aquatic fulvic acid. *Anal. Chim. Acta* **169**, 87–98.

Pearson, R.G. (1967). Hard and soft acids and bases. *Chem. Br.* **3**, 103–107.

Pennell, K.D. & Rao, P.S.C. (1992). Comment on "The Surface Area of Soil Organic Matter". *Environ. Sci. Technol.* **26**, 402–404.

Perdue, E.M. (1978). Solution thermochemistry of humic substances–I. Acid–base equilibria of humic acid. *Geochim. Cosmochim. Acta* **42**, 1351–1358.

Perdue, E.M. (1985). Acidic functional groups of humic substances. In *Humic Substances in Soil, Sediment, and Water*, ed. G.R. Aiken, D.M. McKnight, R.L. Wershaw & P. MacCarthy, pp. 493–526. New York: Wiley.

Perdue, E.M. (1990). Modeling the acid–base chemistry of organic acids in laboratory experiments and in freshwaters. In *Organic Acids in Aquatic Ecosystems*, ed. E.M. Perdue & E.T. Gjessing, pp. 111–126. Chichester: Wiley.

Perdue, E.M. (1998). Chemical properties, structure, and metal binding. In *Aquatic Humic Substances. Ecology and Biogeochemistry*, ed. D.O. Hessen & L.J. Tranvik, pp. 41–61. Berlin: Springer.

Perdue, E.M. & Carreira, L.A. (1997). Modeling competitive binding of protons and metal ions by humic substances. In *Biogeochemistry of Trace Metals*, ed. D.C. Adriano, Z.S. Chen, S.S. Yang & I.K. Iskandar, pp. 381–401. Northwood: Science Reviews.

Perdue, E.M. & Lytle, C.R. (1983). A distribution model for binding of protons and metal ions by humic substances. *Environ. Sci. Technol.* **17**, 654–660.

Perdue, E.M., Beck, K.C. & Reuter, J.H. (1976). Organic complexes of iron and aluminium in natural waters. *Nature* **260**, 418–420.

Perdue, E.M., Reuter, J.H. & Parrish, R.S. (1984). A statistical model of proton binding by humus. *Geochim. Cosmochim. Acta* **49**, 1257–1263.

Petrović, M., Kaštelan-Macan, M. & Horvat, A.J.M. (1999). Interactive sorption of metal ions and humic acids onto mineral particles. *Wat. Air Soil Pollut.* **111**, 41–56.

Piccolo, A. (1995). *Humic Substances in Terrestrial Ecosystems.* Amsterdam: Elsevier.

Pinheiro, J.P., Mota, A.M., Goncalves, M.L.S.S., van der Weijde, M. & van Leeuwen, H.P. (1996). Comparison between polarographic and potentiometric speciation for cadmium/humic systems. *J. Electroanal. Chem.* **410**, 61–68.

Piret, E.L., White, R.G., Walther, H.C. & Madden, A.J. (1960). Some physicochemical properties of peat humic acids. *Sci. Proc. R. Dublin Soc., Ser. A* **1**, 69–79.

Pitzer, K.S. (1991). Ion interaction approach: theory and data correlation. In *Activity Coefficients in Electrolyte Solutions*, ed. K.S. Pitzer, pp. 75–153. Boca Raton, FL: CRC Press.

Plankey, B.J. & Patterson, H.H. (1987). Kinetics of aluminium–fulvic acid complexation in acidic waters. *Environ. Sci. Technol.* **21**, 595–601.

Plechanov, N., Josefsson, B., Dyrssen, D. & Lundquist, K. (1983). Investigations on humic substances in natural water. In *Aquatic and Terrestrial Humic Materials*, ed. R.F. Christman & E.T. Gjessing, pp. 387–405. Ann Arbor, MI: Ann Arbor Publishers.

Pomogailo, A.D. & Wöhrle, D. (1996). Synthesis and structure of macromolecular metal complexes. In *Macromolecule–Metal Complexes*, ed. F. Ciardelli, E. Tsuchida & D. Wöhrle, pp. 11–129. Berlin: Springer.

Posch, M., Forsius, M. & Kämäri, J. (1993). Critical loads of sulphur and nitrogen for lakes I: Model description and estimation of uncertainty. *Wat. Air Soil Pollut.* **66**, 173–192.

Posner, A.M. (1964). Titration curves of humic acid. *Proc. 8th Int. Conf. Soil Sci., Part 2, Bucharest, Romania*, pp. 161–174.

Powell, H.K.J. & Fenton, E. (1996). Size fractionation of humic substances – effect on protonation and metal-binding properties. *Anal. Chim. Acta* **334**, 27–38.

Qualls, R.G. & Haines, B.L. (1991). Geochemistry of dissolved organic nutrients in water percolating through a forest ecosystem. *Soil Sci. Soc. Am. J.* **55**, 1112–1123.

Quentel, F., Madec, C. & Courtot-Coupez, J. (1987) Determination of humic substances in seawater by electrochemistry (mechanisms) *Anal. Lett.* **20**, 47–62.

Radovanovic, H. & Koelmans, A.A. (1998). Prediction of in situ trace metal distribution

coefficients for suspended solids in natural waters. *Environ. Sci. Technol.* **32**, 753–759.
Ragle, C.S., Engebretson, R.R. & von Wandruszka, R. (1997). The sequestration of hydrophobic micropollutants by dissolved humic acids. *Soil Sci.* **162**, 106–114.
Rainville, D.P. & Weber, J.H. (1982). Complexing capacity of soil fulvic-acid for Cu^{2+}, Cd^{2+}, Mn^{2+} and Zn^{2+} measured by dialysis titration – a model based on soil fulvic-acid aggregation. *Can. J. Chem.* **60**, 1–5.
Randhawa, N.S. & Broadbent, F.E. (1965). Soil organic matter–metal complexes: 6. Stability constants of zinc–humic acid complexes at different pH values. *Soil Sci.* **99**, 362–366.
Rate, A.W., McLaren, R.G. & Swift, R.S. (1992). Evaluation of a log-normal distribution 1st-order kinetic model for copper(II)–humic acid complex dissociation. *Environ. Sci. Technol.* **26**, 2477–2483.
Rate, A.W., McLaren, R.G. & Swift, R.S. (1993). Response of copper(II) humic acid dissociation kinetics to factors influencing complex stability and macromolecular conformation. *Environ. Sci. Technol.* **27**, 1408–1414.
Ravichandran, M., Aiken, G.R., Ryan, J.N. & Reddy, M.M. (1999). Inhibition of precipitation and aggregation of metacinnabar (mercuric sulfide) by dissolved organic matter isolated from the Florida Everglades. *Environ. Sci. Technol.* **33**, 1418–1423.
Reid, P.M., Wilkinson, A.E., Tipping, E. & Jones, M.N. (1990). Determination of molecular weights of humic substances by analytical (U.V. scanning) ultracentrifugation. *Geochim. Cosmochim. Acta* **54**, 131–138.
Reuss, J.O. (1983). Implications of the calcium–aluminum exchange system for the effect of acid precipitation on soils. *J. Env. Qual.* **12**, 591–595.
Reuss, J.O., Cosby, B.J. & Wright, R.F. (1987). Chemical processes governing soil and water acidification. *Nature* **329**, 27–32.
Reuter, J.H. & Perdue, E.M. (1981). Calculation of molecular weights of humic substances from colligative data: application to aquatic humus and its molecular size fractions. *Geochim. Cosmochim. Acta* **45**, 2017–2022.
Richards, E.G. (1980). *An Introduction to the Physical Properties of Large Molecules in Solution.* Cambridge: Cambridge University Press.
van Riemsdijk, W.H. & Koopal, L. (1992). Ion binding by natural heterogeneous particles. In *Environmental Particles Vol.* 1, ed. J. Buffle & H.P. van Leeuwen, pp. 455–495. Boca Raton, FL: Lewis.
van Riemsdijk, W.H., de Wit, J.C.M., Mouse, S.L.J., Koopal, L.K. & Kinniburgh, D.G. (1996). An analytical isotherm equation (CONICA) for nonideal mono- and bidentate competitive ion adsorption to heterogeneous surfaces. *J.Coll. Int. Sci.* **183**, 35–50.
Riley, J.P. & Chester, R. (1971). *Introduction to Marine Chemistry.* London: Academic.
Rimmer, D. (1998). Ultimate interface. *New Scientist*, 14 November.
Robertson, A.P. (1996). *Goethite/Humic Acid Interactions and Their Effects on Copper Binding.* Ph.D. Thesis, Stanford University.
Robertson, A.P. & Leckie, J.O. (1999). Acid/base, copper binding, and Cu^{2+}/H^+ exchange properties of a soil humic acid, an experimental and modeling study. *Environ. Sci. Technol.* **33**, 786–795.
Robinson, R.A. & Stokes, R.H. (1959). *Electrolyte Solutions.* 2nd edn. London: Butterworth.
Rosenqvist, I.T. (1978). Alternative sources for acidification of river water in Norway. *Sci. Tot. Environ.* **10**, 39–49.
Rowell, D.L. (1994). *Soil Science. Methods and Applications.* Harlow: Longman.
Rozan, T.F., Lassman, M.E., Ridge, D.P. & Luther III, G.W. (2000). Evidence for iron,

copper and zinc complexation as multinuclear sulphide clusters in oxic rivers. *Nature* **406**, 879–882.

Ryan, D.K. & Weber, J.H. (1982). Copper(II) complexing capacities of natural waters by fluorescence quenching. *Environ. Sci. Technol.* **16**, 866–872.

Saar, R.A. & Weber, J.H. (1979). Complexation of cadmium(II) with water- and soil-derived fulvic acids: effect of pH and fulvic acid concentration. *Can. J. Chem.* **57**,1263–1268.

Saar, R.A. & Weber, J.H. (1980a). Comparison of spectrofluorometry and ion-selective electrode potentiometry for determination of complexes between fulvic acid and heavy-metal ions. *Anal. Chem.* **52**, 2095–2100.

Saar, R.A. & Weber, J.H. (1980b). Lead(II)–fulvic acid complexes. Conditional stability constants, solubility, and implications for lead(II) mobility. *Environ. Sci. Technol.* **14**, 877–880.

Sanchez, A.L., Schell, W.R. & Thomas, E.D. (1988). Interactions of ^{57}Co, ^{85}Sr and ^{137}Cs with peat under acidic precipitation conditions. *Health Phys.* **54**, 317–322.

Santore, R.C., Driscoll, C.T. & Aloi, M. (1995). A model of soil organic matter and its function in temperate forest soil development. In *Carbon Forms and Functions in Forest Soils*, ed. W.W. McFee & J.M. Kelly, pp. 275–298. Madison, WI: Soil Science Society of America.

Santos, E.B.H., Esteves, V.I., Rodrigues, J.P.C. & Duarte, A.C. (1999). Humic substances' proton-binding equilibria: assessment of errors and limitations of potentiometric data. *Anal. Chim. Acta* **392**, 333–341.

Sauvé, S., McBride, M., Norvell, W.A. & Hendershot, W.H. (1997). Copper solubility and speciation of in situ contaminated soils: effects of copper level, pH and organic matter. *Wat. Air Soil Pollut.* **100**, 133–149.

Sauvé, S., McBride, M. & Hendershot, W. (1998). Soil solution speciation of lead(II): effects of organic matter and pH. *Soil Sci. Soc. Am. J.* **62**, 618–621.

Sauvé, S., Norvell, W.A., McBride, M., Norvell, W.A. & Hendershot, W.H. (2000). Speciation and complexation of cadmium in extracted soil solutions. *Environ. Sci. Technol.* **34**, 291–296.

Sawhney, B.L. (1972). Selective sorption and fixation of cations by clay minerals: a review. *Clays Clay Min.* **20**, 93–100.

Scatchard G. (1949). The attraction of proteins for small molecules and ions. *Ann. N.Y. Acad. Sci.* **51**, 660–672.

Schecher, W.D. & Driscoll, C.T. (1988). An evaluation of the equilibrium calculations within acidification models: the effect of uncertainty in measured chemical components. *Wat. Resour. Res.* **24**, 533–540.

Schecher, W.D. & Driscoll, C.T. (1995). ALCHEMI: a chemical equilibrium model to assess the acid–base chemistry and speciation of aluminium in dilute solutions. In *Chemical Equilibrium and Reaction Models*, ed. R. Loeppert, A.P. Schwab & S. Goldberg, pp. 325–356. Madison, WI: Soil Science Society of America.

Schindler, D.W. & Curtis, P.J. (1997). The role of DOC in protecting freshwaters subjected to climatic warming and acidification from UV exposure. *Biogeochem.* **36**, 1–8.

Schindler, P.W. & Stumm, W. (1987). The surface chemistry of oxides, hydroxides and oxide minerals. In *Aquatic Surface Chemistry. Chemical Processes at the Particle–Water Interface*, ed. W. Stumm, pp. 83–110. New York: Wiley.

Schlautman, M.A. & Morgan, J.J. (1993). Effects of aqueous chemistry on the binding of polycyclic aromatic hydrocarbons by dissolved materials. *Environ. Sci. Technol.* **27**, 961–969.

Schlesinger, W.H. (1997). *Biogeochemistry: An Analysis of Global Change*, 2nd edn. San

Diego: Academic Press.
Schnitzer, M. (1985). Nature of nitrogen in humic substances. In *Humic Substances in Soil, Sediment, and Water*, ed. G.R. Aiken, D.M. McKnight, R.L. Wershaw & P. MacCarthy, pp. 303–325. New York: Wiley.
Schnitzer, M. (1990). Selected methods for the characterisation of soil humic substances. In *Humic Substances in Soil and Crop Sciences: Selected Readings*, ed. P. MacCarthy, C.E. Clapp, R.L. Malcolm & P.R. Bloom, pp. 65–89. Madison, WI: American Society of Agronomy, Soil Science Society of America.
Schnitzer, M. (1991). Soil organic matter – the next 75 years. *Soil Sci.* **151**, 41–58.
Schnitzer, M. & Khan, S.U. (1972). *Humic Substances in the Environment*. New York: Dekker.
Schubert, J. (1948). The use of ion exchangers for the determination of physical-chemical properties of substances, particularly radiotracers, in solutions: 1. *Theor. J. Phys. Colloid. Chem.* **53**, 340–350.
Schulten, H.R. (1996). A new approach to the structural analysis of humic substances in water and soils. In *Humic and Fulvic Acids: Isolation, Structure and Environmental Role, ACS Symposium Series No.* 651, ed. J.S. Gaffney, N.A. Marley & S.B. Clark, pp. 42–56. Washington, DC: American Chemical Society.
Schwarzenbach, R.P., Gschwend, P.M. & Imboden, D.M. (1993). *Environmental Organic Chemistry*. New York: Wiley.
Schwertmann, U., Carlsson, L. & Fechter, H. (1984). Iron oxide formation in artificial ground waters. *Schweiz. Z. Hydrol.* **46**, 185–191.
Schwertmann, U., Kodama, H. & Fischer, W.R. (1986). Mutual interactions between organics and iron oxides. In *Interactions of Soil Minerals with Natural Organics and Microbes*, ed. P.M. Huang & M. Schnitzer, pp. 223–250. Madison, WI: Soil Science Society of America.
Scott, M.J., Jones, M.N., Woof, C. & Tipping, E. (1998). Concentrations and fluxes of dissolved organic carbon in drainage water from an upland peat system. *Environ. Int.* **24**, 537–546.
Scrutton, M.C. (1973). Metal enzymes. In *Inorganic Biochemistry Vol. 1*, ed. G.L. Eichhorn, pp. 381–437. Amsterdam: Elsevier.
Sedlacek, J., Källqvist, T. & Gjessing, E. (1983). Effect of aquatic humus on uptake and toxicity of cadmium to *Selenastrum capricornutum* Printz. *In Aquatic and Terrestrial Humic Materials*, ed. R.F. Christman & E.T. Gjessing, pp. 495–516. Ann Arbor, MI: Ann Arbor Science.
Sekaly, A.L.R., Back, M.H., Chakrabarti, C.L., Grégoire, D.C., Lu, J.Y. & Schroeder, W.H. (1998). Measurements and analysis of dissociation rate constants of metal–fulvic acid complexes in aqueous solutions. Part II: measurement of decay rates by inductively-coupled plasma mass spectrometry and determination of rate constants for dissociation. *Spectrochim. Acta* **53**, 847–858.
Senesi, N. (1990). Molecular and quantitative aspects of the chemistry of fulvic acid and its interactions with metal ions and organic chemicals. Part I. The electron spin resonance approach. *Anal. Chim. Acta* **232**, 51–75.
Senesi, N. (1993). Nature of interactions between organic chemicals and dissolved humic substances and the influence of environmental factors. In *Organic Substances in Soil and Water: Natural Constituents and Their Influence on Contaminant Behaviour*, ed. A.J. Beck, K.C. Jones, M.H.B. Hayes & U. Mingelgrin, pp. 73–101. Cambridge: Royal Society of Chemistry.
Senesi, N., Griffith, S.M., Schnitzer, M. & Townsend, M.G. (1977). Binding of Fe^{3+} by humic materials. *Geochim. Cosmochim. Acta* **41**, 969–976.
Senesi, N., Sposito, G. & Martin, J.P. (1986). Copper(II) and iron(III) complexation by soil

humic acids: an IR and ESR study. *Sci. Tot. Environ.* **55**, 351–362.
Senesi, N., Rizzi, F.R., Dellino, P. & Acquafredda, P. (1997). Fractal humic acids in aqueous suspensions at various concentrations, ionic strengths, and pH values. *Coll. Surf. A.* **127**, 57–68.
Shapiro, J. (1966). The relation of humic color to iron in natural waters. *Verh. Internat. Verein. Limnol.* **16**, 477–484.
Sharma, V.S. & Schubert, J. (1969). Statistical factor in the formation and stability of ternary and mixed ligand complexes. *J. Chem. Ed.* **46**, 506–507.
Shaw, D.J. (1978). *Introduction to Colloid and Surface Chemistry.* London: Butterworth.
Shen, Y.-H. (1999). Sorption of natural dissolved organic matter on soil. *Chemosphere* **38**, 1505–1515.
Shevchenko, S., Bailey, G.W. & Akim, L.G. (1999). The conformational dynamics of humic polyanions in model organic and organo-mineral aggregates. *J. Mol. Struct. (Theochem)* **460**, 179–190.
Shin, H.S., Lee, B.H., Yang, H.B., Yun, S.S. & Moon, H. (1996). Bimodal normal distribution model for binding of trivalent europium by soil fulvic acid. *J. Radioanal. Nucl. Chem.* **209**, 123–133.
Sholkovitz, E.R. (1976). Flocculation of dissolved organic and inorganic matter during the mixing of river water and seawater. *Geochim. Cosmochim. Acta* **40**, 831–845.
Sholkovitz, E.R., Boyle, E.A. & Price, N.B. (1978). The removal of dissolved humic acids and iron during estuarine mixing. *Earth Planet. Sci. Lett.* **40**, 130–136.
Shuman, M. (1990). Carboxyl acidity of aquatic organic matter: possible systematic errors introduced by XAD extraction. In *Organic Acids in Aquatic Ecosystems*, ed. E.M. Perdue & E.T. Gjessing, pp. 97–109. Chichester: Wiley.
Shuman, M., Collins, G.J., Fitzgerald, O.J & Olson, D.L. (1983). Distribution of stability constants and dissociation rate constants among binding sites on estuarine copper–organic complexes: Rotated disk electrode studies and an affinity spectrum analysis of ion-selective electrode and photometric data. In *Aquatic and Terrestrial Humic Materials*, ed. R.F. Christman & E.T. Gjessing, pp. 349–370. Ann Arbor, MI: Ann Arbor Science.
Skogerboe, R.K. & Wilson, S.A. (1981). Reduction of ionic species by fulvic acid. *Anal. Chem.* **53**, 228–232.
Skyllberg, U. (1999). pH and solubility of aluminium in acidic forest soils: a consequence of reactions between organic acidity and aluminium alkalinity. *Eur. J. Soil Sci.* **50**, 95–106.
Smith, D.G. (1981). *The Cambridge Encyclopaedia of Earth Sciences.* Cambridge: Cambridge University Press.
Smith, D.S. & Kramer, J.R. (1998). Multi-site aluminium speciation with natural organic matter using multiresponse fluorescence data. *Anal. Chim. Acta* **363**, 21–29.
Smith, D.S. & Kramer, J.R. (2000). Multisite metal binding to fulvic acid determined using multiresponse fluorescence. *Anal. Chim. Acta* **416**, 211–220.
Smith, R.M., Martell, A.E. & Motekaitis, R.J. (1993). Critical Stability Constants Database 46 Version 4.0. Gaithersburg, MD: National Institute of Standards and Technology.
Sparks, D.L. (1989). *Kinetics of Soil Chemical Processes.* San Diego: Academic Press.
Sposito, G. (1981). Trace metals in contaminated waters. *Environ. Sci. Technol.* **15**, 396–403.
Sposito, G. (1986). Sorption of trace metals by humic materials in soils and natural waters. *CRC Crit. Rev. Environ. Cntrl.* **16**, 193–229.
Sposito, G. (1989). *The Chemistry of Soils.* New York: Oxford University Press.
Sposito, G. & Holtzclaw, K.M. (1977). Titration studies on the polynuclear, polyacid

nature of fulvic acid extracted from sewage sludge-soil mixtures. *Soil Sci. Soc. Am. J.* **41**, 330–336.

Sposito, G., Holtzclaw, K.M. & Keech, D.A. (1977). Proton binding in fulvic acid extracted from sewage sludge–soil mixtures. *Soil Sci. Soc. Am. J.* **41**, 1119–1125.

Steelink, C. (1985). Implications of elemental characteristics of humic substances. In *Humic Substances in Soil, Sediment, and Water*, ed. G.R. Aiken, D.M. McKnight, R.L. Wershaw & P. MacCarthy, pp. 457–476. New York: Wiley.

Steelink, C., Wershaw, R.L., Thorn, K.A. & Wilson, M.A. (1989). Application of liquid-state NMR spectroscopy to humic substances. In *Humic Substances II. In Search of Structure*, ed. M.H.B. Hayes, P. MacCarthy, R.L. Malcolm & R.S. Swift, pp. 281–308. Chichester: Wiley.

Stevenson, F.J. (1976). Stability constants of Cu^{2+}, Pb^{2+}, and Cd^{2+} complexes with humic acids. *Soil Sci. Soc. Am. J.* **40**, 665–672.

Stevenson, F.J. (1994). *Humus Chemistry: Genesis, Composition, Reactions*, 2nd edn. New York: Wiley.

Stumm, W. (1992). *Chemistry of the Solid–Water Interface*. New York: Wiley.

Stumm, W. & Furrer, G. (1987). The dissolution of oxides and aluminium silicates; examples of surface-coordination-controlled kinetics. In *Aquatic Surface Chemistry. Chemical Processes at the Particle–Water Interface*, ed. W. Stumm, pp. 197–219. New York: Wiley.

Stumm, W. & Morgan, J.J. (1996). *Aquatic Chemistry*, 3rd edn. New York: Wiley.

Sun, L., Perdue, E.M. & McCarthy, J.F. (1995). Using reverse osmosis to obtain organic matter from surface and ground waters. *Wat. Res.* **29**, 1471–1477.

Susetyo, W., Dobbs, J.C., Carreira, L.A., Azarraga, L.V. & Grimm, D.M. (1990). Development of a statistical model for metal–humic interactions. *Anal. Chem.* **62**, 1215–1221.

Susetyo, W., Carreira, L.A., Azarraga, L.V. & Grimm, D.M. (1991). Fluorescence techniques for metal–humic interactions. *Fres. J. Anal. Chem.* **339**, 624–635.

Sutheimer, S.H. & Cabaniss, S.E. (1995). Aqueous Al(III) speciation by high performance cation exchange chromatography with fluorescence detection of the aluminium–lumogallion complex. *Anal. Chem.* **67**, 2342–2349.

Sutheimer, S.H. & Cabaniss, S.E. (1997). Aluminium binding to humic substances determined by high performance cation exchange chromatography. *Geochim. Cosmochim. Acta* **61**, 1–9.

Sverdrup, H. (1990). *The Kinetics of Base Cation Release Due to Chemical Weathering*. Lund: Lund University Press.

Sverdrup, H., Warfvinge, P., Blake, L. & Goulding, K. (1995). Modeling recent and historic soil data from the Rothamsted Experimental Station, UK using SAFE. *Agric. Ecosyst. & Environ.* **53**, 161–177.

Swift, R.S. (1985). Fractionation of soil humic substances. In *Humic Substances in Soil, Sediment, and Water*, ed. G.R. Aiken, D.M. McKnight, R.L. Wershaw & P. MacCarthy, pp. 387–408. New York: Wiley.

Swift, R.S. (1989a). Molecular weight, size, shape, and charge characteristics of humic substances: some basic considerations. In *Humic Substances II. In Search of Structure*, ed. M.H.B. Hayes, P. MacCarthy, R.L. Malcolm & R.S. Swift, pp. 449–465. Chichester: Wiley.

Swift, R.S. (1989b). Molecular weight, shape, and size of humic substances by ultracentrifugation. In *Humic Substances II. In Search of Structure*, ed. M.H.B. Hayes, P. MacCarthy, R.L. Malcolm & R.S. Swift, pp. 467–495. Chichester: Wiley.

Tanford, C. (1961). *Physical Chemistry of Macromolecules*. New York: Wiley.

Tanford, C. (1980). *The Hydrophobic Effect: Formation of Micelles and Biological Membranes*, 2nd edn. New York: Wiley.

Tanford, C. & Hauenstein, J.D. (1956). Hydrogen ion equilibria of ribonuclease. *J. Am. Chem. Soc.* **78**, 5287–5291.

Tate, K.R. & Theng, B.K.G. (1980). Organic matter and its interactions with inorganic soil constituents. In *Soils with Variable Charge*, ed. B.K.G. Theng, pp. 225–249. Palmerston North: New Zealand Society of Soil Science.

Tate, R.L. (1987). *Soil Organic Matter. Biological and Ecological Effects.* New York: Wiley.

Teasdale, R.D. (1987). Copper-induced indefinite aggregation of humic substances: theoretical consequences for copper-binding behaviour. *J. Soil Sci.* **38**, 433–442.

Temminghoff, E.J.M., van der Zee, S.E.A.T.M. & De Haan, F.A.M. (1998). Effects of dissolved organic matter on the mobility of copper in contaminated sandy soil. *Eur. J. Soil Sci.* **49**, 617–628.

Temminghoff, E.J.M., Plette, A.C.C., van Eck, R. & van Riemsdijk, W.H. (2000). Determination of the chemical speciation of trace metals in aqueous systems by the Wageningen Donnan Membrane Technique. *Anal. Chim. Acta* **417**, 149–157.

Templeton, G.D. & Chasteen, N.D. (1980). Vanadium–fulvic acid chemistry: Conformational and binding studies by electron spin probe techniques. *Geochim. Cosmochim. Acta* **44**, 741–752.

Tessier, A. & Turner, D.R. (1995). *Metal Speciation and Bioavailability in Aquatic Systems.* New York: Wiley.

Tessier, A., Couillard, Y., Campbell, P.G.C. & Auclair, J.C. (1993). Modeling Cd partitioning in oxic lake sediments and Cd concentrations in the freshwater bivalve *Anodonta grandis. Limnol. Oceanogr.* **38**, 1–17.

Tessier, A., Fortin, D., Belzile, N., DeVitre, R.R. & Leppard, G.G. (1996). Metal sorption to diagenetic iron and manganese oxyhydroxides and associated organic matter: narrowing the gap between field and laboratory measurements. *Geochim. Cosmochim. Acta* **60**, 387–404.

Theng, B.K.G. & Scharpenseel, H.W. (1975). The adsorption of ^{14}C-labelled humic acid by montmorillonite. In *Proceedings of the International Clay Conference, Mexico City 1975*, ed. G.W. Bailey, pp. 643–653. Barking: Applied Science Publishers.

Thomason, J.W., Susetyo, W. & Carreira, L.A. (1996). Fluorescence studies of metal–humic complexes with the use of lanthanide ion probe spectroscopy. *Appl. Spect.* **50**, 401–408.

Thurman, E.M. (1985). *Organic Geochemistry of Natural Waters.* Dordrecht: Martinus Nijhoff/Dr. W. Junk Publishers.

Tiller, C.L. & O'Melia, C.R. (1993). Natural organic matter and colloidal stability: models and measurements. *Coll. Surf. A* **73**, 89–102.

Tipping, E. (1981). The adsorption of aquatic humic substances by iron oxides. *Geochim. Cosmochim. Acta* **45**, 191–199.

Tipping, E. (1990). Interactions of organic acids with inorganic and organic surfaces. In *Organic Acids in Aquatic Ecosystems*, ed. E.M. Perdue & E.T. Gjessing, pp. 209–221. New York: Wiley.

Tipping, E. (1993). Modelling ion binding by humic acids. *Coll. Surf. A* **73**, 117–131.

Tipping, E. (1994). WHAM – A chemical equilibrium model and computer code for waters, sediments and soils incorporating a discrete site/electrostatic model of ion-binding by humic substances. *Comp. Geosci.* **20**, 973–1023.

Tipping, E. (1996a). CHUM: a hydrochemical model for upland catchments. *J. Hydrol.* **174**, 305–330.

Tipping, E. (1996b). Hydrochemical modelling of the retention and transport of metallic radionuclides in the soils of an upland catchment. *Environ. Pollut.* **94**, 105–116.

Tipping, E. (1998a). Modelling the properties and behaviour of dissolved organic matter in soils. *Mitteil. Deutsch. Boden. Gesell.* **87**, 237–252.

Tipping, E. (1998b). Humic Ion-Binding Model VI: an improved description of the interactions of protons and metal ions with humic substances. *Aq. Geochem.* **4**, 3–48.

Tipping, E. & Cooke, D. (1982). The effects of adsorbed humic substances on the surface charge of goethite (α-FeOOH) in freshwaters. *Geochim. Cosmochim. Acta* **46**, 75–80.

Tipping, E. & Heaton, M.J. (1983). The adsorption of aquatic humic substances by two oxides of manganese. *Geochim. Cosmochim. Acta* **47**, 1393–1397.

Tipping, E. & Higgins, D.C. (1982). The effect of adsorbed humic substances on the colloid stability of haematite particles. *Coll. Surf.* **5**, 85–92.

Tipping, E. & Hurley, M.A. (1988). A model of solid-solution interactions in acid organic soils, based on the complexation properties of humic substances. *J. Soil Sci.* **39**, 505–519.

Tipping, E. & Hurley, M.A. (1992). A unifying model of cation binding by humic substances. *Geochim. Cosmochim. Acta* **56**, 3627–3641.

Tipping, E. & Ohnstad, M. (1984). Aggregation of aquatic humic substances. *Chem. Geol.* **44**, 349–357.

Tipping, E. & Woof, C. (1991). The distribution of humic substances between the solid and aqueous phases of acid organic soils; a description based on humic heterogeneity and charge-dependent sorption equilibria. *J. Soil Sci.* **42**, 437–448.

Tipping, E., Griffith, J. & Hilton, J. (1983). The effect of adsorbed humic substances on the uptake of copper(II) by goethite. *Croat. Chim. Acta* **56**, 613–621.

Tipping, E., Hilton, J. & James, B. (1988a). Dissolved organic matter in Cumbrian lakes and streams. *Freshwat. Biol.* **19**, 371–378.

Tipping, E., Backes, C.A. & Hurley, M.A. (1988b). The complexation of protons, aluminium and calcium by aquatic humic substances: a model incorporating binding-site heterogeneity and macroionic effects. *Wat. Res.* **22**, 597–611.

Tipping, E., Thompson, D.W. & Woof, C. (1989). Iron oxide particulates formed by the oxygenation of natural and model lakewaters containing Fe(II). *Arch. Hydrobiol.* **115**, 59–70.

Tipping, E., Reddy, M.M. & Hurley, M.A. (1990). Modelling electrostatic and heterogeneity effects on proton dissociation from humic substances. *Environ. Sci. Technol.* **24**, 1700–1705.

Tipping, E., Woof, C. & Hurley, M.A. (1991). Humic substances in acid surface waters; modelling aluminium binding, contribution to charge-balance, and control of pH. *Wat. Res.* **25**, 425–435.

Tipping, E., Thompson, D.W., Woof, C. & Longworth, G. (1993). Transport of haematite and silica colloids through sand columns eluted with artificial groundwaters. *Environ. Technol.* **14**, 367–372.

Tipping, E., Berggren D., Mulder J. & Woof C. (1995a). Modelling the solid-solution distributions of protons, aluminium, base cations and humic substances in acid soils. *Eur. J. Soil Sci.* **46**, 77–94.

Tipping, E., Fitch, A. & Stevenson, F.J. (1995b). Proton and copper binding by humic acid: application of a discrete-site/electrostatic ion-binding model. *Eur. J. Soil Sci.* **46**, 95–101.

Tipping, E., Woof, C., Kelly, M., Bradshaw, K. & Rowe, J.E. (1995c). Solid-solution distributions of radionuclides in acid soils: application of the WHAM chemical speciation model. *Environ. Sci. Technol.* **29**, 1365–1372.

Tombacz, E. & Meleg, E. (1990). A theoretical explanation of the aggregation of humic substances as a function of pH and electrolyte concentration. *Org. Geochem.* **15**, 375–381.

Torres, R.A. & Choppin, G.R. (1984). Europium(III) and americium(III) stability constants with humic acid. *Radiochim. Acta* **35**, 143–148.

Town, R.M. & Filella, M. (2000). Dispelling the myths: Is the existence of L1 and L2 ligands necessary to explain metal ion speciation in natural waters? *Limnol. Oceanogr.* **45**, 1341–1357.

Truitt, R.E. & Weber, J.H. (1981). Determination of complexing capacity of fulvic acid for copper(II) and cadmium(II) by dialysis titration. *Anal. Chem.* **53**, 337–342.

Turner, D.R. (1995). Problems in trace metal speciation modeling. In *Metal Speciation and Bioavailability in Aquatic Systems*, ed. A. Tessier & D.R. Turner, pp. 149–203. Chichester: Wiley.

Turner, D.R., Varney, M.S., Whitfield, M., Mantoura, R.F.C. & Riley, J.P. (1986). Electrochemical studies of copper and lead complexation by fulvic acid I. Potentiometric measurements and a critical comparison of binding models. *Geochim. Cosmochim. Acta* **50**, 289–297.

Underdown, A.W., Langford, C.H. & Gamble, D.S. (1985). Light scattering studies of the relationship between cation binding and aggregation of a fulvic acid. *Environ. Sci. Technol.* **19**, 132–136.

Ure, A.M. & Davidson, C.M. (1995). *Chemical Speciation in the Environment*. London: Blackie.

Vance, G.F. & David, M.B. (1992). Dissolved organic carbon and sulfate sorption by spodosol mineral horizons. *Soil Sci.* **154**, 136–144.

Vandenbroucke, M., Pelet, R. & Debyser, Y. (1985). Geochemistry of humic substances in marine sediments. In *Humic Substances in Soil, Sediment, and Water*, ed. G.R. Aiken, D.M. McKnight, R.L. Wershaw & P. MacCarthy, pp. 249–273. New York: Wiley.

Vanselow, A.P. (1932). Equilibria of the base-exchange reactions of bentonites, permutites, soil colloids, and zeolites. *Soil Sci.* **33**, 95–113.

Vermeer, A.W.P. (1996). *Interactions between Humic Acid and Hematite and their Effects on Metal Ion Speciation.* Ph.D. Thesis, Wageningen Agricultural University, The Netherlands.

Visser, S.A. (1964). A physico-chemical study of the properties of humic acids and their changes during humification. *J. Soil Sci.* **15**, 202–219.

Voelker, B.M., Morel, F.M.M. & Sulzberger, B. (1997). Iron redox cycling in surface waters: Effects of humic substances and light. *Environ. Sci. Technol.* **31**, 1004–1011.

de Vries, W. & Breeuwsma, A. (1987). The relation between soil acidification and element cycling. *Wat. Air Soil Pollut.* **35**, 293–310.

Vulava, V.M., Kretzschmar, R., Rusch, U., Grolimund, D., Westall, J.C. & Borkovec, M. (2000). Cation competition in a natural subsurface material: modeling sorption equilibria. *Environ. Sci. Technol.* **34**, 2149–2155.

Vulkan, R., Zhao, F.-J., Barbosa-Jefferson, V., Preston, S., Paton, G.I., Tipping, E. & McGrath, S.P. (2000). Copper speciation and impacts on bacterial biosensors in the pore water of copper-contaminated soils. *Environ. Sci. Technol.* **34**, 5115–5121.

Wafica, M.A.N. (1994). Enrichment of Fe, Mn, Cu, Zn, Pb and Cd in humic acid sediment cores, Lake Edku, Egypt. *Arab Gulf J. Scient. Res.* **12**, 417–431.

Walker, W.J., Cronan, C.S. & Bloom, P.R. (1990). Aluminium solubility in organic soil horizons from northern and southern forested watersheds. *Soil Sci. Soc. Am. J.* **54**, 369–374.

Warwick, P. & Hall, T. (1992). High-performance liquid chromatographic study of nickel complexation with humic and fulvic acids in an environmental water. *Analyst*

117, 151–156.

Wells, M.L., Kozelka, P.B. & Bruland, K.W. (1998). The complexation of 'dissolved' Cu, Zn, Cd and Pb by soluble and colloidal organic matter in Narragansett Bay, RI. *Marine Chem.* **62**, 203–217.

Wershaw, R.L. (1986). A new model for humic materials and their interactions with hydrophobic organic chemicals in soil–water or sediment–water systems. *J. Contam. Hydrol.* **1**, 29–45.

Westall, J.C., Jones, J.D., Turner, G.D. & Zachara, J.M. (1995). Models for the association of metal ions with heterogeneous environmental sorbents. 1. Complexation of Co(II) by leonardite humic acid as a function of pH and $NaClO_4$ concentration. *Environ. Sci. Technol.* **29**, 951–959.

Wetzel, R.G. (1975). *Limnology*, 1st edn. Philadelphia: Saunders.

White, A.F. & Brantley, S.L. (1995). *Chemical Weathering Rates of Silicate Minerals. Rev. Mineral. Vol. 31.* Washington DC: Mineralogical Society of America.

White, R.E. (1987). *Introduction to the Principles and Practice of Soil Science.* Oxford: Blackwell.

Wilkinson, A.E., Hesketh, N., Higgo, J.J.W., Tipping, E. & Jones, M.N. (1993). The determination of the molecular mass of humic substances from natural waters by analytical ultracentrifugation. *Coll. Surf. A* **73**, 19–28.

Wilkinson, K.J., Nègre, J.-C. & Buffle, J. (1997). Coagulation of colloidal material in surface waters: the role of natural organic matter. *J. Contam. Hydrol.* **26**, 229–243.

Wilson, D.E. & Kinney, P. (1977). Effects of polymeric charge variations on the proton–metal ion equilibria of humic materials. *Limnol. Oceanogr.* **22**, 281–289.

Wilson, M.A. (1987). *NMR Techniques and Applications in Geochemistry and Soil Chemistry.* Oxford: Pergamon.

Wilson, M.A., Collin, P.J., Malcolm, R.L., Perdue, E.M. & Cresswell, P. (1988). Low molecular weight species in humic and fulvic fractions. *Org. Geochem.* **12**, 7–12.

Wilson, S.A. & Weber, J.H. (1977). A comparative study of number-average dissociation-corrected molecular weight of fulvic acids isolated from water and soil. *Chem. Geol.* **19**, 285–293.

Wimmer, H., Kim, J.I. & Klenze, R. (1992). A direct speciation of Cm(III) in natural aquatic systems by time-resolved laser-induced fluorescence spectroscopy (TRLFS). *Radiochim. Acta* **58/59**, 165–171.

de Wit, H. (1992). *Proton and Metal Ion Binding to Humic Substances.* Ph.D. Thesis, Wageningen Agricultural University, The Netherlands.

de Wit, J.C.M., van Riemsdijk, W.H., Nederlof, M.M., Kinniburgh, D.G. & Koopal, L.K. (1990). Analysis of ion binding on humic substances and the determination of intrinsic affinity distributions. *Anal. Chim. Acta* **232**, 189–207.

de Wit, J.C.M., van Riemsdijk, W.H. & Koopal, L.K. (1993a). Proton binding to humic substances. 1. Electrostatic effects. *Environ. Sci. Technol.* **27**, 2005–2014.

de Wit, J.C.M., van Riemsdijk, W.H. & Koopal, L.K. (1993b). Proton binding to humic substances. 2. Chemical heterogeneity and adsorption models. *Environ. Sci. Technol.* **27**, 2015–2022.

de Wit, H.A., Kotowski, M. & Mulder, J. (1999). Modeling aluminium and organic matter solubility in the forest floor using WHAM. *Soil Sci. Soc. Am. J.* **63**, 1141–1148.

Witter, A.E., Mabury, S.A. & Jones, A.D. (1998). Copper(II) complexation in northern California rice field waters: an investigation using differential pulse anodic and cathodic stripping voltammetry. *Sci. Tot. Environ.* **212**, 21–37.

Woolard, C.D. & Linder, P.W. (1999). Modelling of the cation binding properties of fulvic acids: An extension of the RANDOM algorithm to include nitrogen and sulphur donor sites. *Sci. Tot. Environ.* **226**, 35–46.

Wu, Q., Apte, S.C., Batley, G.E. & Bowles, K.C. (1997). Determination of the mercury complexation capacity of natural waters by anodic stripping voltammetry. *Anal. Chim. Acta* **350**, 129–134.

Xia, K., Bleam, W. & Helmke, P.A. (1997a). Studies of the nature of Cu^{2+} and Pb^{2+} binding sites in soil humic substances using X-ray absorption spectroscopy. *Geochim. Cosmochim. Acta* **61**, 2211–2221.

Xia, K., Bleam, W. & Helmke, P.A. (1997b). Studies of the nature of binding sites of first row transition elements bound to aquatic and soil humic substances using X-ray absorption spectroscopy. *Geochim. Cosmochim. Acta* **61**, 2223–2235.

Xia, K., Skyllberg, U.L., Bleam, W.F., Bloom, P.R., Nater, E.A. & Helmke, P.A. (1999). X-ray absorption spectroscopic evidence for the complexation of Hg(II) by reduced sulfur in soil humic substances. *Environ. Sci. Technol.* **33**, 257–261.

Xue, H.B. & Sigg, L. (1993). Free cupric ion concentration and Cu(II) speciation in a eutrophic lake. *Limnol. Oceanogr.* **38**, 1200–1213.

Xue, H.B. & Sigg, L. (1994). Zinc speciation in lake waters and its determination by ligand exchange with EDTA and differential pulse anodic stripping voltammetry. *Anal. Chim. Acta* **284**, 505–515.

Xue, H.B. & Sigg, L. (1998). Cadmium speciation and complexation by natural organic ligands in fresh water. *Anal. Chim. Acta* **363**, 249–259.

Xue, H.B. & Sigg, L. (1999). Comparison of the complexation of Cu and Cd by humic and fulvic acids and by ligands observed in lake waters. *Aq. Geochem.* **5**, 313–335.

Xue, H.B., Kistler, D. & Sigg, L. (1995). Competition of copper and zinc for strong ligands in a eutrophic lake. *Limnol. Oceanogr.* **40**, 1142–1152.

Xue, H.B., Oestreich, A., Kistler, D. & Sigg, L. (1996). Free cupric ion concentrations and Cu complexation in selected Swiss lakes and rivers. *Aq. Sci.* **58**, 69–87.

Yin, Y., Allen, H.E., Huang, C.P. & Sanders, P.F. (1997). Interaction of Hg(II) with soil-derived humic substances. *Anal. Chim. Acta* **341**, 73–82.

Young, L.B. & Harvey, H.H. (1992). The relative importance of manganese and iron oxides and organic matter in the sorption of trace metals by surficial lake sediments. *Geochim. Cosmochim. Acta* **56**, 1175–1186.

Young, S.D. & Bache, B.W. (1985). Aluminium–organic complexation: formation constants and a speciation model for the soil solution. *J. Soil Sci.* **36**, 261–269.

Zachara, J.M., Resch, C.T. & Smith, S.C. (1994). Influence of humic substances on Co^{2+} sorption by a subsurface mineral separate and its mineralogic components. *Geochim. Cosmochim. Acta* **58**, 553–566.

Zhang, H. & Davison, W. (1999). Diffusional characteristics of hydrogels used in DGT and DET techniques. *Anal. Chim. Acta* **398**, 329–340.

Zobrist, J. & Stumm, W. (1981). Chemical dynamics of the Rhine catchment area in Switzerland: extrapolation to the "pristine" Rhine river input into the ocean. *Proc. Review and Workshop on River Inputs to Ocean Systems.* Rome: Food and Agriculture Organisation.

Zsolnay, A. (1996). Dissolved humus in soil waters. In *Humic Substances in Terrestrial Ecosystems*, ed. A. Piccolo, pp. 171–223. Amsterdam: Elsevier.

Index

Printed in the United States
By Bookmasters